SSADM & GRAPES

SSADM & GRAPES

Two Complementary
Major European Methodologies for
Information Systems Engineering

Edited by
R. Duschl and N.C. Hopkins

Authors:
A. Aue, R. Haggenmüller, B. Knuth, M. Pfeiffer, K. Robinson

Springer-Verlag
Berlin Heidelberg New York
London Paris Tokyo
Hong Kong Barcelona Budapest

Dipl.-Ing. Rudolf Duschl
Direktor der Siemens Nixdorf Informationssysteme AG
München, Germany

Nicholas Charles Hopkins, BA (Hons), MBCS
Head of Systems and Advanced Technology Group of CCTA
Norwich, United Kingdom

ISBN 3-540-55380-0 Springer-Verlag Berlin Heidelberg New York
ISBN 0-387-55380-0 Springer-Verlag New York Berlin Heidelberg

This work is subject to copyright. All rights are reserved, whether the whole or part of the material is concerned, specifically the rights of translation, reprinting, reuse of illustrations, recitation, broadcasting, reproduction on microfilm or in any other way, and storage in data banks. Duplication of this publication or parts thereof is permitted only under the provisions of the German Copyright Law of September 9, 1965, in its current version, and permission for use must always be obtained from Springer-Verlag. Violations are liable for prosecution under the German Copyright law.

© British Crown / Siemens Nixdorf / Springer-Verlag 1992
Printed in Germany

This report, SSADM & GRAPES, Two Complementary Major European Methodologies for Information Systems Engineering, is subject to the joint copyright of the British Crown, Siemens Nixdorf and Springer-Verlag. No part of this publication may be reproduced, stored in a retrieval system, or transmitted in any form or by any means, without prior written permission.

In addition, the Hi-Ho case study material (Primarily Annexes 1, 2 and 4) may not be used for training purposes or for the purpose of subsequent comparisons without the written permission of Model Systems Ltd.

The use of general descriptive names, registered names, trademarks, etc. in this publication does not imply, even in the absence of a specific statement, that such names are exempt from the relevant protective laws and regulations and therefore free for general use.

Cover design: K. Lubina, Schöneiche
Typesetting: Camera-ready by editors; Printing: Mercedes-Druck, Berlin; Binding: B. Helm, Berlin
62/3020 5 4 3 2 1 0 Printed on acid-free paper

To Our Readers

The systematic development of large-scale distributed application systems is one of the major challenges facing software engineering and computing science. Only adequate description formalisms including graphical representations and well-adapted methodologies can help cope with the enormous quantitative and qualitative complexity involved. In a system development ranging from requirement engineering to efficient implementations a lot of experimenting is necessary to make sure that the envisaged concepts are adequate. Experimenting with methodologies and with case studies is of high value here.

The study of the development of the Hi-Ho employment agency and the comparison of the advanced methodologies and formal description techniques SSADM and GRAPES is a valuable contribution along this line. It will help to explain, demonstrate, and understand the needs of systematic development and the virtues of advanced methodologies for large-scale distributed application systems.

Munich, January 1992

Manfred Broy
Professor of Computer Science

Foreword of the German Editor

The worldwide fundamental innovation in information technology (IT) is the key to economic productivity and international competitiveness. The basic trends of this innovation, very-large-scale integration, digital telecommunication and international standardization, embody the strategic potential for complex system solutions for office and factory automation, communication, transport, environment, and health care.

The utilization of the IT potential for long-term commercial and corporate strategies tackles new dimensions of complexity and technical architecture. In this process, new attitudes and approaches are emerging with more and more clarity: systems engineering is evolving into a uniform engineering discipline for overall planning, developing and maintaining of integrated systems. Systems engineering methodologies will soon be as important as technology and standards.

Future systems engineering at all stages of overall planning, developing and maintaining must

- guarantee methodological precision and consistency
- link customer and contractor in a uniform transnational understanding.

SSADM, a life cycle-based method with worldwide reputation and DOMINO/GRAPES, a modelling method for technically complex problems applied all over the world, ideally complement each other in meeting the demands for the future. The present study demonstrates that SSADM and DOMINO/GRAPES are complementary and combined will form a sound basis for a standardized and manufacturer-independent methodology for complex IT solutions for administration, commerce, and trade. This methodology based on the traditional European strength in system technology will give Europe a leading position in the world market for system integration.

Munich, January 1992

Rudolf Duschl

Foreword of the British Editor

Information system engineers in the 1970's and 1980's have made great progress in moving towards systematic techniques which improve productivity and quality in IS development. In the 1990's we must begin to overcome three types of barriers to further development.

- **national**, where mother tongues inhibit information exchange
- **proprietary**, where vendor environments inhibit data transfer
- **cultural**, where different procedures and skills inhibit our cooperation on international projects

There are two major European needs for Information Systems Engineering to overcome these barriers. We desperately need a common technical specification for a management framework for methodologies which provides a consistent European approach to the job of systems development. Euromethod will help here. But as well as this we need a common specification language to **help** us bring our different techniques closer together.

This study, represents a significant first step towards such a language. I invite methodologists of all backgrounds to analyse this study. I hope they will propose ways of comparing the rigour and expressiveness of other system specification languages to help us toward the goal of harmonization. If this process alone moves on we will break through significant barriers to our common Engineering culture.

Norwich, January 1992

Nic Hopkins

Contents

1	**Introduction**	11
1.1	Purpose	11
1.2	What is Being Compared?	11
1.3	Intended Audience	12
1.4	Study Assumptions	12
1.5	The Task	12
1.6	The Comparison Study	12
1.7	Overview of the Case Study	13
1.8	Summary	13
2	**Overview of Both Methods**	16
2.1	Overview of SSADM	16
2.2	Overview of DOMINO/GRAPES	20
3	**Motivation, Background and Philosophy**	28
3.1	SSADM	28
3.2	GRAPES History and Development	29
3.3	Comparison	30
4	**Methodology Style**	31
4.1	SSADM Style	31
4.2	GRAPES Style	32
4.3	Comparison of SSADM and GRAPES Styles	34
5	**Architecture of the Developed Systems**	35
5.1	SSADM Architecture	36
5.2	GRAPES Processing Architecture	38
5.3	Examples of the SSADM Architecture in the Hi-Ho Case Study	38
5.4	Examples of GRAPES's Processing Architecture in the Hi-Ho Case Study	45
5.5	Comparison of the Architecture of End-Products of GRAPES & SSADM	48
6	**Life Cycles**	49
6.1	The SSADM Life Cycle	49
6.2	The DOMINO Life Cycle	51
6.3	Coverage of the SSADM Life Cycle in the Hi-Ho Case Study	55
6.4	Coverage of the GRAPES Life Cycle in the Hi-Ho Case Study	56
6.5	Comparison of the GRAPES and SSADM Life Cycles	56
7	**Fact Representation & Syntax**	57
7.1	Fact Representation & Syntax in SSADM	57
7.2	Fact Representation & Syntax in GRAPES	59
7.3	Examples of the SSADM Fact Representation & Syntax	60
7.4	Examples of the GRAPES Fact Representation & Syntax	70
7.5	Comparison of the GRAPES & SSADM Fact Representation & Syntax	77
8	**The Knowledge Collection Process**	81
8.1	Knowledge Collection in SSADM	81
8.2	Knowledge Collection in GRAPES	83
8.3	Examples of the SSADM Knowledge Collection Process	85
8.4	Examples of the GRAPES Knowledge Collection Process	91
8.5	Comparison of the GRAPES and SSADM Knowledge Collection Processes	96
9	**Transformations**	97
9.1	Transformations in SSADM	97
9.2	Transformations in GRAPES	100
9.3	Examples of the SSADM Transformations	103
9.4	Examples of the GRAPES Transformations	109
9.5	Comparison of the GRAPES and SSADM Transformations	115
10	**Conclusions**	116
10.1	Answers to Specific Questions	116
10.2	Opportunities for Harmonization of GRAPES and SSADM	119
Annex 1		
The Hi-Ho Recruitment Case Study		121
	Introduction	122
	The current Hi-Ho system	122
	Requirements for the new Hi-Ho system	123
	Sample forms in use in the current system	124

Annex 2
Hi-Ho SSADM Development 129
    Data flow diagrams 130
    Logical data model 133
    Logical data model-data store
    cross-references 134
    Enquiry access paths 136
    LDM enhanced with RDA results 139
    Results of ELH analysis, 1st pass 140
    Further DFD & LDM documentation 147
    Entity life histories after 2nd pass 152
    ECDs after 2nd pass 159
    After specification of deletion strategy 165
    Update process models 175
    Enquiry process models 188
    Function DFD equivalents 192
    Function input structures 200
    Probable function processing from stage 6 ... 205
    Input/output data 211

Annex 3
Hi-Ho GRAPES Development 221
    Model of Required System 223
    Static structure 224
    Communication 226
    Behaviour 236
    Model hierarchy 249
    Reusable units 250
    Model of Current Situation 253
    Static structure 254
    Communication 256
    Behaviour 265

Annex 4
**Hi-Ho SSADM Development
Using Super-Events** 277

Annex 5
Hi-Ho GRAPES Intermediate Diagrams 305

References 321

Index 323

1 Introduction

1.1 Purpose

The CCTA of the UK government and Siemens Nixdorf (SNI) have undertaken a joint project to compare the CCTA's methodology SSADM and Siemens Nixdorf's product GRAPES®. SSADM is an analysis and design methodology for computer systems. GRAPES is a graphical language used within the framework provided by the DOMINO® process technology of Siemens Nixdorf for analysis and design of information systems. (In the 1991 Statement of Direction it was declared the Siemens Nixdorf main line for modelling technology.) The study was undertaken in February and March 1991.

The goal of the study was to compare SSADM and DOMINO/GRAPES to discover whether both approaches elicit the same knowledge about a system.

The motivation for this comparison is to assess the feasibility of harmonizing both organisations' efforts in the progress to a European standard methodology, EUROMETHOD.

This publication describes how the comparison study was undertaken and the conclusions reached. The workings and findings of the study are documented by giving the following details:

- a statement of the scope, nature and application of SSADM version 4 and GRAPES;
- the parameters of the comparison, including a description of their purpose;
- areas of similarity and difference, pointing out especially: overlaps, gaps, incompatibilities and differences in perspective or purpose;
- the general semantics of the languages of each;
- issues for future harmonization raised at technics, tools, training, usability and standardization levels;
- issues relevant to extending the approaches to cope with knowledge-based systems.

1.2 What is Being Compared?

SSADM

SSADM is a method which helps the analyst to construct a framework within which to document a clear understanding of the business requirement. The SSADM documentation is constantly refined as detailed understanding of the requirement increase.

For project management purposes the SSADM life cycle is defined in terms of work packages, modules, stages and steps. Each work package is completed when a specified set of documentation is produced and satisfies specified quality criteria. All activities undertaken at each point are those necessary to produce the required documentation.

The analysis is undertaken concentrating on three key perspectives:

- functions (users' views of system processing to react to events)
- events (real-world business events, e.g. "receipt of application"; or system-generated, e.g. end-of-month trigger for management reports)
- data (the system manipulates and maintains data to deliver the system's functionality)

An overview of SSADM is given in chapter 2.1.

DOMINO/GRAPES

The DOMINO process technology includes board on a life cycle consisting of phases and milestones. DOMINO also covers the following aspects of a project: quality management, configuration management and project management.

The DOMINO process technology provides a specification of deliverables at the end of each life cycle phase. To ensure a uniform acceptance procedure over all projects, criteria are provided for the quality of documentation and products.

To support the quality assurance of DOMINO milestone end-products it was recognized that a formal graphical representation was required.

GRAPES is a formal graphical engineering language capable of representing various views of system models. The three main structuring criteria are:

- system structure
- structure of information
- structure of information processing

Chapter 2.2 provides an overview of DOMINO and GRAPES.

1.3 Intended Audience

The SSADM/GRAPES comparison study was undertaken in an objective way to form impartial conclusions about the strengths and weaknesses of the two approaches. The results, based on a particular case study, are valid for wider application of the approaches.

The results of the study are documented in such a way as to reflect the professional and intellectual content. This makes the study interesting to a wide audience, not only SSADM and GRAPES practitioners, but also the wider methodology community. In particular the following readership groups can be identified:

- SSADM practitioners
- GRAPES practitioners
- European methodologies standardization bodies: e.g. EUROMETHOD, CEN ISE Working Group
- Academic community interested in methodologies, software engineering and object orientation

SSADM and GRAPES practitioners will be particularly interested in the report conclusions which cover the topics:

- identification of the strengths and weaknesses of the methodologies;
- how to address shortcomings, using techniques in the other approach;
- what aspects of systems analysis and design are not covered – possibly highlighting a need to use an alternative approach.

1.4 Study Assumptions

The study was undertaken based on the following assumptions:

- Potential for a system is present in some business area and the relevant knowledge is available for requirements analysis.
- It is possible to map deliverables and life cycle coverages of SSADM and GRAPES; and they have the same requirements with regard to capture and analyses transformation.
- Comparing classes of facts (of method/system) is possible given the differing natures of the approaches and will show something about each method's view of the world.

- The Zachman framework (see references) can be used to make an objective comparison of the use of techniques within the methodologies based on differing approaches.

1.5 The Task

There were several objectives which the study team had to address, as follows:

- to compare the high-level DOMINO milestones and SSADM version 4 products
- to classify facts about the methods and about systems as described by the methods
- to illustrate abstract facts against actual case studies
- to use the illustrations and comparisons as the basis for a statement concerning the knowledge about systems captured by the methods
- to provide a statement of the issues and conclusions about meta-models, tools and each method's view of the world, by examining the knowledge collection process
- to provide a statement of the issues and conclusions about skills, understanding, method strengths and weaknesses, and training
- to highlight where further work can be undertaken, in particular on the harmonization of SSADM and DOMINO/GRAPES

1.6 The Comparison Study

The work plan was composed of four work packages:

1. Development of the selected case study in SSADM
2. Development of the selected case study in GRAPES
3. Comparison of case study results
4. General comparison

Development of the selected case study in SSADM

The study team considered the development of a SSADM solution through stages 3 and 5 of SSADM version 4. Since the selected case study had already unearthed the SSADM logical view of the problem it was felt that no purpose would be served by back-forming the current system data flow models, so stage 1 was omitted.

1.7 Overview of the Case Study

Development of the selected case study in GRAPES

The study team decided on development of a GRAPES solution primarily from milestones A10 to A30 of the DOMINO master process.

Comparison of case study results

The study team took an initial examination of the study results to ensure that the relevant information had been acquired. This was done by assessing the results from the following viewpoints:

- How complete are the results?
- How consistent are the results?
- What further effort would be needed for implementation?
- How understandable are the results?
- How easy is it to make changes?
- What tool support was needed?

General comparison

Having interpreted the study results the team than assessed them to show how the conclusions drawn during the case study relate to the wider world of systems development.

- Is the same knowledge about a system captured?
- Where do the methods have strengths and weaknesses?
- How comprehensive are the consistency checks?
- What support is given for software re-use?
- What support is given for structuring?
- Do the methods suggest the same view of the world?
- Does any difference in world view have an useful outcome?
- Where are the areas of application of the methods?
- Are both methods targeting the same set of systems to be described?
- What skill levels are necessary to use the method?
- What training is necessary to use the methods?
- What tool support is given, and what is required?
- In which areas would both systems benefit from a joint add-on development?
- What are the limits of this study?

1.7 Overview of the Case Study

The topic of the case studies was modelling an employment agency, called Hi-Ho, which specialized in professional and management placements.

The main service provided by Hi-Ho is the arrangement of interviews between employers and applicants on the basis of the reported vacancies of the employers and the skills of the applicants.

Information on the companies structure, the profiles of the jobs, and the policies in handling employer and applicant requests had to be expressed in terms of the models.

Additionally a model for an intended change in the interview arrangement procedures had to be modelled. Further details are given in the Annex 1.

1.8 Summary

This part gives a brief summary of the purpose and main conclusions of each of the following chapters.

Overview of Both Methods

Chapter 2 gives short overviews of the two methods compared in this study. This chapter is intended for those who aren't familiar with one or both methods.

Motivation, Background and Philosophy

Chapter 3 is a brief description of the historical development of the two approaches. This identifies that DOMINO/GRAPES has largely been developed "top-down", i.e. driven by the desire to have a standard notation for concepts prescribed for documentation of modelling results by the DOMINO engineering process.

Interestingly the development of SSADM has largely been "bottom-up", i.e. by incorporating different modelling techniques into a combined approach.

This difference in history and motivation might be the reason why SSADM is richer in techniques, whereas GRAPES is very formal and produces executable model specifications.

Methodology Style

Chapter 4 describes the style of modelling of the two approaches in terms of

- modelling of systems
- structure of the models
- emphasis of concepts

This reveals that GRAPES is really an integrated language intended for formal representation of a system, whereas SSADM is a collection of techniques for discovering the structure of a system.

Architecture of the Developed Systems

During the study it became apparent that there are aspects of the basic systems architecture which warranted investigation. The findings for this are documented in chapter 5. To illustrate these aspects an analogy to the ANSI-SPARC 3-schema database architecture was drawn.

This reveales that SSADM produces processing models which also show a 3-schema architecture with internal, conceptual and external scenarios; and also uses slightly different techniques for producing and documenting each of the different schemas.

GRAPES does not explicitly recognize such a structuring in its models, i.e. has no means of expressing affilation to a special schema. But due to its object-oriented approach to modelling objects can be identified that show the characteristics of a multi-schema architecture.

Life Cycles

In chapter 6 it is shown that the two approaches define life cycles which are broadly similar. They use the same principles of refinement and transformation from a physical view of the current system to an abstract, logical view of the required system through to a new physical design in terms of user environment as well as database and program design.

Mapping each method onto the Zachman framework reveales that some SSADM techniques bridge the lines between cells. This signifies strong knowledge collection capabilities. GRAPES covers these aspects primarily in the network distribution column of the grid.

During the study the life cycles of SSADM and GRAPES were not fully covered because of constraints laid down on the scoping and resourcing of the study.

Fact Representation and Syntax

Chapter 7 describes how information is recorded within both methods.

In the area of system dynamics SSADM tends to represent more facts, more directly, about the system being developed than GRAPES does.

On the other hand there are significant limitations in SSADM's fact representation which are addressed by GRAPES e. g. time events or common processing structures.

The Knowledge Collection Process

Chapter 8 describes how information about a system is gathered in the two approaches, i.e. which techniques are applied to obtain knowledge from the application domain.

It is shown where the methods put special emphasis in the knowledge collection. The main difference here is the analysis of processing. GRAPES first examines the overall processing in a system whereas SSADM starts from the life histories of single entities.

It can be seen that SSADM is stronger than GRAPES in the knowledge acquisition process. GRAPES on its own can be characterized as a "language to describe models" in a formal and consistent way. SSADM is a "method for discover-ing models".

Transformations

Both methods are concerned with taking some information and refining it to describe the desired system. Chapter 9 describes how the knowledge is changed over time. SSADM and GRAPES support similar transformations in the production of required system data flow models and communication diagrams and in building entity relationship diagrams.

Interestingly, in the production of processing descriptions the transformation direction of GRAPES is exactly opposite to that of these SSADM end-products. Again SSADM proved to be stronger in giving rules for transformations, whereas the strength of GRAPES lies in the possibility of automatic consistency checks.

Conclusious

Chapter 10 provides the answers to the questions raised at the beginning. It also addresses the topics:

- similarities in the methods
- further considerations
- opportunities for harmonization

The most important conclusion is that what seemed to be totally different at first glance turned out to be very similar in the end.

Of course there are differences but these merely highlight possible ways of harmonizing the methods. SSADM can profit by adapting some of the formality of GRAPES. GRAPES can benefit from SSADM knowledge acquisition.

Annexes – The Case Study

The annexes provide the description of the case study. The documentation produced by each approach is included to support the detailed conclusions of the study.

2 Overview of Both Methods

2.1 Overview of SSADM

SSADM (the Structured Systems Analysis and Design Methodology) has been the UK Central Government standard information systems development methodology since May 1981, and has been mandatory for administrative data processing systems since January 1983. It is also used by many companies in the private sector. It is perhaps the leading systems methodology in Europe.

SSADM was designed to tackle a number of traditional development problems, for example, that systems have:

- taken too long, and cost too much, in development
- cost too much to run
- not met the users' requirements
- been difficult to adapt to meet changing requirements
- depended on a few essential personnel for their running

To achieve its design objectives SSADM uses a number of software engineering techniques, some, but not all of which, are common to other development approaches. SSADM's philosophy is an outgrowth of many conventional techniques but is consistent with state-of-the-art techniques such as object-oriented design. However, SSADM version 4 (the latest version) aims for a conventional implementation.

The starting point for SSADM is a brief statement of requirements agreed with the system users. The output of SSADM is a system specification including

- database and/or file specifications
- program specifications
- a system test plan
- clerical / manual interface specifications
- an operating schedule

An SSADM project is divided into a number of modules during which knowledge is acquired about the system being studied. This knowledge is captured in the form of intermediate products such as a logical data model (LDM), data flow models (DFMs) and entity life histories (ELHs), representing the facts which have been discovered. These intermediate products are transformed via other intermediate products into the end products listed above. All of the products have quality criteria assigned to them. Once produced there is a review of the product to ensure that it is of the appropriate quality.

The high-level structure of an SSADM project is as shown in figure 2.1.

The two most critical products are the program specifications and the physical data design, since these two are directly used in the development stage of the project, and the contents of the others are, to a great extent, determined by these two (figure 2.2 and figure 2.3).

The contents of these end-products is important, but so is their structure, and, in particular, their modularity. In order to show their modularity and structure their development is explained here in reverse order.

Physical design

In SSADM stage 6 the final designs for the database and major programs are obtained by transforming a set of logical requirements into a physical implementation with an adequate level of performance in a particular hardware and software environment. How the logical design is produced by earlier stages of SSADM is discussed later, but the fact of its pre-existence means

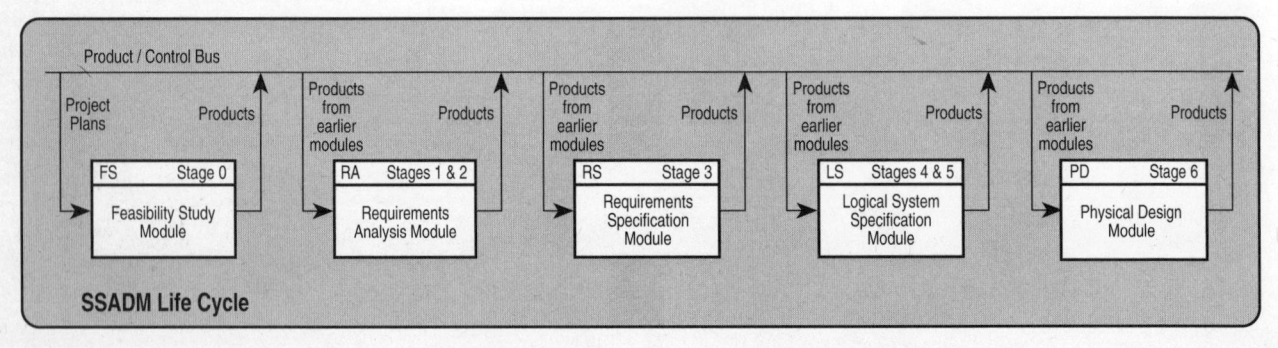

Figure 2.1

2.1 Overview of SSADM

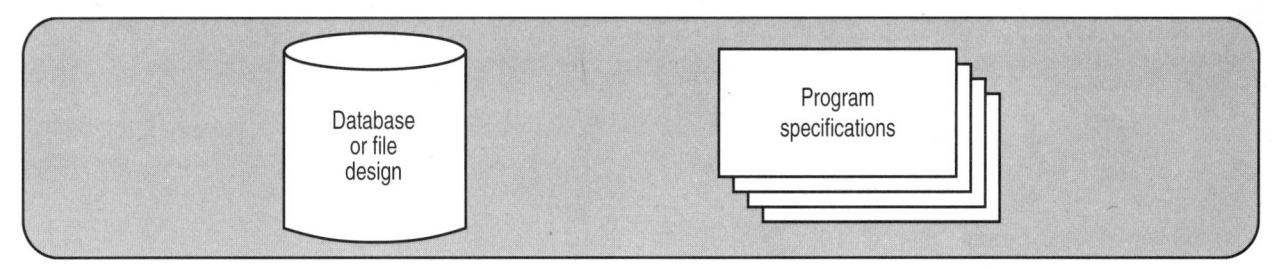

Figure 2.2

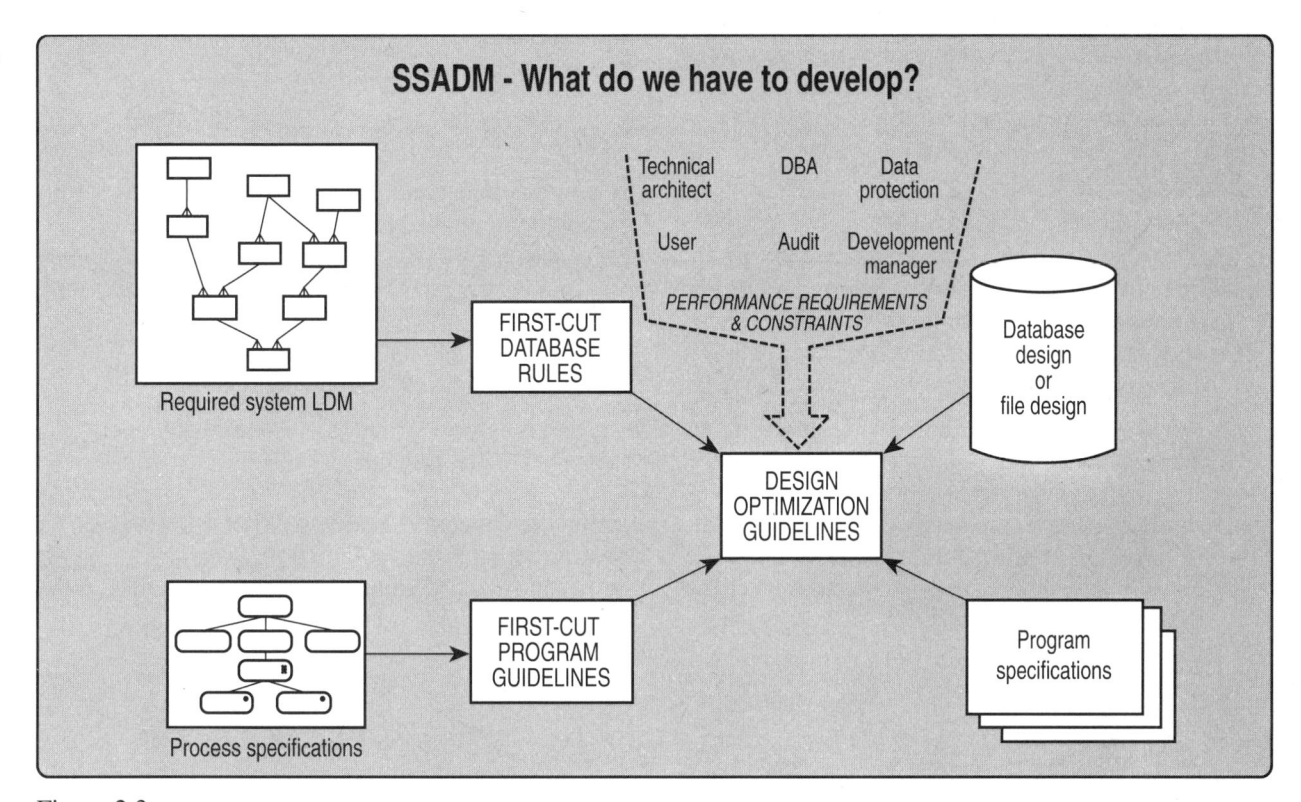

Figure 2.3

that the final SSADM design is not totally locked into a particular implementation environment; the logical design is portable to other environments.

The fact that the physical design is produced by transforming the logical design (ideally by *adding* to the logical design) in order to meet a set of performance objectives means that performance issues are addressed before implementation and that optimization activities which do not provide specifically requested benefits are not undertaken.

An SSADM project produces an optimized database design and program specifications by optimizing a trial design – a "**first-cut** design" – according to a procedure set out in stage 6 of SSADM version 4.

The optimization step is in two parts:

- identify problem areas by estimating the resources required to run the major transactions in the system (major batch runs, high-volume interactive transactions, critical-performance real time transactions etc.)
- manipulate the design (reorganize the database structure and space planning, change access methods, restructure the programs etc.) in order to meet the required performance objectives.

SSADM includes a procedure to follow for optimization, standard values for estimating performance of most of the major DBMSs, and suggestions on what corrective action to try when particular performance problems occur.

Performance criteria and constraints are set by a number of different people involved with the project, including:

- end-users
- security, finance, audit
- operations manager
- systems and programming manager

The **first-cut** designs for the database and major programs are created by applying simple design rules to a logical data model and a set of logical process specifications. These design rules produce designs that are legal (in the sense that they could be implemented), but in almost all cases would not perform adequately.

The physical design is developed in this way for three major reasons:

- the first-cut design exists only as a design (on paper or on an automated sizing/timing spreadsheet), not as a real database and programs; hence it is relatively quick and cheap to manipulate;
- if it turns out that the performance objectives cannot be achieved, then the users and management have the option of changing some of them, e.g. slower response, less stored data, larger CPU, more storage, while the system is still at the design stage – there are fewer options after the database has been built and the programs written;
- the "first-cut" rules allow for quick and simple conversion from a logical to a workable physical design, without the complication of performance requirements. This follows the well-established engineering practice of "to build a high-performance engine, first design one that will work, then tune it".

Deriving the required system logical data model

The required system logical data model, completed in SSADM stage 3, is built from three types of input, as shown in figure 2.4.

Logical data model (LDM)

The logical data model is developed and refined from the beginning of the project.

- Stage 1   LDM of data used in current system
- Stage 3   LDM of required system
- Stage 3   relational (TNF) data analysis; enhancement of the LDM
- Stage 3   refinements from ELH development fed back into LDM

The major concern in building the LDM is to define a structure that contains all the data groups required for processing and supports the necessary access paths between them.

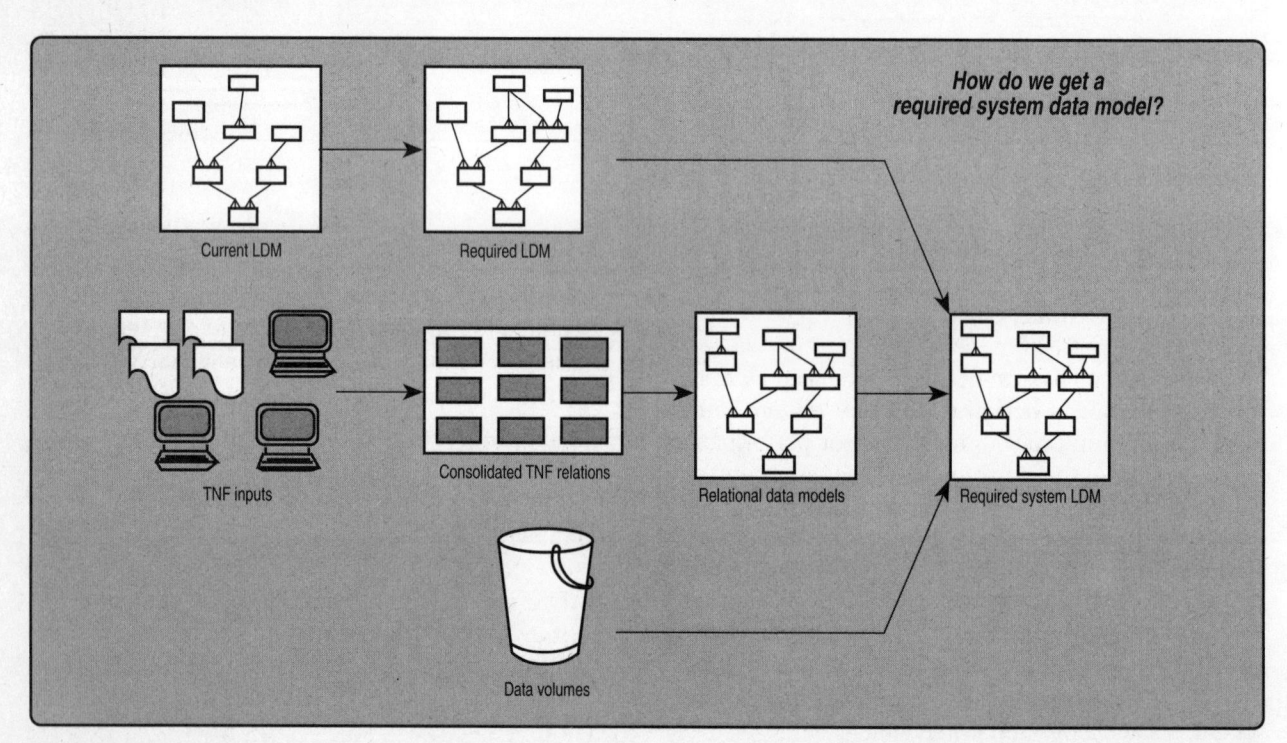

Figure 2.4

2.1 Overview of SSADM

Relational data analysis models (RDA)

The RDA models are built in SSADM stage 3, from the contents of system input forms and screens, output reports and screens and retained data.

The major concern in building the RDA models is to define a structure that encompasses all the retained data in the system, such that no data item (apart from common keys) is duplicated, and every data item has a unique key.

Data volumes and frequencies

These are collected as an ongoing task, starting in stage 1 of SSADM.

Deriving the process specifications

Process specifications also derive from three components as shown in figure 2.5:

Data flow diagrams, leading to functions, prototypes and input/output formats

- Stage 1   DFDs of current physical system
- Stage 3   DFDs of idealized current services
- Stage 3   DFDs of required system
- Stage 3   functions extracted from DFDs
- Stage 3   specification prototypes for selected functions
- Stage 5   user dialogues

Process models fill in the details of processing below the DFDs.

Update process models

- Stage 3   entity life histories (ELH) define updating requirements of LDM entities in terms of the time sequence of external events affecting entity instances
- Stage 3   effect correspondence diagrams (ECDs) collect all processing for the same external event and structure updating accesses caused by the event
- Stage 5   update process models are created by collapsing ECDs into processes

Enquiry process models

- Stage 3   enquiry access paths (EAPs) structure retrievals from the LDM
- Stage 5   enquiry process models are created by transforming EAPs into processes

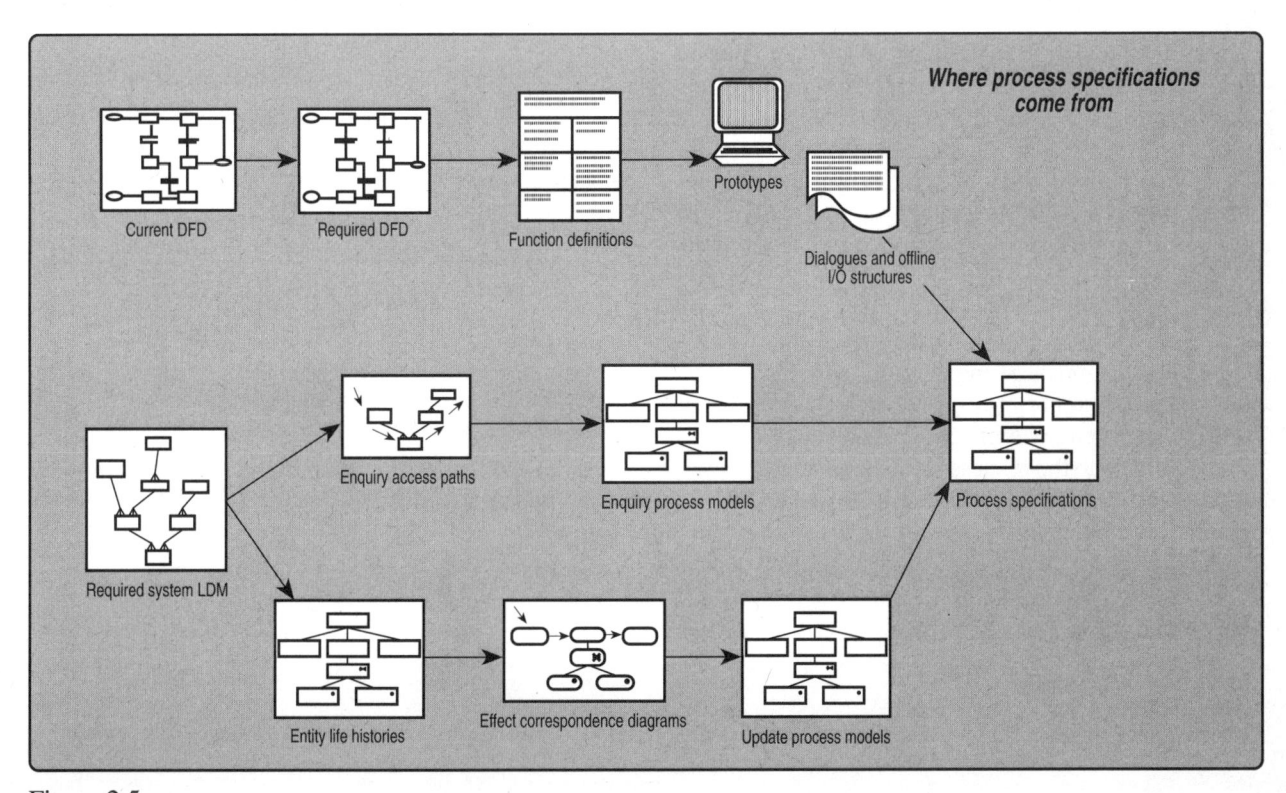

Figure 2.5

2.2 Overview of DOMINO/GRAPES

DOMINO is Siemens Nixdorf's comprehensive process engineering technology. It aims at increasing productivity during the development of information and software systems. DOMINO comprises and integrates tools for the various tasks in systems engineering. The organizational framework is defined in the Process Engineering Handbook (see references), which prescribes the procedures for carrying out projects and for quality assurance.

For capturing analysis and design results the graphical modelling language GRAPES is used. It has a well-defined syntax which allows checks of correctness and consistency. Additionally the notation is executable, which renders animation and simulation of models feasible.

In spring 1991 the revised versions of the Process Engineering Handbook and the GRAPES Language Description were released (see references).

2.2.1 Introduction to DOMINO

An important concern of DOMINO is the systematic integration of essential services in the fields of consultancy, engineering and programming. The consistency and soundness of this systematic approach must be ensured at all stages, from the analysis of the problem to the economical implementation of a solution (making due allowance for the timely provision of the required functions, performance, quality and security). The incorporation and safeguarding of existing investments already made in data processing systems by means of reverse engineering play an important role here.

Components

The DOMINO process engineering technology integrates three different engineering fields: business modelling, systems engineering and software engineering. Supplementary to the process engineering technology a broad range of tools is provided by the DOMINO system development environment (SPU).

The DOMINO process engineering technology covers four levels:

1. *The project life cycle addresses four process segments:*
   - analysis
   - system design
   - technical realization
   - maintenance/operation

   Each process segment begins and ends with a well defined milestone.

2. *Quality management refers to:*
   - project organization
   - project structure
   - risk analysis

   It addresses especially reviews and final tests for product quality and deliverable quality.

3. *Configuration management secures that:*
   - deliverables and products can be identified definitively;
   - versions of deliverables and products can be distinguished clearly;
   - changes to deliverables and products can be executed only in a controlled manner.

4. *Project management:*
   - addresses four components:
     - project structure and configuration planning;
     - scheduling and supervising;
     - demand of working stock and capacity planning;
     - cost budgeting;
   - specifies what management activities have to be executed at which milestone.
   - integrates a quality network into the project management.

Life cycle

A central part of the process engineering technology is the DOMINO life cycle as defined in the master process of the PHB (Process Engineering Handbook). It defines and connects development, management and quality assurance activities.

The master process divides the life cycle into four process segments (see above) according to the activities performed, the results to be achieved, and the techniques used.

Each process segment is divided into a number of process steps, each of which is delimited by milestones. The milestones are checkpoints in the development process. Each milestone defines a certain state of development. At this point all the relevant results and documents are brought together to a milestone deliverable, which then is checked for quality, cost and schedule, according to a fixed set of assessment criteria.

Depending on the result of the quality assurance process, decisions on project progress are taken. In this context, an important feature of DOMINO process engineering is the separation of responsibility for project management and quality assurance. People responsible for assessment of results must not be involved in their production. This guarantees unbiased assessment of the milestone deliverables.

Some of the milestones are also baselines. Baselines provide connection of the development process to the configuration management. Baseline results must not be changed within the development process. Changes to these results have to be requested and approved in a special change request procedure.

Tool support

The DOMINO process and engineering technique provides a broad range of tools. These tools are integrated in Siemens Nixdorf's DOMINO system development environment. It provides for decentralized software development workplaces, e.g. at UNIX workstations, which can be connected to a range of mainframes, e.g. BS2000, MVS, UNIX.

The tools range from editors, dictionaries up to code generators and database interfaces. The SPU covers the main parts of the DOMINO process technology, providing support e.g. for structured interviews, configuration management, cost estimation. A major part of the SPU are the tools that support the usage of the graphical modelling language GRAPES.

2.2.2 Introduction to GRAPES

The graphical engineering language GRAPES can be used in all three engineering fields: business modelling, systems engineering and software engineering. It is the key element of the DOMINO process engineering technology. Its uniform concept allows it to cover the development process comprehensively and consistently, from the very first planning involved at the problem analysis stage right up to a detailed implementation concept.

Components

The GRAPES language combines classical techniques (see figure 2.6). It includes important basic elements of the requirements engineering and design language IORL (Input/Output Requirements Language), SA (Structured Analysis), SADT (Structured Analysis and Design Technique), SDL (Specification and Description Language), and ERM (Entity Relationship Modelling).

GRAPES permits object-oriented modelling, abstraction for data types and processes as well as data modelling using entity relationship diagrams. This makes it a suitable specification language for organization engineers, system engineers and computer scientists.

Main concepts

The main idea behind GRAPES is to build system models, which are structured in such a way as to reproduce the natural divisons within that system and allow simulation of the modelled systems. This is achieved by structuring the system by means of objects which represent the components of the real system. The relationships between the components are represented by communication relationships between the objects. The flow of information is represented by messages.

Structuring a system using objects

The abstraction which forms the basis of system modelling with GRAPES sees the system to be modelled as a structured group of individual parts which interact with each other to achieve a common aim, i.e. the services provided by the complete system. These parts are autonomous entities, each independently performing its allocated task. They can communicate with each other, i.e. they can exchange information and thus affect each other's activities, but are otherwise completely separate. Within the GRAPES model structure, these entities are called objects.

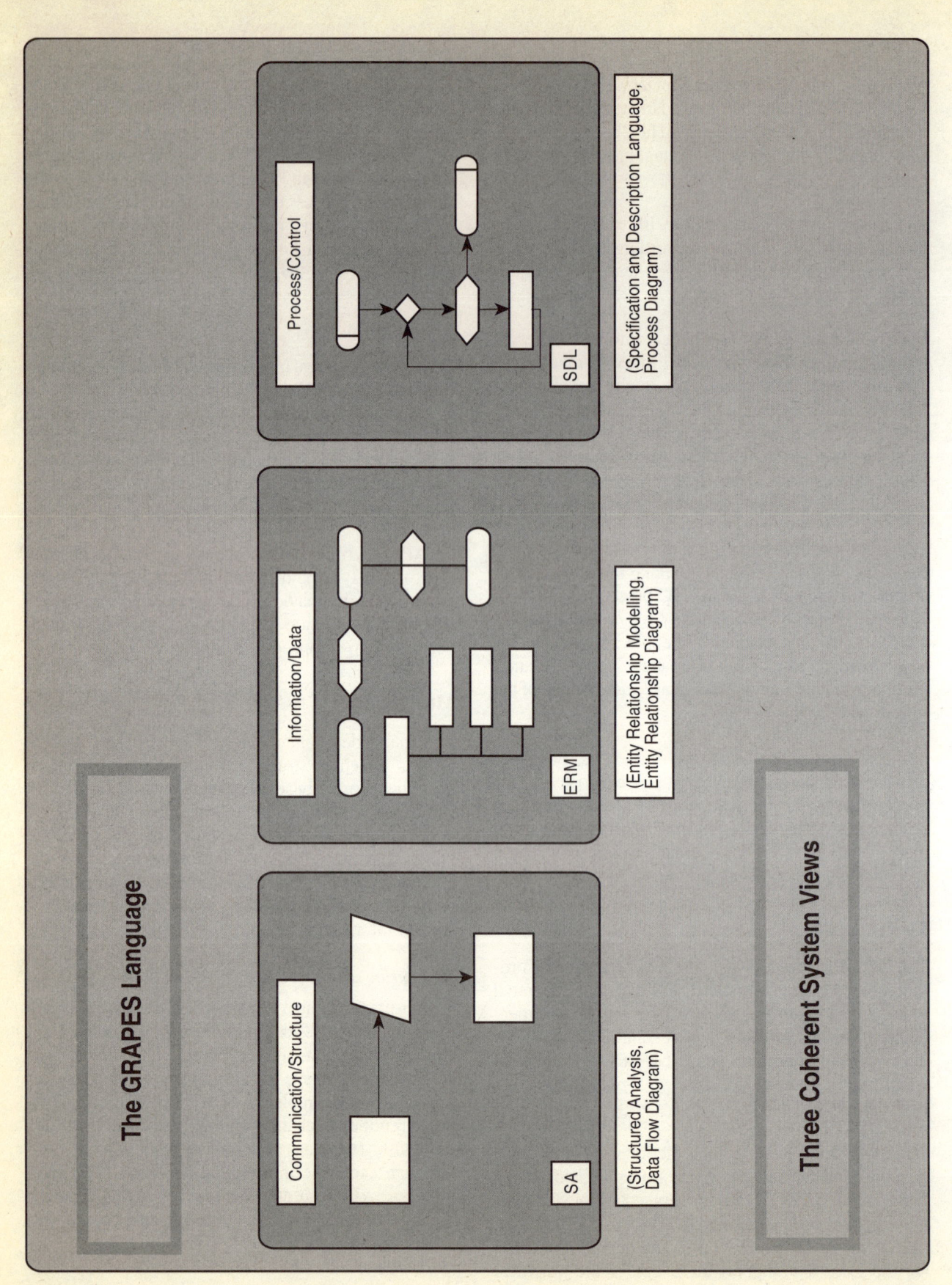

Figure 2.6

2.2 Overview of DOMINO/GRAPES

Individual objects within a system can themselves be seen as systems with more specific tasks, and which are composed of a set of so-called subobjects. Objects of this sort are called structure objects.

Finally, there are objects whose characteristics and activities as seen from outside are so elementary that they are no longer defined in terms of a subobject structure, but rather by a so-called process. Objects of this type are called process objects. A process is a sequence of operations which represent the activities of the object, e.g. its reaction to any messages received.

A GRAPES model of a system is thus a hierarchy of structure objects T, T_1, T_2 (see figure 2.7), which are defined in terms of the communication structure of their subobjects, and process objects $T_{1.1}$, $T_{1.2}$, $T_{2.1}$, $T_{2.2}$, whose behaviour is explicitly defined. Process objects form the bottom of the hierarchy. The process objects of this object hierarchy taken together form the complete system under consideration.

Communication carriers and media

The only communication carriers between objects are messages. These contain the information to be transported from one object to another.

The communication medium used for transporting messages is the channel. Channels are always connected to two objects, thus permitting message transfer. Channels are typecast and direction-specific, i.e. a single channel can only pass messages of a specific type and in a particular direction.

The external view of an object

When seen from the outside, only two things are known about an object, namely its function and its interface. The interface defines the types of channels which can be connected to an object, i.e. the message types it can send or receive.

An object reacts to incoming messages by performing an activity (which may not be visible from the outside) and perhaps sending messages via the connected channels. The type and contents of these messages depend on the life history of the incoming messages. This means that objects have the capacity to store message contents and information concerning communication to date. This capacity is called the memory of an object. The same behaviour characteristics can be observed with structure objects and process objects.

The internal view of an object

A structure object is defined by the subobjects it contains and by the way in which these subobjects are interconnected. The function and interface are known for each of the subobjects. This corresponds to the external view of the subobjects. A structure object communicates with its environment via channels which connect its interface with the interfaces of its subobjects. If a message is sent via one of these channels, it is transferred to the appropriate channel at the interface and relayed from there. If a message arrives on an external channel, it is relayed via the appropriate internal channel to the connected subobject.

The behaviour of the structure object is defined by the interaction of its subobjects. Its current state is defined by the current state of all its subobjects.

The possible states and state changes for a process object are defined explicitly by a process. The object state is defined by the current memory contents, the queued messages which have not yet been evaluated and the current receive state of the process.

When in a receive state, a process accepts a message, if necessary waiting for it to arrive. Depending on the current receive state, the contents of the received message and the current memory contents, the process then performs a transition, thereby switching to a new receive state. During the transition the process can modify the memory contents and send messages to the object interface, which then relays them.

The syntax of process descriptions is defined analogous to the rules of structured programming so that implementation of process descriptions is straightforward if any such programming language is used.

Additional concepts

For further structuring of process descriptions GRAPES provides procedures and modules. Additionally the hierarchy of structure and process objects defines a hierarchy of visibility scopes so that declarations can be hidden in certain scopes. This enables the definition of data types, procedures and modules that may only be used in certain parts of a system and have to be unknown to the rest.

Again these concepts can be implemented straightforwardly if a structured programming language is used.

Graphical notation

GRAPES uses diagrams with an easily understandable set of graphical symbols for representing models. These diagrams allow the main characteristics (the structure, processes and processed information) of a complete system to be represented in the form of models and to be gradually refined. The various diagram types correspond to the various aspects of the system description. The diagram types available are described in brief below.

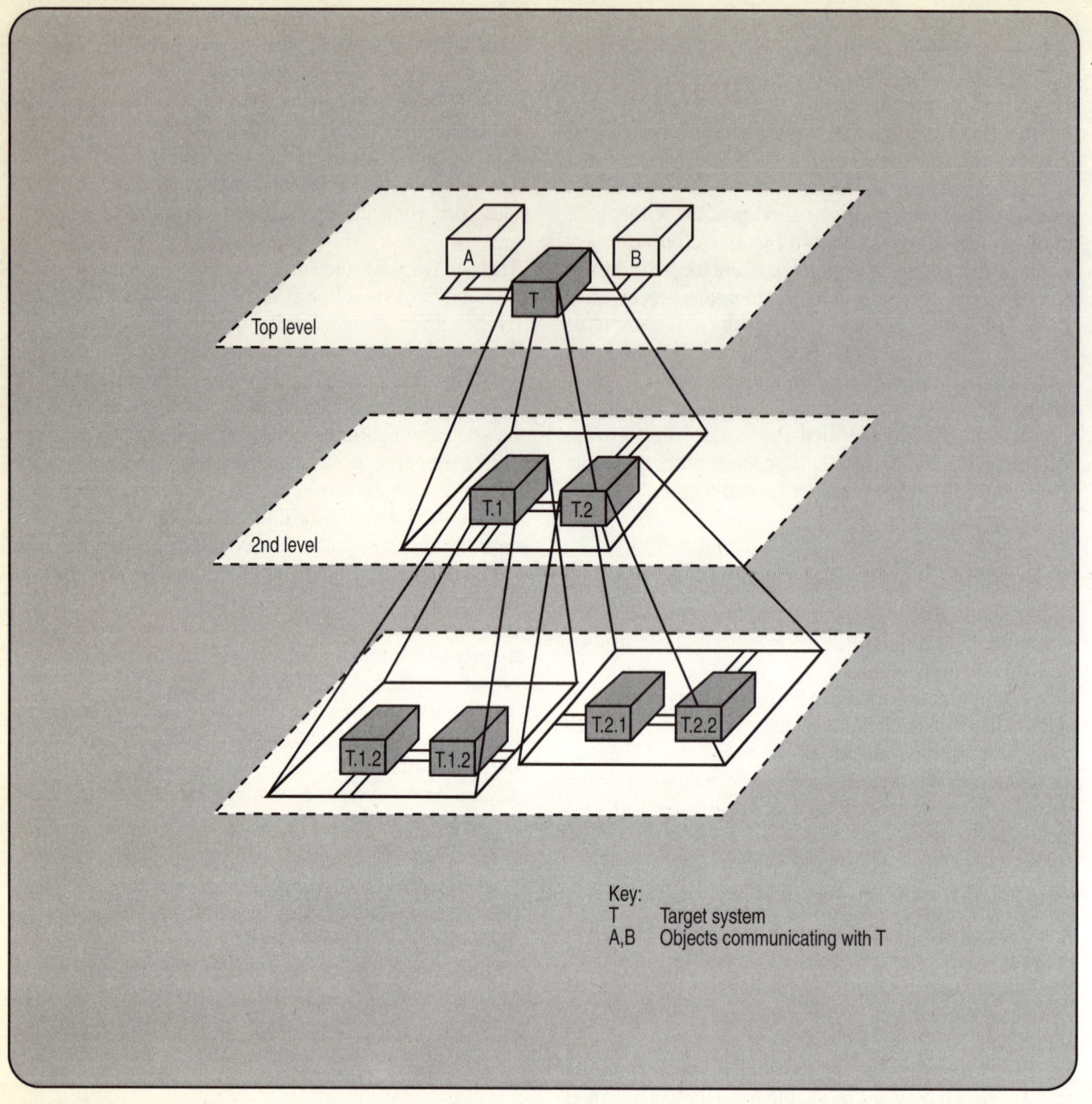

Figure 2.7

- Representation of structure and communication

    CD: **Communication diagrams** are used to describe the way in which an object or object type is made up of subobjects and to represent the communication relationships between the subobjects in the form of communication lines.

    IT: Precisely one **interface table** is assigned to each communication line in a communication diagram. This describes the structure of the communication line in terms of channels and associated data types.

- Representation of behaviour and functionality

    PD: **Process diagrams** are used to define processes (the behaviour of process objects), procedures and functions.

    DT: **Data tables** are used for declaring constants and variables which processes, procedures, functions and modules can exclusively access.

    SD: **Specification diagrams** are used for describing the call interfaces (parameters) of procedures or functions and for describing the export lists of modules.

- Representation of data structures

    DD: **Data diagrams** allow data structures to be modelled. The relationships between sets of data can be described here, as can user-defined data types.

- Representation of the declaration hierarchy

    HD: **Hierarchy diagrams** are used to represent the interrelationships between the documents defined in the model.

See figure 2.8 and 2.9 for the relationships between the diagram types.

Further developments

DOMINO and GRAPES were originally developed to provide a uniform project management procedure and a standard notation for experienced developers.

Extending the range of application, the need for methodical support emerged. Thus the current activities are defining and describing in detail the modelling steps necessary to produce the models required in analysis and design of information and software systems.

GRAPES is being developed further towards knowledge-based systems and further concepts of object-orientation, e.g. inheritance, polymorphism, in order to cover also the field of knowledge engineering and to enhance the support for information systems engineering.

Another impetus for further development is the European challenge of the single market of 1993. With respect to this, DOMINO and GRAPES are being developed to both influence and be consistent with EUROMETHOD, the systems engineering method sponsored by the Commission of European Communities, which is expected to be the European standard of procurers in systems engineering in the second half of the 1990s.

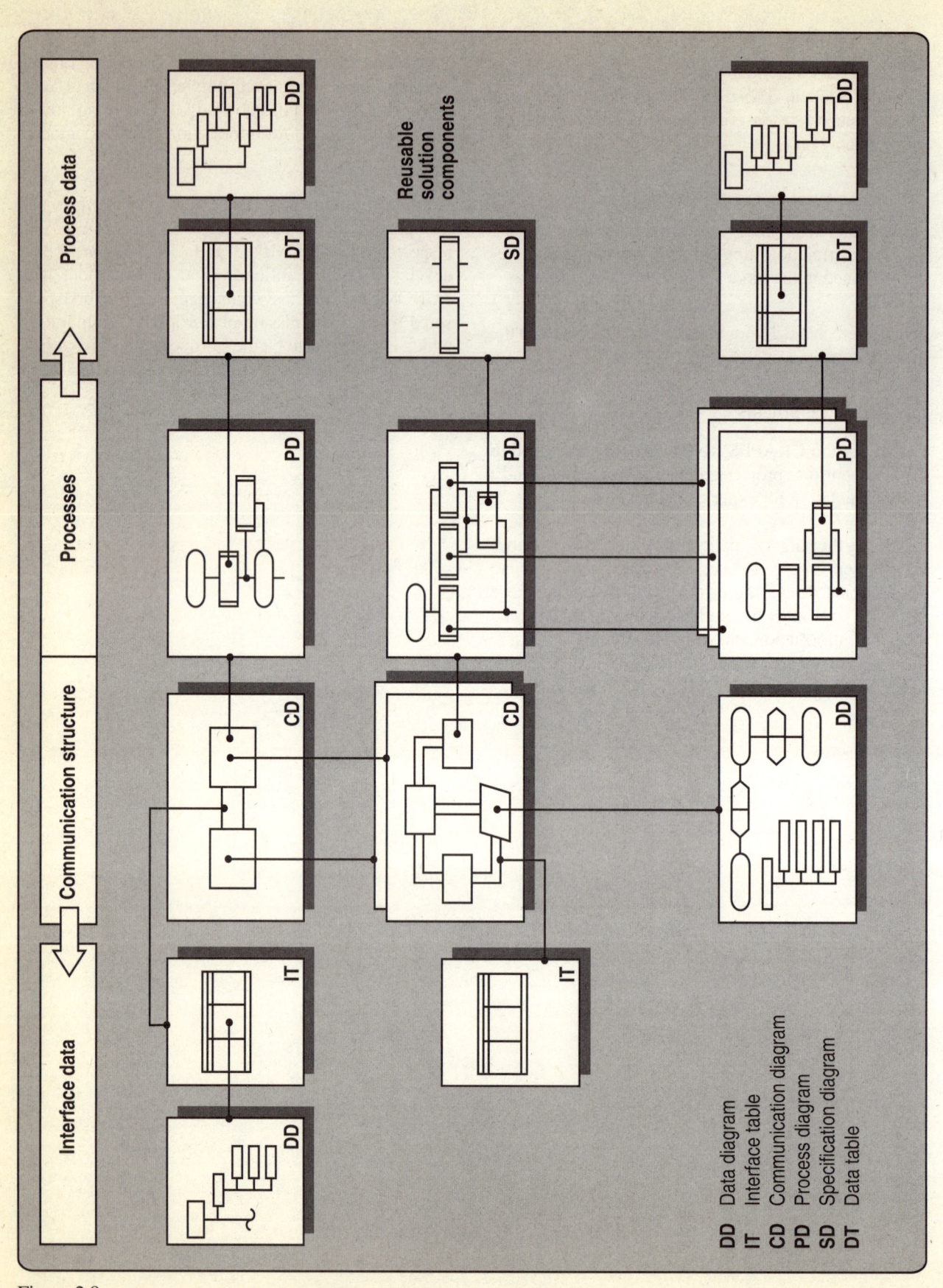

Figure 2.8

2.2 Overview of DOMINO/GRAPES

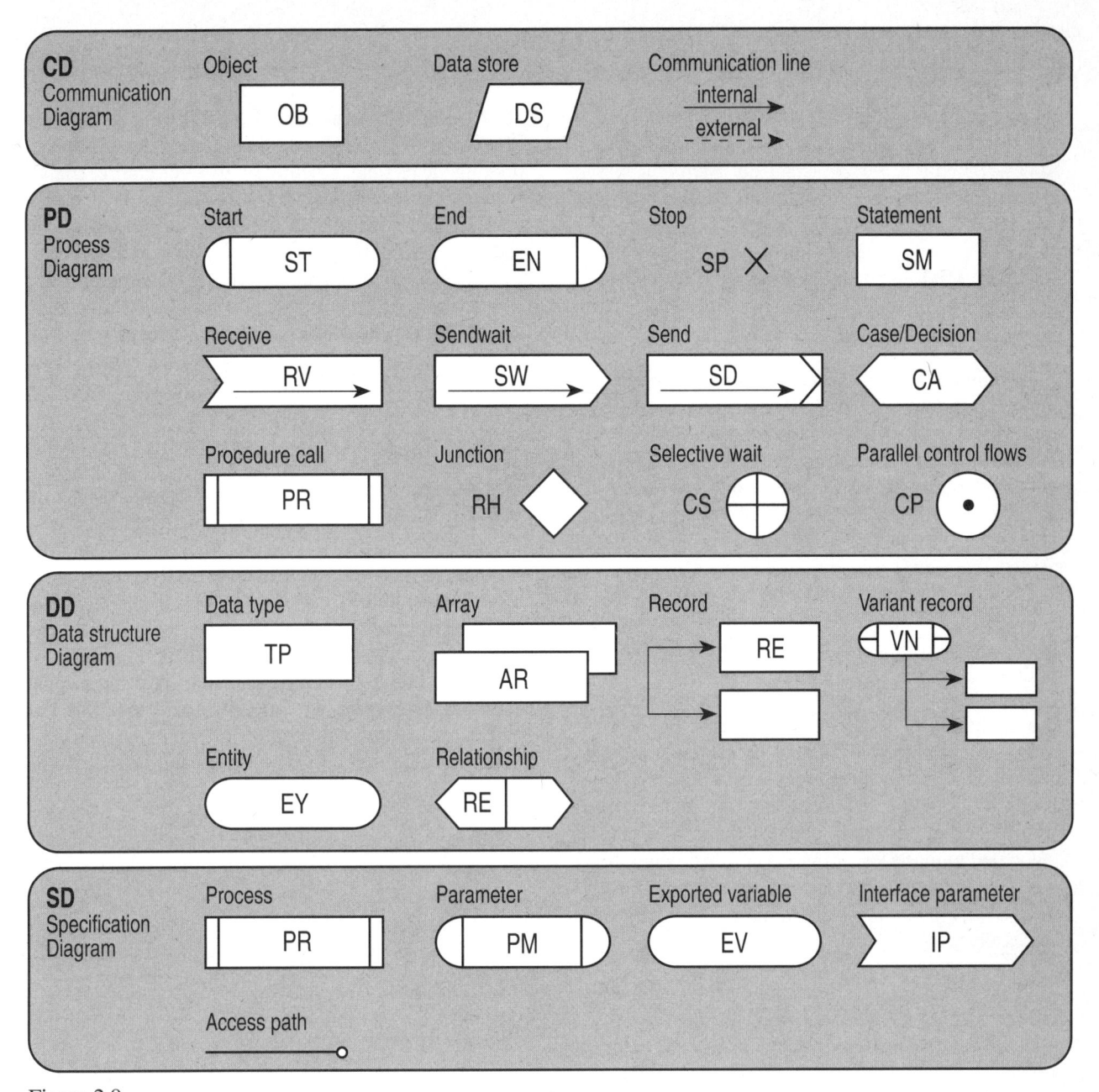

CD Communication Diagram
- Object: OB
- Data store: DS
- Communication line: internal, external

PD Process Diagram
- Start: ST
- End: EN
- Stop: SP
- Statement: SM
- Receive: RV
- Sendwait: SW
- Send: SD
- Case/Decision: CA
- Procedure call: PR
- Junction: RH
- Selective wait: CS
- Parallel control flows: CP

DD Data structure Diagram
- Data type: TP
- Array: AR
- Record: RE
- Variant record: VN
- Entity: EY
- Relationship: RE

SD Specification Diagram
- Process: PR
- Parameter: PM
- Exported variable: EV
- Interface parameter: IP
- Access path

Figure 2.9

3 Motivation, Background and Philosophy

In this chapter we outline the motives of both organizations for defining a formalized software development process. Therefore we will sketch out briefly the history and the philosophy of SSADM and GRAPES.

3.1 SSADM

History & development

The origins of SSADM lie in a time when the UK government has had considerable success in managing the development of computer programs through its Jackson-based programming methodology SDM. At that time there was a perception that this was only tackling one end of the problem of systems development: it was possible to produce good programs which tackled the wrong job. Experience with SDM raised expectations that a similar methodological process (in some people's minds, perhaps even a "cook-book") could be applied to deciding what programs should be produced.

The methodology selection development process was based around objectives of finding proven, objective, analysis and design methods which produced a good database design and produced end-products usable in SDM. Parallel with this activity there was activity concerned with the selection of a compatible approach to project management, which resulted in the selection of PROMPT, and ultimately in the development of its successor, PRINCE.

SSADM was developed by combining elements of two LBMS courses, themselves based on previous techniques developed by BIS and Gane & Sarson. One of the courses contained a primitive use of entity life history analysis and thus was seen to provide the beginnings of a bridge to SDM, which used the same diagramming techniques. In the early trials of SSADM "version 0" it became evident that the entity life history technique needed more formality and that there was still a gap between the SSADM end-products and program specification. This gap was largely plugged in SSADM version 4.

Figure 3.1 shows the ancestry of techniques used in SSADM version 4. Note that LSDM developed out of SSADM, and not the other way round, as is often incorrectly suggested in the computer literature.

As part of the definition of SSADM version 1, an SSADM - PROMPT interface guide was produced, relating the technical methodology to its control framework. The development of SSADM and its project control environment can thus be seen to have been accomplished almost entirely by a bottom-up process. This may explain some of SSADM's strengths and weaknesses. The techniques in SSADM show good fitness for purpose, and are well integrated at the level of purpose. But no formal syntax exists covering all techniques.

SSADM intended implementations

SSADM is directed at 3rd generation languages (mainly COBOL) and 4th generation languages.

SSADM is not currently directed towards object-oriented implementations, although it is capable of being enhanced to map onto those implementations when the underlying technology matures. The construction of update and enquiry process models currently predisposes SSADM towards conventional implementations. For object-oriented implementations it would be advisable to map from effect correspondence diagrams onto the final design.

SSADM is being enhanced to deal with distributed systems. Such enhancement has not been taken into account in this study.

SSADM does not currently cover knowledge-based systems.

SSADM application areas

SSADM is aimed at the production of administrative data processing systems. It can be used for the development of any system which is concerned with managing the behaviour of some portion of the real world, to the extent that the system is structured around a model of the reality it is trying to manage. SSADM could thus be used for other applications such as process control or for discrete event simulation. In practice, however, its domain of application is almost exclusively administrative data processing systems.

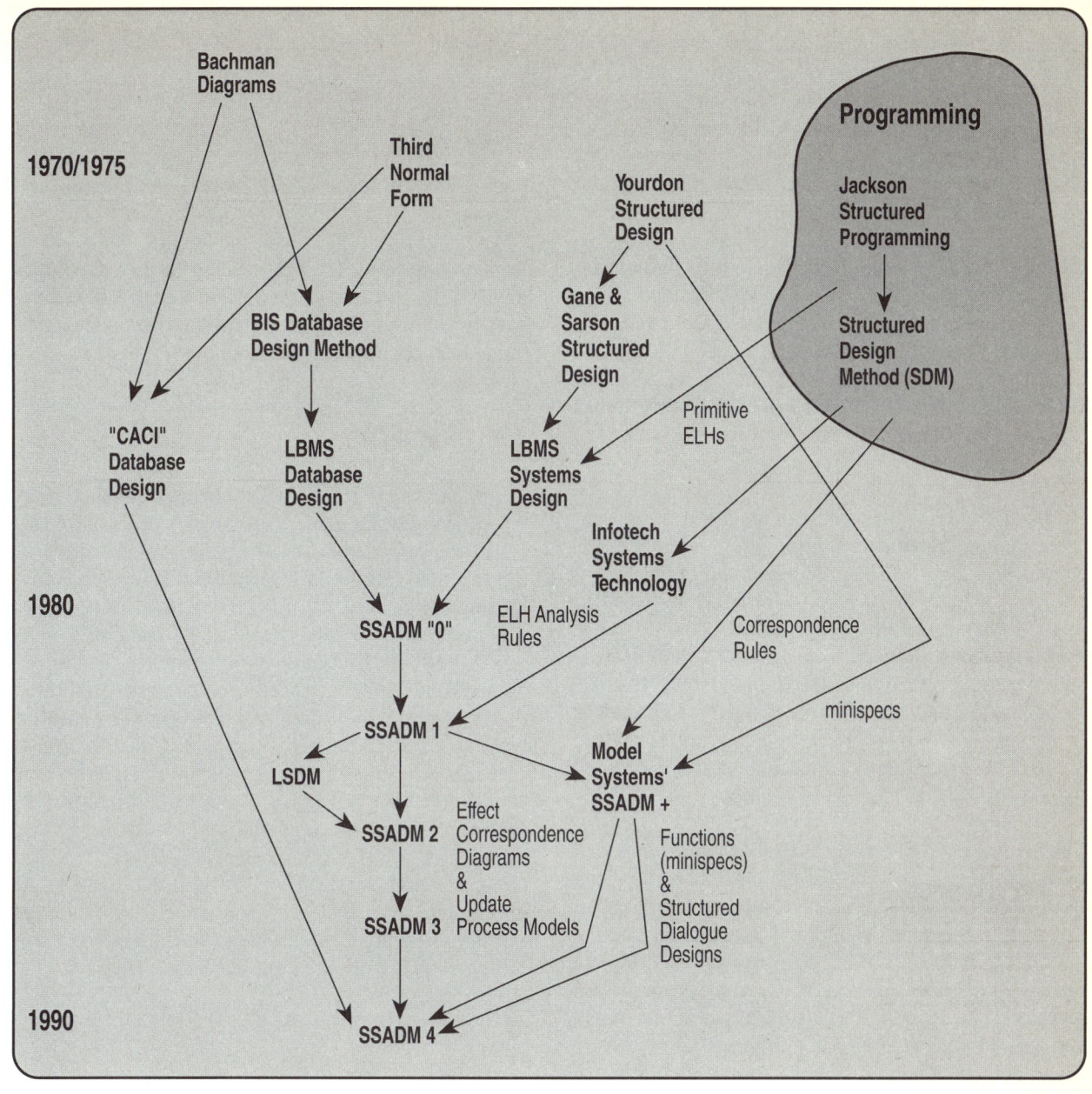

Figure 3.1

3.2 GRAPES History and Development

The roots of the idea of formal software specification and verification within the Siemens company lie in the mid 60's. Since the early 70's two approaches have been in use:

- the graphical approach using Petri nets, state transition diagrams, square charts and Nassi-Shneiderman diagrams;
- the "mathematical" approach using predicate logic specifications, algebraic specifications and formal verification.

Both approaches have been applied in various projects, e.g. "Netz", a large communication software project or the development of a LISP compiler.

A major milestone in this development was the project "Olympia 72", where the complete information system for the Olympic games 1972 in Munich was developed by the Siemens company. During this project it was the first time that principles of modern software engineering were applied on a large scale, e.g. that a structural life cycle process model, the chief programmer team concept and structured programming were used.

In the problem analysis phase of Olympia 72 two "directions" were distinguished. The "vertical analysis" identified the different areas of application, e.g. basketball, hockey, swimming etc. The "horizontal analysis" was done to find identical functionalities in the different areas, e.g. starter lists, result lists etc. So somehow unconsciously the idea of reuse was applied without having elaborated a concept for it.

Based on the experiences collected during Olympia 72, the Siemens Process Technology was defined and a commitment to the graphical approach (to formal software specification) was achieved.
By the way: The Olympia 72 project was performed successfully and punctually. When the Olympic games started at August 24th, 1972 the information system was in full operation.

During the late 70's the concept of graphical system and software specification went through a period of clarification and ripening. So at the beginning of the 80's it became clear that the Siemens modelling methodology of information systems would be based upon the concept of hierarchical nets of communicating automatons. At that time IORL (Input/Output Requirements Language), a modelling technique based on ideas from Carnegie Mellon University and elaborated by the Teledyne Brown Engineering company in a product named TAGS, was adopted by Siemens.

Since 1982 the IORL tool CADOS (Computer-Aided Design for Organizers and Systems engineers) has been used within the AP (Application software & Projects) division of Siemens. In addition to the tool, training material and a CADOS-based software development method have been produced. Experiences in several projects showed that IORL was oriented too much towards old-style FORTRAN and the wish arose to have a modified IORL more suitable for the modelling of commercial applications.

This is when GRAPES was born. Around 1986, based on the experiences with CADOS, the graphical modelling language GRAPES was defined. The basis of GRAPES was and is still the concept of a hierarchical net of communicating automatons. But it has been slightly modified using the ideas of object-orientation. Also the concept of data modelling, which was completely missing in CADOS, has been integrated into GRAPES.

The resulting GRAPES is composed mainly of three components representing:
- system structure: processes (objects) and data flows;
- structure of information: E-R modelling and structured type definitions;
- structure of information processing: process diagrams.

At the moment GRAPES is being further developed towards the capacity to model embedded knowledge-based systems and further object-orientation and additional features for simulation support.

3.3 Comparison

It is interesting to note that GRAPES is a methodology that has largely developed top-down, whereas SSADM has developed bottom-up. Differences in properties of the two methods are largely attributable to these contrasting approaches. Top-down design is strong where the scope of a problem is well understood; it is weak where the boundaries of the problem or its attributes are not well understood and need to be widened. Bottom-up design is strong in dealing with new specific facts; it is weak in integration. Perhaps this is why SSADM is richer in techniques, fact representation and knowledge acquisition than GRAPES but is less formal and its specifications are unexecutable.

GRAPES and SSADM are both aimed at the same application areas. They are both aimed at broadly the same implementations, although GRAPES is more committed to object-oriented implementation.

4 Methodology Style

In this chapter we try to identify the distinctive characteristics of each approach in order to give a high-level understanding of what the approach is, what kind of techniques it uses, and what role those techniques play in the particular approach.

4.1 SSADM Style

Three models of systems

SSADM sees information systems as models of real-world systems. In the SSADM literature three models, or views of data, are referred to:
- logical data model
- data flow diagrams
- entity life histories.

Logical data model
This diagram cuts the system up into a collection of entities (which represent the real-world objects being modelled) and relationships between those entities.

Data flow diagrams
These diagrams give a hierarchical decomposition of a system into a set of processes. They provide a kind of static view of the system dynamics in that they show the main data processing activities and the connections between them. They don't show the individual dynamics of the activities, but rather their relationships.

Entity life histories
These diagrams provide a description of the dynamics or "behaviour" of the system, which may be thought of in this way:
- A system is a collection of related entities.
- The behaviour of a system is the co-ordinated sum of the behaviours of the entities making up the system.
- The behaviour of an entity is the organised collection of events affecting that entity; the organisation of such a collection of events is described by a regular expression, called the entity life history.

The co-ordination of the behaviours of the entities is managed in an "effect correspondence diagram".

Temporal view

Entity life histories give SSADM the capability of representing in their most direct form all the fundamental temporal sequences of activity required in the system.

Hierarchical decomposition is used for system scoping & interface definition, not for detailed processing

Although SSADM develops data flow diagrams, the final processing description does not come directly out of the data flow descriptions. Nor does the data flow diagram decomposition necessarily define the final structure of the system components. The data flow diagrams are used to define the end-user interfaces. Then the functionality associated with an end-user playing a particular "user role" is extracted from the data flow diagram into a "function" which is the main packaging unit of physical implementation.

The detailed processing is developed by identifying the business events carried by the data flows and using those events (supplemented by events identified from the logical data model) to build entity life histories.

When the entity/event analysis is complete the effect correspondence diagrams supplant those parts of the data flow diagrams which are concerned with the relationships between entity-level units of processing.

3-schema architecture

In effect, an SSADM development recognizes that there are different types of what other methodologies call "logical" processing descriptions: those concerned with end-user functionality packaging, and those concerned with the fundamental underlying processing. These different types of processing description are described more fully in chapter 7.

Because SSADM cuts the logical processing up into two components whereas most other methodologies do not, when we add the physical processing we can think of SSADM as a 3-schema architecture as distinct from the older style 2-schema architectures.

Detailed knowledge acquisition & semantic consistency checking

Because the underlying fundamental processing is more objectively based than the end-user processing, and because of its direct temporal viewpoint SSADM can support very detailed knowledge acquisition techniques. It can support a strong focus on semantic consistency and completeness because of implications in one entity life history about what can happen in another. In a way, what is supported is a kind of semantic "double-entry book-keeping".

Cross-checking by double development

A technique proposed by Gilb is to develop systems twice and compare the end-products. SSADM uses this technique without the overheads of developing all of the end-products twice. Different techniques are used to develop different but slightly overlapping aspects of the system, from different perspectives, and an "intersection" product is developed. For example, a data model is prepared intuitively and by analyzing access requirements; this model is cross-checked against models produced by TNF analysis. Similarly, updating requirements are discovered from DFD analysis and entity life history analysis, and can be cross-checked.

4.2 GRAPES Style

Hierarchical decomposition of systems

Models are abstractions, i.e. generalized representations of systems in the real world. A system in the real world is a combination of individual components which interrelate with each other and together provide the range of functions assigned to the system.

GRAPES models are abstractions of real-world systems where the components of a system are represented by objects that communicate with each other. An object can in turn be viewed as a system. The services that an object provides to the outside are provided by the subsystems the system is composed of and their interaction.

Iterating this process of refinement, a pyramid of objects in hierarchical layers emerges. This hierarchy represents the system on different levels, each of which shows a different degree of structuring and detail.

Relationship between subsystems

An object at one level is an abstraction of an independent subsystem. The relationships between subsystems are represented by communication relationships between the objects. The flow of information is represented by message passing. This means the only communication carriers between objects are messages. These contain the information to be transported from one object to another.

Parallelism of subsystems

The objects are autonomous entities, each independently performing its allocated task. They can communicate with each other, i.e. they can exchange information and thus affect each other's activities, but are otherwise completely separate. This modularisation is enforced in GRAPES by describing an object, as seen from the outside, solely by its functions and its interface.

The latter defines the message types which an object can send or receive. The encapsulation of the information stored in an object and the hiding of its internal structure is the reason why objects in GRAPES can act independently and fully in parallel. There is no restriction on the parallel action of objects whatsoever, except their need to communicate synchronously with other objects or the need to receive information from other objects.

Message passing

The communication medium used for transporting messages is the channel. Channels are always connected to two objects, thus permitting message transfer. Channels are typecast and direction-specific, i.e. a single channel can only pass messages of a specific type and in a particular direction. An object communicates with its environment via channels which connect its interface with the interfaces of its subobjects. If a message is sent via one of these channels, it is transferred to the appropriate channel at the interface and relayed from this. If a message arrives on an external channel, it is relayed via the appropriate internal channel to the connected subobject.

Consistency and completeness

In GRAPES the consistency of the object interfaces and the communication paths, i.e. the succession of channels a particular message has to pass in order to get from the sending object to the receiving one, is automatically checked. These consistency checks include e.g. the "syntactic" level of connectivity of channels and the "semantic" level expressed by classification of messages by types.

The focus on the static structure of the system together with the consistency checks of the connection network on each level helps the GRAPES users to achieve completeness of system components in the model. For example, if a channel is not connected on one side this is a hint that there is a missing object. Objects or subsystems which have no connection to the rest of the system, i.e. no means to communicate with the outside, indicate missing channels.

Elementary objects

Objects whose characteristics and activities as seen from the outside are so elementary that they are no longer defined in terms of a subobject structure are in GRAPES described by a process. A process is a sequence of operations which represents the activities of the object, e.g. its reaction to any messages received. The description of this process is based on regular expressions. The GRAPES user sees a graphical flow representation of these regular expressions.

Objects whose behaviour is explicitly defined by processes are called process objects. Process objects form the bottom of the object hierarchy. The upper level objects are called structure objects. They are defined in terms of the communication structure of their subobjects. The behaviour of a structure object results from the interaction of its subobjects and its current state is defined by the current state of all its subobjects.

Process states

A process explicitly defines the possible states and state changes of the process object it describes. The object state is defined by the current memory contents, the queued messages which have not yet been evaluated and the current receive state of the process.

When in a receive state, a process accepts a message, if necessary waiting for it to arrive. Depending on the current receive state, the contents of the received message and the current memory contents, the process then performs a transition, thereby switching to a new receive state. During the transition, the process can modify the memory contents and send messages to the object's interface, which then relays them.

Again the consistency of send and receive operations in process descriptions is checked with the external interface description of the corresponding object.

Reusability

In GRAPES the identification of reusable components at different levels plays an important role. It starts with the identification of classes of data that elicit some kind of similar behaviour. This abstraction process results in typing of data structures and data items.

The basic form of reusability of processing is obtained by declaring common parts of processing as procedures, as known from numerous programming languages. From the outside only the parameters and the functionality of a procedure are known, which makes it reusable in all the places where such functionality is required. Both reuse within one single object and reuse in several objects are possible.

GRAPES supports a higher level of reusability of processing by providing means to declare modules together with their export interfaces. Modules are in principle a higher level of process abstraction, allowing local definition of model elements, i.e. data types, variables, procedures, functions and further local modules.

GRAPES provides a third kind of reusability, which is at the object level. During the modelling process it is not unusual for a number of objects with the same behaviour pattern to occur in the same system. For this reason, the GRAPES language provides the facility for defining object types. If an object description (behaviour pattern) is to be used more than once, an object type can be defined and referenced in the definition of several objects.

4.3 Comparison of SSADM and GRAPES Styles

The differences in style are:
- GRAPES aims for consistent representation of the discovered facts.
- Due to the previous point GRAPES is strong on consistency, type and scope checking, whereas SSADM is strong on knowledge acquisition. This means that SSADM is more methodology-oriented, whereas GRAPES is more description-oriented.
- The structure of SSADM models is "dynamic-structure-biased", i.e. each object symbol represents one single processing unit.
- The detailed analysis of GRAPES is based on message passing, which SSADM it is based on structuring atomic business events.
- In both approaches process descriptions are based on structured constructs, but use different notations. SSADM uses tree representations; GRAPES uses flow representations.

GRAPES is really a language intended for the formal representation of system structures. SSADM is a collection of techniques and rules for discovering the structure of a system.

5 Architecture of the Developed Systems

During this study we tried to explain why, in SSADM, some processing specifications arose out of the entity event analysis (and out of access path analysis) and why some processing specifications arose out of function definition and dialogue design, whereas in GRAPES process specification proceeded by a uniform activity of decomposition. We characterized the first set of SSADM processing specifications as describing that set of processing which was necessarily truly independent of the implementation environment, and the second set of the processing specifications as being about the embedding of the first set of processing in the user's environment. We could see that the code defining the final system would contain three elements:

- code which implements the user interface structure
- code which implements the necessary processing semantics
- code which implements the physical storage and processing environment.

This separation of concerns closely resembled the separation of concerns in the ANSI-SPARC 3-schema database architecture.

The ANSI-SPARC 3-schema database architecture

In the mid 1970's the ANSI DBMS standards programme resulted in the production of a report on the architecture of database management systems which supplanted the older, more naive, notion of a division between the logical and physical views of data. The notion of physical data, i.e. data as actually stored, was retained in the notion of an "internal schema" which defines the storage structure of the data. However, data that was previously thought of as logical data was now categorized as being described either in an "external schema" or in a "conceptual schema".

Broadly speaking, the role of the conceptual schema is to define the meaning and interdependences of the data, whereas the role of the external schema is to define how the data is organised to suit a particular end-user purpose (as shown in figure 5.1).

The architecture was originally proposed in an earlier joint report by the IBM user groups GUIDE and SHARE. At the time of publication of the GUIDE-SHARE report there was considerable lack of understanding about the role of the conceptual schema or "entity data". Since then, however, the idea of having a central entity model of the user's business has gained wide acceptance as a way of, among other things:

- understanding the user's business
- gathering data about volumes and promoting the physical data design
- mediating between different, co-operating DBMSs
- gaining some portability of design
- providing building blocks from which the external views can be constructed

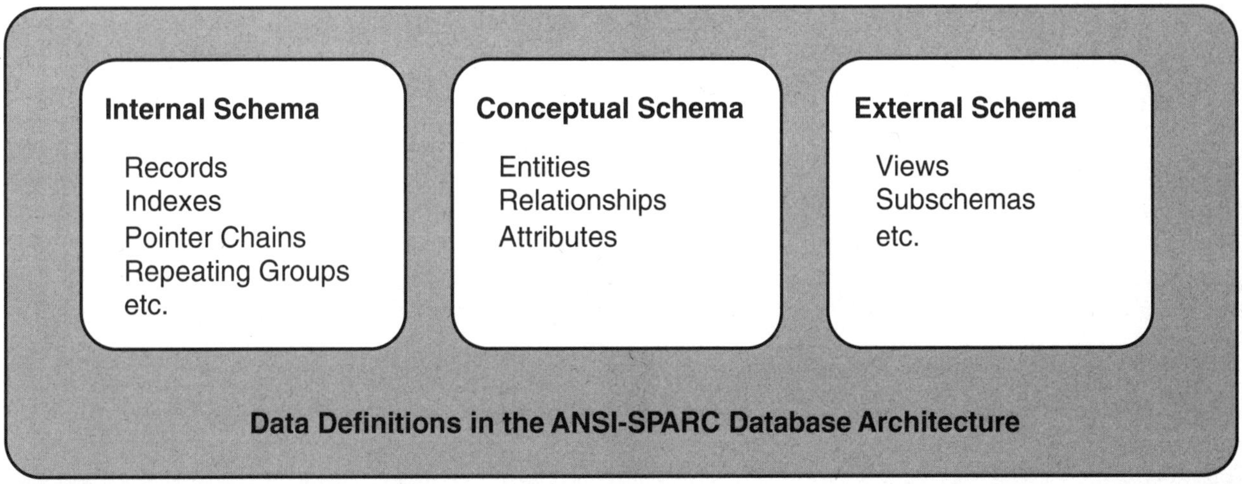

Internal Schema
Records
Indexes
Pointer Chains
Repeating Groups
etc.

Conceptual Schema
Entities
Relationships
Attributes

External Schema
Views
Subschemas
etc.

Data Definitions in the ANSI-SPARC Database Architecture

Figure 5.1

5.1 SSADM Architecture

3-schema processing architectures

The idea of extending the 3-schema architecture from data into processing was raised (by Jim Lucking of ICL, among others) at the time of publication of the ANSI-SPARC report, principally to extend the conceptual schema to cover a "conceptual scenario", but no-one knew how to do this at the time.

The three areas of concern covered in the SSADM processing products, however, closely resemble the areas of concern covered by the ANSI-SPARC 3-schema database architecture. This led us to suggest that the processing specifications produced by SSADM conformed to a 3-schema framework.

- A conceptual scenario would be both a static and a dynamic model of the user's world and capture all of the semantics, i.e. all of the necessary business knowledge, about the system to be built.
- An external scenario would be a model of how the user of a data processing system interacted with that system.
- An internal scenario would describe the processing necessary to map a conceptual view onto an implementation.

The advantages of recognizing a conceptual scenario as a system component distinct from an external scenario are extensions of the advantages of recognizing a conceptual schema:

- Collecting in a recognizable place all of the knowledge about the business.
- Creating code which is portable both between different user environments and different physical implementations.
- Separating that part of the system which is objectively constrained and so must be *discovered* (the conceptual view) from those parts whose properties are subjectively constrained and so must be *designed* (the internal and external views).

The SSADM 3-schema processing architecture

It is quite clear to which of the three levels of processing description almost all of the SSADM end-products relate. The only products which are difficult to classify are the entity access paths and enquiry process models. To the extent that these products describe necessary feed-back information in the implemented system, independent of user processing environment, we have included them in the conceptual view which is shown in figure 5.2.

Internal Scenario	Conceptual Scenario	External Scenario

Records
Indexes
Pointer Chains
Repeating Groups
etc.
--- 	
Program / Data Interface
	Entities
	Relationships
	Attributes
	--- 	
	Entity Life Histories
	Events / Commit Units
	Effect Correspondence
	Entity Access Paths
	Update Process Models
	Enquiry Process Models
		Screen Layouts
		Printer Layouts
		etc.
		--- 	
		Functions
		  Dialogues
		  Batch Runs
		Scheduling

Data & Process Definitions in the SSADM Architecture

Figure 5.2

The 3-schema processing architecture explains a great deal about the history of SSADM, its relationship to other methodologies, the role of prototyping in SSADM and other systems, where SSADM might relate to knowledge engineering, and the possibilities that exist for developing "cook-book" design methods:

- The history of SSADM can be seen as the gradual recognition and emergence of the 3-schema processing architecture. In the early days, there was reluctance even to acknowledge that the LDM was an entity model; it was commonplace to refer to "data groups" rather than "entities". There was continuous tension between those who thought the data flow diagrams were the most important element and who sometimes wanted to scrap entity life histories, and those who thought that entity life histories were the only objective way to describe system dynamics and that data flow diagrams were so subjective as to be almost useless.

  The somewhat unsatisfactory rules for the production of first-cut programs in versions of SSADM up to and including version 3 were based on criteria about processing cycles which are quite clearly external scenario considerations.

  In SSADM version 4 it is now clear that the prime focus of entity life history analysis is the conceptual scenario and that the prime focus of data flow diagram analysis and the resulting function definition is the external scenario.

- Hard methodologies like IE can now be seen as failing to address adequately the discovery of the conceptual scenario.

- Cultural methodologies such as Mumford's can now be seen as addressing the performance and acceptability of the external scenario.

- In SSADM prototyping is used for dialogue design (the external scenario) and for improvement of a first-cut design into an optimised design (the internal scenario). We believe that prototyping methods are more appropriate to design than to discovery, and have practical experience which seems to justify this belief. If this belief is true SSADM makes exactly the right use of prototyping.

- The conceptual scenario holds all of the significant knowledge about the application area. If SSADM is to be extended into the knowledge engineering domain the CCTA's interest should be focussed on the conceptual scenario techniques (LDM, RDA, entity-event analysis).

- Because the conceptual scenario is objectively constrained it is possible to believe that there is in some sense a "right answer". This may not be quite so formally "right" that it is a completely canonical representation of the knowledge such that we could guarantee that two analysts discovering exactly the same knowledge would produce exactly the same representation. However, we know that through the use of stereotypes and a disciplined approach to analysing the entity life histories based on the logical data model we can produce a highly objective and procedural approach to entity-event analysis. A "cook-book" method requires two things: a more or less canonical solution and an objective method for developing that solution. Both of these preconditions are present in the tools for developing the conceptual scenario.

- The external scenario depends on a number of different things: organisation structure, end-user input/output device technology, ergonomics, arbitrary preferences of particular users, audit principles, security, user politics. There is certainly no "right answer" to the trading-off of these different requirements. This implies that there will always be a strong subjective element in the development of a solution. Thus there is little point in searching for an objectively correct method for building DFDs or for an objectively correct method for designing screens. Any method for building the external scenario must be a design method, i.e. must invent solutions; heuristic approaches like prototyping clearly have a strong role to play here.

- The internal scenario depends on tradeoffs between a number of things whose relative importance is subjectively defined: time objectives, space objectives, maintainability. Once again, this implies that there is no "right answer" and that a heuristic, prototyping approach is needed.

- Prototyping and heuristic methods are needed in systems development. But they are not the whole of the story. Just because they are needed in some areas of systems development, this does not imply that they are needed everywhere. The conceptual scenario defines an area where prototyping is less appropriate (and, we believe, counter-productive).

5.2 GRAPES Processing Architecture

GRAPES produces both logical and physical models of the required system. The GRAPES logical models do not distinguish between external and conceptual processes. In fact, using the hierarchical decomposition approach for the development of GRAPES logical models (the method taught on the GRAPES courses) it is not clear that the external scenario and conceptual scenario aspects of a GRAPES logical model would appear in separate GRAPES processes.

However, the GRAPES logical models developed for Hi-Ho quite clearly show components which can be classified as external processes and conceptual processes. This decomposition of the GRAPES logical model into components which have on the one hand an entity-life-history-like appearance and, on the other hand, a dialogue-like appearance is probably due to the fact that GRAPES is partially object-oriented and is being developed into a much more object-oriented product. Object-oriented thinking comes naturally to the GRAPES Hi-Ho design team and would predispose them to the recognition of conceptual scenario processes as separate object types. The GRAPES Hi-Ho design team used this object-oriented approach rather than the hierarchical decomposition method.

Although conceptual scenario objects can be distinguished by inspection of the GRAPES Hi-Ho solution, GRAPES itself does not know that there is a special distinction to be made between objects which represent conceptual scenario processes and objects which represent external scenario processes.

There is no way at present of recording the external/conceptual object distinction in GRAPES. For this reason, and for the reason that a GRAPES analysis using the hierarchical decomposition method would probably lead to a different decomposition.

Nevertheless, the trend in GRAPES is to recognize separate objects which are potentially classifiable as conceptual or external. For this reason we think it is fair to suggest that GRAPES has the potential to encompass a 3-schema processing architecture.

5.3 Examples of the SSADM Architecture in the Hi-Ho Case Study

Neither the SSADM nor the GRAPES case-study developments went as far as physical design. But this is unimportant because the distinguishing difference between SSADM and other methods, including GRAPES, is the splitting of the logical view of processing into an external view and a conceptual view.

Example of an external scenario component

The "arrange interviews for applicants" function results in a new applicant dialogue which has the logical structure as shown in figure 5.3.

In some implementation environments (e.g. Application Master) this function structure might be implemented directly; in others (e.g. CICS) the function might be dismembered into separate processes for each screen.

Examples of conceptual scenario components

The update process models invoked from the "new applicant" dialogue are shown in figures 5.4 and 5.5.

But the update process models are not the basic conceptual processes. Figures 5.6 – 5.9 show four entity life histories which are the basic conceptual processes in applicant registration.

An effect correspondence diagram as shown in figure 5.10 gives the conceptual view of an event, showing how it feeds through the entity life history processes.

The update process model is an implementation of the effect correspondence diagram in a non-distributed, non-object-oriented environment.

The "new applicant" dialogue also invokes an enquiry process model as shown in figure 5.11.

This enquiry process model is included in the conceptual scenario because it expresses an important business rule: vacancies for which the applicant has already rejected a job offer are not eligible for further interview arrangement for the same applicant.

The conceptual processes may be invoked by many external processes. For example, the interview arrangement process may also be invoked as a result of the way in which Hi-Ho is organised to process vacancies, as shown in figure 5.12.

5.3 Examples of the SSADM Architecture in the Hi-Ho Case Study

F1 Online   New Applicant Dialogue

Operations list
1. Get Input
2. i: = 1
3. i: = i + 1
4. Invoke Applicant Registration
5. Invoke Suitable Vacancies Query
6. Invoke Interview Arrangement
7. Put Skill (i) to Applicant Registration

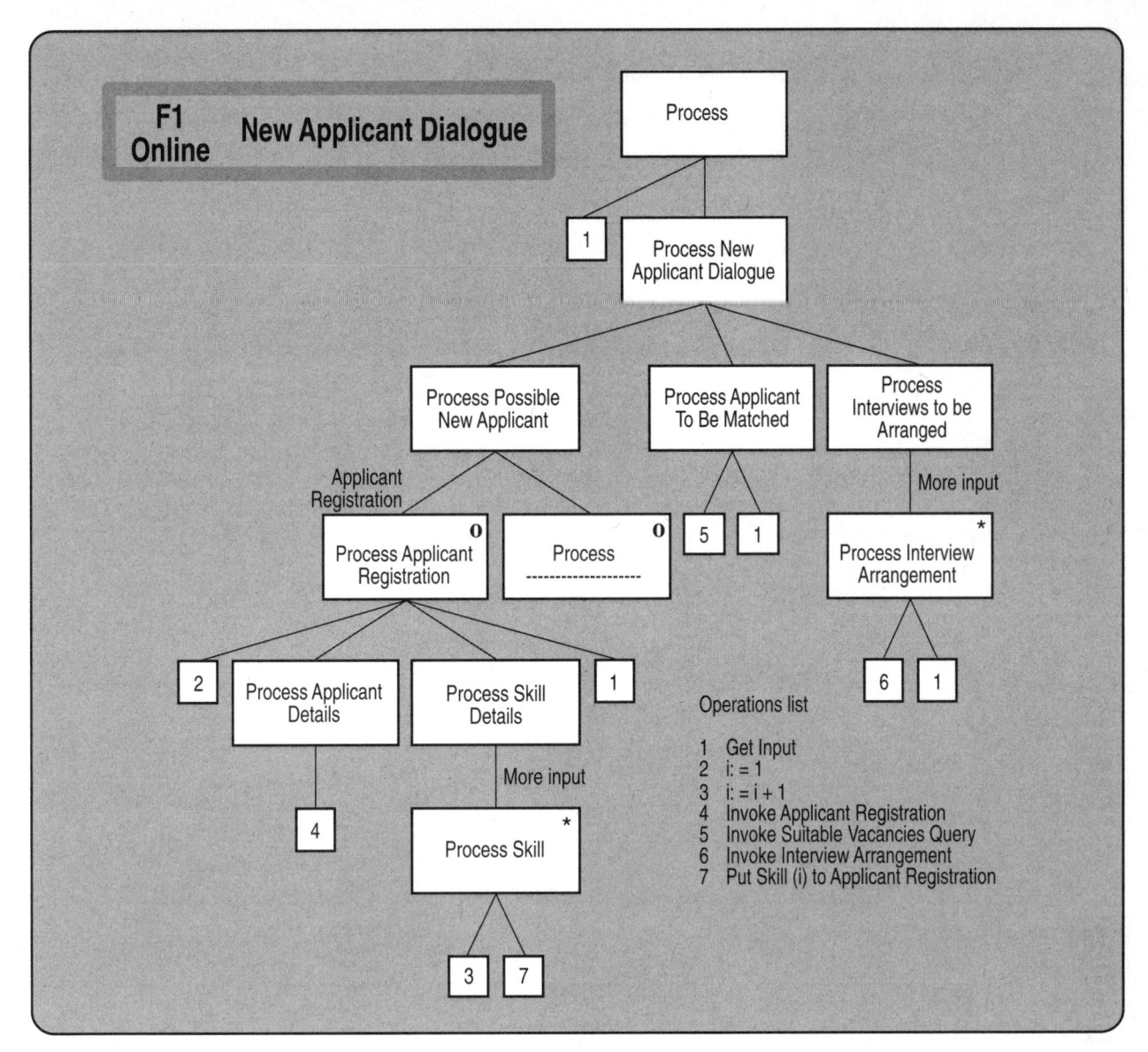

Figure 5.3

Applicant Registration

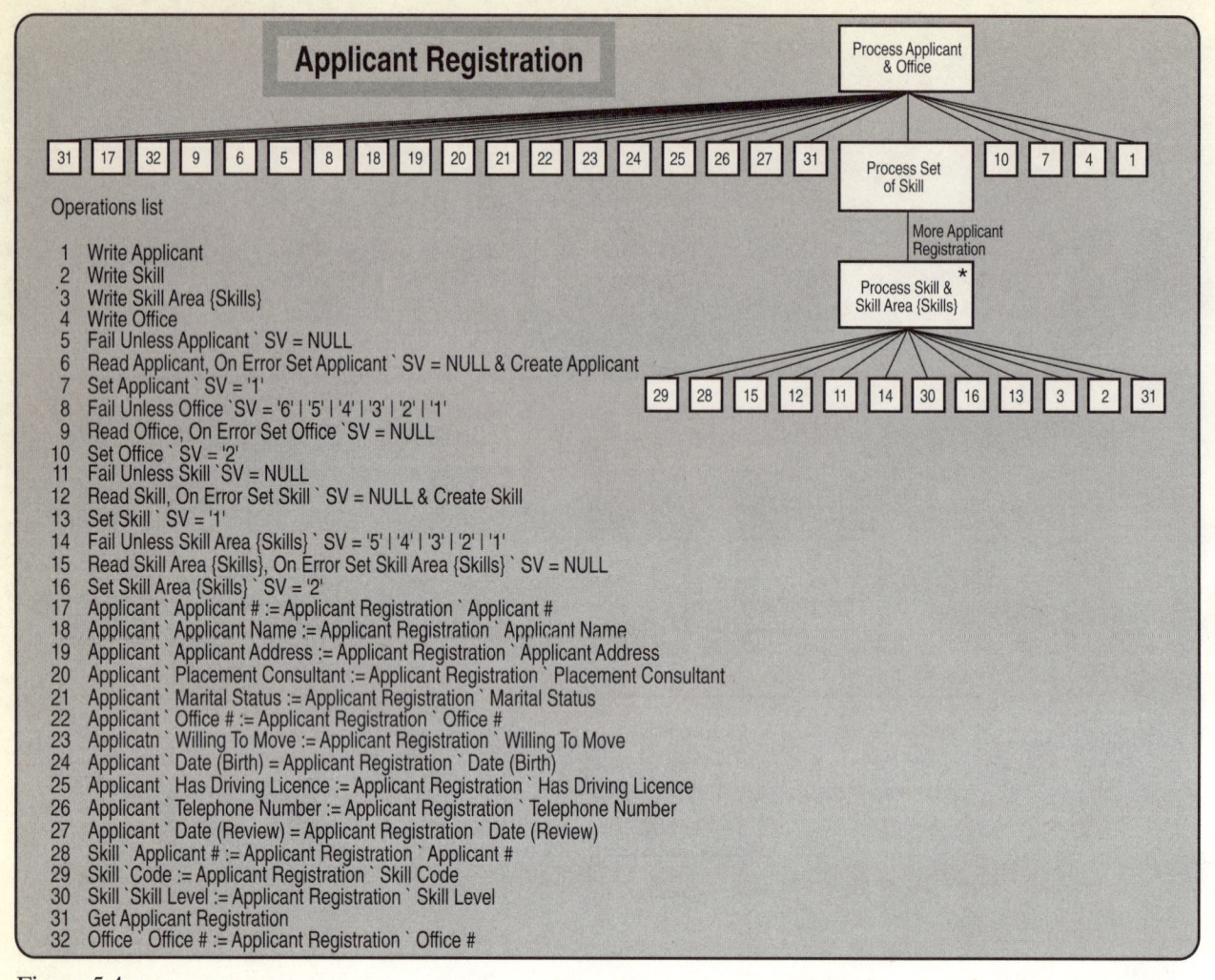

Boxes (tree structure): Process Applicant & Office → 31, 17, 32, 9, 6, 5, 8, 18, 19, 20, 21, 22, 23, 24, 25, 26, 27, 31, Process Set of Skill, 10, 7, 4, 1

More Applicant Registration → Process Skill & Skill Area {Skills} * → 29, 28, 15, 12, 11, 14, 30, 16, 13, 3, 2, 31

Operations list

1. Write Applicant
2. Write Skill
3. Write Skill Area {Skills}
4. Write Office
5. Fail Unless Applicant ` SV = NULL
6. Read Applicant, On Error Set Applicant ` SV = NULL & Create Applicant
7. Set Applicant ` SV = '1'
8. Fail Unless Office `SV = '6' | '5' | '4' | '3' | '2' | '1'
9. Read Office, On Error Set Office `SV = NULL
10. Set Office ` SV = '2'
11. Fail Unless Skill `SV = NULL
12. Read Skill, On Error Set Skill ` SV = NULL & Create Skill
13. Set Skill ` SV = '1'
14. Fail Unless Skill Area {Skills} ` SV = '5' | '4' | '3' | '2' | '1'
15. Read Skill Area {Skills}, On Error Set Skill Area {Skills} ` SV = NULL
16. Set Skill Area {Skills} ` SV = '2'
17. Applicant ` Applicant # := Applicant Registration ` Applicant #
18. Applicant ` Applicant Name := Applicant Registration ` Applicant Name
19. Applicant ` Applicant Address := Applicant Registration ` Applicant Address
20. Applicant ` Placement Consultant := Applicant Registration ` Placement Consultant
21. Applicant ` Marital Status := Applicant Registration ` Marital Status
22. Applicant ` Office # := Applicant Registration ` Office #
23. Applicatn ` Willing To Move := Applicant Registration ` Willing To Move
24. Applicant ` Date (Birth) = Applicant Registration ` Date (Birth)
25. Applicant ` Has Driving Licence := Applicant Registration ` Has Driving Licence
26. Applicant ` Telephone Number := Applicant Registration ` Telephone Number
27. Applicant ` Date (Review) = Applicant Registration ` Date (Review)
28. Skill ` Applicant # := Applicant Registration ` Applicant #
29. Skill ` Code := Applicant Registration ` Skill Code
30. Skill `Skill Level := Applicant Registration ` Skill Level
31. Get Applicant Registration
32. Office ` Office # := Applicant Registration ` Office #

Figure 5.4

Interview Arrangement

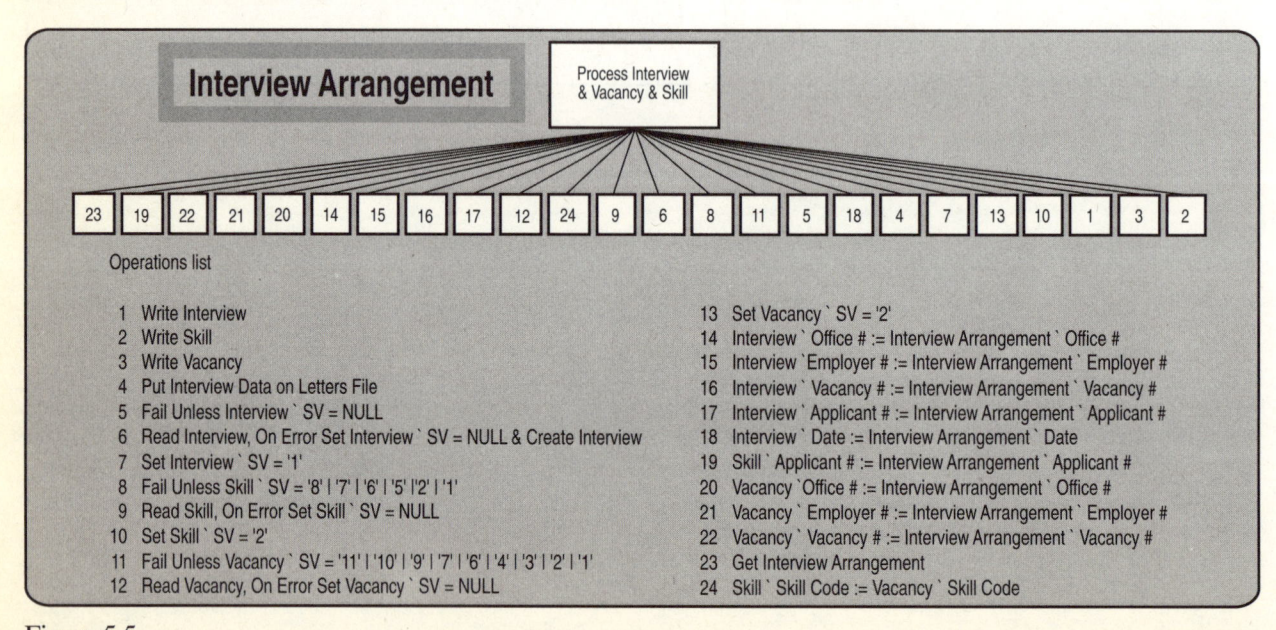

Process Interview & Vacancy & Skill → 23, 19, 22, 21, 20, 14, 15, 16, 17, 12, 24, 9, 6, 8, 11, 5, 18, 4, 7, 13, 10, 1, 3, 2

Operations list

1. Write Interview
2. Write Skill
3. Write Vacancy
4. Put Interview Data on Letters File
5. Fail Unless Interview ` SV = NULL
6. Read Interview, On Error Set Interview ` SV = NULL & Create Interview
7. Set Interview ` SV = '1'
8. Fail Unless Skill ` SV = '8' | '7' | '6' | '5' | '2' | '1'
9. Read Skill, On Error Set Skill ` SV = NULL
10. Set Skill ` SV = '2'
11. Fail Unless Vacancy ` SV = '11' | '10' | '9' | '7' | '6' | '4' | '3' | '2' | '1'
12. Read Vacancy, On Error Set Vacancy ` SV = NULL
13. Set Vacancy ` SV = '2'
14. Interview ` Office # := Interview Arrangement ` Office #
15. Interview `Employer # := Interview Arrangement ` Employer #
16. Interview ` Vacancy # := Interview Arrangement ` Vacancy #
17. Interview ` Applicant # := Interview Arrangement ` Applicant #
18. Interview ` Date := Interview Arrangement ` Date
19. Skill ` Applicant # := Interview Arrangement ` Applicant #
20. Vacancy `Office # := Interview Arrangement ` Office #
21. Vacancy ` Employer # := Interview Arrangement ` Employer #
22. Vacancy ` Vacancy # := Interview Arrangement ` Vacancy #
23. Get Interview Arrangement
24. Skill ` Skill Code := Vacancy ` Skill Code

Figure 5.5

5.3 Examples of SSADM Architecture in the Hi-Ho Case Study

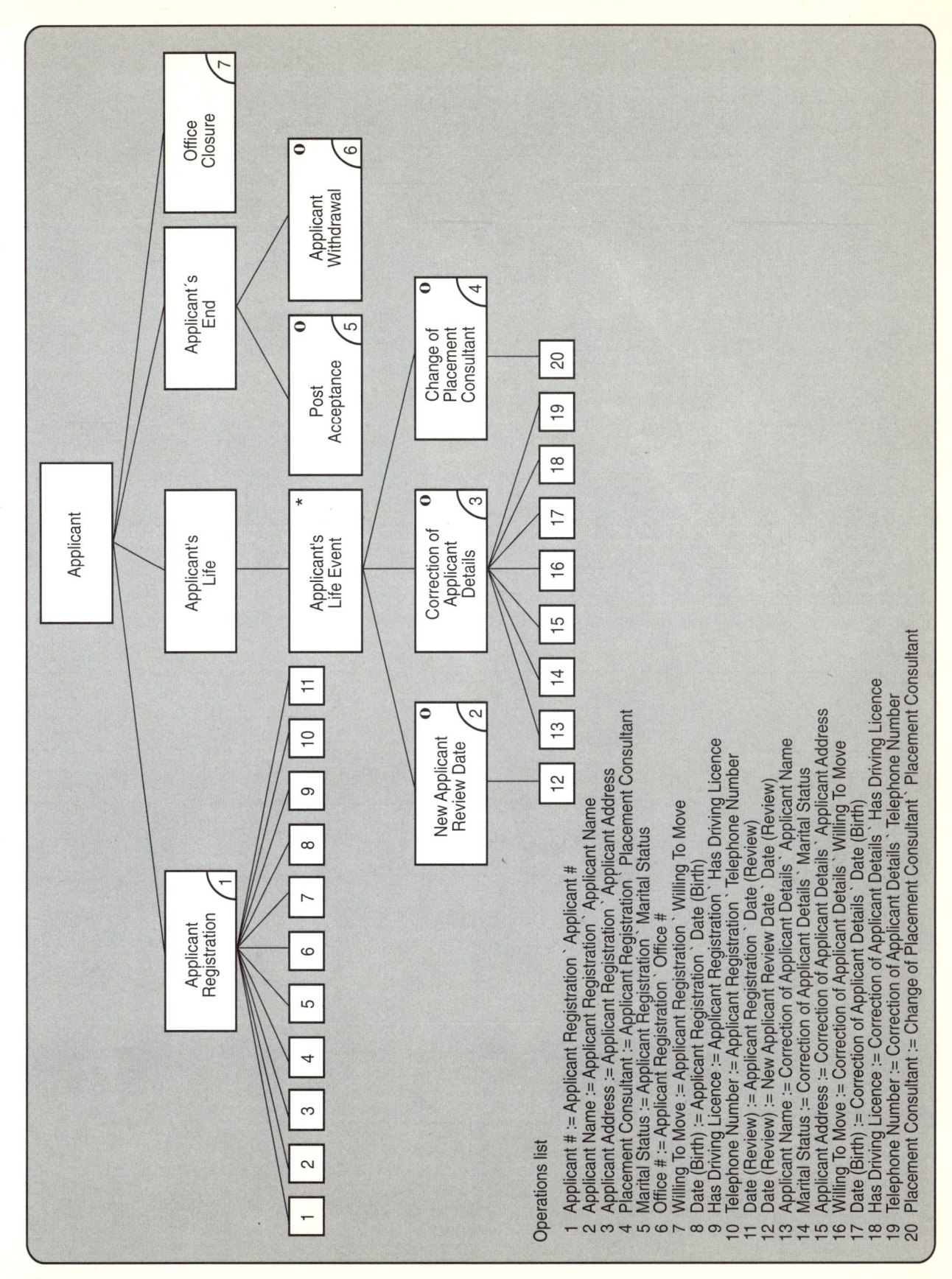

Operations list

1. Applicant # := Applicant Registration ` Applicant #
2. Applicant Name := Applicant Registration ` Applicant Name
3. Applicant Address := Applicant Registration ` Applicant Address
4. Placement Consultant := Applicant Registration ` Placement Consultant
5. Marital Status := Applicant Registration ` Marital Status
6. Office # := Applicant Registration ` Office #
7. Willing To Move := Applicant Registration ` Willing To Move
8. Date (Birth) := Applicant Registration ` Date (Birth)
9. Has Driving Licence := Applicant Registration ` Has Driving Licence
10. Telephone Number := Applicant Registration ` Telephone Number
11. Date (Review) := Applicant Registration ` Date (Review)
12. Date (Review) := New Applicant Review Date ` Date (Review)
13. Applicant Name := Correction of Applicant Details ` Applicant Name
14. Marital Status := Correction of Applicant Details ` Marital Status
15. Applicant Address := Correction of Applicant Details ` Applicant Address
16. Willing To Move := Correction of Applicant Details ` Willing To Move
17. Date (Birth) := Correction of Applicant Details ` Date (Birth)
18. Has Driving Licence := Correction of Applicant Details ` Has Driving Licence
19. Telephone Number := Correction of Applicant Details ` Telephone Number
20. Placement Consultant := Change of Placement Consultant ` Placement Consultant

Figure 5.6

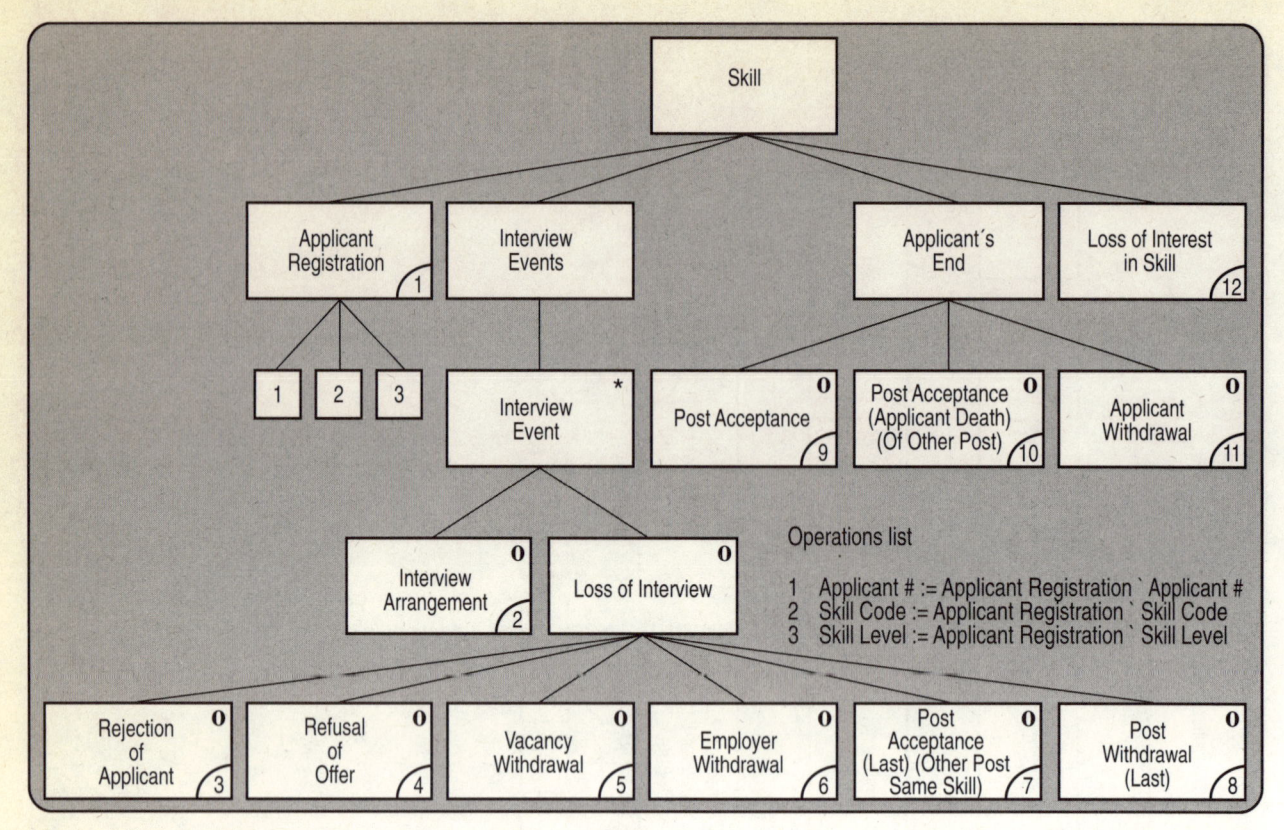

Figure 5.7

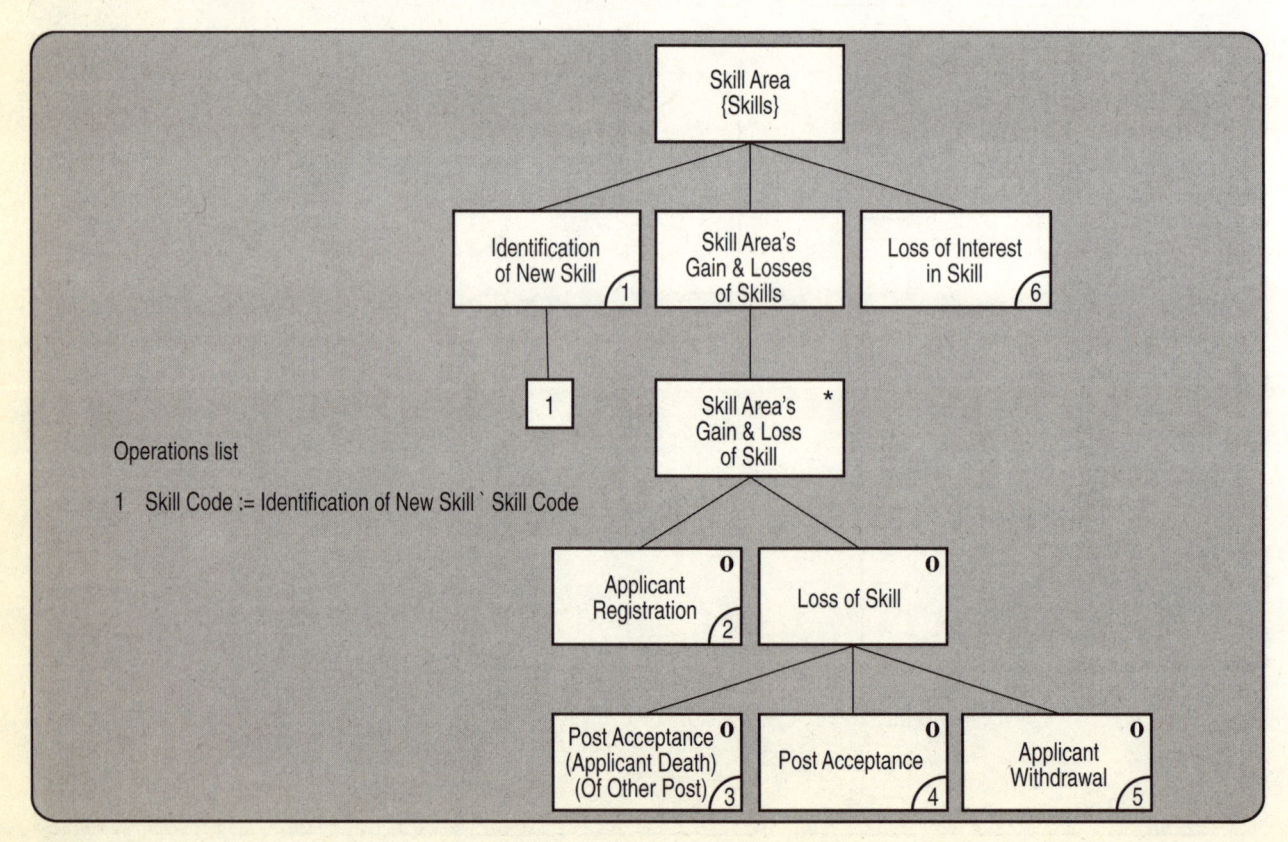

Figure 5.8

5.3 Examples of SSADM Architecture in the Hi-Ho Case Study

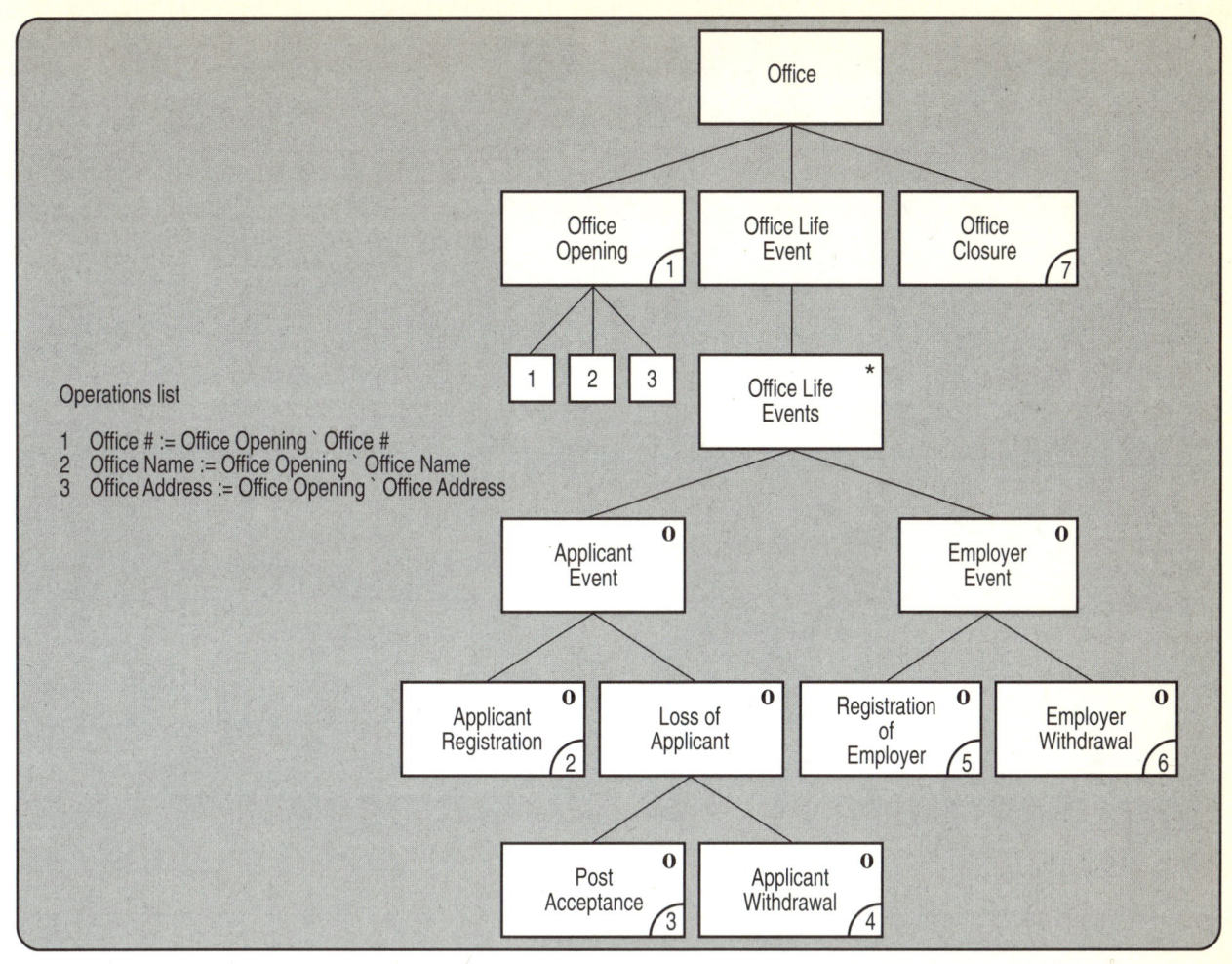

Operations list

1   Office # := Office Opening ` Office #
2   Office Name := Office Opening ` Office Name
3   Office Address := Office Opening ` Office Address

Figure 5.9

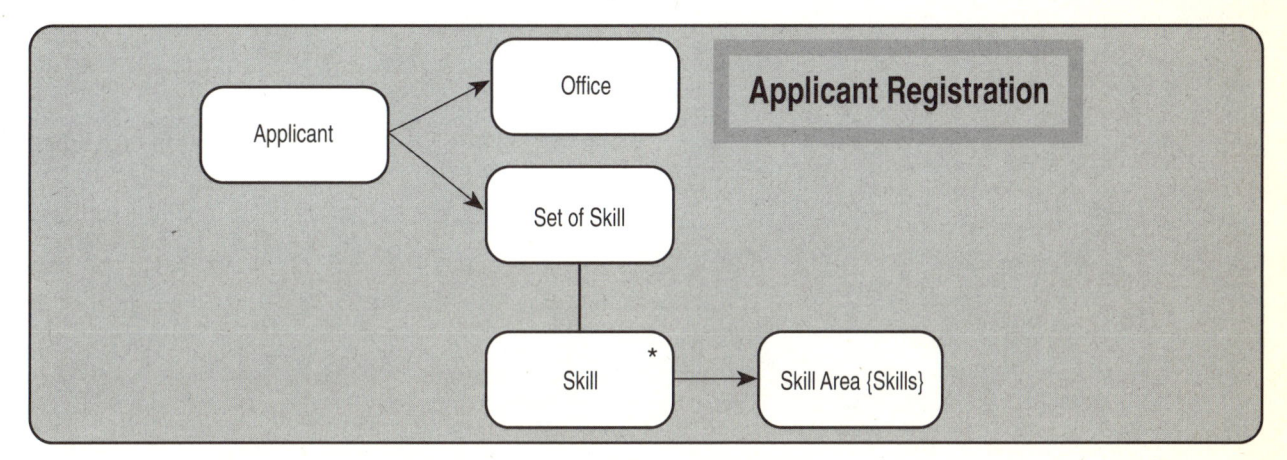

Figure 5.10

Suitable Vacancies Query

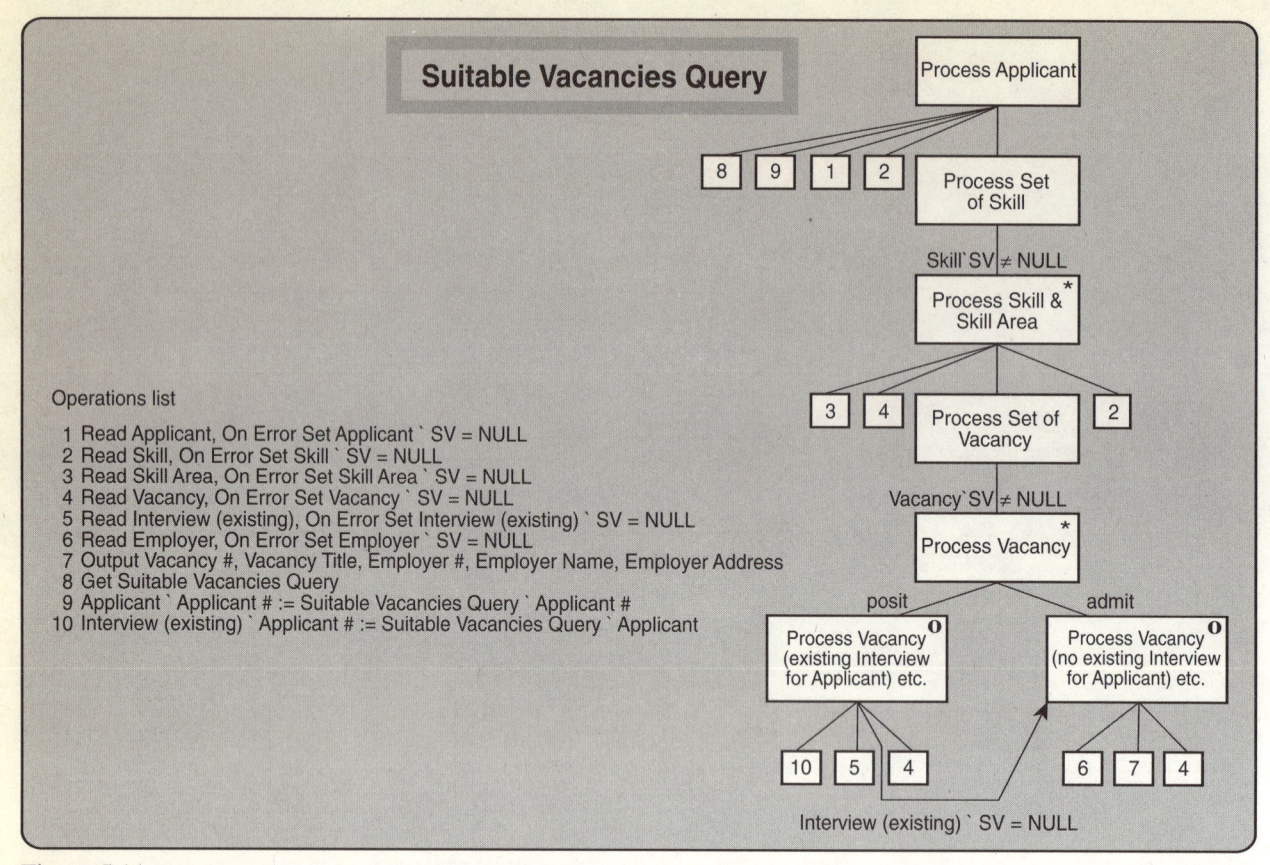

Operations list

1. Read Applicant, On Error Set Applicant ` SV = NULL
2. Read Skill, On Error Set Skill ` SV = NULL
3. Read Skill Area, On Error Set Skill Area ` SV = NULL
4. Read Vacancy, On Error Set Vacancy ` SV = NULL
5. Read Interview (existing), On Error Set Interview (existing) ` SV = NULL
6. Read Employer, On Error Set Employer ` SV = NULL
7. Output Vacancy #, Vacancy Title, Employer #, Employer Name, Employer Address
8. Get Suitable Vacancies Query
9. Applicant ` Applicant # := Suitable Vacancies Query ` Applicant #
10. Interview (existing) ` Applicant # := Suitable Vacancies Query ` Applicant

Figure 5.11

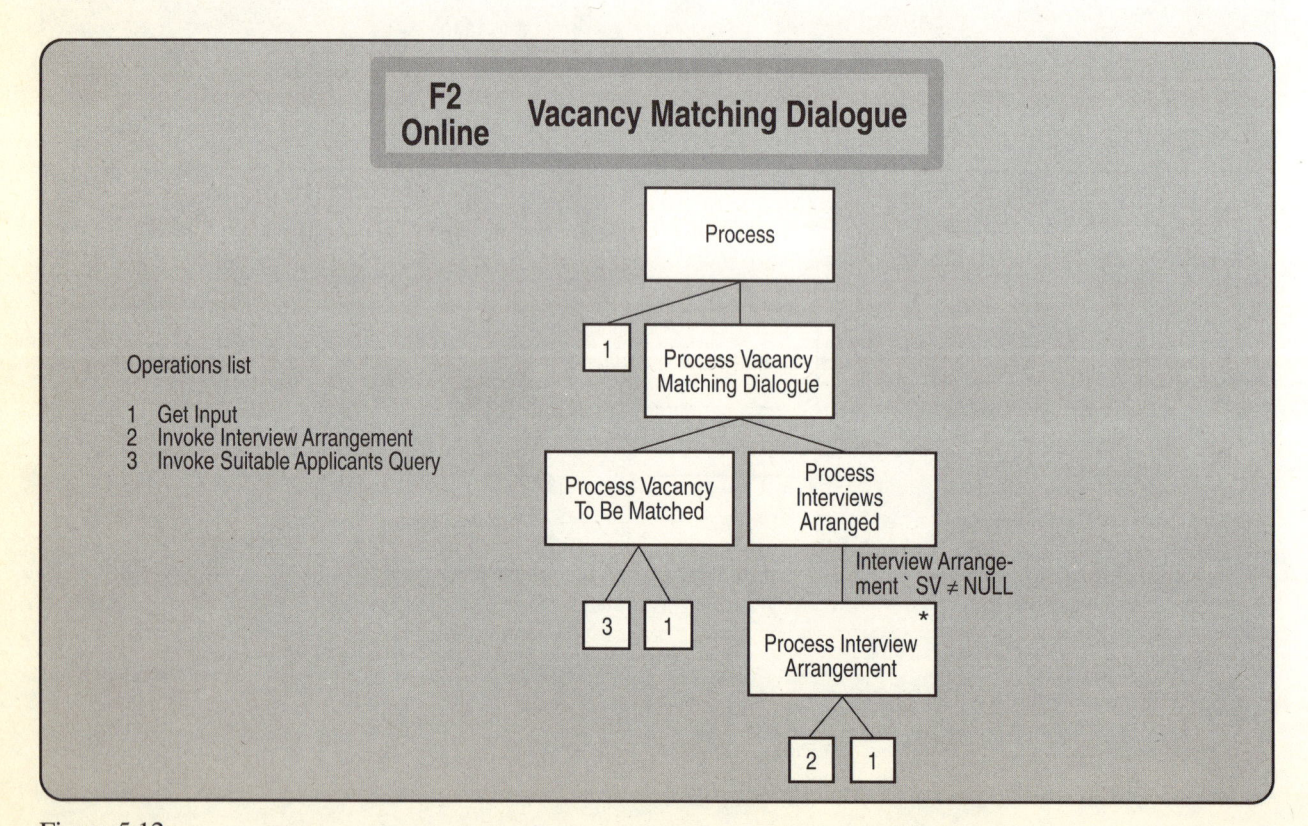

F2 Online — Vacancy Matching Dialogue

Operations list

1. Get Input
2. Invoke Interview Arrangement
3. Invoke Suitable Applicants Query

Figure 5.12

5.4 Examples of GRAPES's Processing Architecture in the Hi-Ho Case Study

As mentioned earlier, GRAPES makes no distinction between external and conceptual scenario components of the logical processes. However, one can easily find processes in the case study solution which play the different roles.

Example of a quasi-external scenario component

The process on figure 5.13 and 5.14 describes the interaction of the placement consultants with the required system, while figure 5.15 documents the behaviour of the applicant as an example of a quasi-conceptual scenario.

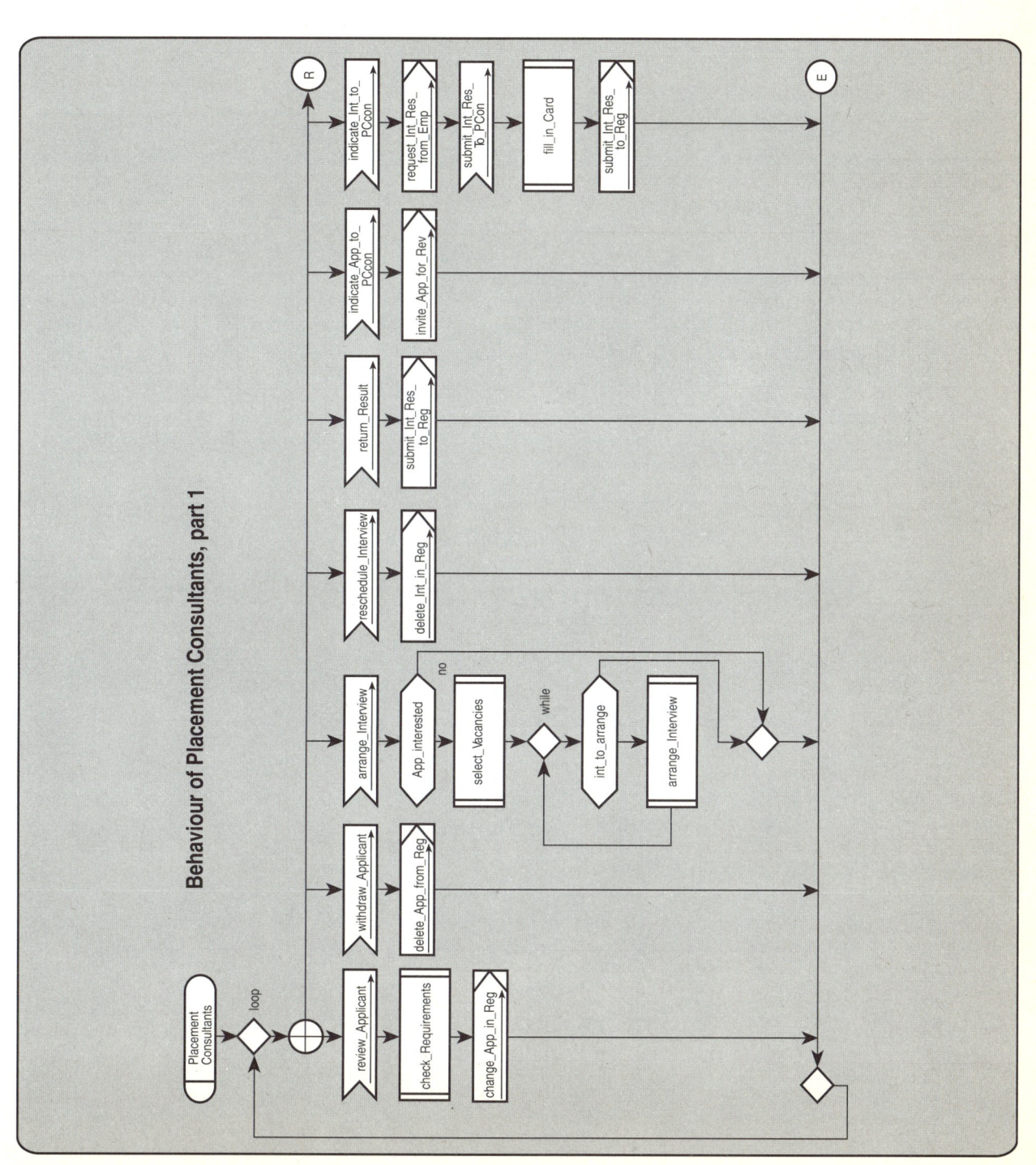

Figure 5.13

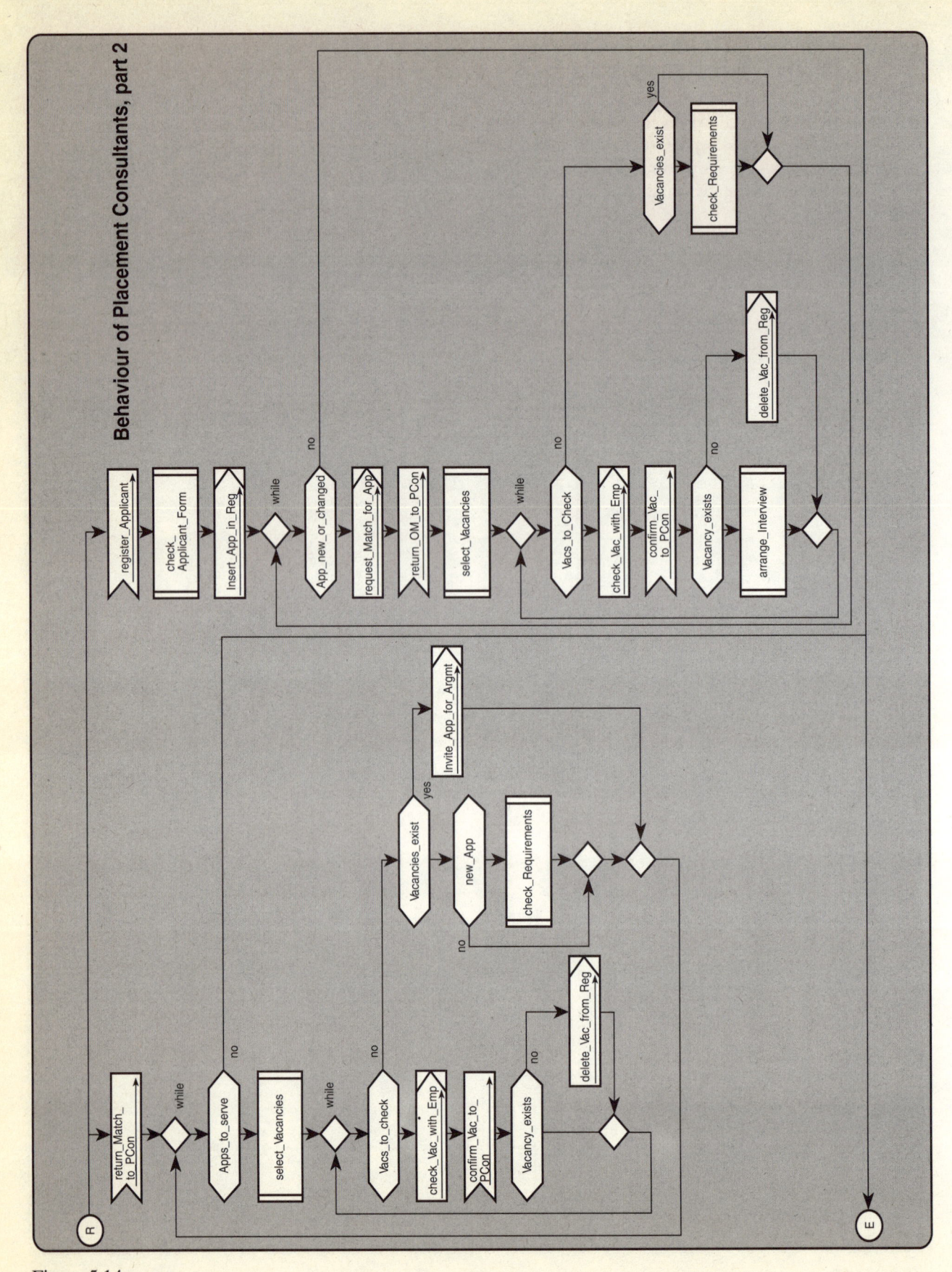

Figure 5.14

5.4 Examples of GRAPES's Processing Architecture in the Hi-Ho Case Study

Example of a quasi-conceptual scenario component

Behaviour of Applicant Register

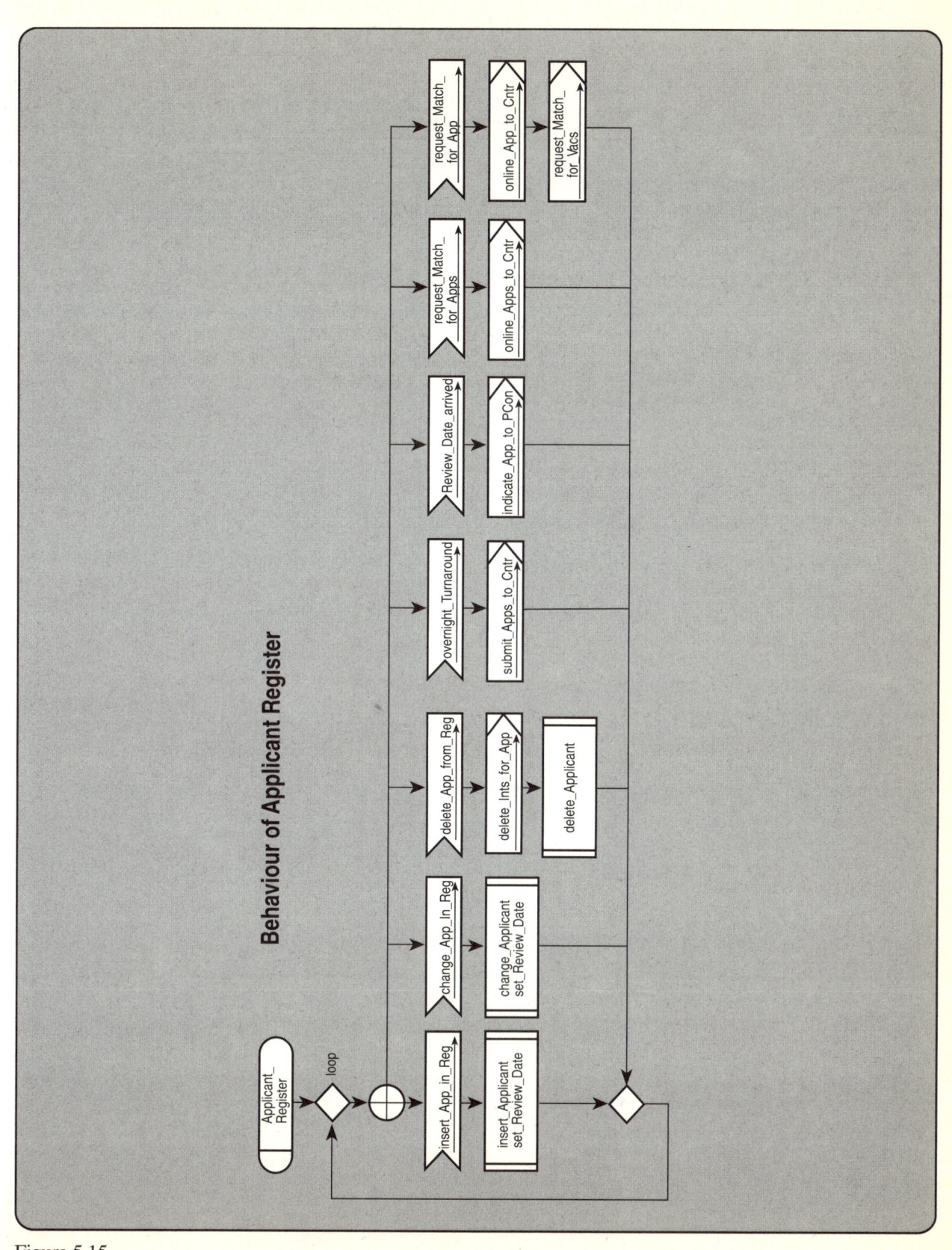

Figure 5.15

5.5 Comparison of the Architecture of End-Products of GRAPES & SSADM

SSADM is clearly 3-schema. The different schemas are represented by different end-products. Different techniques are used to build products in the different schemas.

In GRAPES no distinction is drawn between processes in the conceptual and external scenarios. The same techniques are used to build both types of products; in fact, the GRAPES designer is not conscious of the difference between the two schemas. Although the GRAPES Hi-Ho case study has components which are clearly recognisable as belonging to one or the other of the schemas it is not necessarily true that a different design team, using a top-down approach, would have produced a decomposition which reflected the split between the two schemas.

GRAPES, however, doesn't contradict a position where the 3-schema view would be embedded in. The GRAPES case study decomposition has been influenced by factors other than prior knowledge of the SSADM solution. The most significant of these factors, a commitment to further development of the object-oriented view in GRAPES, would influence the GRAPES design in the direction of the SSADM solution.

The **external scenarios** represent different views of the Hi-Ho organization.

The SSADM solution recognizes a number of different functions for the placement consultant based on the different roles which he is playing. The GRAPES solution encapsulates all placement consultant/system interactions in a single process.

The reason for the differences are:
- SSADM encourages the identification of user roles and the building of functions around them, whereas GRAPES does not;
- the absence of a process replication mechanism in GRAPES at present encourages the building of a process to model the behaviour of the set of placement consultants, rather than an instance;
- the solutions represent different views about the best way for the placement consultant to interact with the system.

Since no dialogue prototyping could be carried out it is impossible to say whether one solution would be preferred by the users to the other, or whether both are inadequate in some way.

The **conceptual scenarios** are as close as they could possibly be, given the differences in the ways the two approaches represent the facts.

There are really only two significant differences between the two conceptual scenarios:
- The SSADM solution represents directly the time sequences in which things happen, at the level of a single applicant. The GRAPES solution can't represent an individual applicant's time sequences directly because it describes the behaviour of the *set* of applicants; this is due to the current absence of a process array construct.
- The GRAPES solution encapsulates the retrievals as well as the updates in the applicant register behaviour. The SSADM solution ignores retrievals in the entity life histories, except where those retrievals influence the outcome of an updating event. The SSADM semantics are thus poorer than GRAPES (or, rather, poorer than those of a GRAPES enhanced by a process replication mechanism) if it becomes necessary to describe the interaction of retrievals and updates.

The closeness of the two conceptual views can be regarded as evidence for the desirability of recognising the conceptual scenario as a distinct part of the architecture of the developed system.

6 Life Cycles

In this chapter we describe and compare the SSADM and GRAPES life cycles. One of the ways in which we compare the life cycle coverage of the two approaches is done by mapping them onto the architectural framework described by John Zachman, see references.

6.1 The SSADM Life Cycle

Overview of the SSADM life cycle

The SSADM life cycle comprises six stages[1]:

1 *Investigation of current environment*

   Here a requirements catalogue is built, a data flow model of the current system is constructed, a logical data model is constructed and used to modify the current system data flow to produce a logical data flow model of the current services.

2 *Business systems options* ***

   The logical model of the current services (LDM and LDFM) is used together with the requirements catalogue to generate (typically up to 3) business system options which solve the requirements to one degree or another with different cost / benefit metrics. An option is chosen or constructed.

3 *Definition of requirements* ***

   The selected business system option is developed into a detailed requirements specification. The logical data model is extended to deliver the functionality required by the selected business system option, and so is the logical data flow model. The logical data flow model is then used to derive "functions" (packages of user-invocable functionality).

The I/O structures of the functions are subjected to relational data analysis and the results are used to enhance the required system logical data model. The I/O structures are also used to derive specification prototypes for the system dialogues.

The events identified in the functions and the structural components of the required system logical data model are then used to build entity life histories and effect correspondence diagrams to define the system behaviour. Enquiry access paths are produced to define the production of system outputs.

All of this is then assembled into the requirements specification.

4 *Technical systems options*

   Here a technical option is chosen.

5 *Logical design* ***

   Here update and enquiry process models are produced from the entity life histories, effect correspondence diagrams, and entity access paths.

6 *Physical design* ***

   A first-cut physical database design is produced. A function component implementation strategy is decided upon. Finally the design is sized, timed and optimized.

---

[1] The stages marked with asterisks (***) above represent the ends of SSADM "modules". Module ends may be thought of as "milestones".

SSADM and the Zachman framework

In the SSADM mapping to the Zachman framework it is noteworthy how some SSADM techniques bridge the lines between cells. This signifies strong knowledge collection and semantic cross-checking capabilities, as it is shown in figure 6.1.

SSADM in the Zachman Framework

	Data Description	Process Description	Network Description
Scope	Requirements Catalogue	Soft Systems Guide? Requirements Catalogue	
Business Model	Overview LDM?	Resource & Service Flows	↑ ?

Information Model	LDM:
	  LDS
	  Entity & Relationship
	    Descriptions
	  TNF Contents
	  Data Catalogue
	
	ELHs ECDs EAPs
		Data Flow Diagrams
		Functions
		I/O Structures
		Dialogues
		
		Update Process Models
		Enquiry Process Models
			Guide on distributed systems to be produced

Technology Model	Database Block Planning & Relationship Implementation Plans	Process-Data Interface Function Component Implementation Map 3 GL Interface Guide	↓ ?
Detailed Description			
Actual System			

Figure 6.1

6.2 The DOMINO Life Cycle

The DOMINO life cycle is based on the principle of structuring the development process for information systems into discrete phases. Precisely defined and monitored results, known as milestones, are associated with each of the phases.

The milestones not only facilitate the logical structuring of work processes, they are also used in project management and monitoring. The scheduling of the phases and the milestone results produced are of decisive importance for the quality of an information processing system and for productivity during its manufacture. The primary consideration at all stages must be the degree to which the final product is of use to the customer. To ensure that the product is of full benefit to the customer, quality must be guaranteed from the very start of the project.

In DOMINO quality is defined as the overlapping of intended purpose, requirements and product characteristics. The aim of DOMINO is the improvement of the quality of the products step by step. To achieve this goal, DOMINO integrates three different engineering fields: organization engineering, systems engineering and software engineering. This integration leads to a common understanding of the tasks and is thus the basis for consistent transfer of results from one phase to the next.

The focus of this comparison is on the technical aspects, therefore project management and quality assurance aspects of the DOMINO life cycle model are not pursued further below.

The life cycle model is divided into the phases: problem analysis, requirements definition, technical implementation, and operation and maintenance. The graphical language GRAPES is used in the first three phases. A series of models of the system under consideration is produced to emphasize different system aspects. The milestones A10 to T20 are covered by GRAPES models (see figure 6.2).

The milestone A10

At A10 a logical model of the current system together with a rough logical model of the required system are to be delivered.

We have to distinguish clearly the physical view of the current system and the logical view of it. It is no goal of DOMINO to construct a complete, consistent physical model of the current system. In an information system project the information about the current system is gained by interviews describing its physical reality. But when we derive a GRAPES model from those results we have to abstract from the physical reality and to concentrate on the essential aspects of the system to manage the complexity. So even if the interview results are a physical model of the current system, at this milestone a logical model of the current system has to be delivered.

The second deliverable at A10 is a rough logical model of the required system. Its purpose is to identify and clearly define:

- the system environment in terms of the project scope and system boundary
- external entities that interact with the system
- data flows that carry inputs and outputs across the system boundary to and from the external entities.

In addition, the system's reaction to external events is outlined in process diagrams.

The rough logical model describes the system as it is seen by the environment, showing in detail the possible communication with the system and a sketch of its behaviours. No information is contained about the internals of the system, its components, subsystems and states.

The functionality of the model of the required system is determined by the requirements placed on the system by its environment.

The milestone A20

The deliverable at A20 is a rough logical model of the target system. The rough logical model of the required system (A10) is refined until the target system is identified. The contents of the deliverable are the internal structure and components of the required system, the communication and the data flows within the required system boundary and an outline of the reaction of the target system components to events.

During the refinement process described above, it is necessary to ensure that the selected requirements are fulfilled. The selection of requirements to be fulfilled is done together with the customer. Alternatives are discussed. The feasibility of fulfilling the critical requirements is evaluated, in particular, with respect to performance. The recommendations of the financial and technical controllers are evaluated with due regard to the clarity, completeness and consistency of the model. The final decision is shared by the customer and the financial and technical controllers.

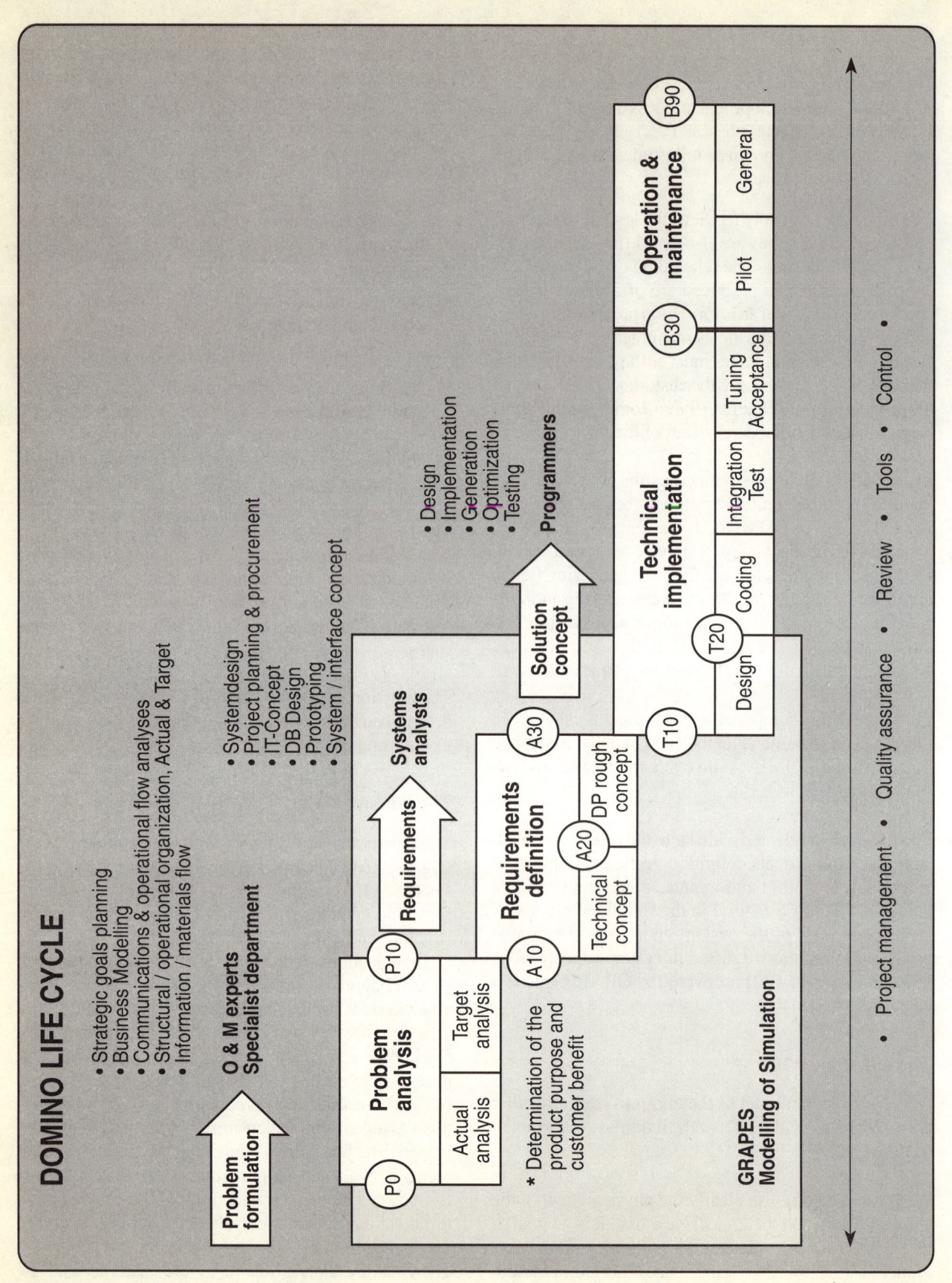

Figure 6.2

The milestone A30

At A30 a detailed logical model of the target system has to be delivered together with a rough physical model.

The detailed logical model of the target system is a refinement of the A20 deliverable in which the processes that make up the behaviour and communication of the elementary components of the system and operate on the data have been specified.

The detailed logical model provides the physical design team with an implementation-independent specification of the required system functionality. This includes not only the functionality of the system as seen from the outside, but also the functionality of the subsystems and subcomponents together with their interactions.

The rough physical model of the target system contains essentially the structure of the target system, the software environment and the distribution of software components.

The milestone T20

The deliverable at T20 is a detailed physical model of the target system. It is a refinement of the rough physical model of A30 and specifies the functionality of all components of the target system.

The model incorporates the final structure of the software system with the specification of requirements and interfaces of system elements at the various levels. The functionality of each component with regard to its communication and processing is specified in as much detail as it is necessary to provide everything needed to decide how application development should be taken forward. The detailed physical model of the target system is the basis for implementation decisions and has to ensure that all requirements are fulfilled by an implemented system that conforms to it.

Recent development

The DOMINO-GRAPES milestone definitions have recently been reworked. The new-style milestones which correspond to the SSADM life cycle are as follows:

P20

At this milestone, the relevant product is an extract from the strategic model of the overall current system.

P30

At this milestone, the relevant product is the strategic model of the overall required logical system.

A10

At this milestone, the relevant product is an extract from the strategic model of the overall required logical system defining the top-level view of the development project's required logical system.

A20

This milestone produces the first refinement of the project's required logical system.

A30

This milestone produces the final refinement of the project's required logical system. It also produces a rough model of the required physical system.

T20

This milestone produces the final refinement of the project's required physical system.

GRAPES and the Zachman framework

The mapping of GRAPES onto the Zachman framework revealed, quite unexpectedly – because this had not been covered in previous work during the study, that GRAPES gave some significant coverage to column 3 (see figure 6.3). This resulted in the identification of the GRAPES process type/process instance capability and its role in representing system distribution.

GRAPES in the Zachman Framework

	Data Description	Process Description	Network Description
Scope	DOMINO Tool: MOSAIK (out of the scope of GRAPES)		
Business Model	**Entity-Relationship Diagrams** **Data Diagrams**	Communication Diagrams Interface Tables Data Tables Process Diagrams Specification Diagrams	**GRAPES Object Types & Instances ?**
Information Model	**Entity-Relationship Diagrams** **Data Diagrams**	Communication Diagrams Interface Tables Data Tables Process Diagrams Specification Diagrams	**GRAPES Object Types & Instances ?**
Technology Model	**Entity-Relationship Diagrams** **Data Diagrams**	Communication Diagrams Interface Tables Data Tables Process Diagrams Specification Diagrams	**GRAPES Object Types & Instances ?**
Detailed Description			
Actual System	Techniques & tools of the DOMINO SE Environment (not part of the GRAPES functionality)		

Figure 6.3

6.3 Coverage of the SSADM Life Cycle in the Hi-Ho Case Study

The case study covers stages 3 and 5 of SSADM version 4, largely excluding dialogue design activities. A small foray was made into stage 6 to produce a more-or-less plausible dialogue structure for a friendly implementation environment. The coverage of SSADM projects is shown in figure 6.4.

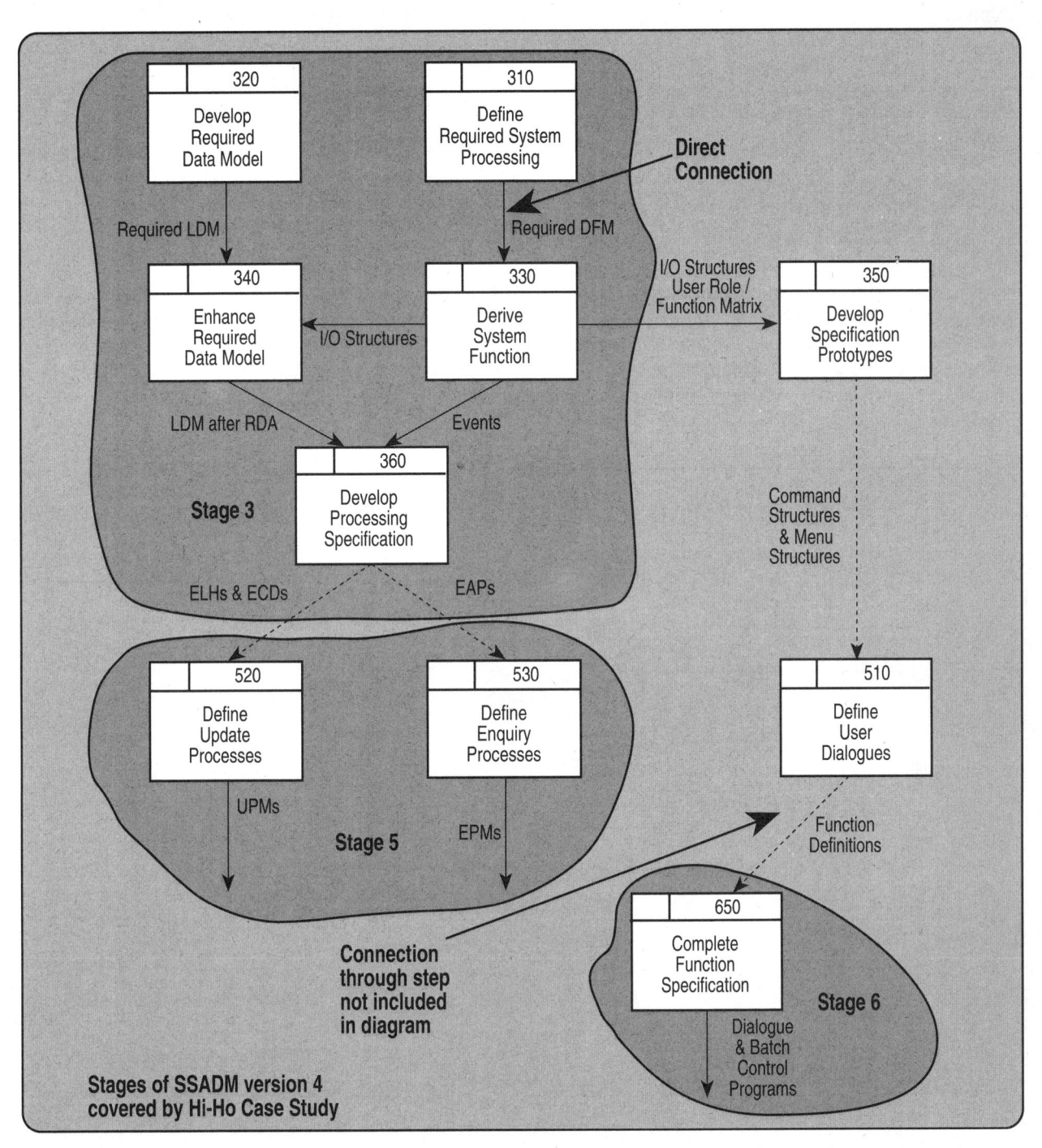

Figure 6.4

6.4 Coverage of the GRAPES Life Cycle in the Hi-Ho Case Study

As shown in figure 6.5 the delivered case study covers:

- Part of the A10 milestone. (The strategic aspect is missing.)
- The A20 milestone.
- Part of the A30 milestone. (The detailed logical model of the system is included; the rough physical model is missing.)

6.5 Comparison of the GRAPES and SSADM Life Cycles

The two life cycles are broadly similar. They use the same principles of refinement and transformation from a physical view of the current system to an abstract, logical view of the required system through to a new physical design in terms of user environment and database / program design.

The main differences between the two life cycles are as follows:

- Different interaction with strategic modelling. The GRAPES approach envisages continuity with an existing IT strategy. The SSADM approach is uncommitted to the strategic environment.
- Different boundaries set by the milestones. But the activities encompassed by the milestones are compatible.
- SSADM proceeds by user selection from options. This is intended to give the users a bigger stake in the developed system, but also to make them understand what they are committing themselves to.

The mapping of the two life cycles' products onto the Zachman framework revealed a capability in GRAPES which the rest of the study had not revealed (and one that might help in SSADM's extensions into the domain of distributed processing). It also revealed that SSADM was strong on bridging techniques.

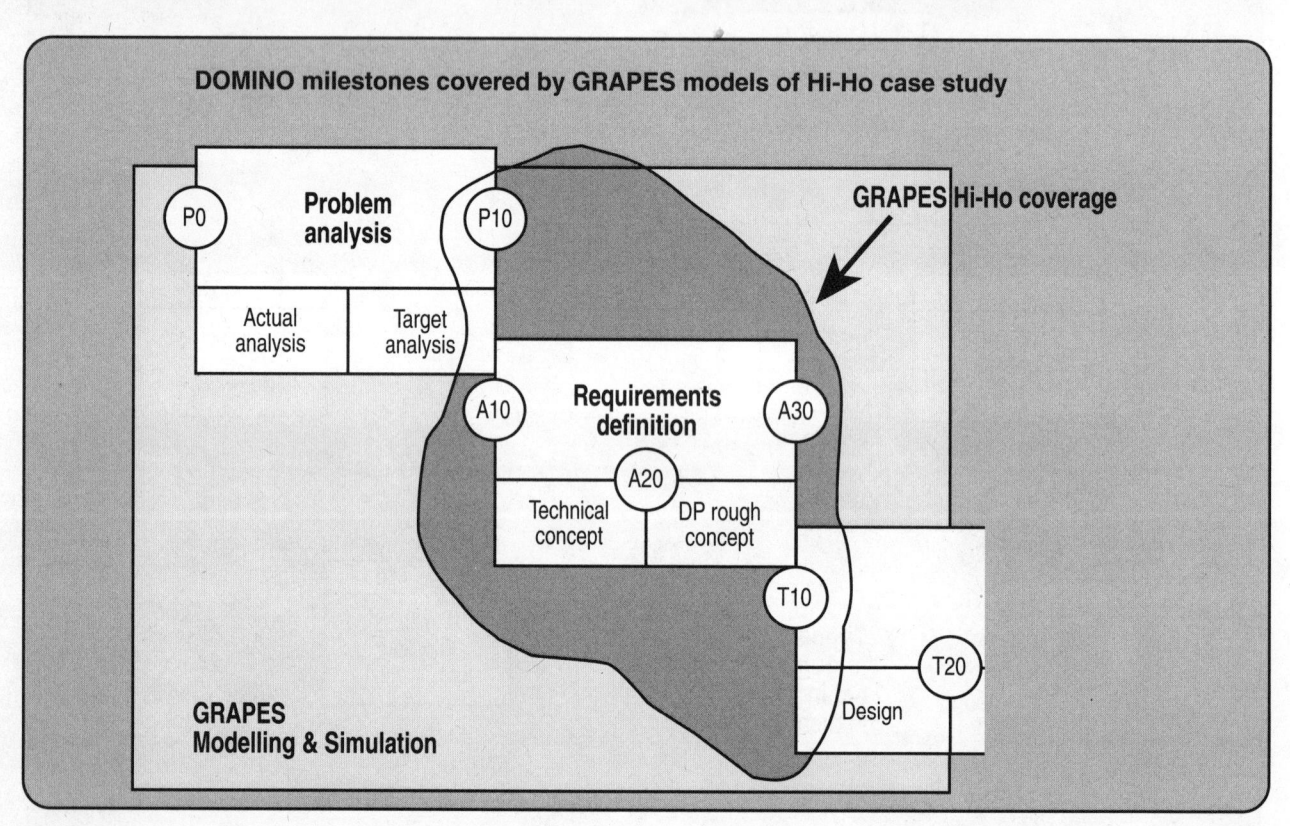

Figure 6.5

7 Fact Representation & Syntax

We investigated the techniques in which facts were represented in the two methodologies and formed the following classification scheme:

Implicit / Explicit

We can distinguish between whether a fact has to be actually stated or whether it is implicit in a methodology's documentation scheme. Implicitness can be a good feature or a bad feature depending on how well it fits with the world-view, or paradigm, of the end-user of the methodology.

In methodologies like SSADM and GRAPES, which have some object-oriented features, implicit representation of facts about those features are likely to be difficult for methodology users who have not acquired the paradigm shift which comes with object-orientation.

On the whole, though, we take it that implicitness is a good thing because it reduces the number of facts that must be kept in the methodology user's short-term memory when considering a problem.

Direct / Indirect

We can distinguish between whether an explicit fact is directly stated in a methodology's documentation scheme, or whether it can be deduced from other directly stated facts. If a fact is of occasional interest and does not figure strongly in some interaction with other facts in a knowledge collection or validation process, it does not particularly matter whether the fact is recorded indirectly.

But if the fact needs to be deduced at the same time as several other facts need to be deduced, and compared with those other facts in some way, the capacity of the methodology user's short-term memory is likely to be exceeded; this increases the difficulty of the knowledge collection or validation process by one or, possibly, several orders of magnitude.

7.1 Fact Representation & Syntax in SSADM

Facts represented in SSADM

Data flow diagrams / functions / I/O structures

Data flow diagrams show how services are organized and processing is undertaken. Functions show the final packaging of the functionality. I/O structures document the inputs to and outputs from a function.

Explicit & direct
- System scope
- Packages of user-invocable functionality and end-user / function interactions
- Detailed external interfaces
- Communication between functions with different scheduling requirements via transient data stores

Implicit
- Each external entity represents a potential array of external entities of the defined type. However, there is no way, at the moment, of defining which external entities have a singular instantiation.

Logical data models and access paths

Logical data models show the overall relationship between the data elements. Access paths contain the information for navigating through the logical data model.

Explicit & direct
- Entity types
- Relationship types: attributed relationships defined as entities
- Optionality and mutual exclusion
- Attributes
- Data typing. But the support for this is weak. There is no real knowledge of type constructors or of abstract data types. There is no scoping of where types can be used as in GRAPES.
- Volumetrics
- Entry point to database for access path.
- Navigations through data model to support events and enquiries.
- Entity / relationship usage
- Access volumetrics

Explicit & indirect
- Entity subtyping. This can be deduced from a combination of inspection of relationship cardinality limitations and mutual exclusion.

Entity life histories & effect correspondence diagrams

Entity life histories show the evolution of single entities in terms of processing triggered by events. Effect correspondence diagrams show the relationship of events and entity life histories.

Explicit & direct
- Update success units. The events define the atomic units of business processing. An event must succeed or fail as a whole.
- Access paths for updates
- Instance-level control flow of processing for each entity
- Preconditions for events
- Expectations about likely further processing.
- Catastrophic changes required when expectations are unjustified
- Sequence, selection, iteration & parallelism within an entity life history.
- Behaviour cycles. Iterations of events or groups of events which have some well-defined business functions or are subject to some well-defined business constraints.

Explicit & indirect
- Side-effects of previous event processing having been done or omitted in error (by following event sequences and cycles through life histories).
- State transitions. We have classified these as indirect because they can be derived from the entity life history control structures. However, since state transitions are documented explicitly once they have been derived, we could equally well have listed them as directly represented facts. Alternatively, since the entity life history approach enables the specification of systems behaviour without the direct identification of states, we might think of state handling as implicit.

Implicit
- Array parallelism between entity life histories of the same type.

SSADM syntax

Data flow diagrams
- Modified Gane & Sarson.
- Three consumer / producers of data:
  - external entity
  - process
  - data store
- Data flow must have a process at at least one end
- Decomposed into multi-level hierarchy
- Asymmetric consistency (low level must represent everything at upper level)

Logical data model
- Modified Bachman diagram
- Two-name relationships
- Only one-to-many relationships in final design
- Attributes of relationships not allowed
- Informal notion of kernel, characteristic and associate entities (simple keys, hierarchical keys and compound keys)
- Mutual exclusion construct
- Indirect subtyping

I/O structures
- Jackson structures (well-structured expressions of sequence, selection and iteration).

Entity life histories, update and enquiry process models
- Modified Jackson structures (well-structured expressions)
- Implicit arrays of processes
- Backtracking / escape constructs
- Parallel "sequence" constructs
- Operation list (i.e. read, write, assignment statements etc.) and allocation of operations to structure

Effect correspondence diagrams
- Modified Jackson structures (well-structured expressions)
- Explicit arrays of processes
- Jackson correspondence
- Path expressions

Weaknesses
- Weak data typing
- Poor domain support
- Informal operations on types
- No formal name scoping, leading to more cumbersome representation of names to avoid ambiguity, or to actual ambiguity if the analyst abbreviates names
- Weak cross-checking between some techniques (e.g. DFD processes and entity life histories / effect correspondence diagrams)
- Informal syntax for operations in entity life histories etc.

7.2 Fact Representation & Syntax in GRAPES

Facts represented in GRAPES

Communication diagrams

Communication diagrams show the overall structure and the communication relationships of the system components.

Explicit & direct
- System scope
- The structure of the system and its components (decomposition into subobjects)
- Communication relationships between parts of the system (communication lines)
- External interface of the system (communication lines in higher level CD)

Implicit
- Parallelism between objects

Interface tables

Interface tables contain the details about the communication.

Explicit & direct
- External interface of the system (ports and message types)
- Communication relationships (channels and message types)
- Interfaces of the system components (ports and message types)
- Communication mode (asynchronous or synchronous)

Implicit
- Control flow (message passing triggers activities)

Process diagrams

Process diagrams describe the functionality of the system in terms of processing and flow of control.

Explicit & direct
- Control flow, preconditions and likely further processing at the object class level (in contrast to the object instance level)
- Representation of repetitive behaviour at the class level
- Reaction to messages and timer events
- Timer objects
- Use of common procedures by several inputs
- Parallelism within an object

Indirect
- States & state transitions at the class level
- Access paths through data when events / inputs occur

Entity relationship diagrams

Entity relationship diagrams show the overall relationship between the data elements.

Explicit & direct
- Entities and relationships. Relationships can be attributed
- Optionality, mutual exclusion, cardinality
- Role names

Data structure diagrams

Data structure diagrams contain the details about the data.

Explicit & direct
- Structure of stored data – attributes of entities and relationships
- Structure of flowing data
- Data type definition, scope definition and domain support

Data tables

Data tables establish the relationship between processing and data.

Explicit & direct
- Local variables of processes, modules and procedures
- Initial values for variables
- Symbolic constants

Specification diagrams

Specification diagrams describe the interfaces of reusable components.

Explicit & direct
- Formal parameters of procedures
- Exported procedures, data types and variables

Implicit
- Hiding of information

Hierarchy diagram

Hierarchy diagrams give an overview of the model structure.

Explicit & direct
- Hierarchical object structure. (Redundant representation of structure and process.)
- Scopes of declarations. (Redundant representation of declarations in PDs, SDs and DDs about "parent".)

GRAPES syntax

Communication diagrams

- Modified structured analysis data flow diagrams
  – objects
  – information flows between objects
- Subjected to information hiding principles
- Decomposed into multi-level hierarchy
- Symmetric consistency (low level must represent everything at upper level and vice versa.

Entity relationship diagrams

- Modified Chen diagrams
  – entities
  – relationships, which may have attributes
- Cardinalities including 1-1 are specifiable
- Role names
- Mutual exclusion

Data structure diagrams

- PASCAL-like record and type constructors

Process diagrams

- Modified SDL (CCITT / Telecom standard)
- Well-formed process diagrams represent well-structured expressions
- Although well-formed process diagrams look like flowcharts they are drawn from definitions which strictly represent sequences, selections and iterations
- Parallel "sequences" can be defined
- Can define procedure types and then instantiate them

Weaknesses

- No backtracking / escape construct in well-formed process diagram
- Multiple instances of processes of the same type cannot be defined at present
- Information (e.g. a name) cannot be associated with a loop itself, as opposed to the things contained in the loop
- No way of describing success units

7.3 Examples of the SSADM Fact Representation & Syntax

Figures 7.1 and 7.2 show the SSADM data flow diagram decomposition. Notice how extra detail, not corresponding to anything on the top level, can be added at second level.

Figure 7.3 shows the SSADM logical data model and illustrates most of the types of relationships which are possible.

Figure 7.4 is not a standard SSADM diagram, but it has been included here to provide an example of the definition of the data which underlies the entities.

Notice how reusable types can be defined in SSADM by affixing role names to data names. The role name facility was originally introduced to provide a limited amount of domain support during relational data analysis, originally only used on keys.

Figure 7.5 shows an entity life history, illustrating the main constructs, and shows how the use of backtracking constructs enable the normal processing to be kept clearly visible whilst embedding a control structure which also handles exceptions. The backtracking constructs used are of the "disciplined" kind recently recommended by the CCTA.

Figures 7.6 – 7.8 are included to show how an event (in this case applicant withdrawal) appears in several entity life histories.

Notice, in the entity life histories, how the events are sequenced. For example at the level of an individual applicant, applicant registration always comes before applicant withdrawal.

When the entity life histories are constructed the applicant withdrawal event propagates from the applicant to all of his skills. Notice how the deletion of the applicant's entity also propagates in the same way.

Then notice, in figure 7.8, how the two events both propagate into interview.

The propagation of an event through the system is recorded in an effect correspondence diagram, as in figure 7.9.

But this direct representation of the effects of an event is gained at the expense of some duplication of process description. Notice how part of the post acceptance effect correspondence diagram (figure 7.10) is identical to the applicant withdrawal effect correspondence diagram.

Finally, there is the conceptual scenario of processing required for applicant withdrawal in a non-object-oriented, undistributed implementation, as shown in figure 7.11.

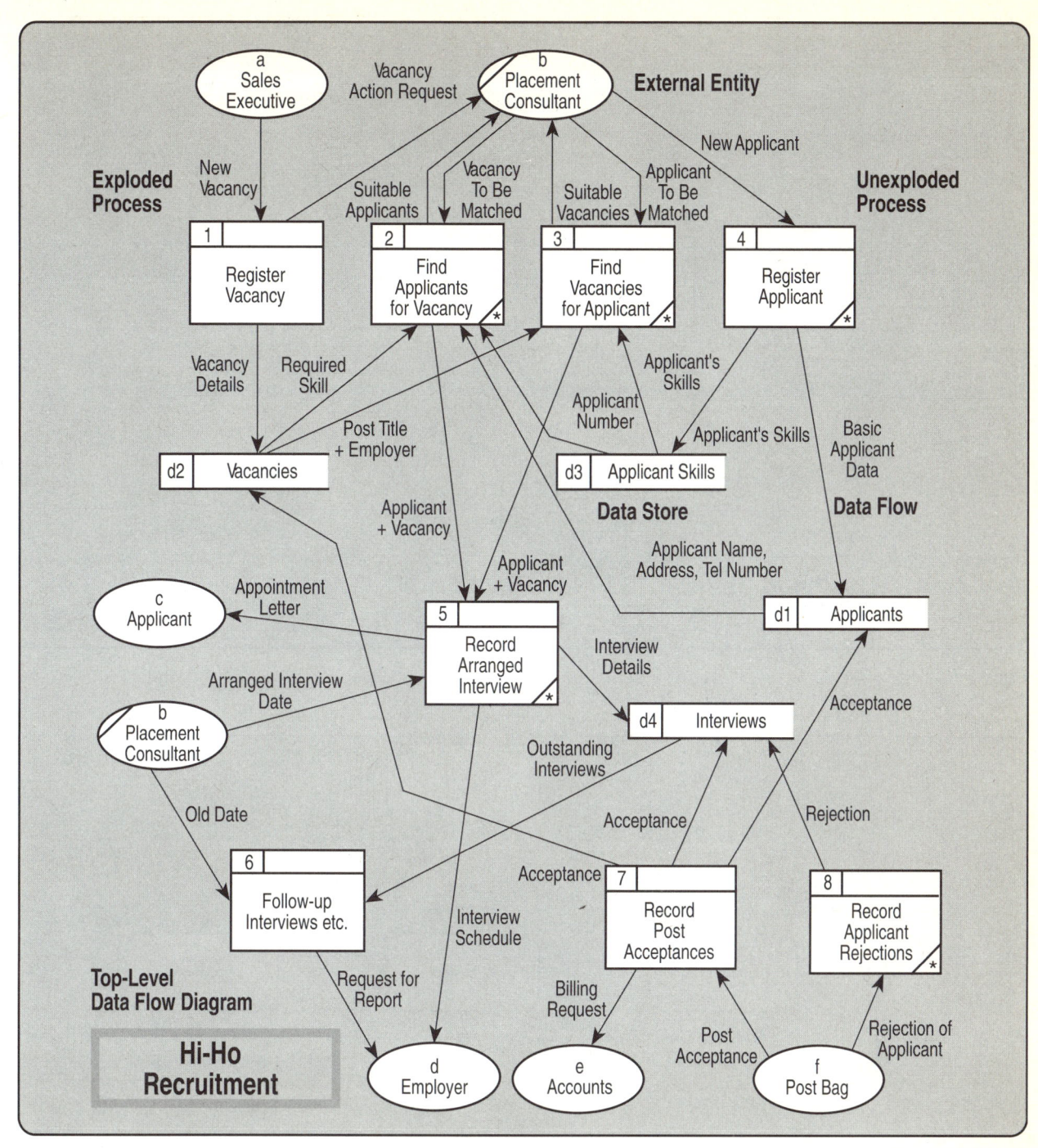

7.3 Examples of the SSADM Fact Representation & Syntax 61

Figure 7.1

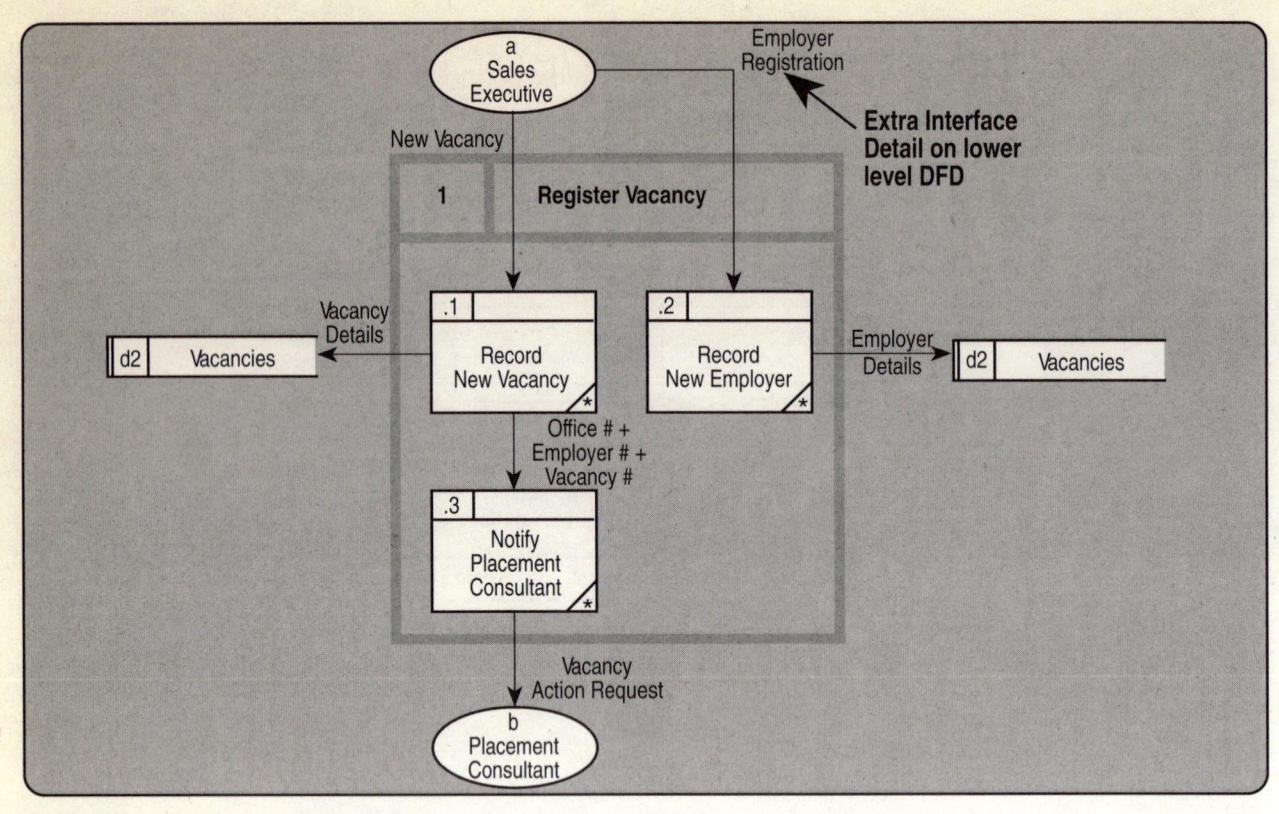

Figure 7.2

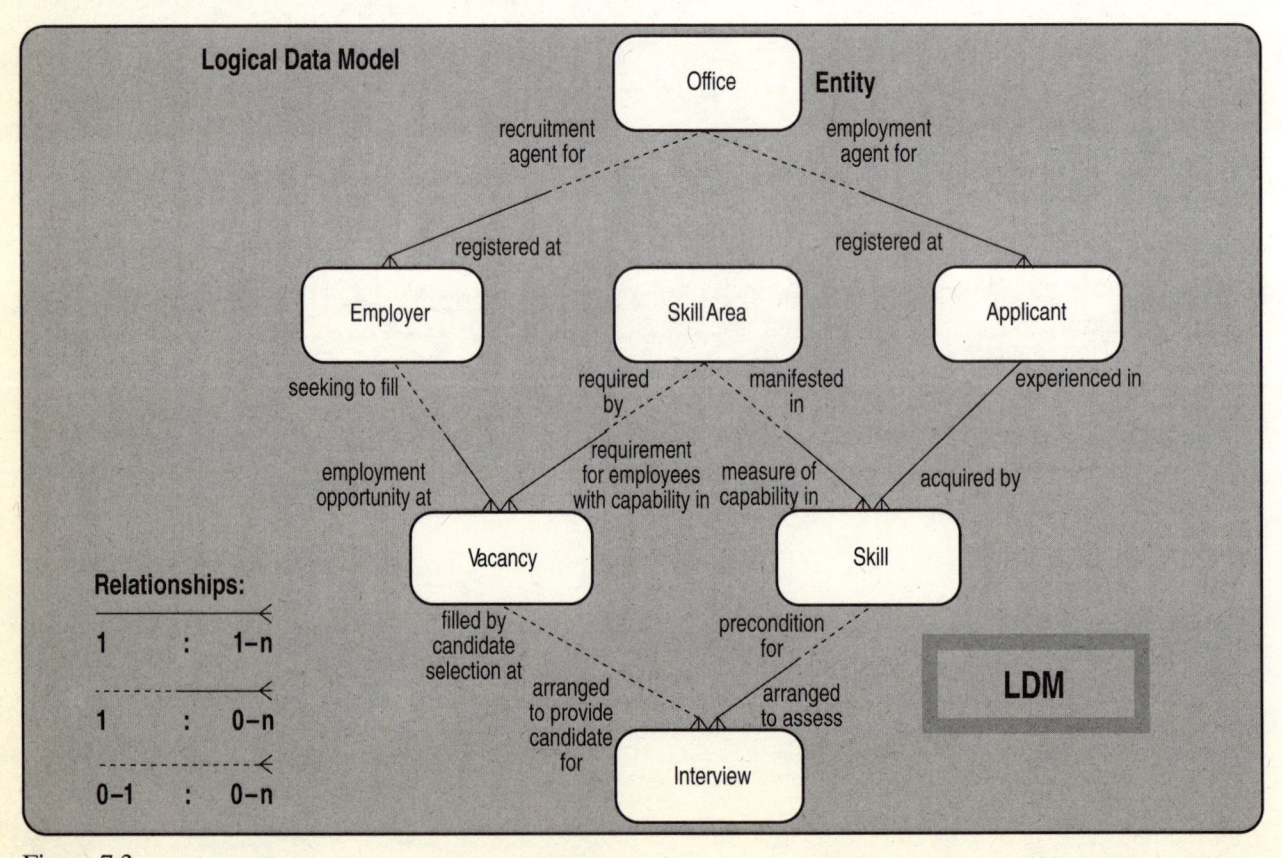

Figure 7.3

7.3 Examples of the SSADM Fact Representation & Syntax

Logical Data Model populated with Attributes

Office
- Office #
- SV
- Office Address
- Office Name

Applicant
- Applicant #
- SV
- Applicant Name
- Applicant Address
- Placement Consultant
- Marital Status
- Office #
- Willing To Move
- Date (Birth)
- Has Driving Licence
- Telephone Number
- Date (Review)

Skill Area
- Skill Code
- SV (Skills)
- SV (Vacancies)
- Skill Description

Employer
- Employer #
- Office #
- SV
- Employer Address
- Employer Name

Skill
- Skill Code
- Applicant #
- SV
- Skill Level

Vacancy
- Vacancy #
- Employer #
- Office #
- SV
- Date (Review)
- Skill Code
- Vacancy Title

Interview
- Applicant #
- Vacancy #
- Employer #
- Office #
- SV
- Date (Interview)

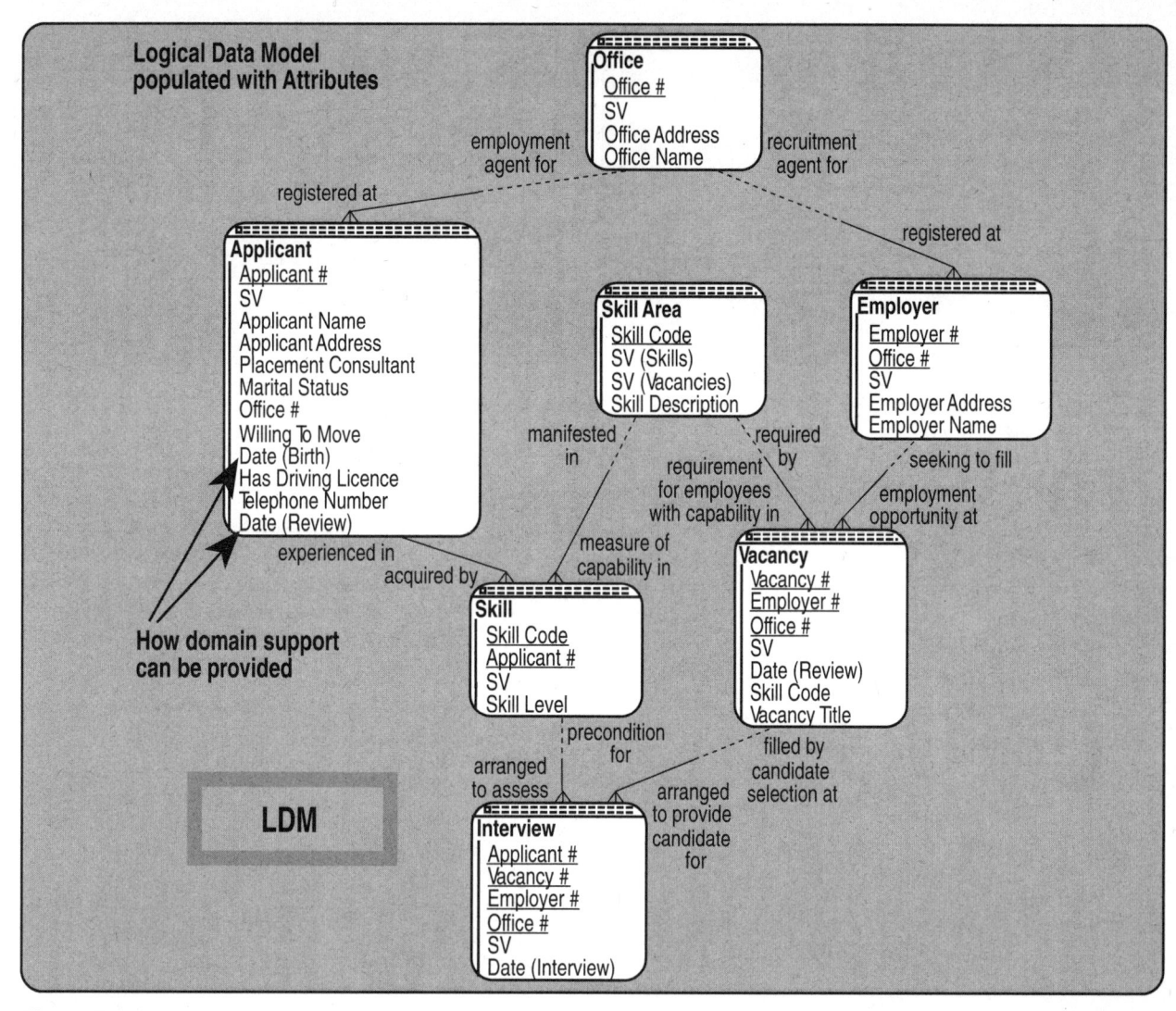

Relationships:
- Office — employment agent for / registered at — Applicant
- Office — recruitment agent for / registered at — Employer
- Applicant — experienced in / acquired by — Skill
- Skill Area — manifested in — Skill
- Skill Area — measure of capability in — (Skill)
- Skill Area — requirement for employees with capability in / required by — Vacancy
- Employer — seeking to fill / employment opportunity at — Vacancy
- Skill — precondition for / arranged to assess — Interview
- Vacancy — filled by candidate selection at / arranged to provide candidate for — Interview

How domain support can be provided (arrows pointing to Date (Birth) and Has Driving Licence)

LDM

Figure 7.4

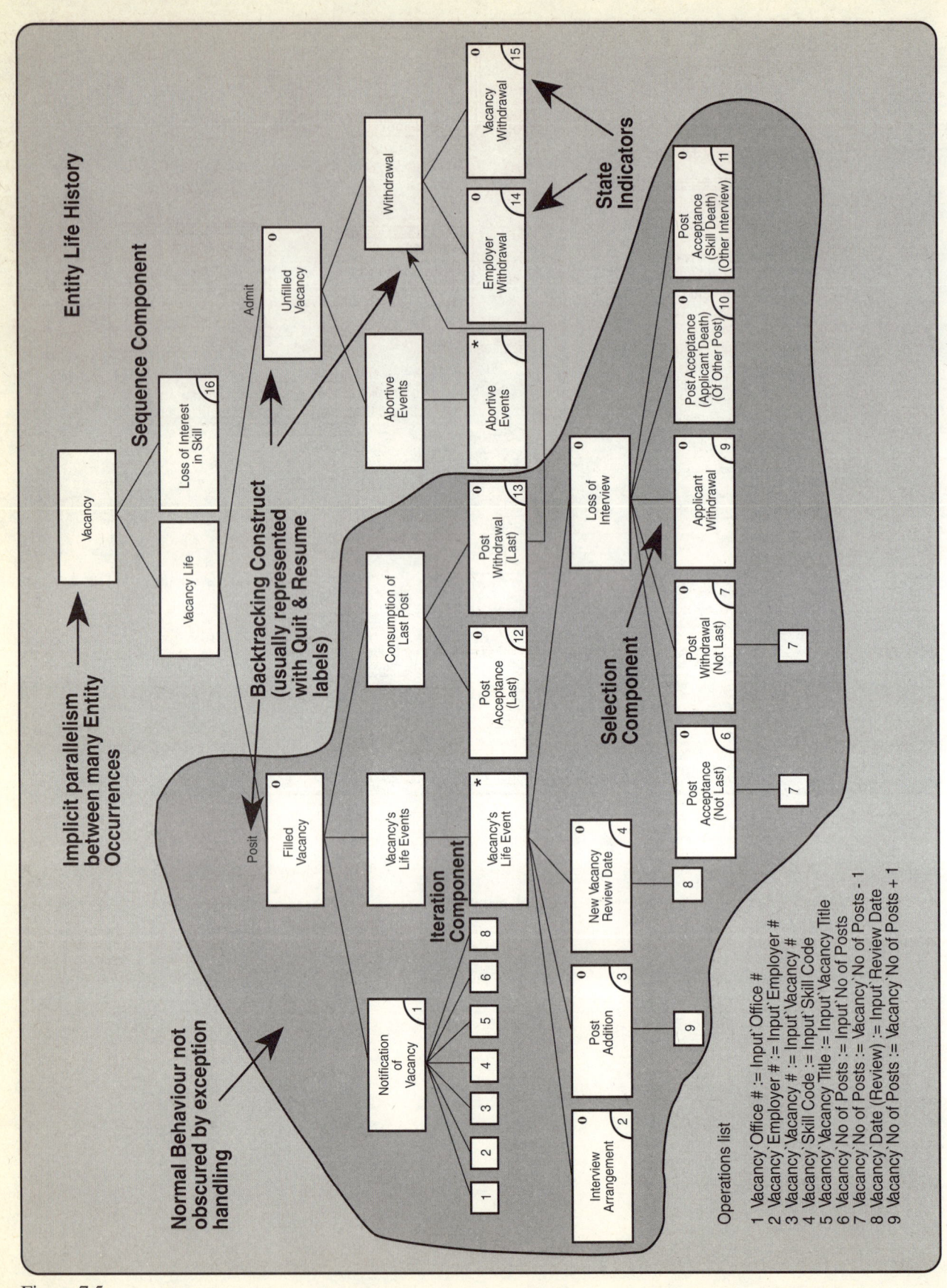

Figure 7.5

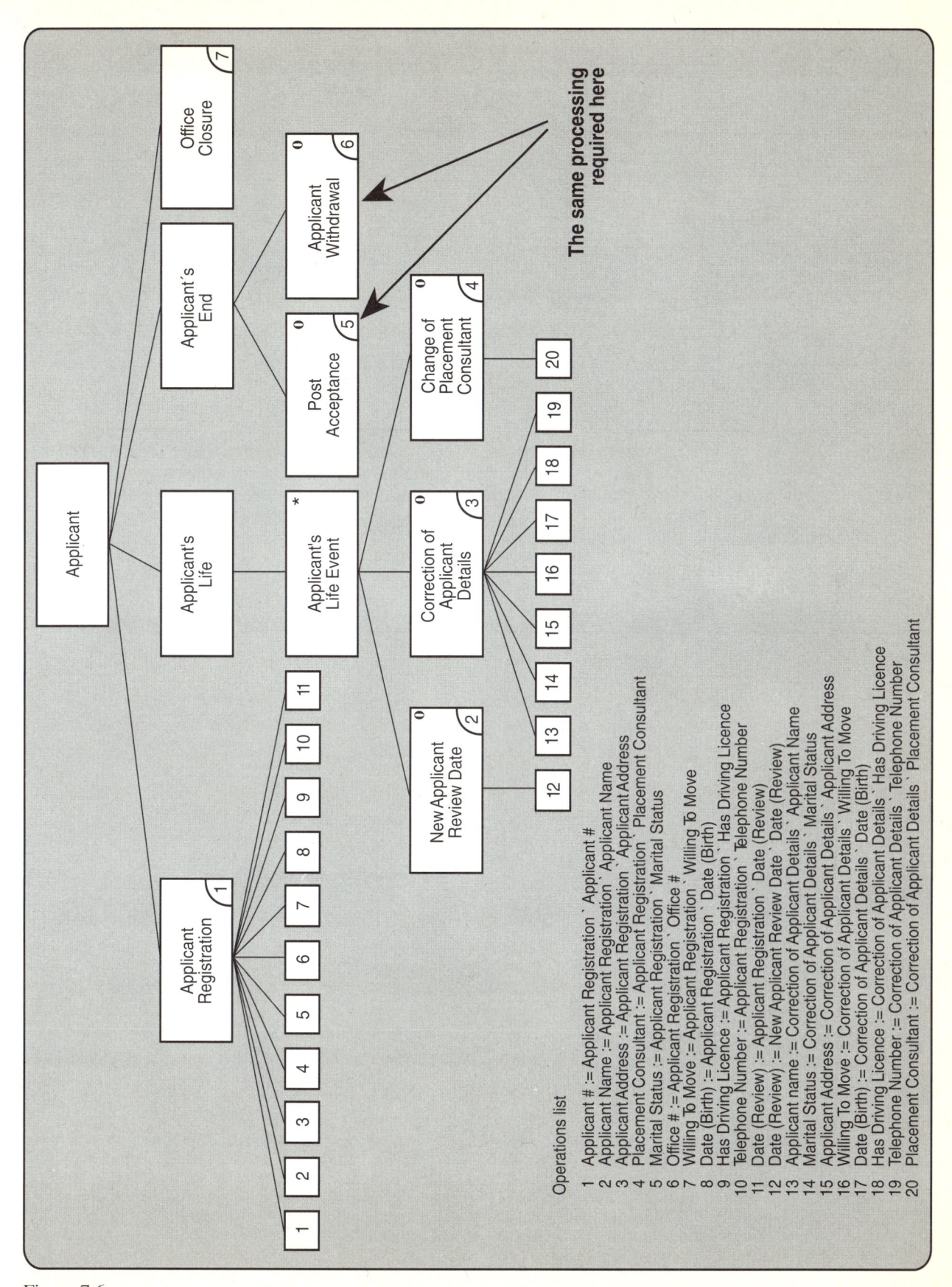

7.3 Examples of the SSADM Fact Representation & Syntax 65

Operations list

1   Applicant # := Applicant Registration ` Applicant #
2   Applicant Name := Applicant Registration ` Applicant Name
3   Applicant Address := Applicant Registration ` Applicant Address
4   Placement Consultant := Applicant Registration ` Placement Consultant
5   Marital Status := Applicant Registration ` Marital Status
6   Office # := Applicant Registration ` Office #
7   Willing To Move := Applicant Registration ` Willing To Move
8   Date (Birth) := Applicant Registration ` Date (Birth)
9   Has Driving Licence := Applicant Registration ` Has Driving Licence
10 Telephone Number := Applicant Registration ` Telephone Number
11 Date (Review) := Applicant Registration ` Date (Review)
12 Date (Review) := New Applicant Review Date ` Date (Review)
13 Applicant name := Correction of Applicant Details ` Applicant Name
14 Marital Status := Correction of Applicant Details ` Marital Status
15 Applicant Address := Correction of Applicant Details ` Applicant Address
16 Willing To Move := Correction of Applicant Details ` Willing To Move
17 Date (Birth) := Correction of Applicant Details ` Date (Birth)
18 Has Driving Licence := Correction of Applicant Details ` Has Driving Licence
19 Telephone Number := Correction of Applicant Details ` Telephone Number
20 Placement Consultant := Correction of Applicant Details ` Placement Consultant

Figure 7.6

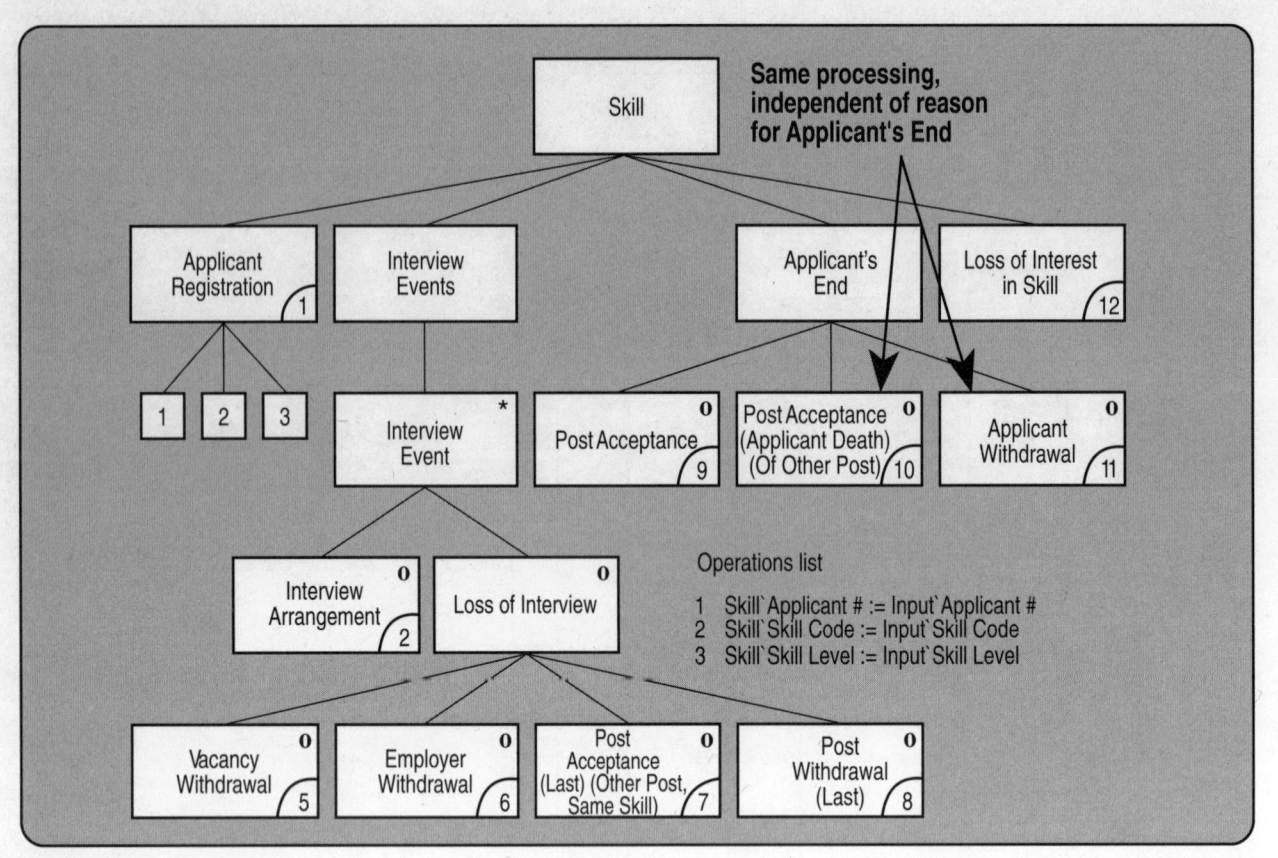

Figure 7.7

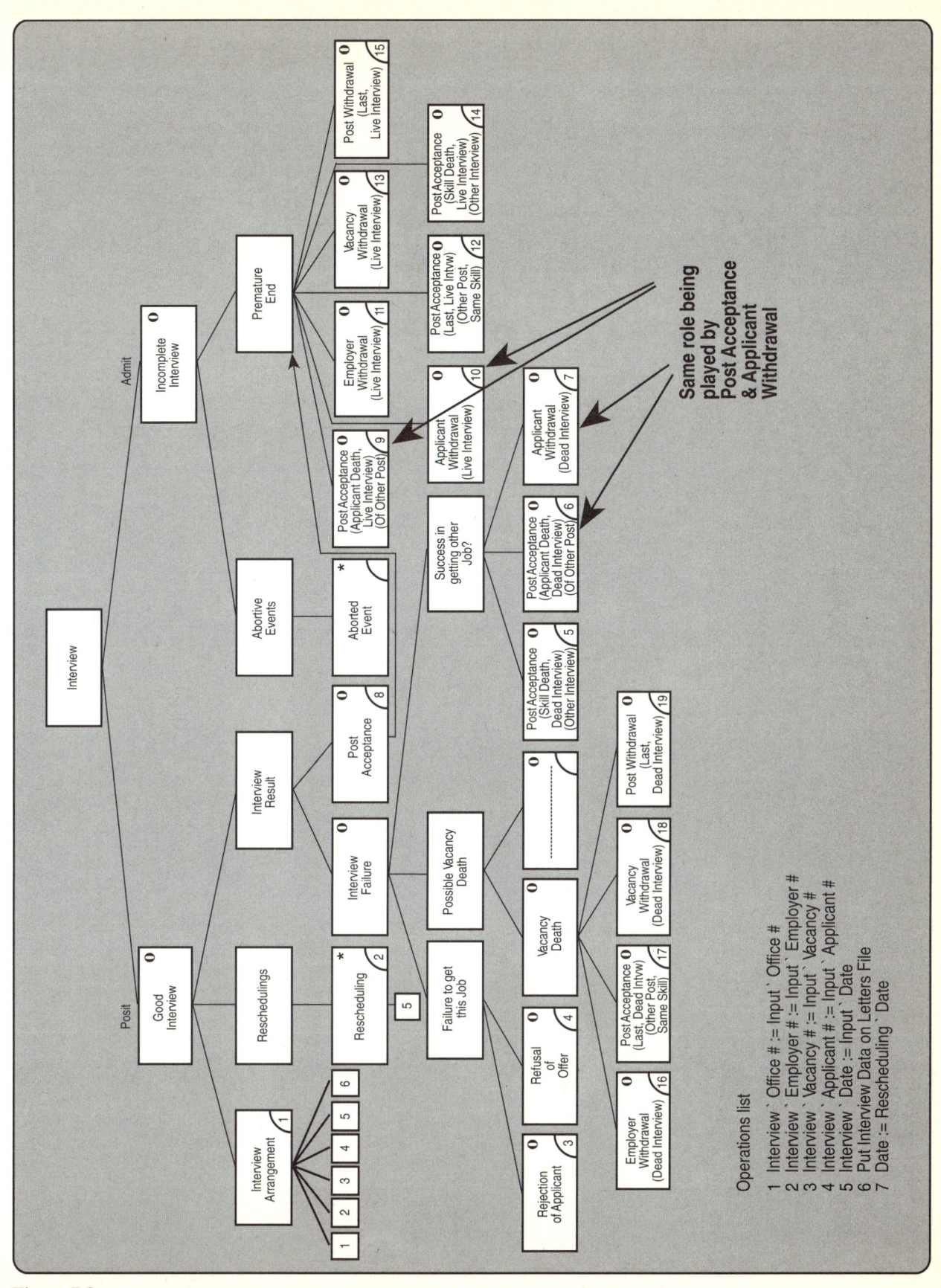

7.3 Examples of the SSADM Fact Representation & Syntax 67

Figure 7.8

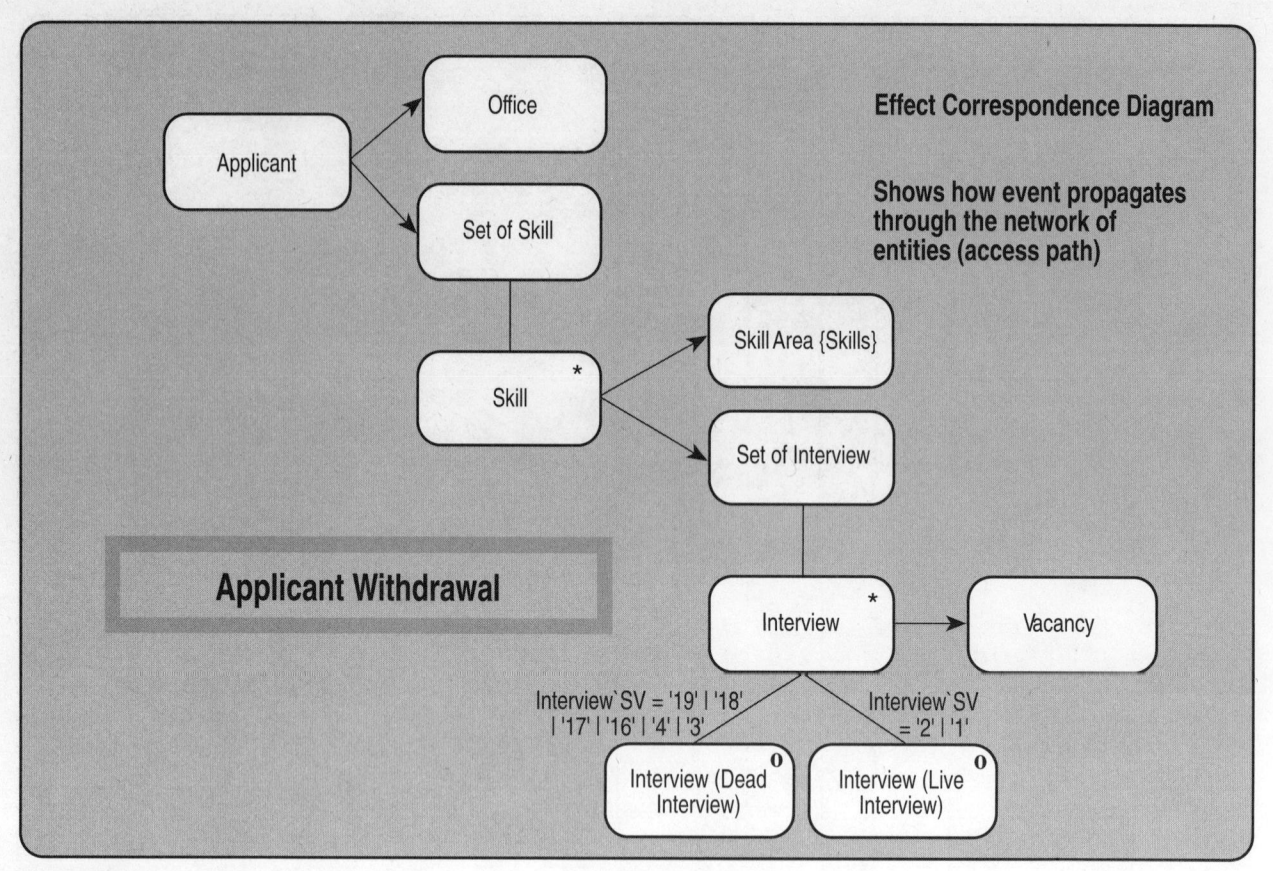

Figure 7.9

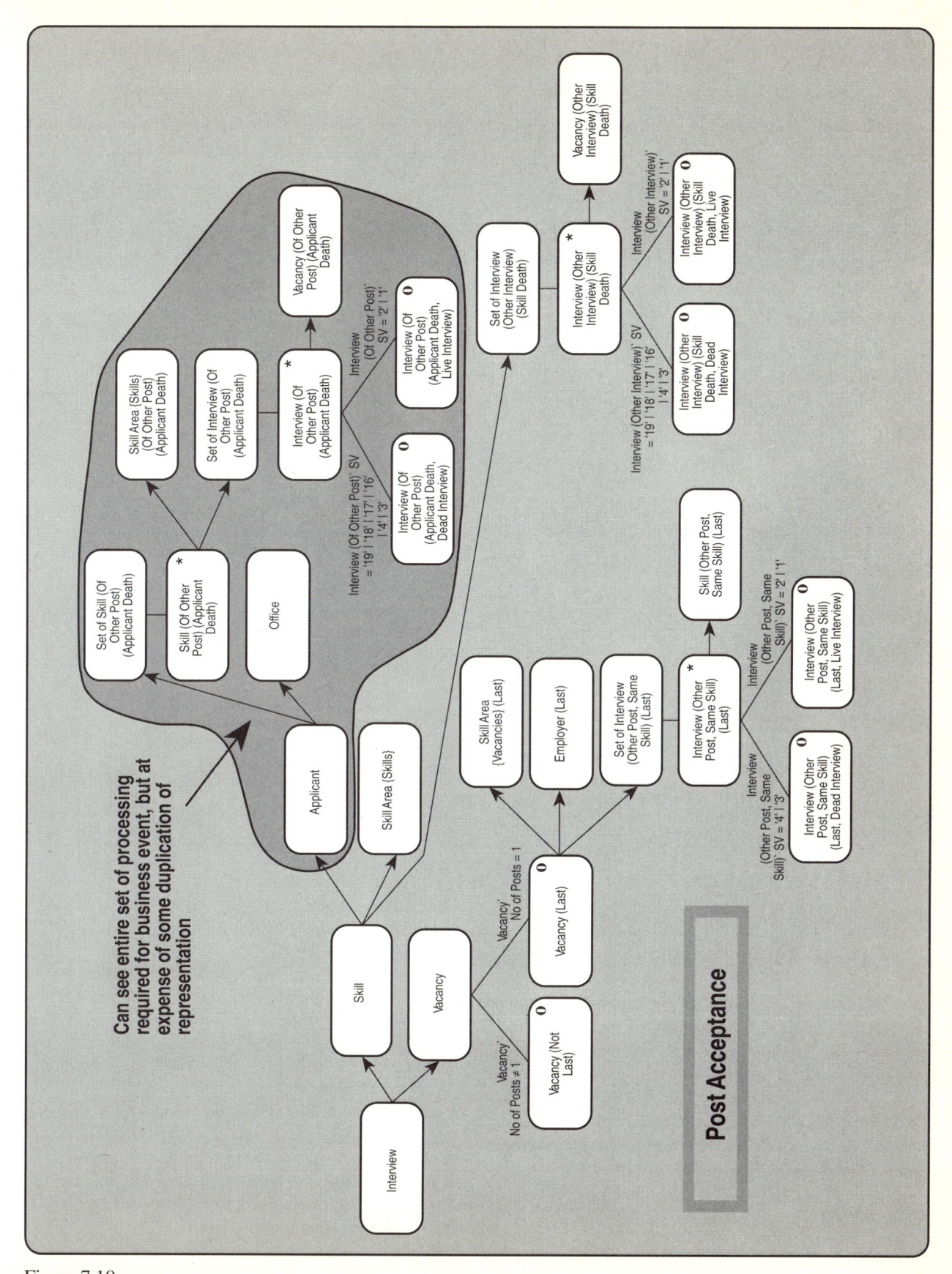

7.3 Examples of the SSADM Fact Representation & Syntax

Figure 7.10

Applicant Withdrawal

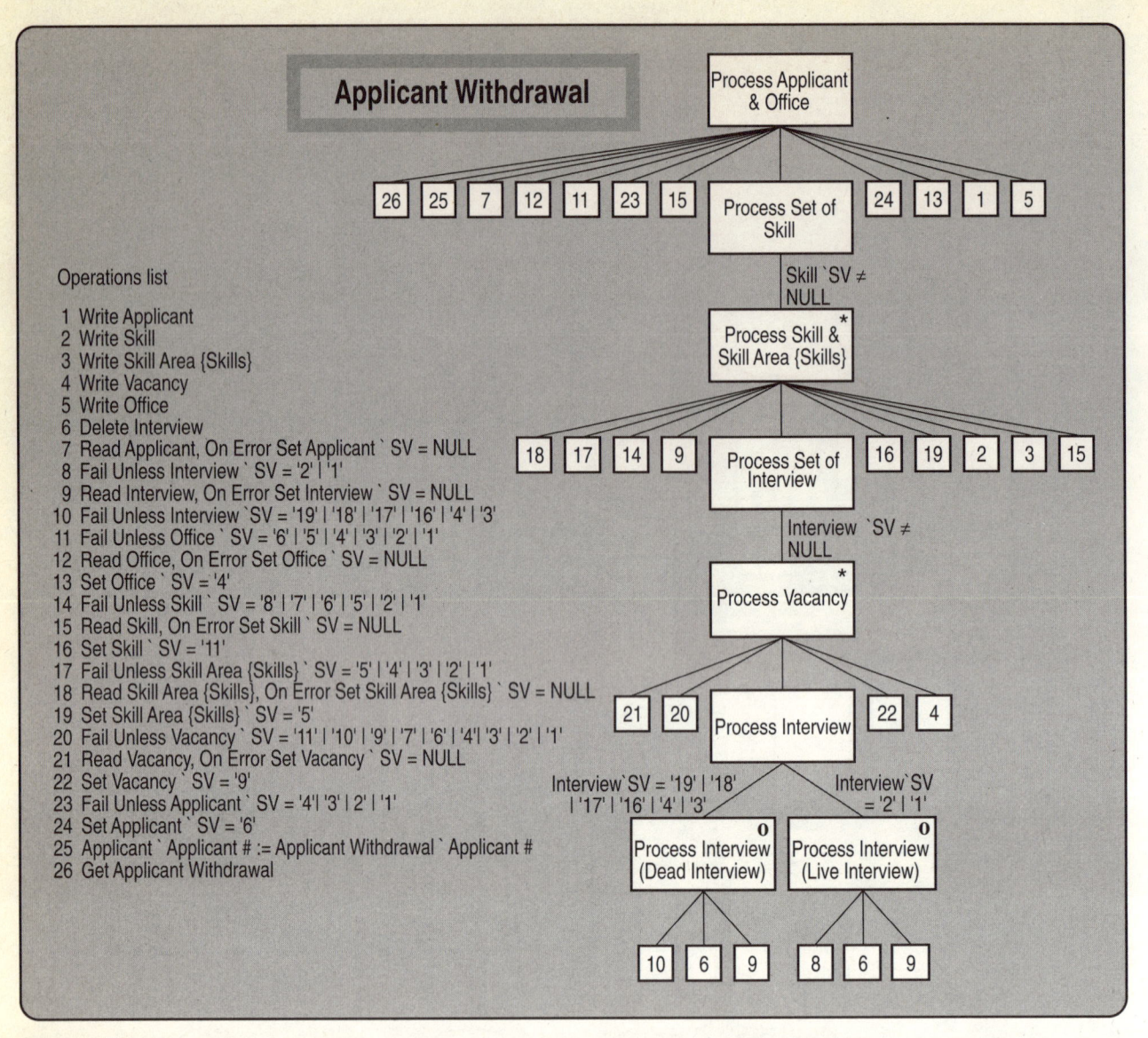

```
                                    Process Applicant
                                        & Office
                                           │
   ┌──┬──┬──┬──┬──┬──┬──┐         ┌──┬──┬──┐
   │26│25│ 7│12│11│23│15│         │24│13│ 1│ 5│
   └──┴──┴──┴──┴──┴──┴──┘         └──┴──┴──┴──┘
                           Process Set of
                               Skill
                                  │  Skill `SV ≠ NULL
                          Process Skill & *
                          Skill Area {Skills}
                                  │
   ┌──┬──┬──┐               Process Set of         ┌──┬──┬──┬──┐
   │18│17│14│ 9│              Interview            │16│19│ 2│ 3│15│
   └──┴──┴──┘                                       └──┴──┴──┴──┘
                                  │  Interview `SV ≠ NULL
                           Process Vacancy *
                                  │
          ┌──┬──┐            Process Interview          ┌──┐
          │21│20│                                        │22│ 4│
          └──┴──┘                                        └──┘
           Interview`SV = '19'|'18'        Interview`SV
           |'17'|'16'|'4'|'3'              = '2'|'1'
           Process Interview  0    Process Interview  0
           (Dead Interview)        (Live Interview)
           ┌──┬──┬──┐              ┌──┬──┬──┐
           │10│ 6│ 9│              │ 8│ 6│ 9│
           └──┴──┴──┘              └──┴──┴──┘
```

Operations list
1 Write Applicant
2 Write Skill
3 Write Skill Area {Skills}
4 Write Vacancy
5 Write Office
6 Delete Interview
7 Read Applicant, On Error Set Applicant ` SV = NULL
8 Fail Unless Interview ` SV = '2' | '1'
9 Read Interview, On Error Set Interview ` SV = NULL
10 Fail Unless Interview `SV = '19' | '18' | '17' | '16' | '4' | '3'
11 Fail Unless Office ` SV = '6' | '5' | '4' | '3' | '2' | '1'
12 Read Office, On Error Set Office ` SV = NULL
13 Set Office ` SV = '4'
14 Fail Unless Skill ` SV = '8' | '7' | '6' | '5' | '2' | '1'
15 Read Skill, On Error Set Skill ` SV = NULL
16 Set Skill ` SV = '11'
17 Fail Unless Skill Area {Skills} ` SV = '5' | '4' | '3' | '2' | '1'
18 Read Skill Area {Skills}, On Error Set Skill Area {Skills} ` SV = NULL
19 Set Skill Area {Skills} ` SV = '5'
20 Fail Unless Vacancy ` SV = '11' | '10' | '9' | '7' | '6' | '4' | '3' | '2' | '1'
21 Read Vacancy, On Error Set Vacancy ` SV = NULL
22 Set Vacancy ` SV = '9'
23 Fail Unless Applicant ` SV = '4' | '3' | '2' | '1'
24 Set Applicant ` SV = '6'
25 Applicant ` Applicant # := Applicant Withdrawal ` Applicant #
26 Get Applicant Withdrawal

Figure 7.11

7.4 Examples of the GRAPES Fact Representation & Syntax

Figure 7.12 is a communication diagram which defines the processing inside the boundary of the Hi-Ho local office. It represents the decomposition of the corresponding object on the higher level.

Two of the three objects in the diagram further described by process diagrams. The third object (Local_Register) is described by a lower-level communication diagram (figure 7.13). The messages from the PC to the Local_Register are defined in interface tables. It compares the flows in each level of detail, as shown in figure 7.14.

Notice in figure 7.14 how the data flow contents are related in this information abstracted from the interface tables for the PC-access arrow in figure 7.12 and some of the arrows in figure 7.13.

Figure 7.15 depicts the "Skill_Entity" corresponds to the SSADM skill area entity. The SSADM skill entity might be represented as an attributed relationship in GRAPES. Note that this kind of representation may occur where` a GRAPES E-R diagram represents a many-to-many relationship.

Underlying the E-R diagram and the interface tables are a set of data diagrams. In fact, in the GRAPES development the applicant's skill information has been represented as a repeating group, as shown in figure 7.16. Notice how data diagrams can be used to construct types from primitives, and other types.

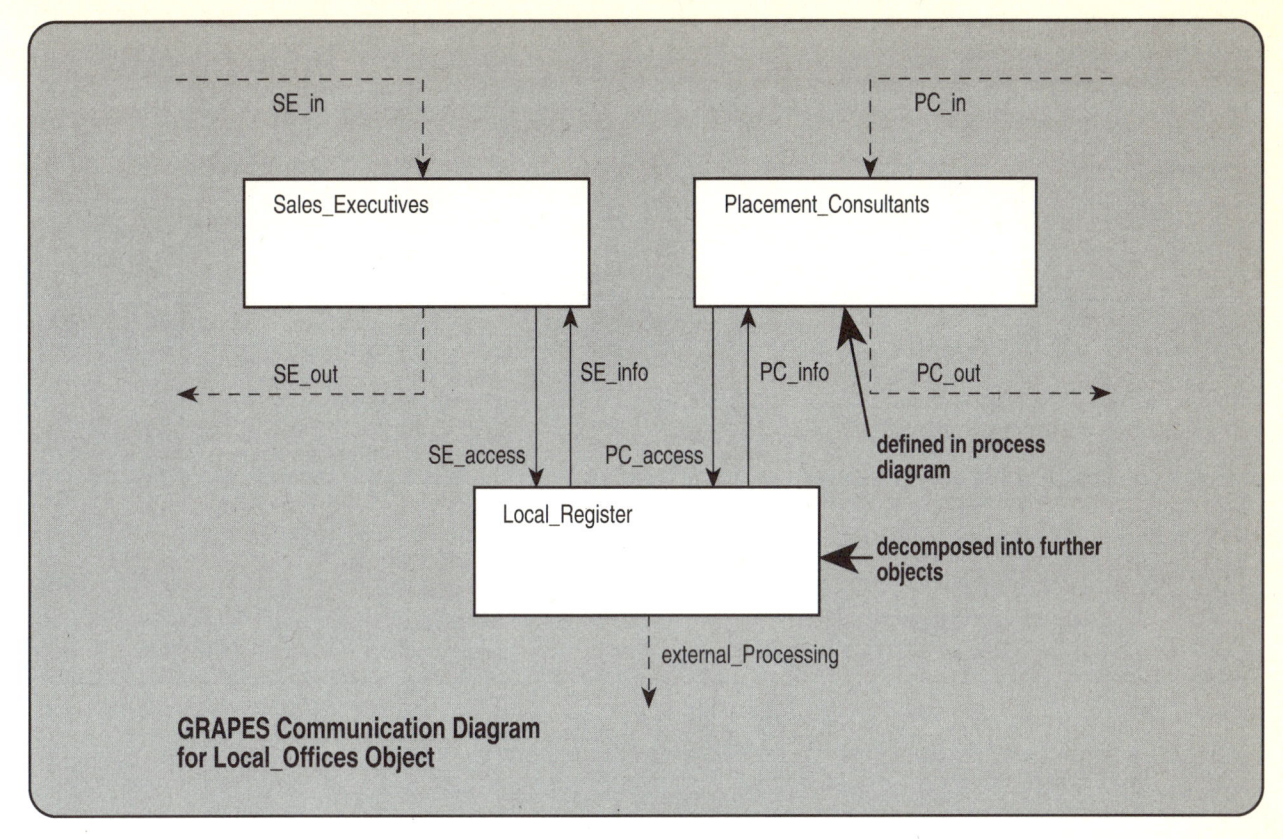

7.4 Examples of the GRAPES Fact Representation & Syntax

Figure 7.12

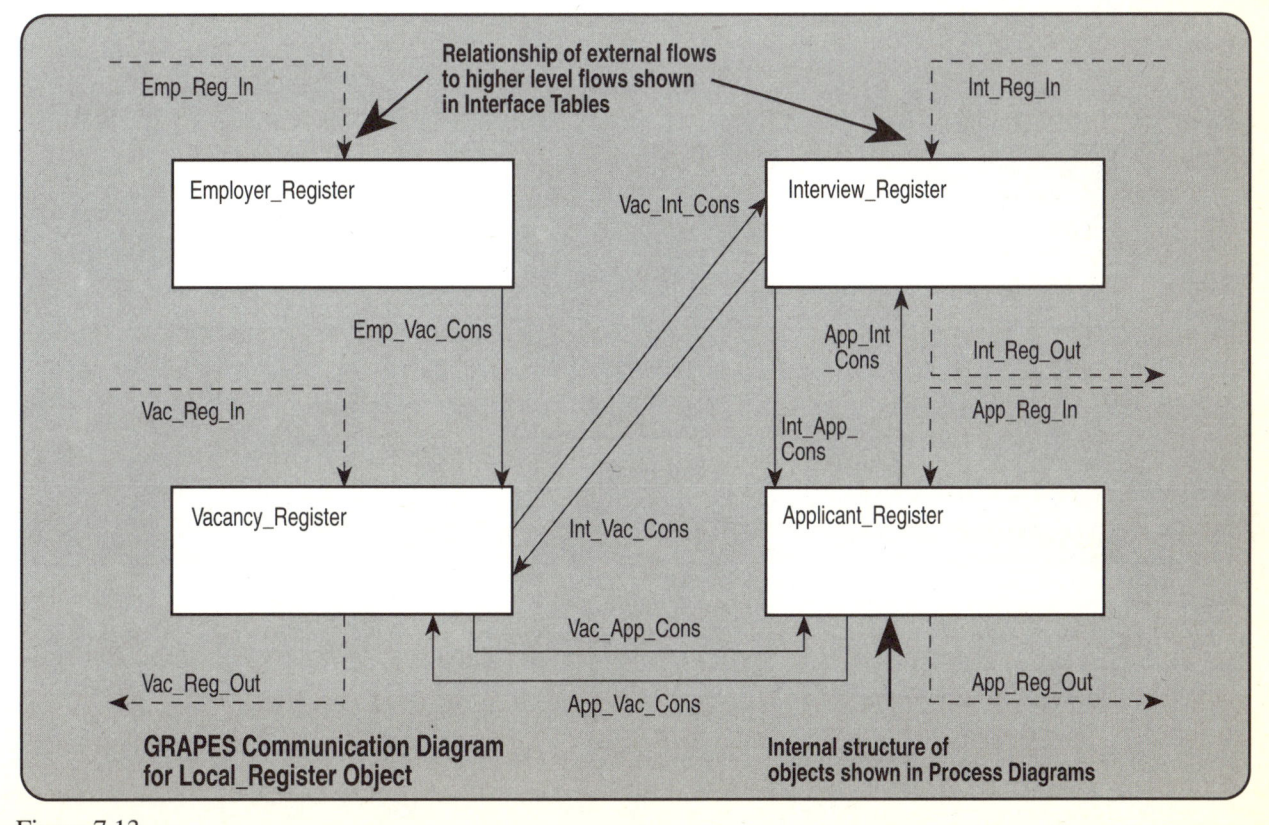

Figure 7.13

Interface Tables

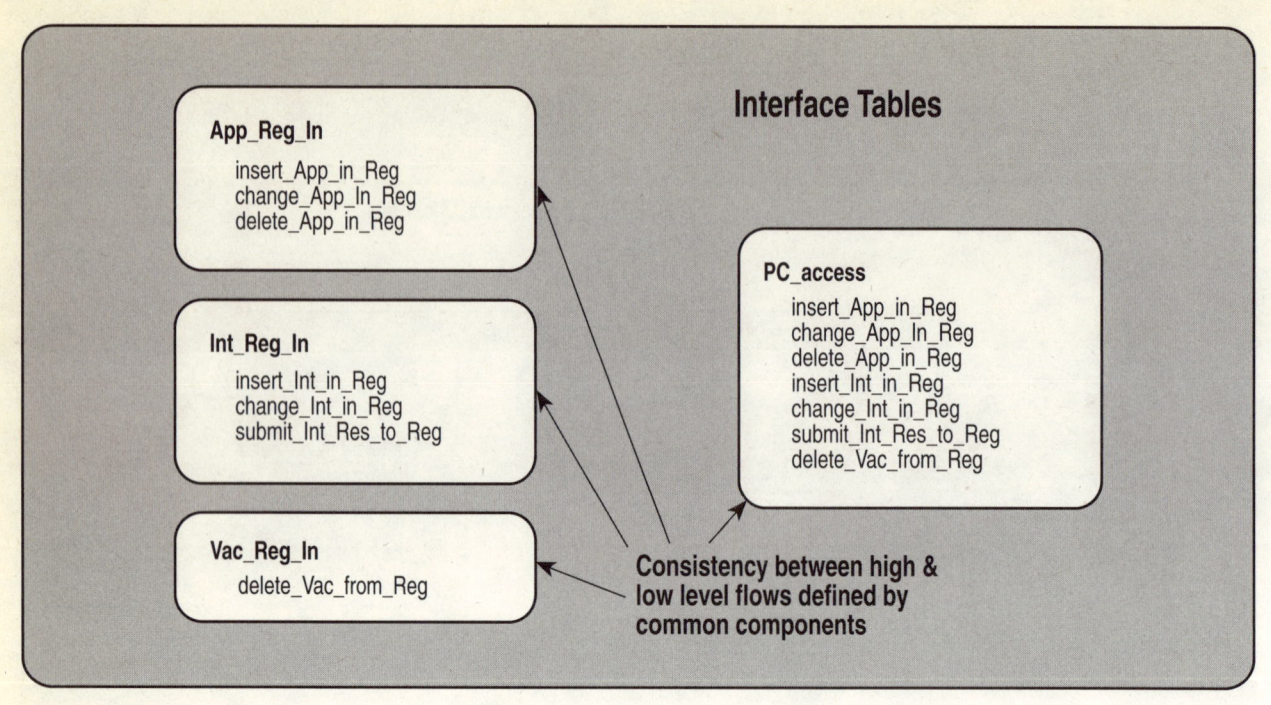

App_Reg_In
- insert_App_in_Reg
- change_App_In_Reg
- delete_App_in_Reg

Int_Reg_In
- insert_Int_in_Reg
- change_Int_in_Reg
- submit_Int_Res_to_Reg

Vac_Reg_In
- delete_Vac_from_Reg

PC_access
- insert_App_in_Reg
- change_App_In_Reg
- delete_App_in_Reg
- insert_Int_in_Reg
- change_Int_in_Reg
- submit_Int_Res_to_Reg
- delete_Vac_from_Reg

Consistency between high & low level flows defined by common components

Figure 7.14

GRAPES E-R Diagram

- Employer_Entity — offers: 0..n / is_offered_by: 1..1 — Vacancy_Entity
- Vacancy_Entity — is_arranged_for: 1..1 — Interview_Entity
- Interview_Entity — is_arranged_for: 1..1 / has: 0..n — Applicant_Entity
- Vacancy_Entity — has: 0..n — (Skill)
- Applicant_Entity — has: 1..n — Skill_Entity
- Vacancy_Entity — requires: 1..1 / is_Requirement_for: 0..n — Skill_Entity
- Skill_Entity — is_Ability_of: 0..n — Applicant_Entity

could be attributed with Applicant's Skill information

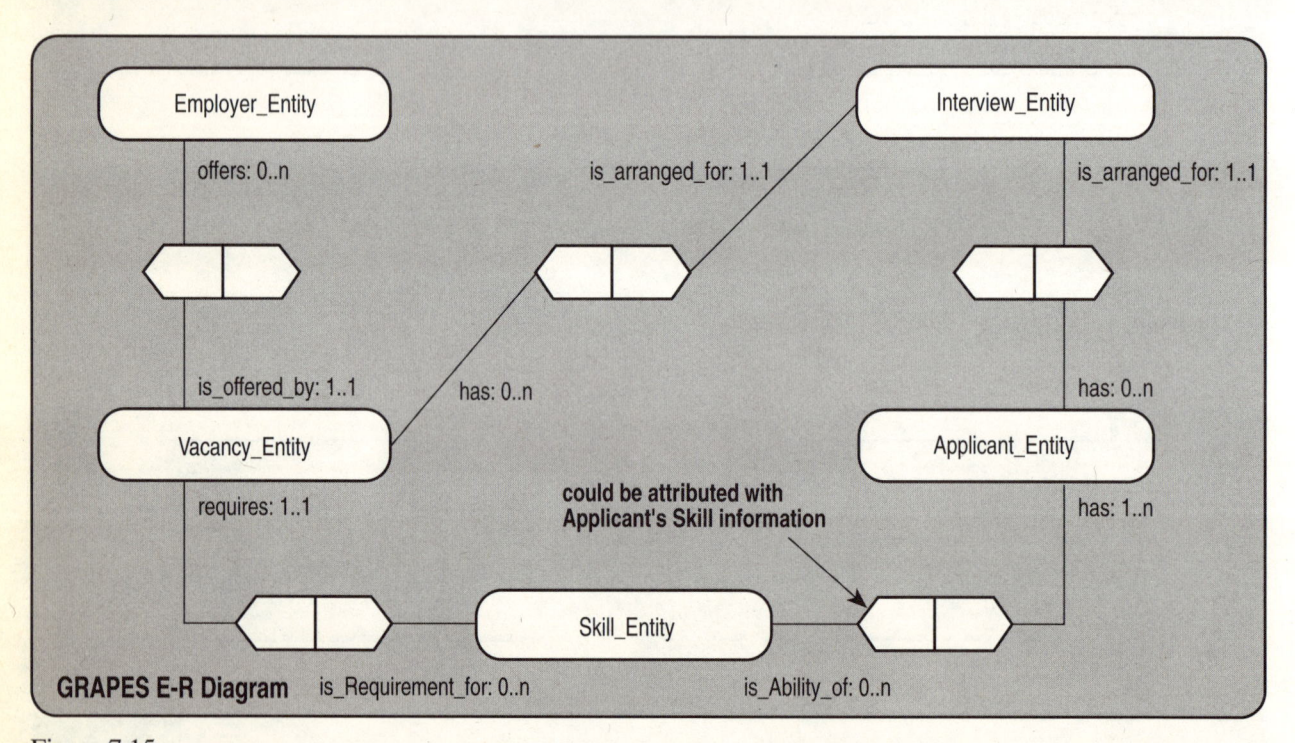

Figure 7.15

7.4 Examples of the GRAPES Fact Representation & Syntax

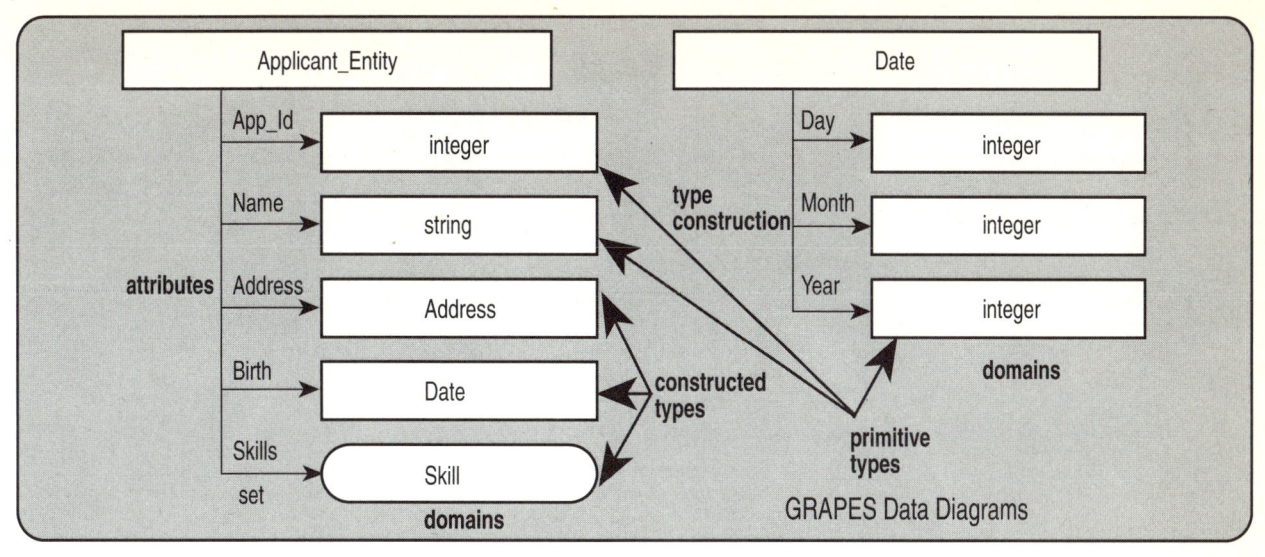

Figure 7.16

Finally, in figures 7.17 – 7.19, we show some GRAPES process diagrams covering approximately the same areas as the entity life histories shown earlier.

Notice, in particular, that:

- Although the diagrams have a flow-like notation, all of the loops are well-formed and it is easy to restructure the diagrams as iterations, selections and sequences. This is no accident. The underlying syntax is based on well-structured expressions.
- The diagrams capture fewer sequence constraints than the corresponding SSADM life histories. This is because they describe the behaviours of sets of entities rather than individuals. At the level of the applicant register it is not possible to specify that all applicant registrations come before all applicant withdrawals.
- Processing which is duplicated in the entity life histories and effect correspondence diagrams occurs only once here.
- Tracing the effects of an external input requires a significant amount of navigating from diagram to diagram and back.

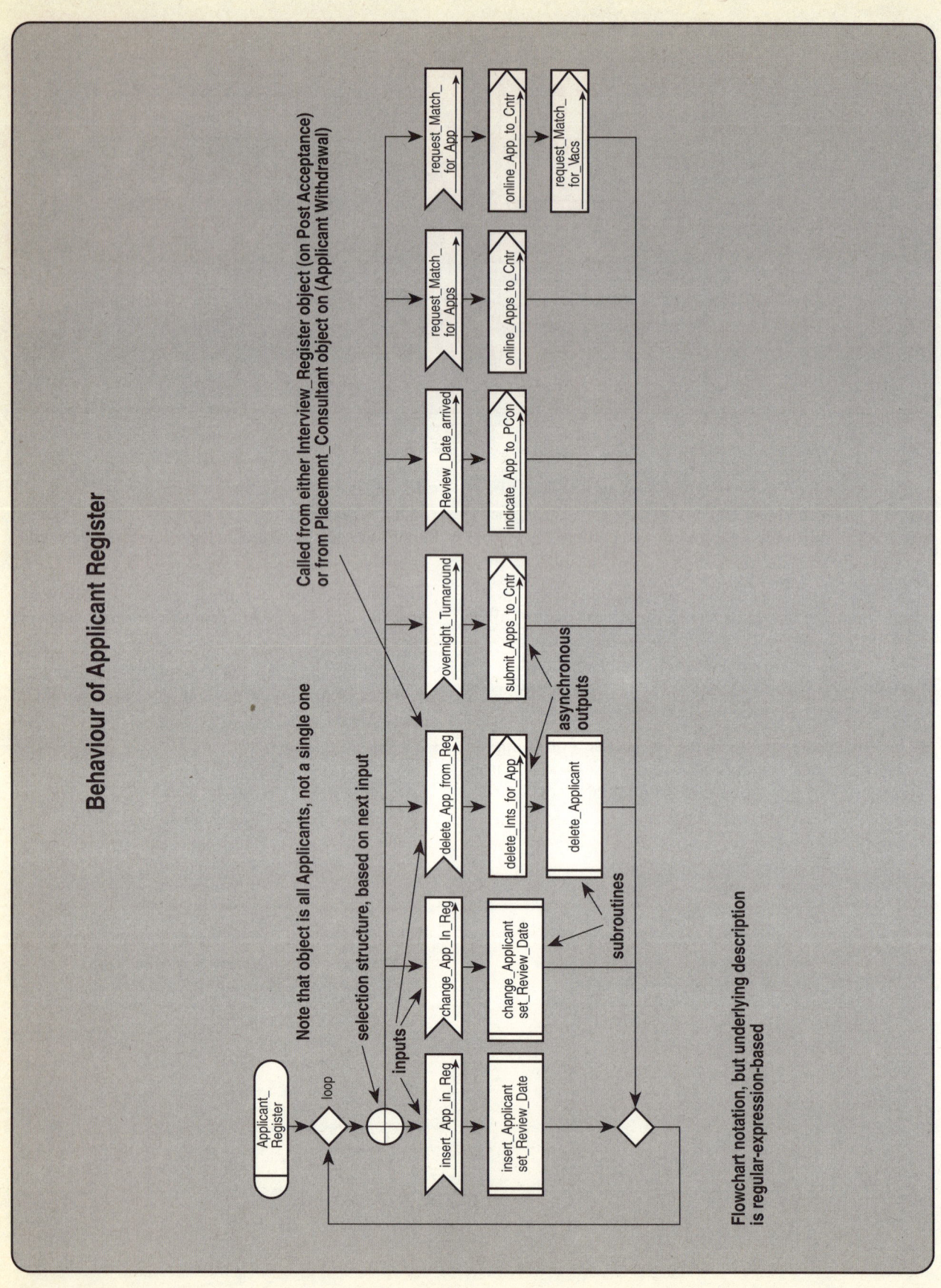

Figure 7.17

Behaviour of Applicant Register

Note that object is all Applicants, not a single one

Called from either Interview_Register object (on Post Acceptance) or from Placement_Consultant object on (Applicant Withdrawal)

selection structure, based on next input

inputs

asynchronous outputs

subroutines

Flowchart notation, but underlying description is regular-expression-based

7.4 Examples of the GRAPES Fact Representation & Syntax

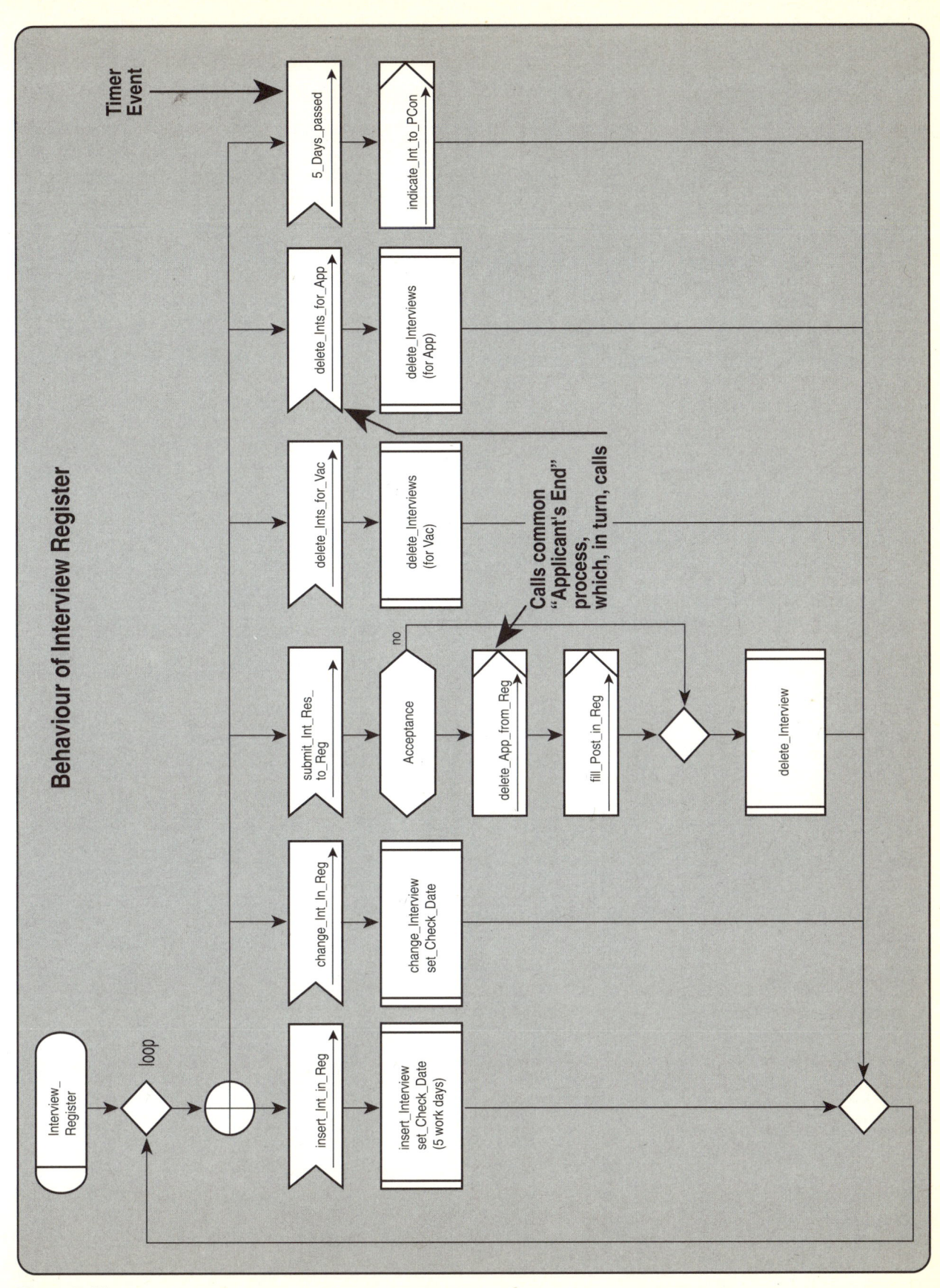

Figure 7.18

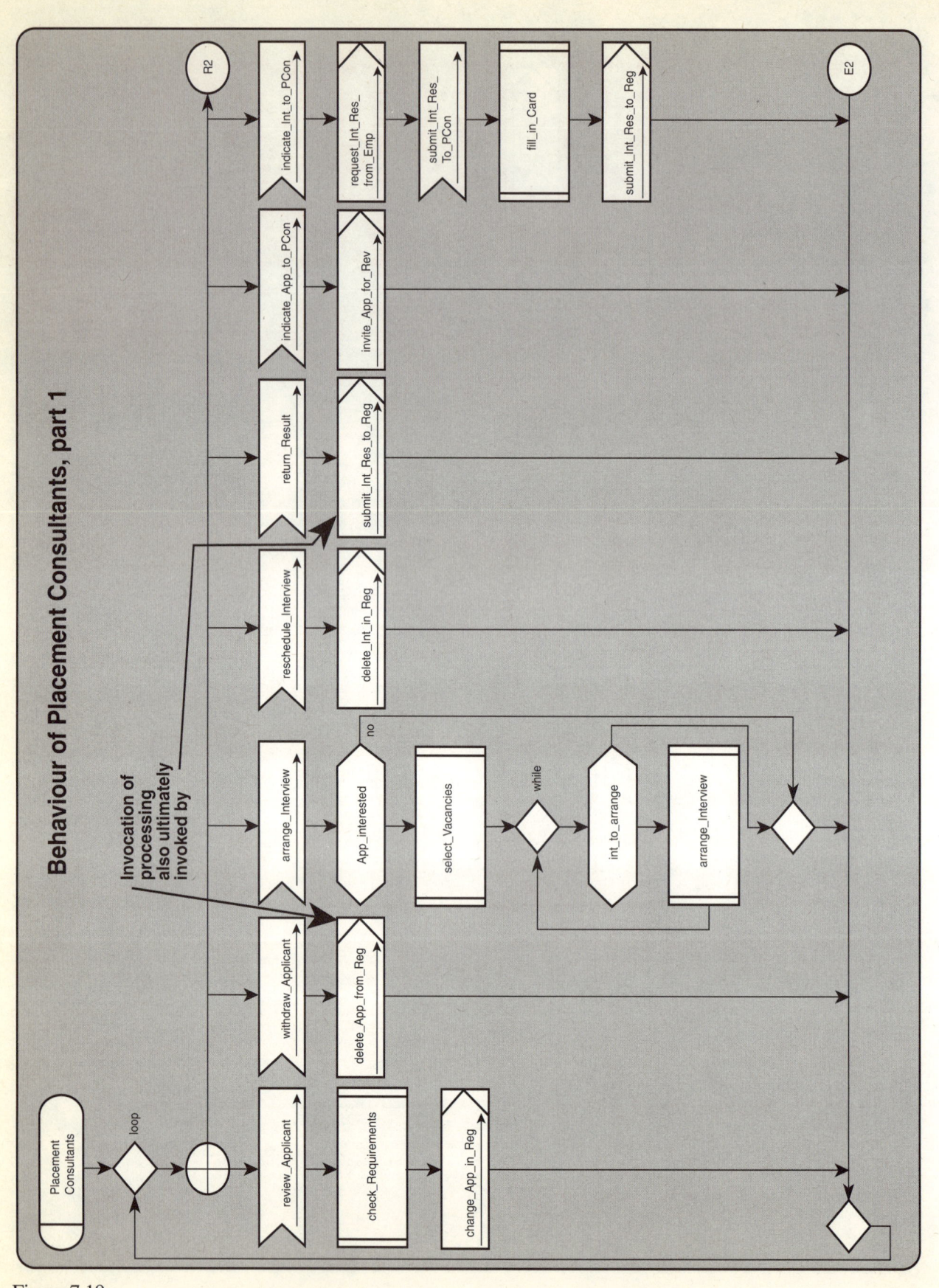

Figure 7.19

7.5 Comparison of the GRAPES & SSADM Fact Representation & Syntax

In the comparisons the following classification scheme is used:

+ directly comparable representations
≈ indirect support, but facts are easily accessible
− no support, or very indirect representation

A second symbol in front of an item indicates that there are a few exceptions from the given classification.

Representation of SSADM facts in GRAPES

Data flow diagrams / functions / I/O structures

Explicit & direct
+ System scope. Communication diagrams.
+ − Packages of user-invocable functionality and end-user / function interactions. Represented in communication diagrams. But because of the absence of multiple process instance construct, function structures are limited to describing interaction at external entity set level, instead of detail of an individual interaction.
+ Detailed external interfaces communication diagrams.
+ Communication between functions with different scheduling requirements via transient data stores. Communication diagrams.

Implicit
− Each external entity represents a potential array of external entities of the defined type. In GRAPES there is currently no way of specifying this.

Logical data models & access paths

Explicit & direct
+ Entity types. Data model.
+ Relationship types. Data model.
+ Optionality and mutual exclusion. Data model.
+ Attributes. Data model.
+ Data typing. GRAPES is stronger than SSADM.
− Volumetrics.
≈ Entry point to database for access path. 1st step in process diagram.
≈ Navigations through data model to support events & enquiries. By tracing messages passed between different process diagrams.
≈ Entity / relationship usage. By tracing messages passed between different process diagrams.
− Access volumetrics.

Explicit & indirect
+ Entity subtyping. This can be deduced from a combination of inspection of relationship cardinality limitations and mutual exclusion in the same way as in SSADM.

Entity life histories & effect correspondence diagrams

Explicit & direct
GRAPES will perform much better here once it has the capability of managing multiple process instances.
− Update success units.
≈ Access paths for updates. By tracing messages passed between different process diagrams.
− Instance-level control flow of processing for each entity
+ − Preconditions for events. Expectations about likely further processing. By setting and testing state variables, but not by process structure.
− Catastrophic changes. GRAPES doesn't provide backtracking or escape constructs.
+ − Sequence, selection, iteration and parallelism – but not within an entity life history.
+ − Behaviour cycles. GRAPES provides no way of naming loops or selections.

Explicit & indirect
− Side-effects of previous event processing done or omitted in error (by following event sequences and cycles through life histories). In theory, this information could be derived in GRAPES by setting and testing states; however, its derivation would be very complex in practice.
+ State transitions; have to be managed by explicit manipulation of state variables.

Implicit
− Array parallelism between entity life histories of same type.

Representation of GRAPES facts in SSADM

Communication diagrams

Explicit & direct
+ System scope. Data flow diagrams.
+ The structure of the system and its components. Data flow diagrams.
+ Communication relationships between parts of the system. DFDs.
+ External interface of the system (message types). Data flow diagrams.

Implicit
+ Parallelism between objects. Also between DFD processes.

Interface tables

Explicit & direct
+ External interface of the system. I/O structures.
− Communication relationships between parts of the system.
  I/O structures are not produced for internal data flows.
− ≈ Communication mode. Not clearly distinguished in SSADM. If anything, SSADM's communication is implicitly synchronous; asynchronous buffered communication is provided through transient data stores.

Implicit
+ Control flow. Triggering is defined in function definition.

Process diagrams

Explicit & direct
+ Control flow, preconditions and likely further processing at the object class level. Can be done at the class level, but normal practice, which captures more detailed facts, is at instance level in entity life history.
+ Representation of repetitive behaviour at the class level. Can be done at the class level, but normal practice, which captures more detailed facts, is at instance level in entity life history.
+ Reaction to messages and timer events. Effect correspondence diagrams.
− ≈ Explicit timer objects. Calendar / timing events are absent from data flow diagrams as triggers, but are often represented by outputs. Timer object entity life histories could be constructed, but the movement away from operational masters makes it less likely that they will be constructed in practice.
+ − Use of common processes by several inputs. SSADM does re-use conceptual scenario processes in more than one function. However, SSADM doesn't recognise that processing is being redundantly defined inside the entity life histories, effect correspondence diagrams and update process models when two or more events cause identical subsets of systems behaviour: SSADM doesn't recognize that the processing carried out, e.g. on post withdrawal, is also carried out on post acceptance.
+ Parallelism within an object. Lower-level DFD processes are implicitly in parallel with each other within the higher-level process. Entity life history syntax contains a parallel "sequence" construct. Entity life history instances implicitly execute their behaviours in parallel with each other.

Indirect
≈ States and state transitions at the class level. Can be done at the class level, but normal practice, which captures more detailed facts, is at instance level in entity life history.
+ Access paths through data when events / inputs occur. Effect correspondence diagrams and entity access paths.

Entity relationship diagrams

Explicit & direct
+ Entities and relationships. But relationships cannot be attributed in SSADM.
+ Optionality, mutual exclusion, cardinality.
+ Role names.

Data structure diagrams

Explicit & direct
+ Structure of stored data – attributes, TNF relations
+ − Structure of flowing data. I/O structures only.
− Data type definition, scope definition & domain support. SSADM is informal here.

Data tables

Explicit & direct
+ − Local variables of processes, modules and procedures. Defined for entity life histories (the TNF record definitions are the local variables of the entity life histories). In principle the same kinds of local variables are needed for dialogues and other processes. SSADM isn't specific about defining these. If the dialogues are built using SDM the local variables are the SDM state-vector.
− Initial values for variables
− Symbolic constants

Specification diagrams

Explicit & direct
− Formal parameters of procedures. Perhaps this could be tackled in the function component implementation map.
− Exported procedures, data types and variables

Implicit
+ Hiding of information

Hierarchy diagram

Explicit & direct
≈ Hierarchical object structure. Present in data flow diagram hierarchy
− + Scopes of declarations. Local data stores

Comparison of GRAPES & SSADM syntax

Data flow diagrams & communication diagrams

The two syntaxes are directly comparable and compatible. A GRAPES communication diagram has objects which function as SSADM external entities (in the top-level diagram only), objects which function as SSADM data stores, as well as objects which function as SSADM processes.

I/O structures in SSADM are represented by interface tables and data definitions in GRAPES. SSADM doesn't require the equivalent of these definitions for data flows to and from data stores, or between processes; however, these could easily be provided and there is at least a case to be made that they ought to be provided. In SSADM no meaning is given to the fact that the same name appears on two different data flows. In GRAPES it is an error to have two arrows with the same name in one diagram.

Consistency between the levels of data flow diagrams has different requirements in GRAPES and SSADM. In GRAPES the consistency is symmetric; as in SSADM, it is invalid to include something in an upper-level diagram which isn't represented in a lower-level one but, unlike SSADM, it is also invalid to include something in a lower-level diagram which isn't represented in an upper-level one. SSADM uses this asymmetry to present simpler, more easily understandable pictures at the upper levels. In SSADM the upper-level pictures are, with minor exceptions like the appearance of upper-level process boxes in the external environment of a lower-level data flow diagram, only there for "comfort". Because the GRAPES interface table information is organised hierarchically, one needs to move up and down levels of communication diagram to track information exchange between process diagrams. So in GRAPES symmetric consistency is important.

GRAPES object type instantiation can be used to represent distribution of the same system functionality to several named end-user locations. SSADM doesn't provide this capability.

Logical data models and entity relationship diagrams

The only significant difference between the two systems (apart from minor details about cardinality) lies in the ability to define relationship attributes. Attributed relationships could be represented easily as entities in SSADM. But how could we recognise those entities in SSADM which ought to become attributed relationships in GRAPES?

At first sight, the solution seems simple: compound key entities ("associative" entities) become attributed relationships. However, the choice of key in SSADM, although governed by some rules of thumb, is arbitrary. An analyst could decide to introduce a hierarchical key to simplify a large compound key.

The rules for choosing keys in SSADM are, in any case, not entirely satisfactory. For example, it is possible to choose a compound key, one of the elements of which represents a relationship which can be "swapped" to a new instance of the same master entity type. Such an update represents an update to the primary key of the detail entity. It is generally agreed by the relational database community that updates to primary keys are undesirable. We recommend some investigation of the rules for choosing keys. Better rules for key choice would then go a long way towards providing a solution to recognise attributed relationships in the SSADM entities.

Entity life histories etc. and process diagrams

Despite the differences between the Jackson diagram appearance of SSADM process descriptions and the flowchart appearance of GRAPES process descriptions, the two representations are both based on well-structured expressions augmented with a parallel "sequence" construct. It should be easy to convert from one representation to the other and back.

The main differences are:

- Implicit recognition in SSADM of multiple process instances of the same type, and the absence of a mechanism for handling this in GRAPES.
- Absence of mechanisms in GRAPES for handling backtracking or escape requirements. This prevents GRAPES process descriptions from clearly separating the description of an expectation about the behaviour of the business system from descriptions of exception handling.

Formality

GRAPES has a formal syntax which covers everything represented. In SSADM formality is generally limited to the formation rules for diagrams. This means that much more detailed syntactic cross-checking can be performed in GRAPES.

On the other hand it may make it more difficult to extend the GRAPES language, because of all the loose ends which need tying up. In practice this may be a good thing or a bad thing. SSADM suffers from time to time from the injection of new concepts whose relationship to existing concepts has not been well thought through (e.g. the adoption of mutual exclusion in SSADM version 3 LDSs without the adoption of a procedure for dealing with them at the first-cut design stage). But SSADM can also experiment with new techniques, not having to worry about ultimate fit until the techniques prove to be useful.

Summary of findings about fact representation

In the area of system dynamics SSADM tends to represent more facts, more directly, about the system being developed than does GRAPES. This is shown in chapter 8 to have some significant consequences in knowledge collection.

However there are some significant limitations in SSADM's fact representation which are addressed by GRAPES:

- Detailed facts captured by GRAPES at the data item level and I/O level are more formal and more susceptible to machine manipulation.

- There is an inability in SSADM to factor out common processing which appears in several entity life histories. This common processing comes from SSADM's redundant representations of the same fact in several diagrams. This is not redundancy in the sense that SSADM is making a translation of a known fact into another format where it is used for a specific purpose, but redundancy in the sense that the SSADM documentation doesn't even "know" that the same fact is being represented several times.

This redundancy is easily correctable by the introduction of "super-events" as shown in annex A4. But then some of the facts directly captured in the entity life histories and effect correspondence diagrams become more indirect as the analyst has to navigate through a classification / inheritance hierarchy to correlate them. GRAPES does capture this common processing but pays the same penalty. A way of gaining both benefits is to maintain a (redundant) representation of the classification / inheritance hierarchy extracted from the basic representations. This is shown at the end of annex A4.

Another way to gain the benefits of both approaches is to use a CASE tool which holds a non-redundant representation and deduces indirect facts from it. If one has such a tool the distinction between direct and indirect facts vanishes, provided, of course, that the tool is programmed to generate all of the required facts from the underlying representation. Tools of this sophistication do not exist at the moment.

- GRAPES manages the generation of "external" time-based events. SSADM tends to leave this to the operator or operating system.

- The content and structure of internal data flows is defined in GRAPES, but not in SSADM. We would expect that this has had some negative effect on the development of the SSADM 3GL Interface Guide.

- SSADM doesn't currently address system distribution.

8 The Knowledge Collection Process

Some common approaches to systems development are thought of as methodologies but actually provide very little in the way of "method". For example, information engineering is defined almost completely in terms of its end products, rather than by how those end products are produced. Some methods, however, claim to assist in the process of producing those end products; thus we believed it was important to study how the different approaches collected the knowledge embedded in the final systems.

8.1 Knowledge Collection in SSADM

The following knowledge acquisition techniques are the major ones used in SSADM:

Knowledge about the scope of an SSADM system

Resource / Service flow analysis

This is a technique for analysing what business activities (as opposed to information handling activities) occur in the organization; for example, a surgeon actually operating on a patient (as opposed to recording that an operation had taken place). The technique can be used to identify the major resource usage in an organisation and those services whose delivery is failing in order to discover the high pay-off areas for information system development (under the assumption that use of IT will improve performance of those business activities).

Corporate data modelling

The data modelling technique discussed below can also be used at the corporate data modelling level. The corporate data model can then be partitioned and the partitions prioritized by end-users.

Knowledge about the static structure of an SSADM system

Logical data modelling

SSADM provides a reasonably objective synthetic process for constructing data models. We use "synthetic" to mean the opposite of "analytic" as discussed in the rational data analysis. The technique does not work by taking concrete things apart and reassembling them; it works by conjuring facts out of a dialogue with end-users. It has a number of substeps:

- Identifying entities:

    The "notional key" technique is used - identifying the kinds of reference numbers the users will need to manage, and then asking what entities those keys stand for. The technique works because keys represent one of the most significant properties of "entityness": distinctness of being: the ability to identify and reidentify an object.

- Identifying relationships:

    Relationships are identified by considering each pair of entities in turn. This step limits our consideration to binary relationships, and to relationships which are direct, and not through other entities. More complex relationships are discovered later, as a result of applying the next two steps.

- Resolution of many-to-many relationships:

    The complete resolution of many-to-many relationships into pairs of one-to-many onto a new, intermediate entity, together with the identification of a good name for that new entity is an important knowledge acquisition step.

- Simplification rules:

    Elimination of relationships by merging entities which are in a strict one-to-one relationship identifies where two entities are really two aspects of the same thing. Elimination of "double-v" structures completes the discovery of tertiary and higher degree relationships. Elimination of "triangles" rids the model of any indirect relationships which have been mistakenly included.

- Validation:

    Ensuring that required outputs can be produced by "walking round" the data model helps discover further facts which have been overlooked.

- Naming:

    Finding good names for entities and relationships is an important knowledge discovery technique. The relationship-naming rules are designed to capture the knowledge about a relationship.

Resource / Service flow analysis

This technique is used to derive a top-level data flow diagram when no current information system exists. The efficacy of the technique is based on the assumption that a new information system must have a similar external structure to that of the business system it supports. This is a "form follows function" argument.

Data flow modelling

This technique follows the following principles (spread across different parts of the life cycle):

- Document flow analysis of the current system

  A picture of the current system is constructed for two reasons: to provide the users with a concrete picture of something they already know to ease their acceptance of later, more abstract systems descriptions; and to provide a baseline from which deficiencies in the current system can be identified. This is not so much knowledge collection as collection of a set of facts from which the knowledge can be deduced.

- Structuring of the current system into a data flow diagram (DFD) hierarchy

  This approach mixes some intuitive and subconscious analysis of coupling and cohesion (not explicitly recognized as SSADM concepts) with principles concerned with isolating exception handling in lower-level diagrams. Here we are structuring the facts collected above and beginning to see what they mean.

- Logicalization of the data flow models

  This technique removes current implementation considerations from the data flow models to reveal the knowledge that they contain about the system structure. The substeps involve:

  – Alignment of system data stores with closely related groups of entities
  – Clustering of processing around "user roles"
  – Removal of unnecessary transient data stores
  – Restructuring of the DFD hierarchy from new bottom level
  – Identification of "bottlenecks" & "common tails" (common processing)

  This technique completes the analysis of the meaning of the facts gathered above. After the system has been enhanced to ensure that the identified requirements have been met, this knowledge forms the basis for identifying the system's "functions" – the packages of processing that particular users will be given the capability of invoking.

Relational data analysis

This technique is an "analytic" technique. It works by taking apart concrete but complex facts and discovering the simple facts on which they are based. Those simple facts (about entities) are then restructured in data modelling notation to reveal the relationships between them and used to enhance or correct the logical data model.

The technique is essentially one for discovering semantics by analysing syntax. It works when (as is the normal case) the formal structure of the data matches the structure of the meaning of the data. It fails otherwise.

The first half of the technique is essentially the "Third Normal Form" (TNF) analysis process discussed widely in computer literature. The second part of the technique uses the appearance of the primary key of one entity as a data item or part of a key in another entity to deduce that a relationship must exist between those two entities.

Knowledge about the behaviour of an SSADM system

Entity life history analysis

The main knowledge acquisition questions are:

- What updating events are implied by the data flow model I/O structures?

  Each triggering input data flow which causes an update must carry information about the occurrence of one or more externally-notified business events. Each output data flow which seems to occur spontaneously, and which originates in some processing which updates a data store, must occur because some timer-based business event has occurred.

- What updating events are implied by the structure of the logical data model?

  Each entity must be created and killed by some business event. Relationships which can change must do so in response to some business event. The events which cause incrementation of a numeric attribute are usually different from those which cause decrementation of that attribute.

- Are the events identified atomic?

  Can part of the processing fail and yet the event still succeed? If so the event needs to be decomposed.

- What's really happening inside an iterated selection?

  Events which appear inside an iterated selection structure in an entity life history are either truly independent of each other or must be related in a more complex way. Dependence can be tested for by checking whether they both affect or use a common attribute. If they are independent the iterated selection masks a parallel structure. If they are dependent there may be one of a number of things going on: a build-to-a-climax cycle, a set-up-and-use cycle, hidden array parallelism necessitating extraction of a new entity and / or requiring some numeric validation test to ensure that hidden entities can't be killed before they're created etc. All of these represent important knowledge about the system which may help to simplify it, prevent it from failing, help it or the user recover from failure, or help it to behave more intelligently.

- Do iterations of iterations hide entities?

  Once more, hidden entities can be discovered, and, after the new entity life histories are constructed, new events can be identified.

- Why does the same substructure occur in many places in an entity life history?

  A reason can be because some parallelism hasn't been recognized. The events really have different effects depending on when they occur. In the latter case the analyst needs to name the effects and to identify differences in operations to be carried out. He also figures out whether only some of the effects propagate into neighbouring entity life histories rather than all of them doing so.

- What effect does a termination event in one entity have on its data model details?

  Iterations in one entity usually contain birth events of its data model details. Termination events may be thought of as those which either kill off an entity or follow an iteration in the entity. The effect of those termination events on the details must be considered. This is analogous to specifying CASCADE or RESTRICT rules in connection with deletion in a relational database. Note, however, that the question is a more general one: the effects must be considered even if the entity is not being deleted (even if we keep applicants around for enquiry purposes after they have got jobs, we still want to ensure that interviews are not left open when the applicant gets a job), and whether we CASCADE or RESTRICT depends not on the entities, but on the events (a customer's contracts may be broken on customer death, but a customer may not withdraw if he has live contracts), or on the states of the entities (vacancy withdrawal kills the interview if no result has been received, but leaves it around for future queries on applicant performance if an offer has been refused or an applicant has been rejected.

- Can an entity be reborn, or revert to an earlier state in its life?

  This might, for example, let us discover that it's not a good idea to delete an applicant until we're confident he won't be coming back in the near future, or that we need a procedure to reinstate an applicant if we've entered a post acceptance against him in error.

- Can anything else happen between event A and event B?
- Do things always happen in this sequence? Do we always find out about them in this sequence?
- Why can't I build a satisfactory entity life history for this entity?

  Because the logical data model is wrong. This is not genuine knowledge acquisition, but knowledge acquisition derived from getting stuck.

8.2 Knowledge Collection in GRAPES

The knowledge collection process in GRAPES is not prescriptive at present. GRAPES modellers are free to use virtually any method to collect systems knowledge. What is prescribed is that at different milestones models of a well-formed nature have to be delivered. The results are checked in a prescribed quality assurance process which is defined in the DOMINO process technology. How the models are produced is, strictly speaking, not the concern of GRAPES. This is because of the Siemens marketing policy, which is designed to maximize the market for GRAPES.

Some techniques are taught on GRAPES courses, for example the top-down approach for decomposing object hierarchies mentioned below. It should be noted that this was not the method used by the Siemens Hi-Ho team who, in effect, decomposed Hi-Ho more or less along the lines of their E-R model.

Knowledge about the scope of a GRAPES system

MOSAIK

To find out about customers' requirements and current situation there exists an interviewing technique within

DOMINO called MOSAIK. It consists of guidelines for questioning the customer and a tool for recording and statistical evaluation of the results.

The main activities are:
- Document flow analysis
- Job analysis for every workplace
- Analysis of sequence and duration of jobs

MOSAIK is at the moment only loosely coupled with GRAPES. For the future it is planned to extend these techniques to make them support the development of the strategic GRAPES models in the early Problem-analysis-phases of the DOMINO life cycle.

Knowledge about the static structure of a GRAPES system

Data modelling

In the area of data modelling with entity-relationship techniques the analysts are advised to apply some commonly accepted rules.

Simplification rules:
- Elimination of 1:1 relationships
- Elimination of triangles and cycles

Naming recommendations:
- Use of meaningful names
- SPO rule (subject-predicate-object rule): relationships should be readable as a sentence

Normalization:
- TNF (third normal form)

Besides these rules, there exists an entity-relationship-based method described in "G. Held: Informations- und Funktionsmodellierung mit GRAPES" (see references; currently only in German).

Rules for entities:
- An entity must have a key, i.e. at least one identifying attribute
- An entity must have at least one characterizing attribute

Rules for attributes:
- Every attribute is assigned to exactly one entity, i.e. the entities must be in third normal form

Rules for keys:
- A key attribute must not have null values
- An identifying attribute must not be a characterizing attribute in another entity
- Composed primary keys should not be classifying
- Composed primary keys must be minimal

Rules for relationships:
- Multiple relationships should be aggregated to one relationship with an attribute if possible
- Redundant relationships (in cycles) should be removed
- 1:1 relationships should be resolved if appropriate
- m:n relationships with attributes should be transformed into entities (introducing 1:n relationships) if appropriate
- Multidimensional relationships should be resolved if appropriate

Object hierarchy / communication structure

The top-down approach is to identify subobjects by looking for subobjects which implement the dynamic behaviour of the parent object. In schematic form the refinement process follows the model, shown in figure 8.1, whose shape gives its name to the approach.

The approach is a modified Myers – Constantine approach. Cohesion (how tightly related the things are inside an object) and coupling (how loose are the connections between objects) metrics are used to evaluate decompositions (tight and loose, respectively, are the best). Information hiding (Parnas) is also used in evaluating decompositions. The approach provides criteria for deciding when a milestone model has been reached, i.e. terminating conditions for the refinement process.

The main steps of the Z-procedure are:

1. Definition of the external partners and interface of the system
2. Definition of a rough process outline of the reactions to the external events
3. Identification of sequential, parallel, alternative and repeated behaviour
4. Definition of the substructure by packaging behavioural units into subobjects and connecting them with communication lines
5. Repetition of steps 2 – 4 for every object until further decomposition is not useful and all objects are described

In addition there are rules for checking a decomposition:

- Every object must have at least one incoming and at least one outgoing communication line, except those representing external partners
- A sequential path through a set of objects is checked to see whether the whole path represents a single object
- Objects that receive and send the same messages or communicate with the same partners are to be checked to see whether they can be amalgamated into one object
- Objects with similar names and interfaces are checked to see whether they can be amalgamated into one object (difficulties in finding meaningful names often indicate a bad decomposition)

Knowledge about the behaviour of a GRAPES system

The processes of objects that have not been decomposed are defined in more detail and some additional questions are asked:

- What happens if expected information does not arrive in time or does not arrive at all?
- What are the timing constraints?
- Is additional information (communication or local storage) required?
- Are there common behaviour patterns (within one process or in different processes) that can be packed into procedures or modules?
- Are there existing procedures and modules that can be used in specifying the process?

8.3 Examples of the SSADM Knowledge Collection Process

The result of applying the initial steps of logical data modelling contains a "double-V" structure, as shown in figure 8.1.

On resolving the "double-V" structure we discover a constraint, which is shown in figure 8.2.

(It is interesting to note that one of the participants in the study, who had seen only this end-product, queried the absence of a relationship between applicant and interview.)

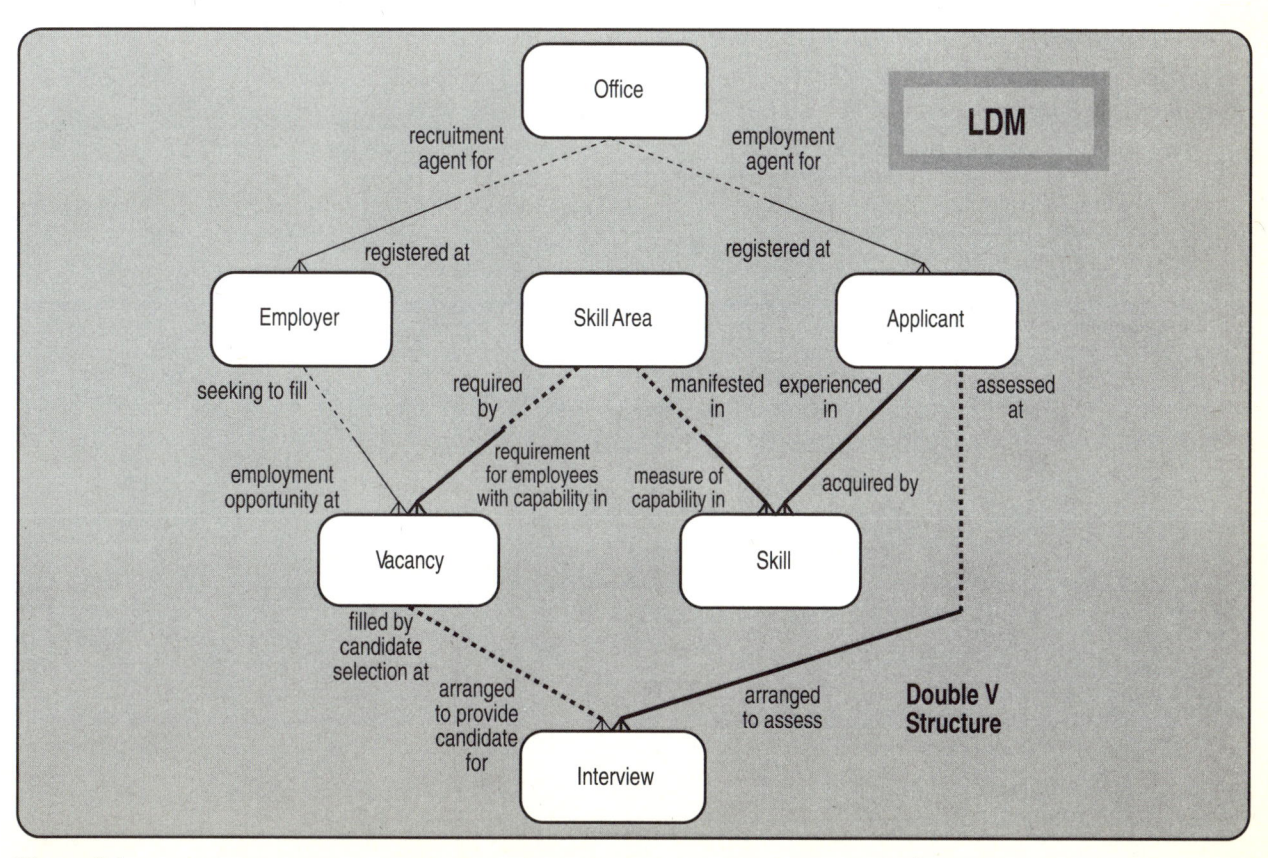

Figure 8.1

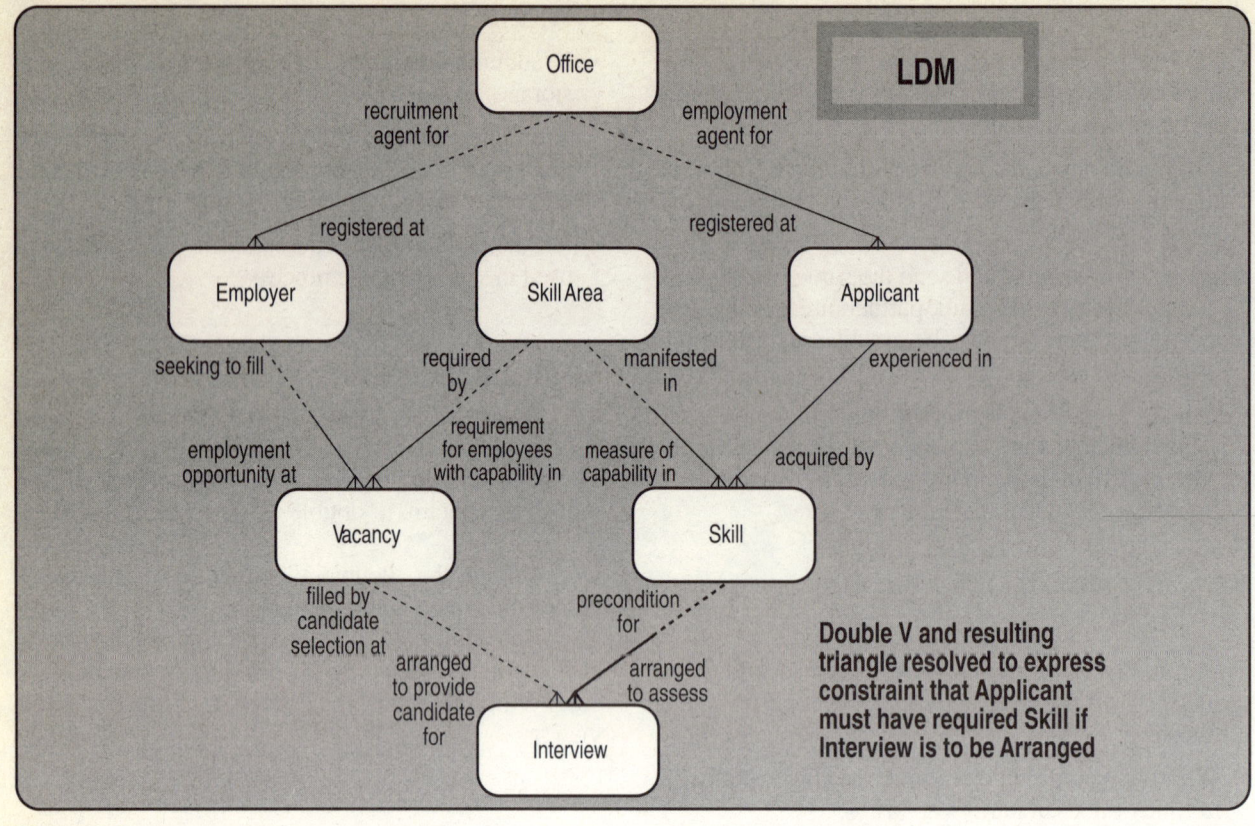

Figure 8.2

Figure 8.3 shows an example of the analysis of iterated selections from the case study:

When we try to define the relationships between the three events under the iterated selection we discover that

- all post acceptances cause skill areas to lose skills
- only the last post acceptance for a vacancy causes a skill area to lose a vacancy.

The entity life history has been split into two in figures 8.4 and 8.5.

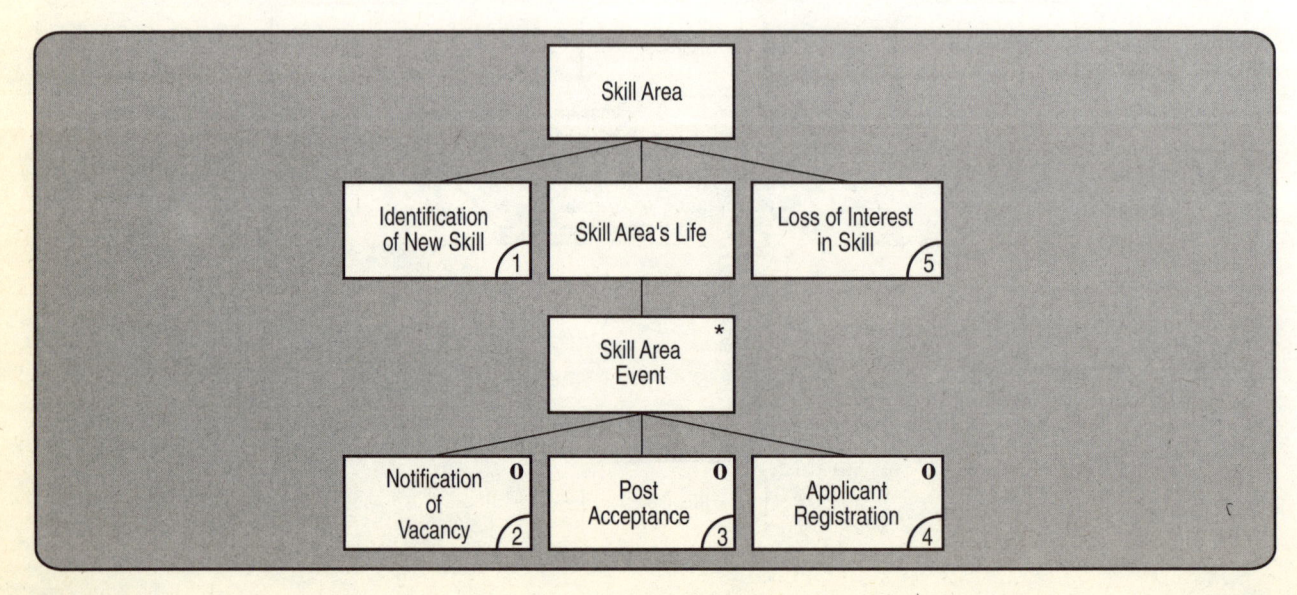

Figure 8.3

8.3 Examples of the SSADM Knowledge Collection Process 87

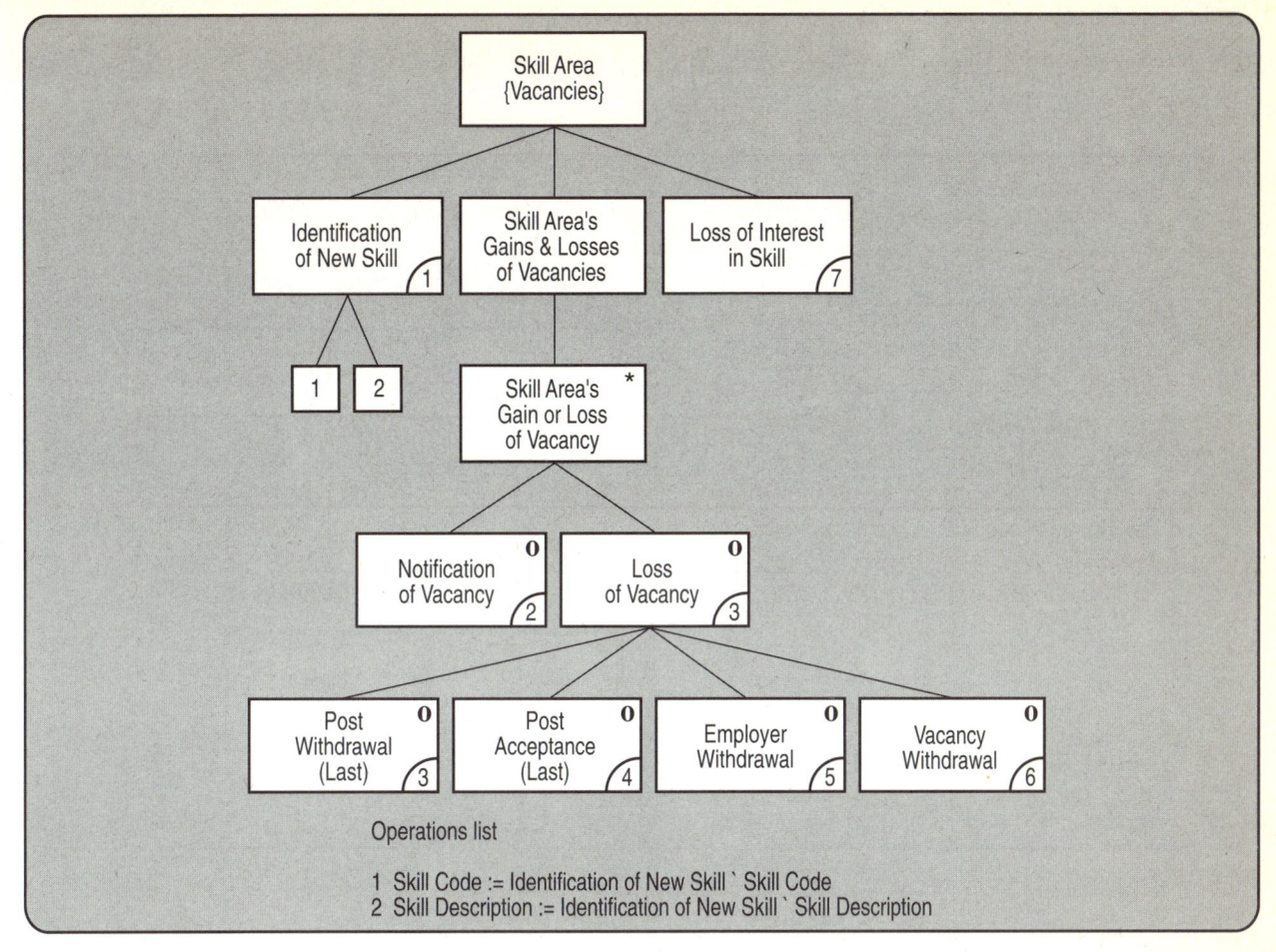

Figure 8.4

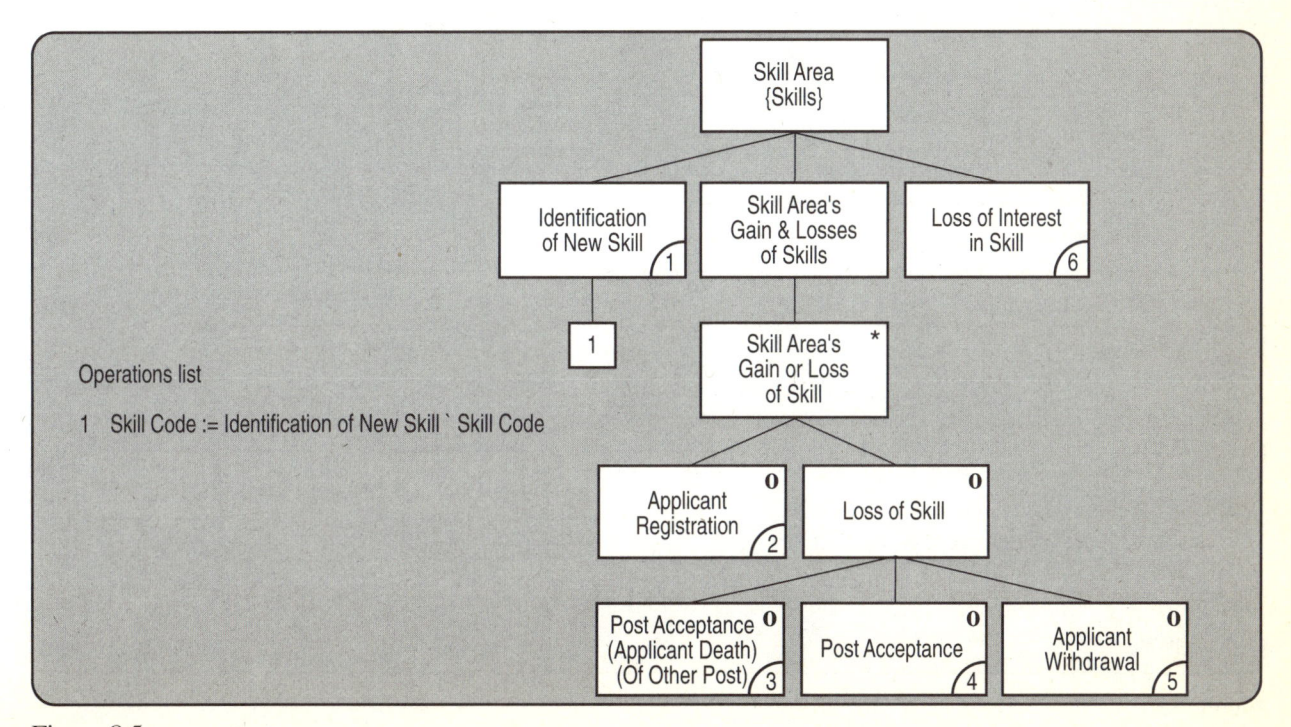

Figure 8.5

In a different case study, figure 8.6 is an example of how iterated selections can hide parallel arrays. The concept of an instalment needs to be recognized and managed. Lateness turns out to be an attribute of the ageing of instalments, and, as is shown in figure 8.7, not an attribute of contracts.

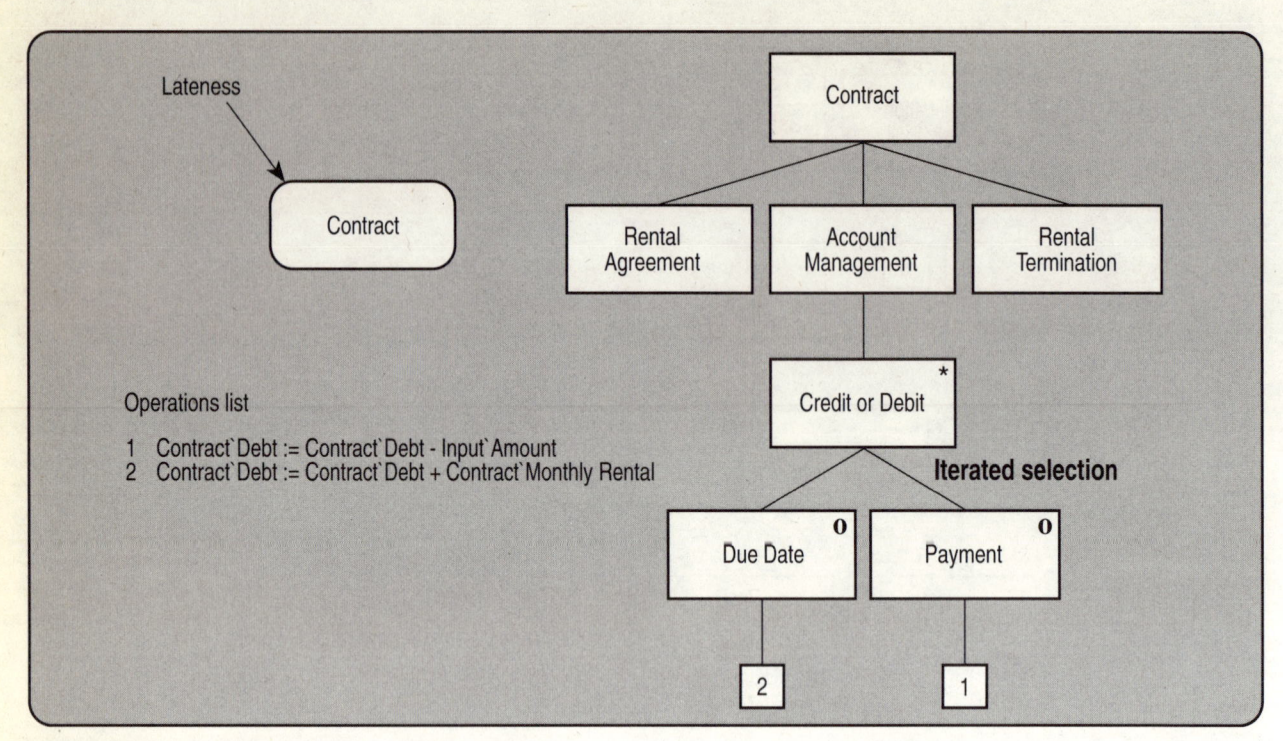

Figure 8.6

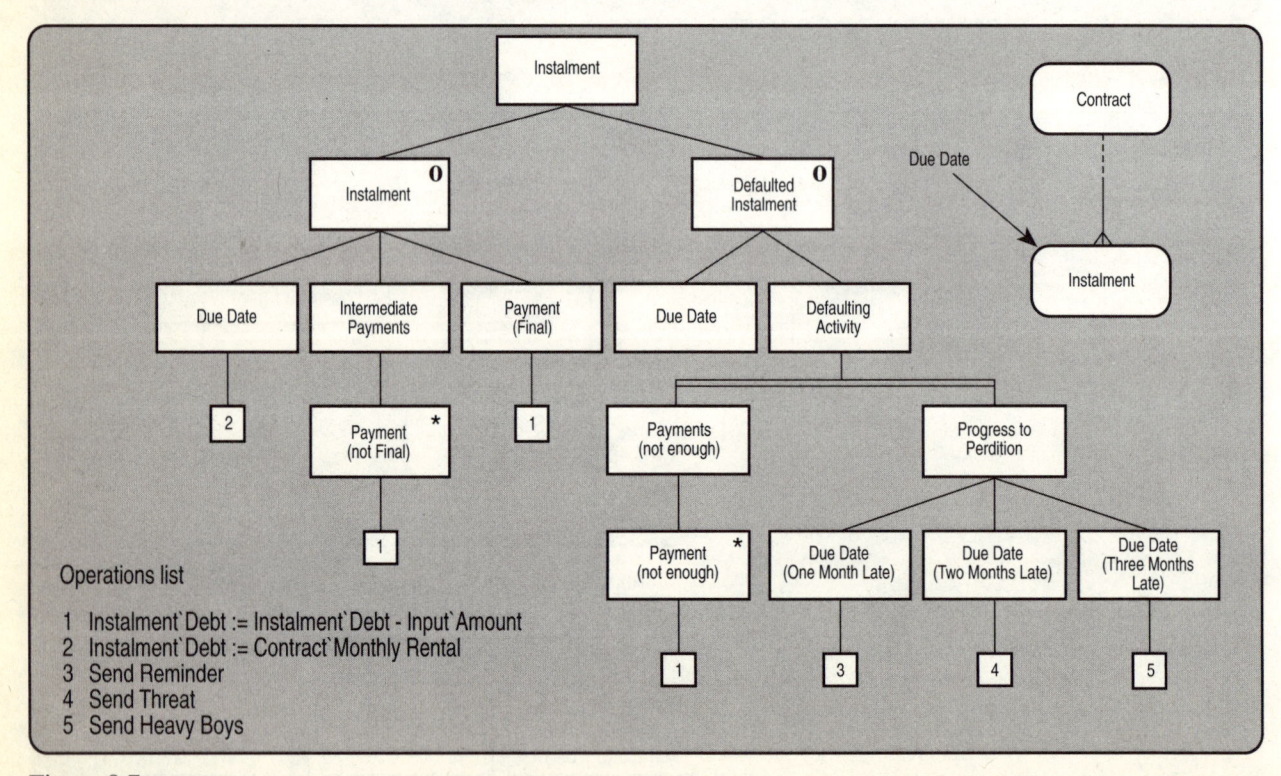

Figure 8.7

8.3 Examples of the SSADM Knowledge Collection Process   89

Figures 8.8 and 8.9 provide an example of the analysis of the consequential effects of termination events:

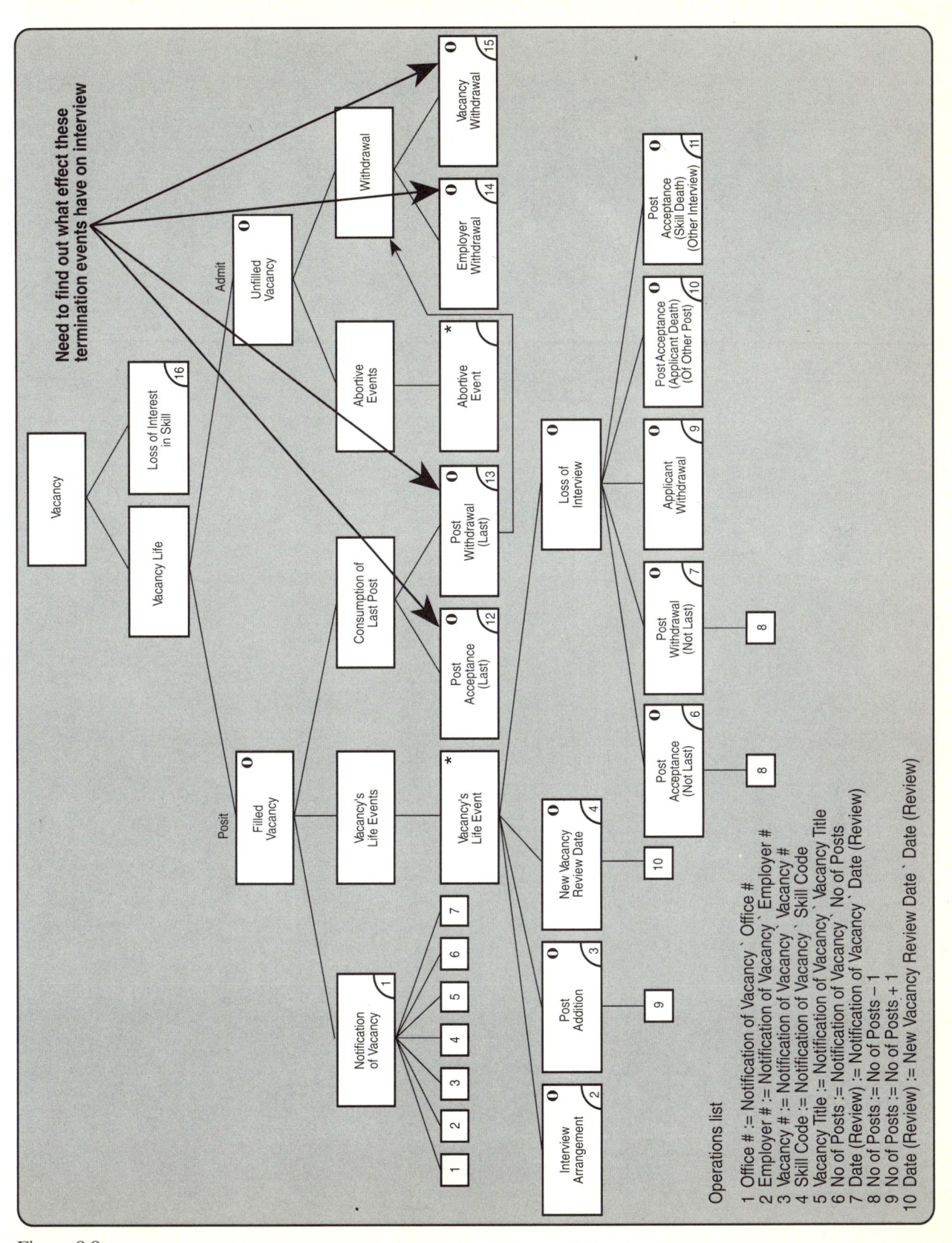

Figure 8.8

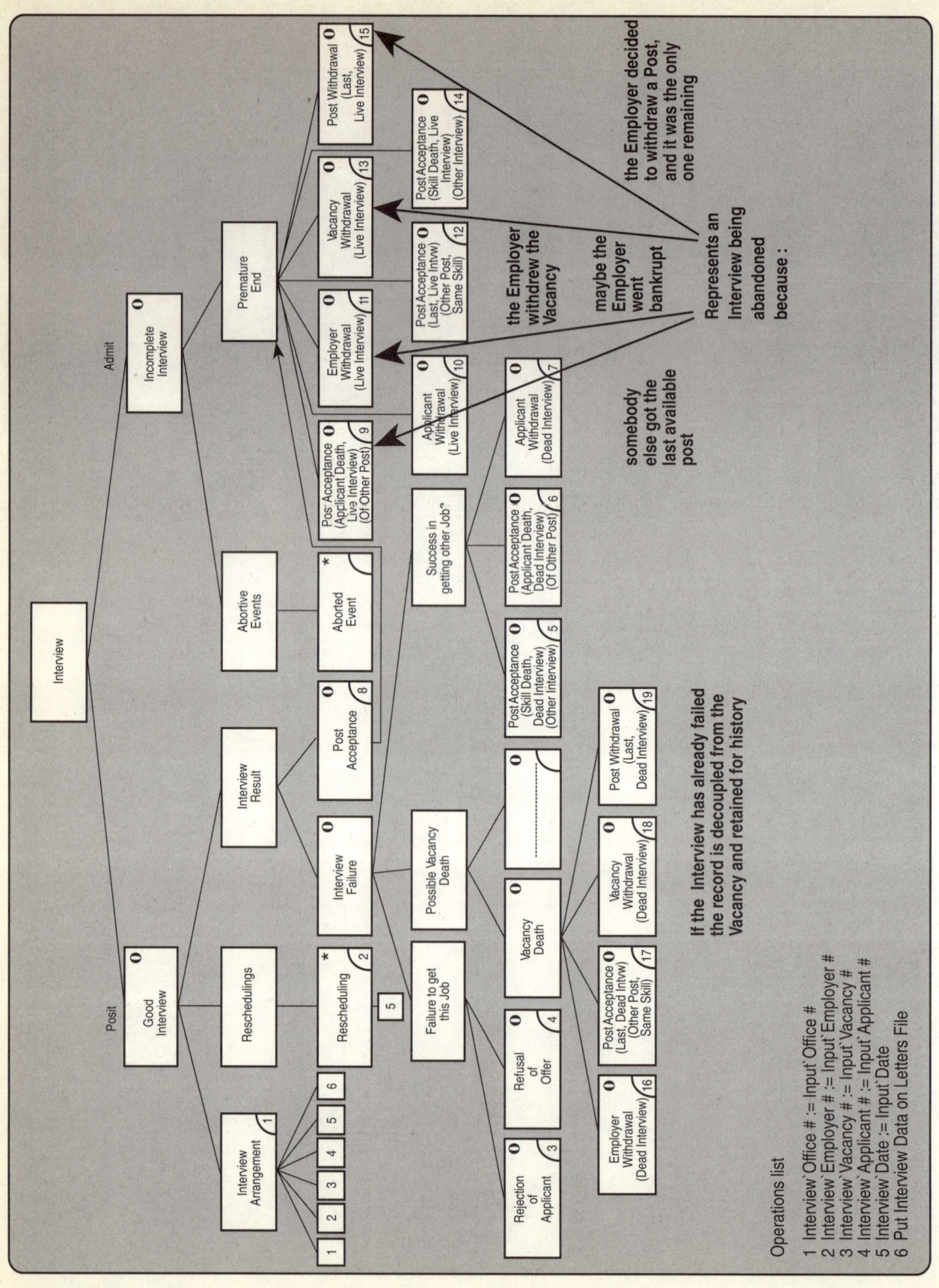

Figure 8.9

8.4 Examples of the GRAPES Knowledge Collection Process

Due to the given information in the Hi-Ho requirements the decomposition procedure was not useful in modelling the case study.

So as a first step we extracted the information about objects and their hierarchical relations from the requirements description. This produced the communication diagrams describing the structure of Hi-Ho (figures 8.10.–8.12).

The next step was to identify the possible messages the external partners of Hi-Ho can send and messages they expect in return. The result of this step was the interface tables corresponding to the top-level communication diagram (figure 8.13).

The structure of the local registers could not be deduced directly from the requirements description. Thus we produced an E-R diagram as the requirements specification described the local register in terms of the stored data. Besides this we specified the attributes of the entities and the forms for external communication as data types in data structure diagrams (figures 8.14, 8.15).

The entities were represented by objects that describe the management of the data (figure 8.16).

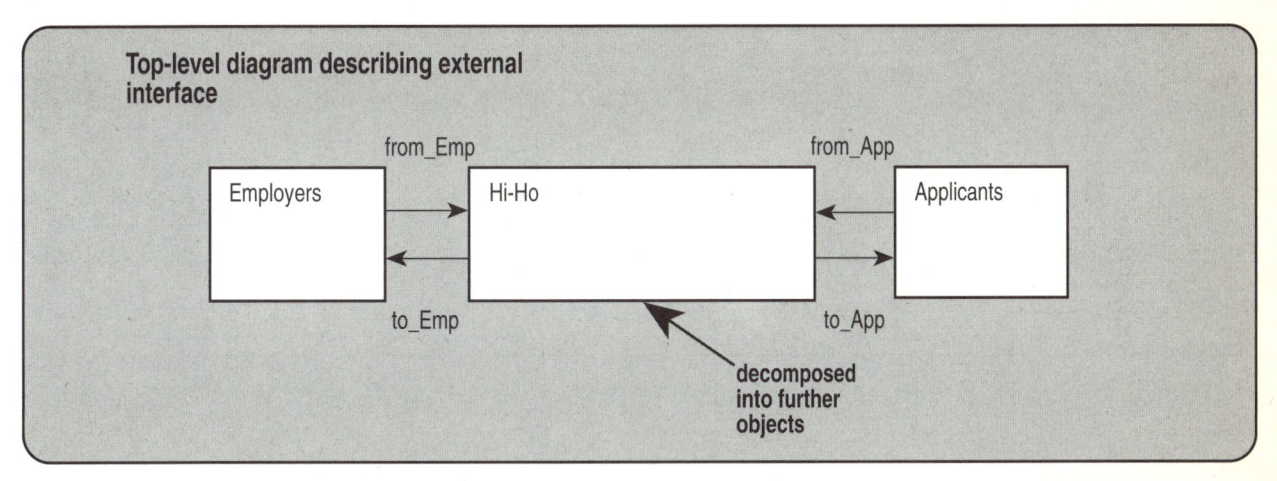

Figure 8.10

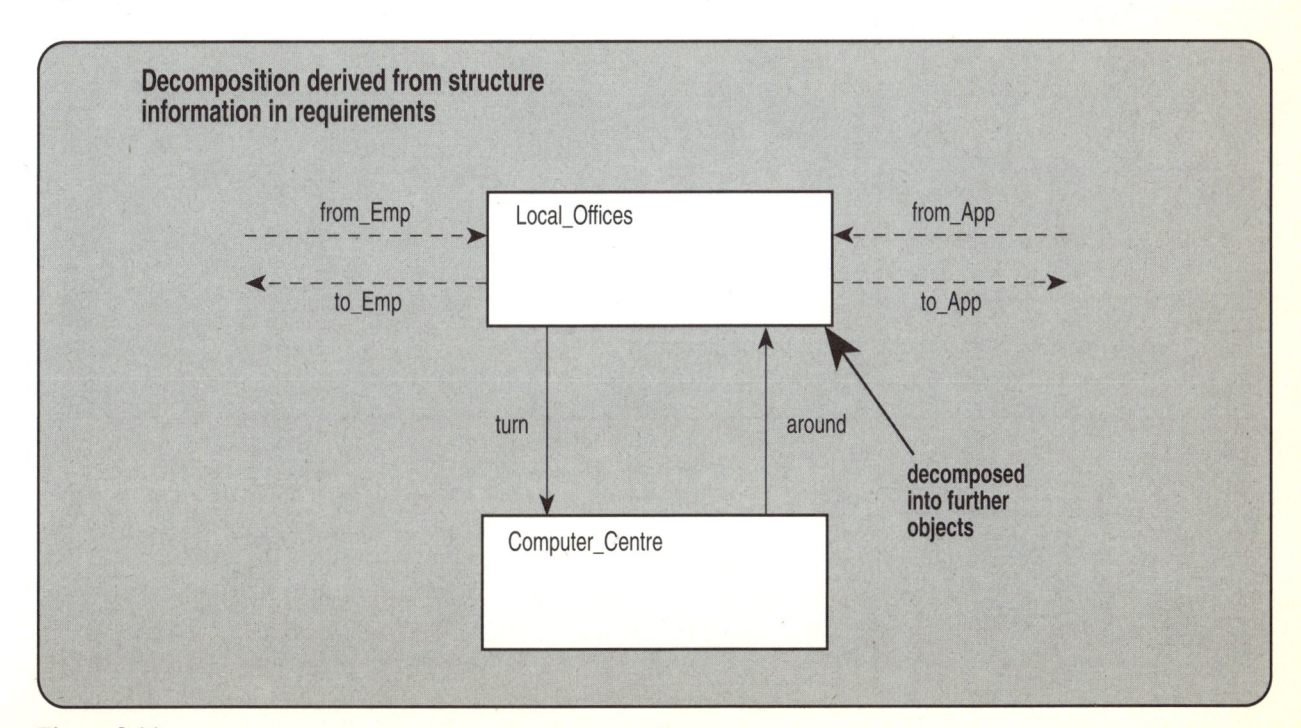

Figure 8.11

GRAPES Communication Diagram for Local_Offices Object

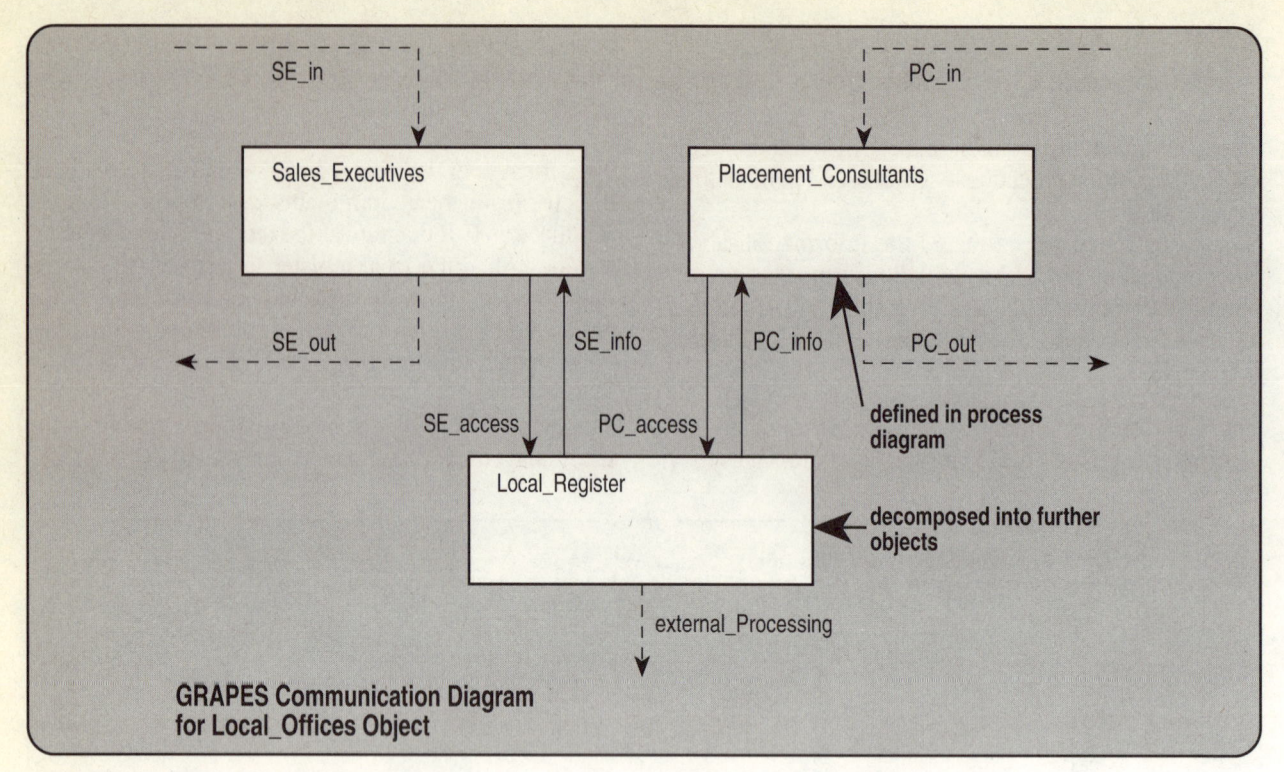

Figure 8.12

Name:	Data type:	Description:
register_Applicant	Applicant_Form	via PC_in to Placement_Consultants
withdraw_Applicant	App_Ident	–"–
review_Applicant	App_Ident	–"–
arrange_Interview	App_Ident	–"–
reschedule_Interview_for_App	Int_Rescheduling	–"–
submit_Reqs_to_PCon	App_Requ_Notification	–"–

GRAPES Interface Table for applicant messages for the communication line from_App

Figure 8.13

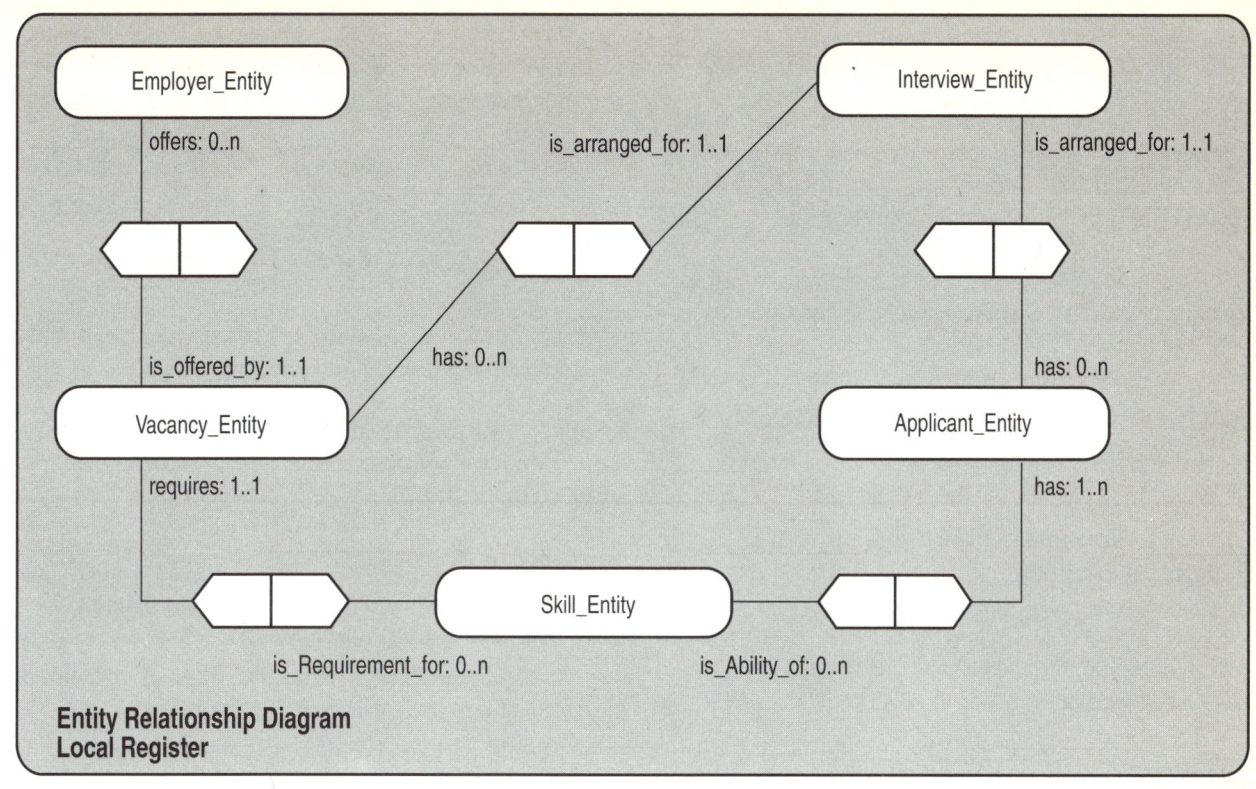

8.4 Examples of the GRAPES Knowledge Collection Process

**Entity Relationship Diagram
Local Register**

- Employer_Entity
  - offers: 0..n
- Interview_Entity
  - is_arranged_for: 1..1
  - is_arranged_for: 1..1
- Vacancy_Entity
  - is_offered_by: 1..1
  - has: 0..n
  - requires: 1..1
- Applicant_Entity
  - has: 0..n
  - has: 1..n
- Skill_Entity
  - is_Requirement_for: 0..n
  - is_Ability_of: 0..n

Figure 8.14

Applicant_Entity
- App_Id → integer
- Name → string
- Address → Address
- Birth → Date
- Skills → Skill

Vacancy_Entity
- Vac_no → integer
- Name → string
- Emp_Id → integer
- Required_Skill → Skill
- Posts → integer

Employer_Entity
- Id → integer
- Address → Address
- Vacancies → integer
- Office_use → Emp_Internals

Data types representing the external forms and local storage.

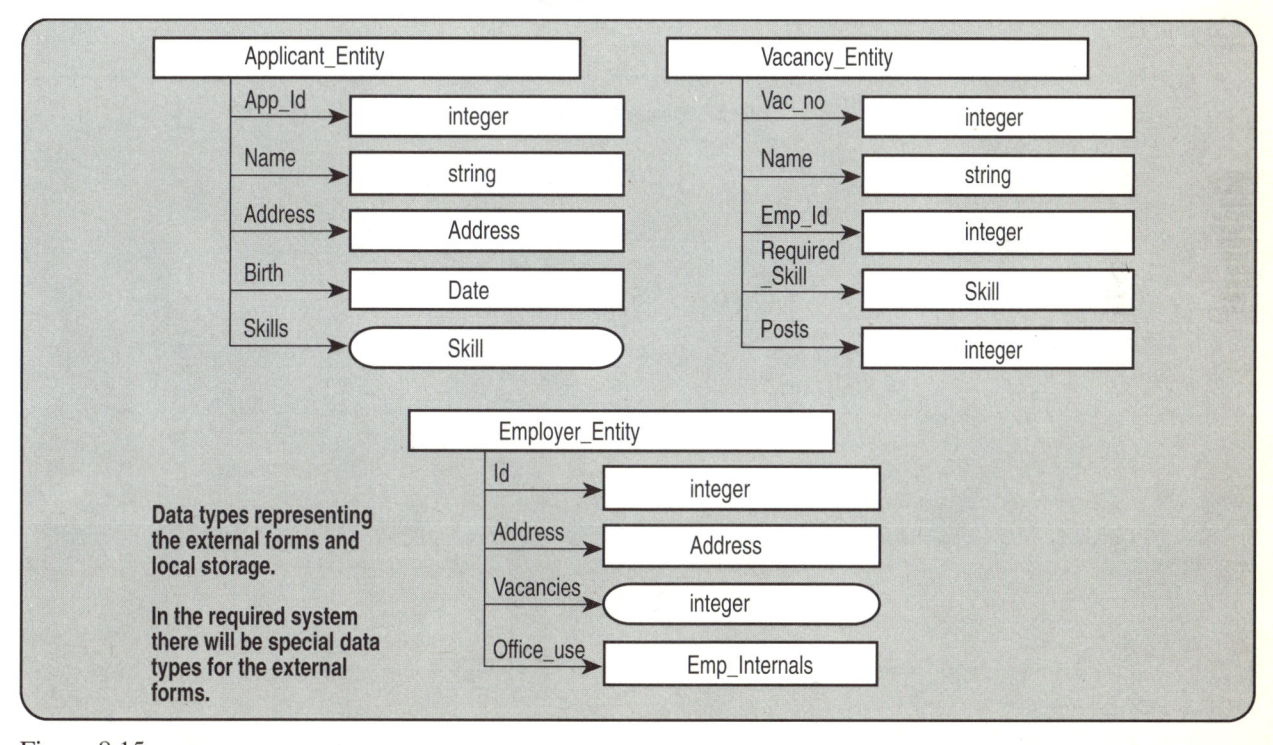

In the required system there will be special data types for the external forms.

Figure 8.15

With the static structure essentially completed we looked at the system's behaviour. The approach used on the case study was to track information flows and processing sequences through the system in a set of "Intermediate Diagrams" and then structure this diagram into well-formed GRAPES process diagrams for the process objects in the systems hierarchy (figure 8.17).

(Note that the creation of such intermediate diagrams is not a standard GRAPES procedure – see Annex).

The sorts of questions being asked were:

– Where does information come from and who is responsible for processing it?
– What is the reaction to the receipt of information?
– What information is produced as the result of some processing activity?
– Who gets the resulting information?
– What information can be transmitted?
– What information needs to be stored?

Due to the shortage of development time in producing the intermediate diagrams we concentrated mainly on the case of correct information flow. Thus questions like the following are not answered by the model.

– What happens if inconsistent information arrives?

   This could be modelled by inserting checking procedures right after the receive operations. These procedures could contain communication and processing for correcting or rejecting inconsistent messages.

– What happens if expected information does not arrive?

   This question was only modelled where explicitly requested in the requirements.

Some questions arose that could not be answered from the requirements description alone, and thus are not treated in the model.

The most important questions are:

– Which is the mechanism for avoiding matches after applicants refusal or rejection?
– Can an employer withdraw when a placement consultant tries to arrange an interview?
– Should partners be informed when an interview is rescheduled?

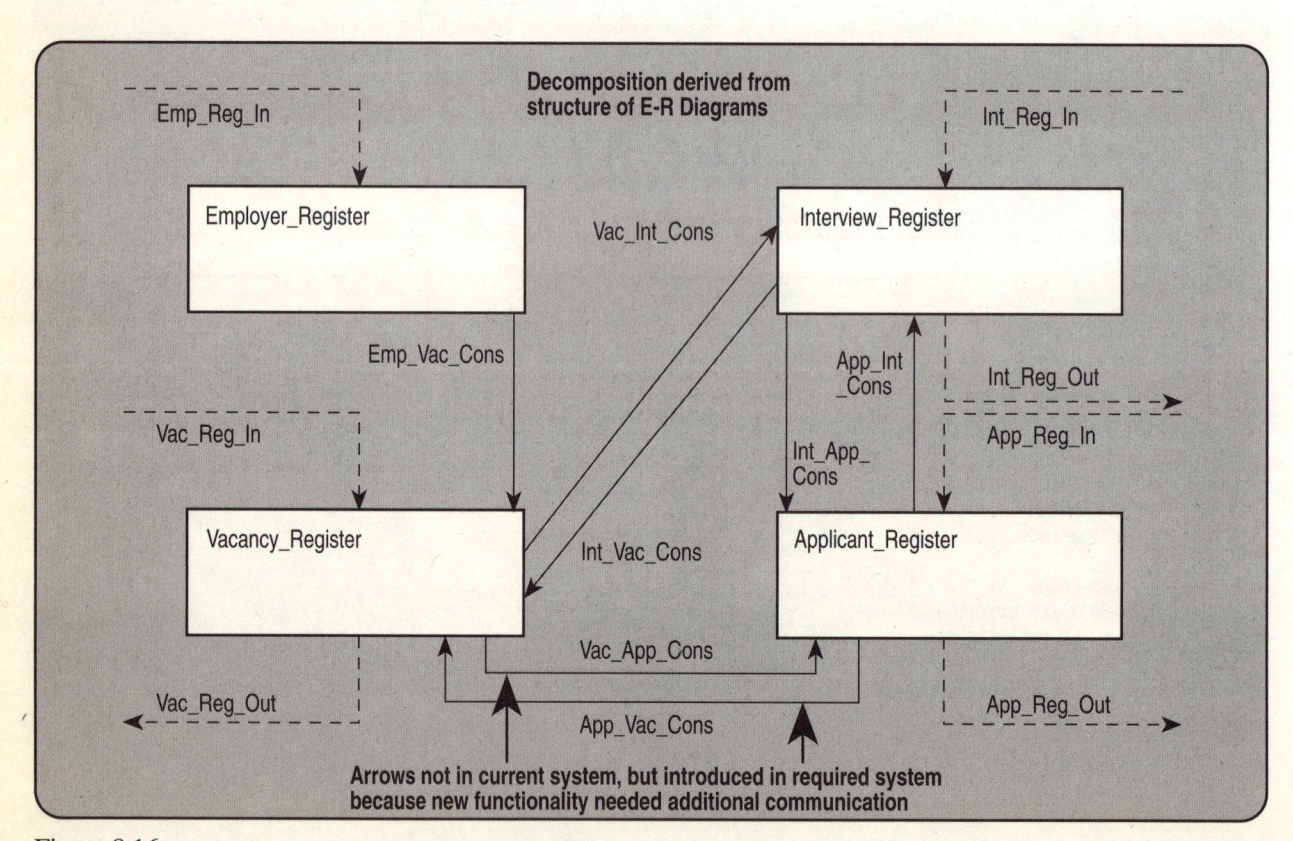

Figure 8.16

8.4 Examples of the GRAPES Knowledge Collection Process

During the project it was suggested that the intermediate diagrams should be retained as an end-product in order to ease the navigation around the diagrams to see what happened when a particular event occurred. If this is done, the intermediate diagrams will have a function analogous to SSADM effect correspondence diagrams.

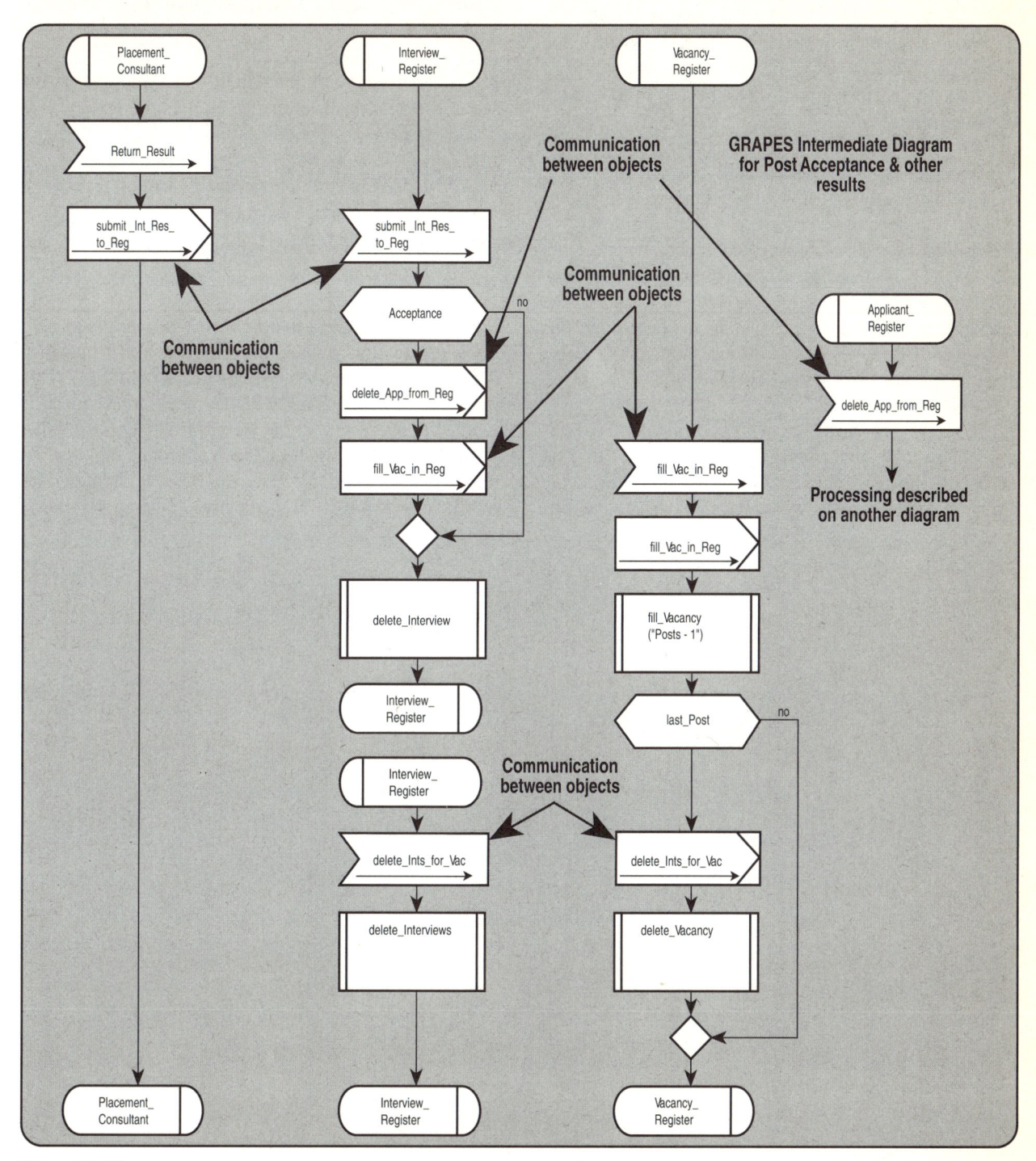

Figure 8.17

8.5 Comparison of the GRAPES and SSADM Knowledge Collection Processes

SSADM is stronger than GRAPES in the knowledge acquisition process. GRAPES can be characterized as a language to describe models. SSADM is a method for discovering models.

It is impossible to say whether a GRAPES development of Hi-Ho, using only the GRAPES standard techniques, would, unaided, have discovered the complexity of processing needed on post acceptance, because the GRAPES Hi-Ho team knew about this detailed processing from earlier discussions of the case study. The GRAPES study team's use of intermediate diagrams was found to have some resemblance to an effect correspondence analysis in SSADM.

What we can say, however, with several years' experience of running this particular case study, is that no student syndicate has ever discovered this processing from SSADM data flow analysis. None of the differences between GRAPES communication diagram production and SSADM data flow diagrams production seems to provide any knowledge acquisition technique which would make GRAPES perform better than SSADM in this area.

It is interesting to note that, in the case study work, the sequence of construction of equivalent products is exactly opposite in GRAPES and SSADM.

SSADM: Entity life histories, before effect correspondence diagrams

GRAPES: Intermediate diagrams before process diagrams

If we had to reduce both approaches to a single knowledge acquisition question, in the context of system behaviour, the GRAPES question would be "What processing has to go on?" whereas the SSADM question would be "What happens to the entities?". This is a rather absurd over-simplification (each approach to some extent also asks the other's question) but it characterizes the flavour of the difference between the approaches.

9 Transformations

As we mentioned earlier, some common approaches to systems development provide very little in the way of "method". But, because some methods claimed to assist in the process of producing those end-products we believed it was important to study how the different approaches collected the knowledge embedded in the final systems. In this chapter we look not just at how the knowledge is collected, but at how the various knowledge collection techniques relate to each other. We are looking to see whether there is an incremental development of end-products via intermediate products, or whether at some points in the methods earlier results are put to one side and new products are created (from facts represented in the earlier results) whose structure is totally new and produced intuitively.

9.1 Transformations in SSADM

From resource/service flow to current system DFD

The resource/service flow models the supply of and demand for services in the user's business system (not in the information system). There should be a direct mapping, almost an isomorphism, between the real system and the information system which supports it.

From resource/service flow to document flow

The resource/service flow shows some elementary information about demands for services. This, together with information accompanying the supply of services, identifies for the analyst a set of documents which he can trace round the current information system. The essential elements are abstracted and a new product is built.

From document flow to top-level of DFD current system

If a resource/service flow has been produced, the elements of the document flow are grouped to match the resource/service flow and a high-level process is abstracted. Otherwise grouping is performed on an intuitive basis.

From document flow to second-level DFD

Details from the document flow which were omitted from the top-level DFD can now be added to the second level DFDs.

From top-level to second-level DFD

Processes in the top-level DFD can be exploded to second level whether or not the explosion detail is present in the document flow. Exception handling can be added.

From second-level DFD to third-level DFD

If the detail of a second-level process is over-complex (more than 10 or so processes), some can be refined into a third-level DFD and grouped in the second-level DFD.

From low-level current DFD to low-level logical DFD

Data stores are restructured to achieve compatibility with intuitively defined subsets of the logical data model. Physical implementation details are abstracted from the low-levels of the current DFDs to create the lower levels of the logical DFDs.

From low-level logical DFD to low-level required DFD

Extra functionality is added to deliver new system requirements.

From low-level to high-level (logical & required DFDs)

Low-level processes are grouped by intuitively discovered areas of interest to provide higher-level views of processing.

From DFDs to functions

Low-level processes are restructured and grouped by common user roles to provide packages of functionality useful to support end-users in their day-to-day work.

SSADM DFD Transformations

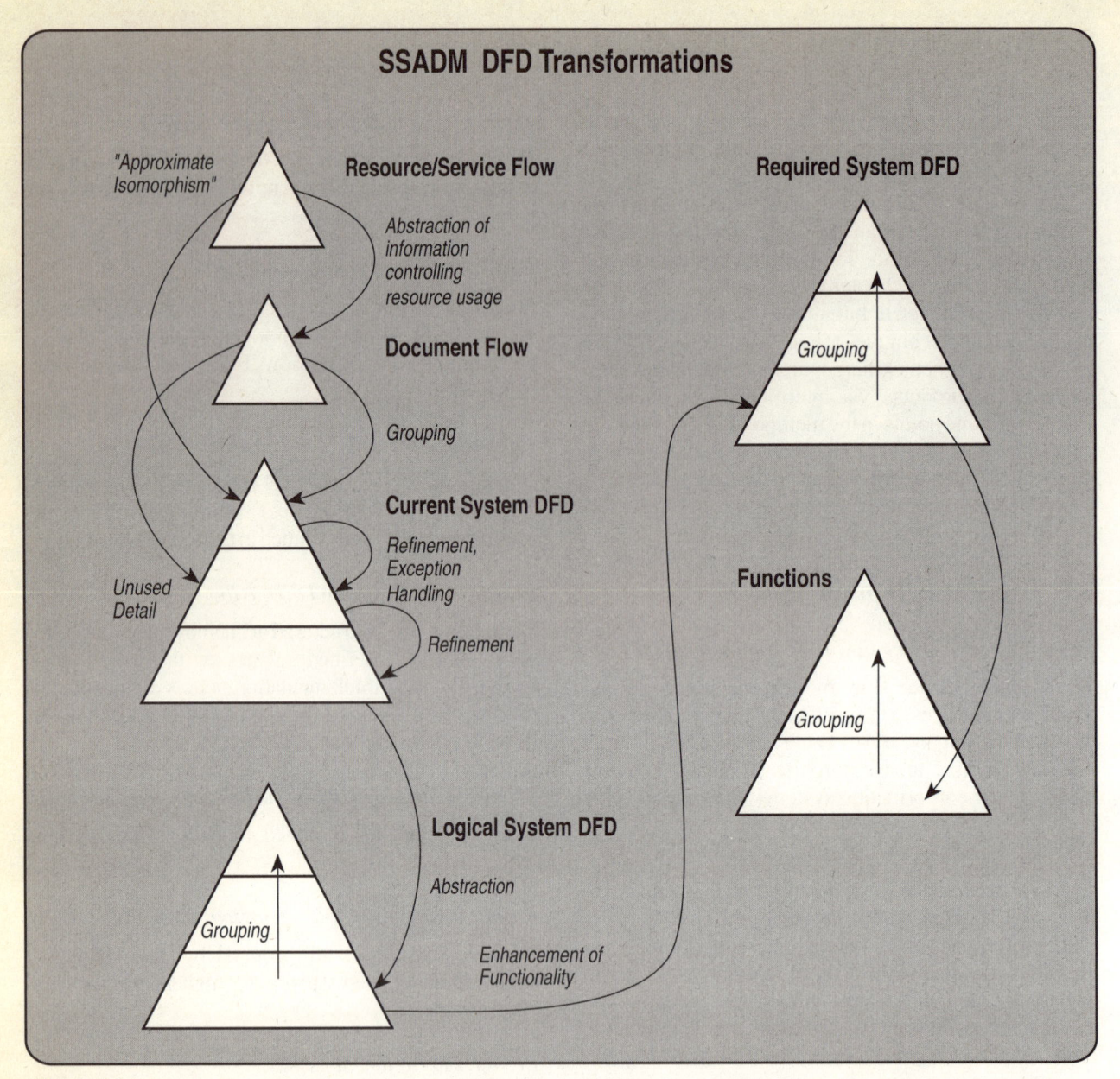

Figure 9.1

Production of a logical data model

Rules are provided to transform an entity list and a grid of relationships into a rough data model. Structure simplification rules further transform that model into one which is simultaneously simpler structurally and richer semantically.

From I/O structures to TNF relations

The input/output data for the end-user functions is normalized.

From TNF relations to required system logical data model

As each I/O structure is normalized, each TNF relation is identified as a data model entity. Key matching rules are used to verify the relationships in the logical data model. The results of multiple TNF results for the same entity are composed to build the entity attribute list.

From required system logical data model to access paths

For each enquiry the entry point into the data model and the entities accessed are identified.

From entity life history to effect correspondence diagram

The behaviours of the entities in the data model are captured in the entity life histories which define the structures of events affecting the entities. Typically, the same event affects more than one entity. The matrix of entities against events is transposed and, for each event, all entities affected by that event are extracted into an effect correspondence diagram.

From access path to enquiry structure

There is an isomorphic transformation between an access path and an effect correspondence diagram-like structure which is an alternative notation for describing the access path.

From effect correspondence diagrams to process models

Groups of correspondences define equivalence classes of effects of events which represent processing done in the same part of the control structure. ECDs and their enquiry equivalents can then be "collapsed" to produce update & enquiry process models.

From logical data model to first-cut design

The required system logical data model is transformed using "first-cut" rules to produce a first-cut database design (figure 9.2).
In addition to these transformations, prototyping activities are used to turn the I/O structures into dialogue designs.

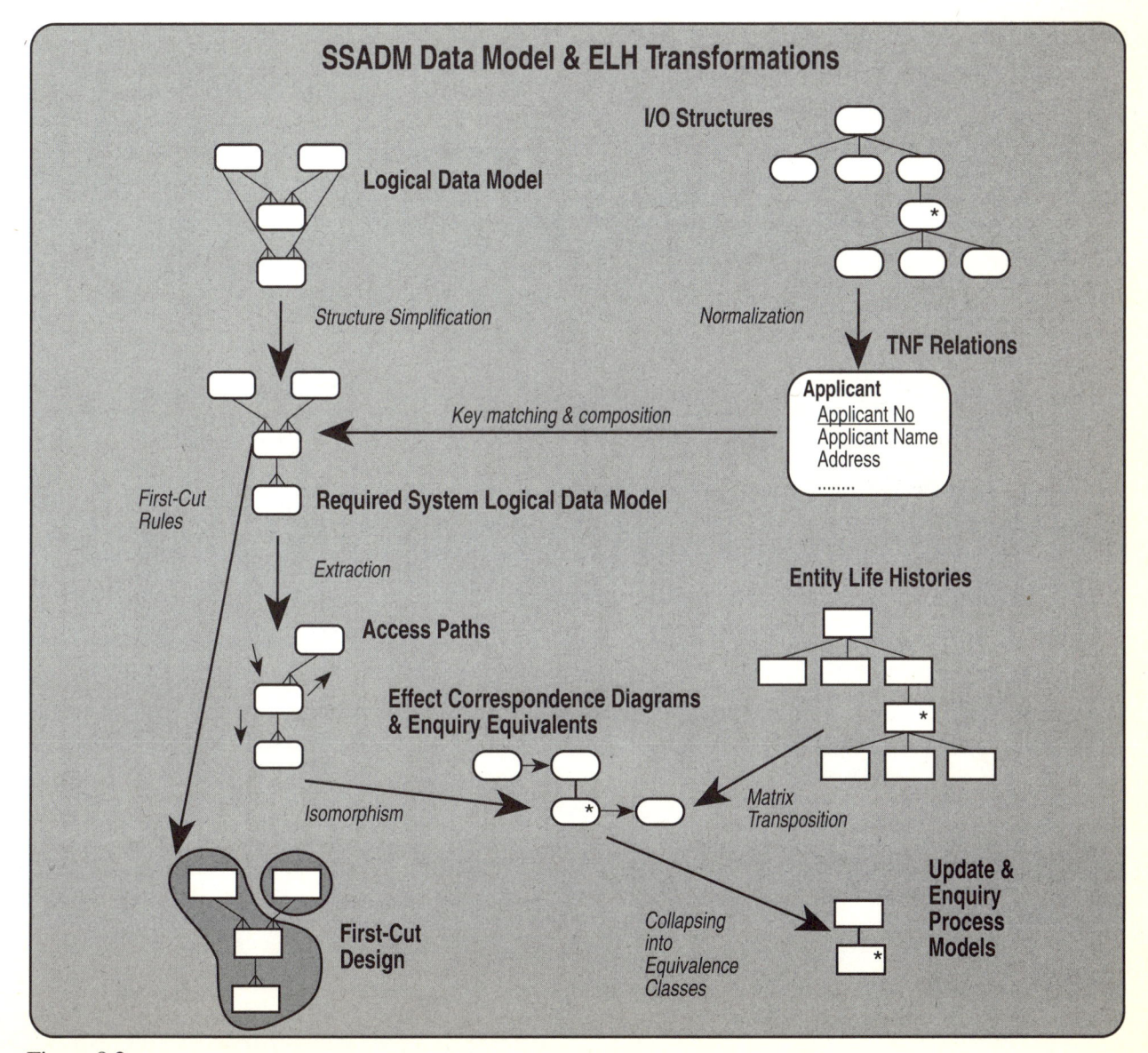

Figure 9.2

9.2 Transformations in GRAPES

In each of the phases we find a different model and different levels of detail, each concentrating on specific aspects of the system. By transformations we understand how to get from one model to the next and from one level of detail to the next during the analysis and design process.

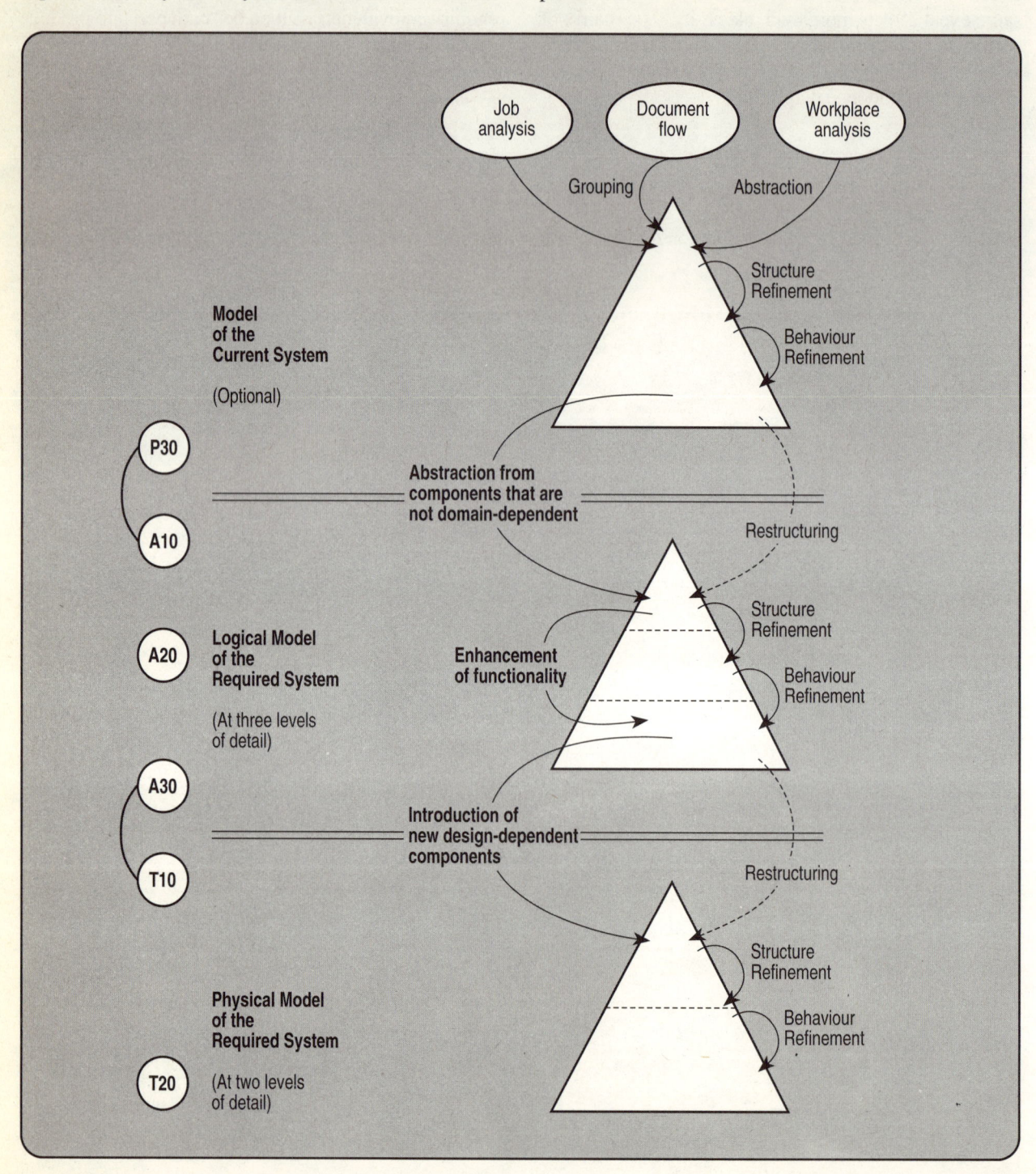

Figure 9.3

9.2 Transformations in GRAPES

According to the DOMINO life cycle we distinguish between the model of the current system which is produced during the problem analysis phase, the logical model of the required system which is developed in the requirements definition phase, and the physical model of the required system which is a product of the technical realization phase.

Transformations between models

Once the model of the current system is finished, the life cycle enters the requirements definition phase, where the logical model of the required system is developed. The transformation from the model of the current system to the logical model of the required system is somewhat

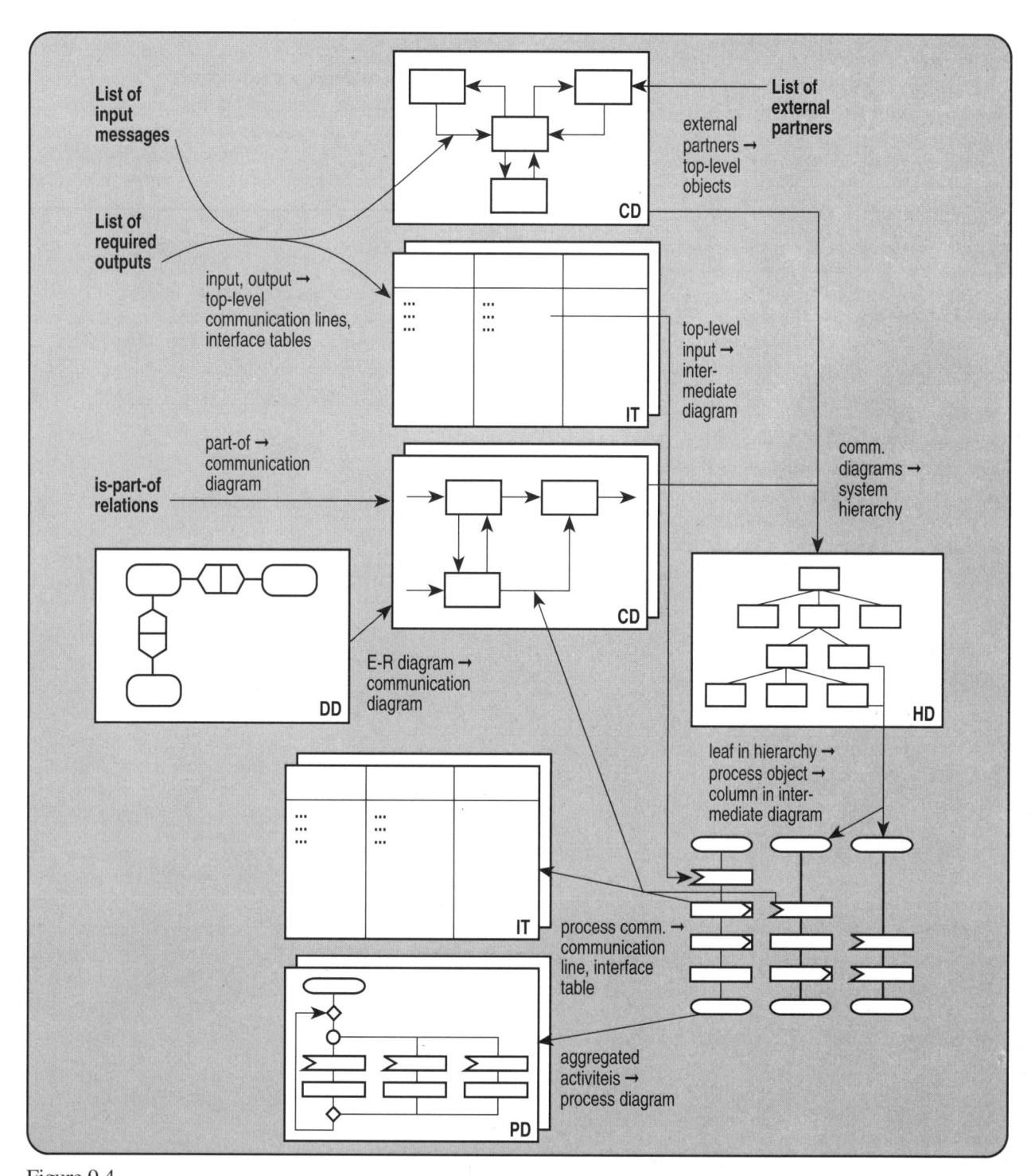

Figure 9.4

problematic, not only within DOMINO/GRAPES but also generally in software engineering methodologies. The difficulty is that a restructuring of the system model has to take place in order to incorporate new requirements into the system. The kind of transformation necessary depends on the new requirements and thus varies from project to project.

GRAPES supports the transformation process by the capability to duplicate parts from the current system model into the model of the required system and to integrate them consistently in the new system structure. The structure of the physical model is constrained by requirements and restrictions placed upon it by the software and hardware environment the system is embedded in. Since the functionality the system has to provide, which is described in the logical model of the target system, has to have its analogue in the physical model, in practice much of the logical systems definition can be used in the physical model, easing the transformation from the logical to the physical view.

Essentially the same consistency and completeness checks are applied as for the logical model of the required system.

A detailed physical model of the target system serves as a specification of the programmers' tasks. Some code may be generated from the specification, for example interfaces of procedures and module frames. The boundaries of the software components and their coupling are derived from the structure of the model. Finally the communication specified in the model serves to identify the information flow between software modules.

Transformations between levels of detail

As already stated, the GRAPES Hi-Ho team did not work with a pure top-down data-flow analysis approach. They integrated data analysis as well as behaviour analysis, documented in the intermediate diagrams, into their knowledge collection process. This led to a set of additional transformations.

Lists of external partners, inputs and outputs → communication diagram, interface tables

The list of external partners is transformed into the top-level communication diagram with an object representing every external partner and one representing the system itself. Each list of messages is transformed into an interface table with a corresponding communication line in the top-level communication diagram. Every input or output is represented by a channel in the appropriate interface table.

Is-part-of relations → communication diagram

Another input to modelling is the analysis of is-part-of relations. These are transformed into a communication diagram which contains the objects that are parts of, or contained in, another one. The diagram itself is located hierarchically below the refined object.

Entity relationship diagram → communication diagram

If is-part-of relations cannot be detected directly, then data analysis for an object can produce an E-R diagram. The considered object is then refined by transforming the E-R diagram into a communication diagram that contains objects, each representing an entity of the E-R diagram.

Communication diagrams → hierarchy diagram

When the structure of the model is finished, or while developing it, the hierarchical relations of the objects in the communication diagrams are transformed into a hierarchy diagram, showing the object hierarchy in a tree representation.

Interface tables, hierarchy diagram → intermediate diagrams

For every channel of the top-level interface tables that represents a system input an intermediate diagram is created. The possible columns of these diagrams are taken from the hierarchy diagram. The objects that are leaves in the hierarchy tree are the process objects of the system; they are represented as columns in the intermediate diagrams.

Intermediate diagrams → process diagrams

For every process object a process diagram is created. The appropriate columns from the intermediate diagrams are collected for every process and integrated to be processing paths within the process description.

Intermediate diagrams → interface tables, communication diagrams

Pairs of communication operations in the intermediate diagrams are transformed into channels. If necessary the required interface tables are created and the corresponding communication lines are inserted in the appropriate communication diagrams.

9.3 Examples of the SSADM Transformations

Functions are extracted from the data flow model, as shown in figure 9.5.

After a function has been extracted, its I/O structure will be defined and used as input to relational data analysis and the dialogue design process (figure 9.6).

Figure 9.7 shows a simplified picture of the structure of the input to the function extracted on the preceding page.

This I/O structure includes the input data for the applicant registration and interview arrangement events. The data will be normalized and the results of normalization will be used to enhance the logical data model with attributes.

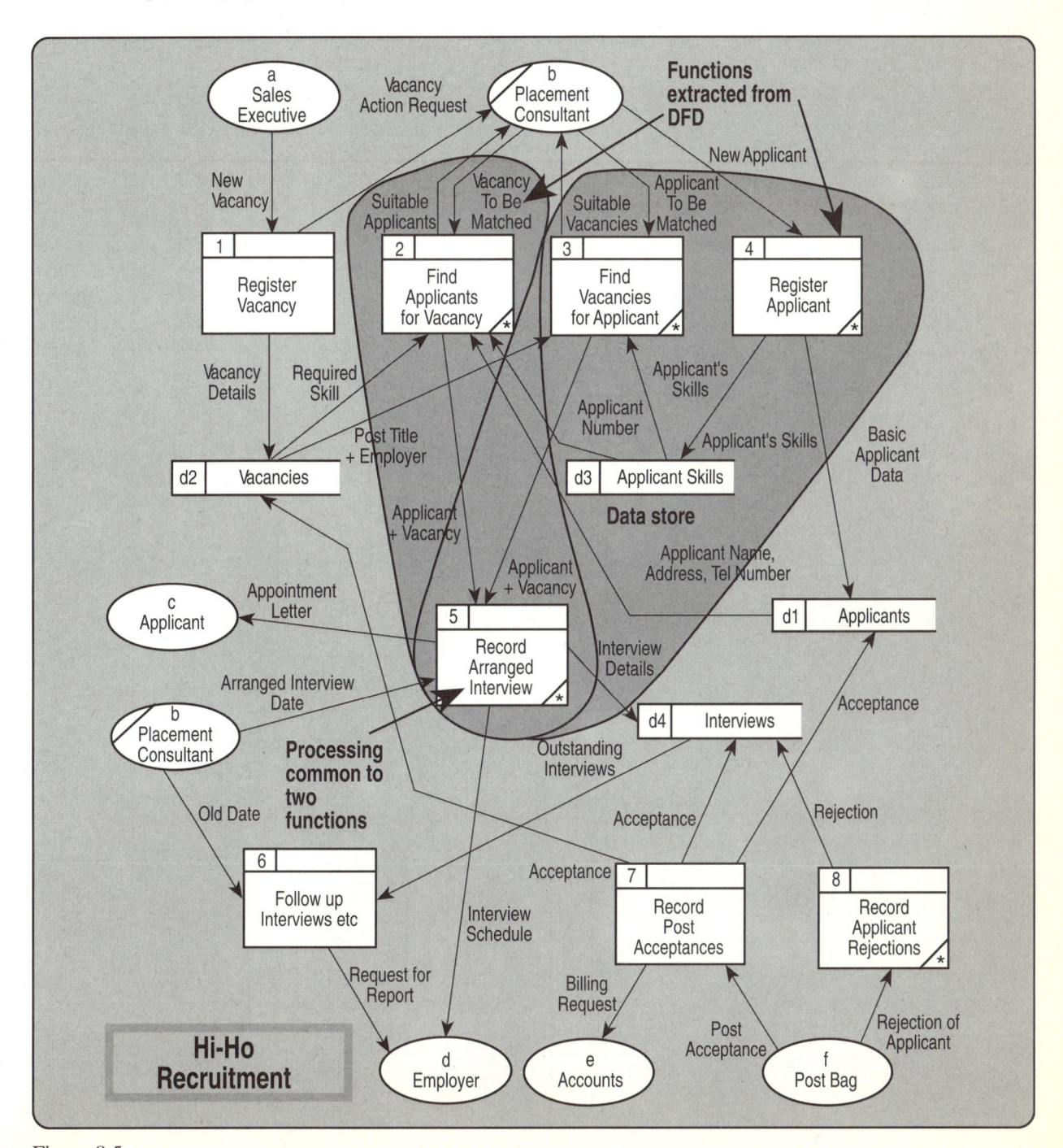

Figure 9.5

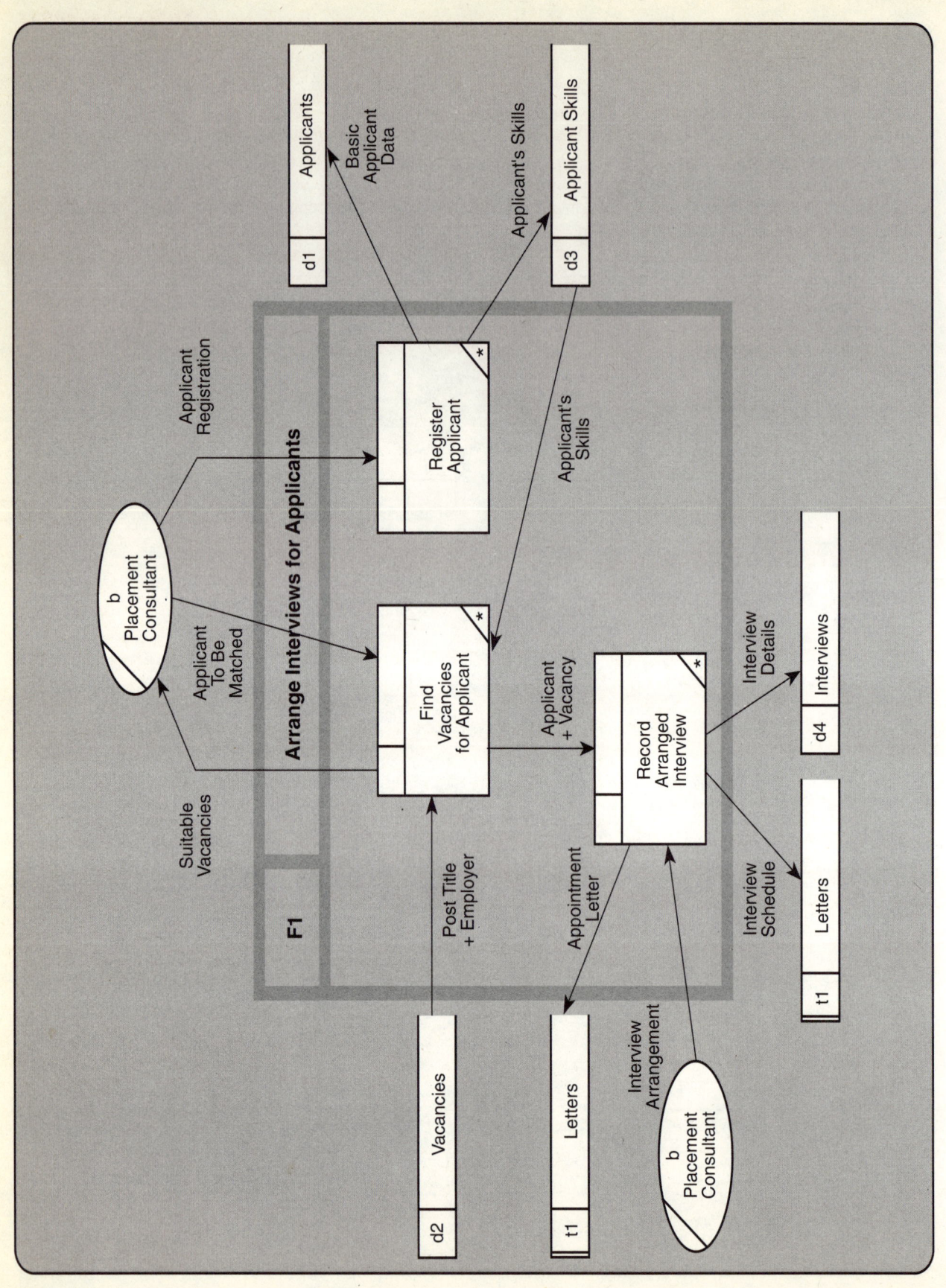

Figure 9.6

9.3 Examples of the SSADM Transformations

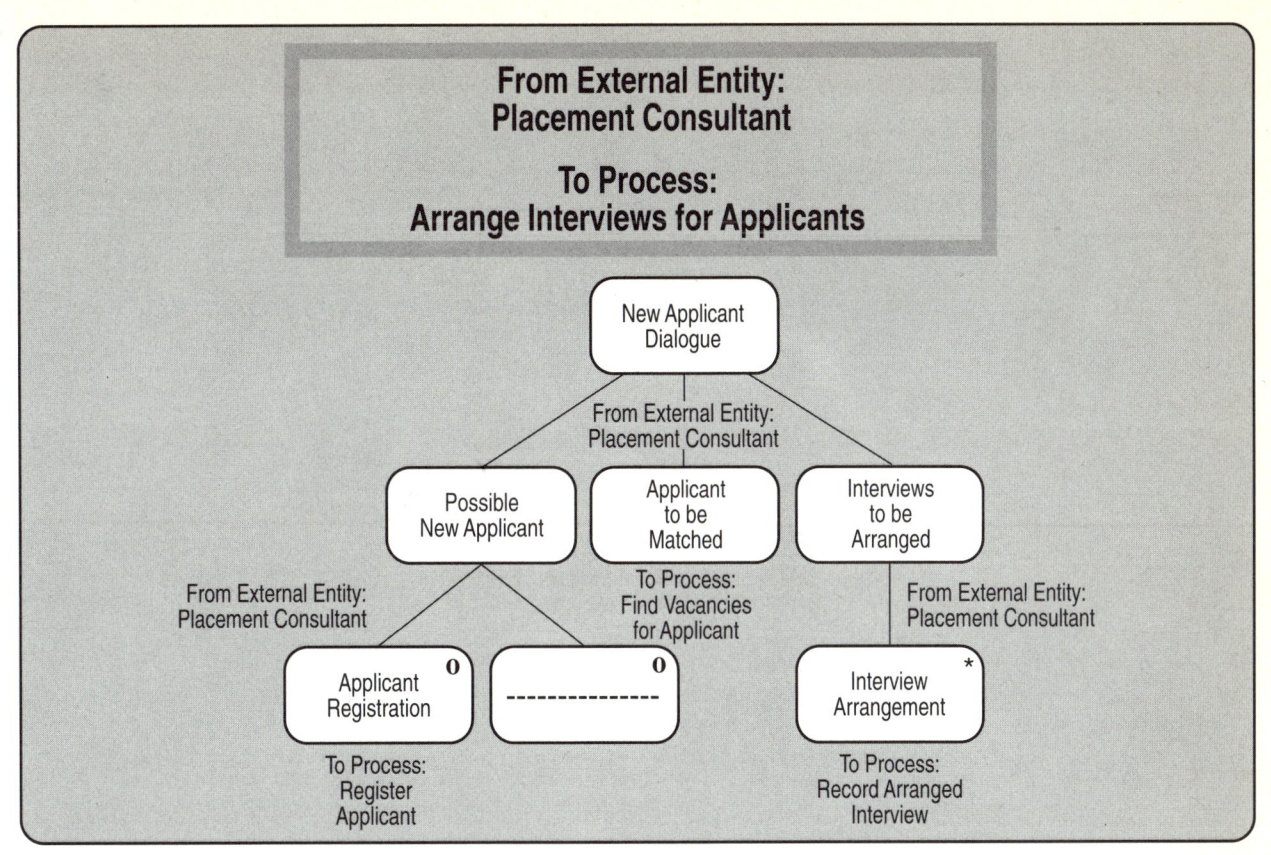

Figure 9.7

After being validated by prototyping, the I/O structure can be transformed into an executable dialogue control structure during stage 6 of SSADM, if the implementation environment is appropriate (see figure 9.8).

F1 Online — New Applicant Dialogue

```
                        Process
                           |
                        1  |
                           |
                    Process New
                    Applicant Dialogue
                    /      |       \
                   /       |        \
          Process Possible  Process Applicant   Process
          New Applicant    to be Matched       Interviews to be
                                                Arranged
            Applicant                              |  More input
            Registration                           |         *
              /    \          5   1          Process Interview
             /      \                        Arrangement
            o        o                          /    \
      Process    Process                       6     1
      Applicant  ---------
      Registration
       /    |    \
      2    |     1
           |
   Process Applicant    Process Skill
   Details              Details
      |                  More | input
      4                       |    *
                         Process Skill
                           /     \
                          2       7
```

Operations list

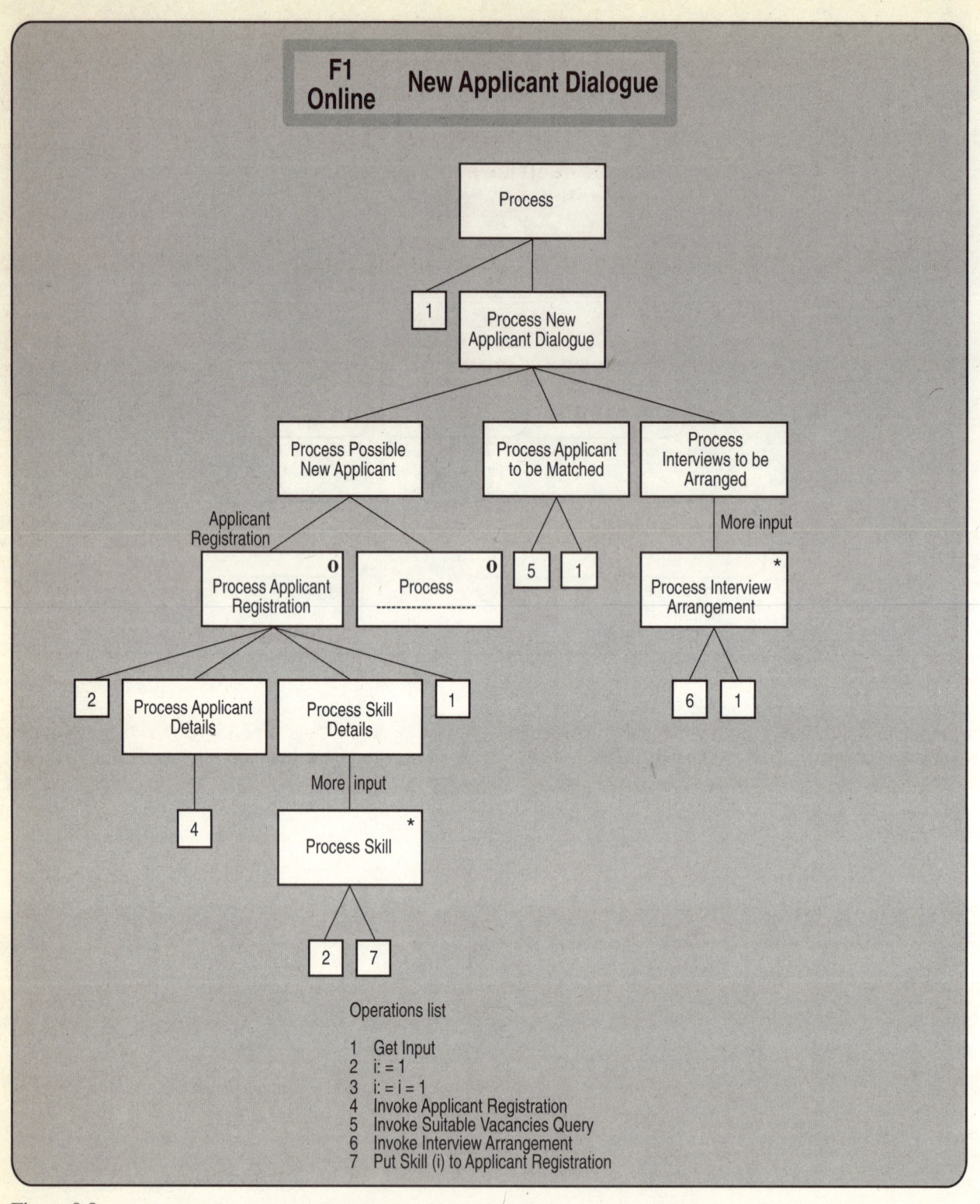

1. Get Input
2. i: = 1
3. i: = i = 1
4. Invoke Applicant Registration
5. Invoke Suitable Vacancies Query
6. Invoke Interview Arrangement
7. Put Skill (i) to Applicant Registration

Figure 9.8

9.3 Examples of the SSADM Transformations

Access paths can be extracted from the logical data model, as shown in figure 9.9.

Access paths can be represented in effect correspondence diagram notation, as in figure 9.10.

Effect correspondence diagrams for update events can be created by transposing the entity life histories and extracting all of the processing for an event from several entity life histories. As was demonstrated during the development of this study, this transformation can be performed almost entirely automatically.

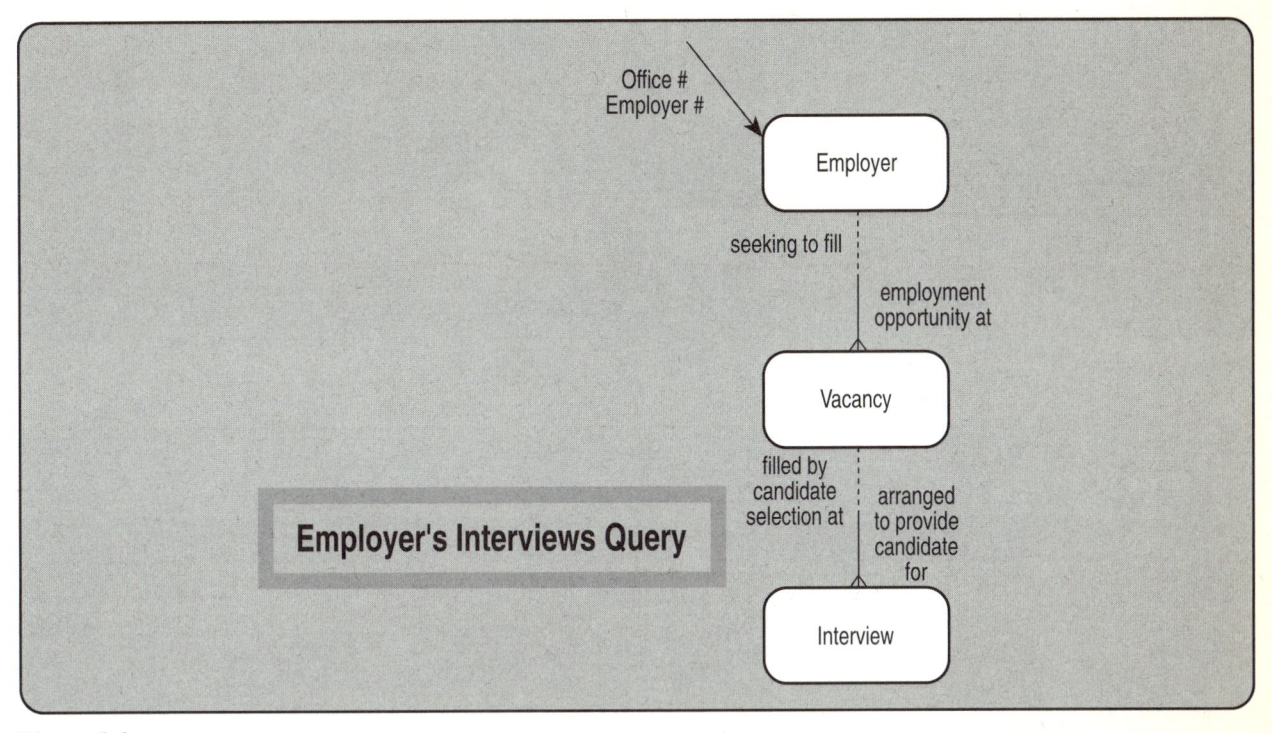

Figure 9.9

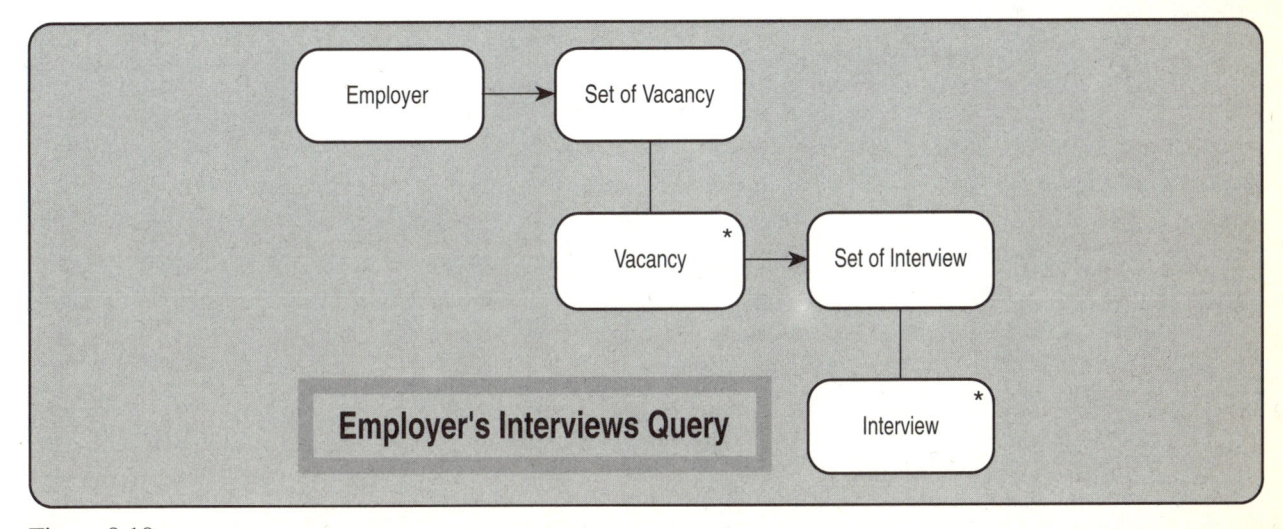

Figure 9.10

The entity life histories which are transformed to produce the effect correspondence diagram shown in figure 9.11 are shown in figures 5.6 – 5.9.

By collapsing components in 1-1 correspondence and generating state testing and setting operations and read and write operations, the update process model in figure 9.12 can be produced.

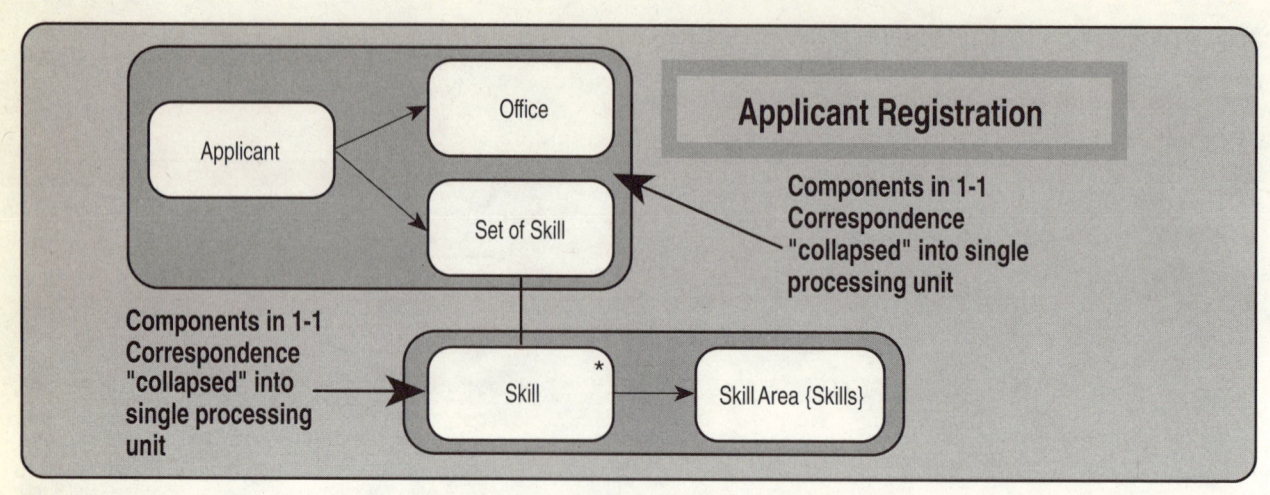

Figure 9.11

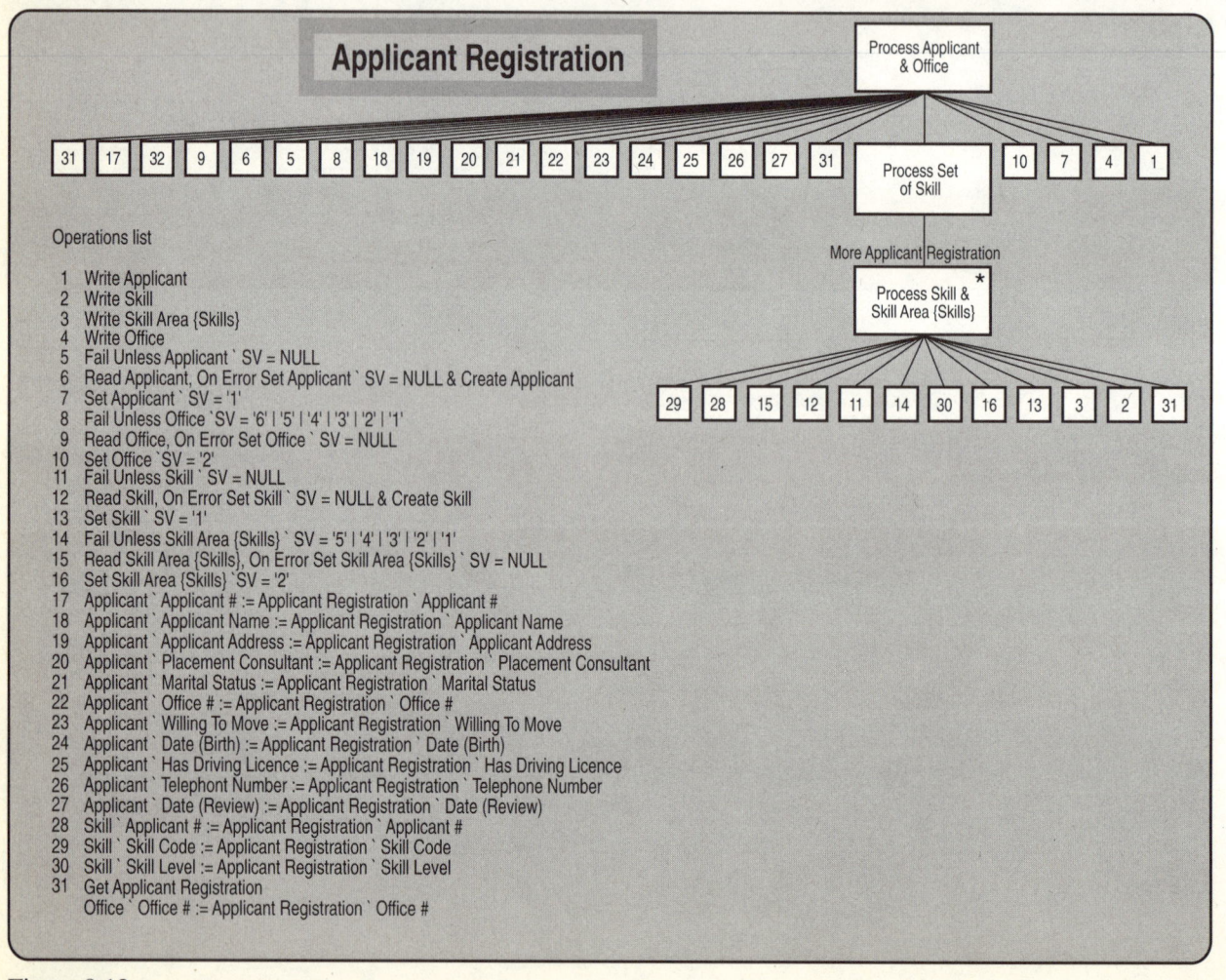

Figure 9.12

Applicant Registration

Operations list

1. Write Applicant
2. Write Skill
3. Write Skill Area {Skills}
4. Write Office
5. Fail Unless Applicant ` SV = NULL
6. Read Applicant, On Error Set Applicant ` SV = NULL & Create Applicant
7. Set Applicant ` SV = '1'
8. Fail Unless Office ` SV = '6' | '5' | '4' | '3' | '2' | '1'
9. Read Office, On Error Set Office ` SV = NULL
10. Set Office ` SV = '2'
11. Fail Unless Skill ` SV = NULL
12. Read Skill, On Error Set Skill ` SV = NULL & Create Skill
13. Set Skill ` SV = '1'
14. Fail Unless Skill Area {Skills} ` SV = '5' | '4' | '3' | '2' | '1'
15. Read Skill Area {Skills}, On Error Set Skill Area {Skills} ` SV = NULL
16. Set Skill Area {Skills} ` SV = '2'
17. Applicant ` Applicant # := Applicant Registration ` Applicant #
18. Applicant ` Applicant Name := Applicant Registration ` Applicant Name
19. Applicant ` Applicant Address := Applicant Registration ` Applicant Address
20. Applicant ` Placement Consultant := Applicant Registration ` Placement Consultant
21. Applicant ` Marital Status := Applicant Registration ` Marital Status
22. Applicant ` Office # := Applicant Registration ` Office #
23. Applicant ` Willing To Move := Applicant Registration ` Willing To Move
24. Applicant ` Date (Birth) := Applicant Registration ` Date (Birth)
25. Applicant ` Has Driving Licence := Applicant Registration ` Has Driving Licence
26. Applicant ` Telephont Number := Applicant Registration ` Telephone Number
27. Applicant ` Date (Review) := Applicant Registration ` Date (Review)
28. Skill ` Applicant # := Applicant Registration ` Applicant #
29. Skill ` Skill Code := Applicant Registration ` Skill Code
30. Skill ` Skill Level := Applicant Registration ` Skill Level
31. Get Applicant Registration
    Office ` Office # := Applicant Registration ` Office #

9.4 Examples of the GRAPES Transformations

Transformations between levels of detail

Lists of external partners, inputs and outputs → communication diagram, interface tables

This transformation produced the communication diagram "Hi-Ho _ Environment" which represents the Hi-Ho company and its external partners. Together with this diagram the interface tables "from _ Emp" and "from _ App" which define the possible inputs, and the interface tables "to _ Emp" and "to _ App" are assembled which define the required outputs (Fig. 9.13 and 9.14).

Is-part-of relations → communication diagram

An example of the transformation of is-part-of relations to communication diagrams can be seen in the CD for local offices. The Hi-Ho requirements state that local offices consist of a placement consultant, a sales executive and a local register. This statement is transformed into a communication diagram containing objects which represent these elements of local offices. Note that the communication lines had not been detected at that point, but came in as a result of a later transformation (Fig. 9.15).

Figure 9.13

Name:	Data Type:	Description:
register_Applicant	Applicant_Form	via PC_in to Placement_Consultants
withdraw_Applicant	App_Ident	–"–
review_Applicant	App_Ident	–"–
arrange_Interview	App_Ident	–"–
reschedule_Interview_for_App	Int_Rescheduling	–"–
submit_Reqs_to_PCon	App_Requ_Notification	–"–

Interface Table : from_App

Figure 9.14

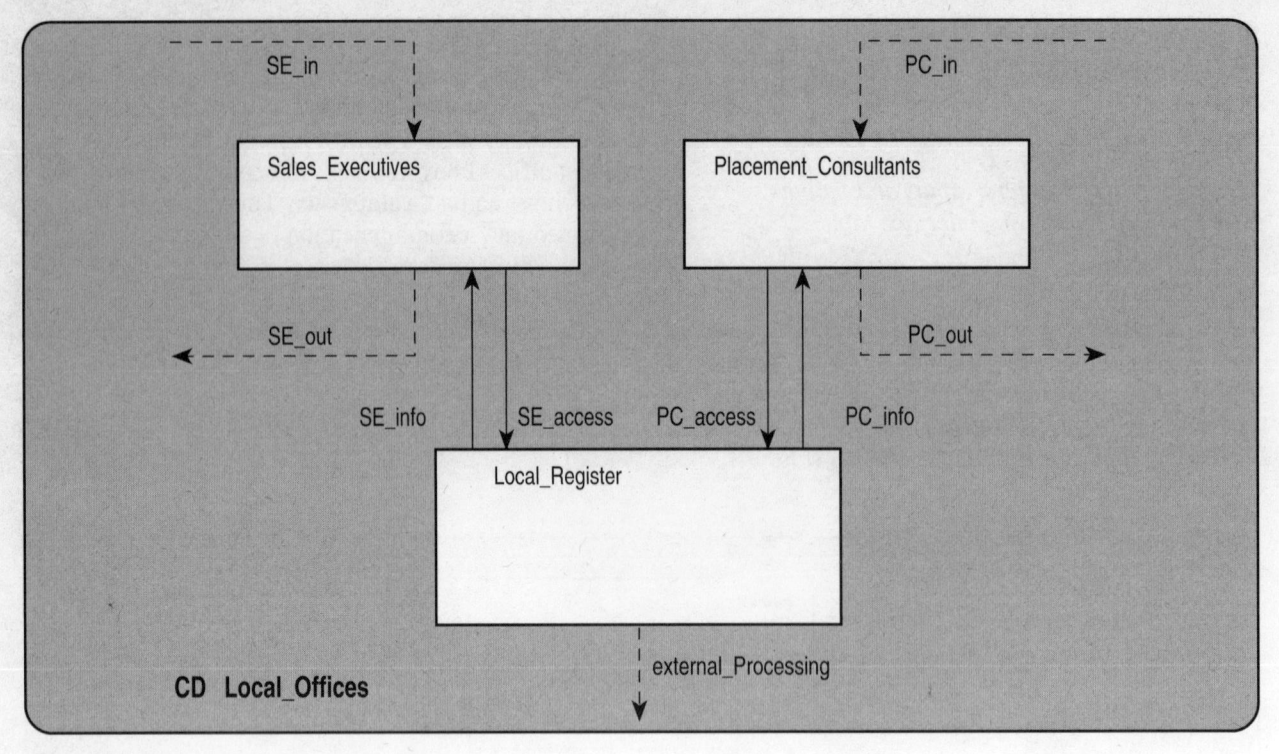

Figure 9.15

Entity relationship diagram → communication diagram

There is only one example in the GRAPES model showing the transformation from an entity relationship diagram to a communication diagram.

The communication diagram defining the local register was derived from an entity relationship diagram. Again the communication lines had not been detected at that point, but came in as a result of a later transformation.

Anyway it is interesting that the structure of the communication lines resembles the structure of the relationships (Fig. 9.16 and 9.17).

Communication diagrams → hierarchy diagram

This is a rather trivial transformation, but it helps in getting an overview of the system and in identifying the process objects, which are the basis of further analysis (Fig. 9.18).

9.4 Examples of the GRAPES Transformations 111

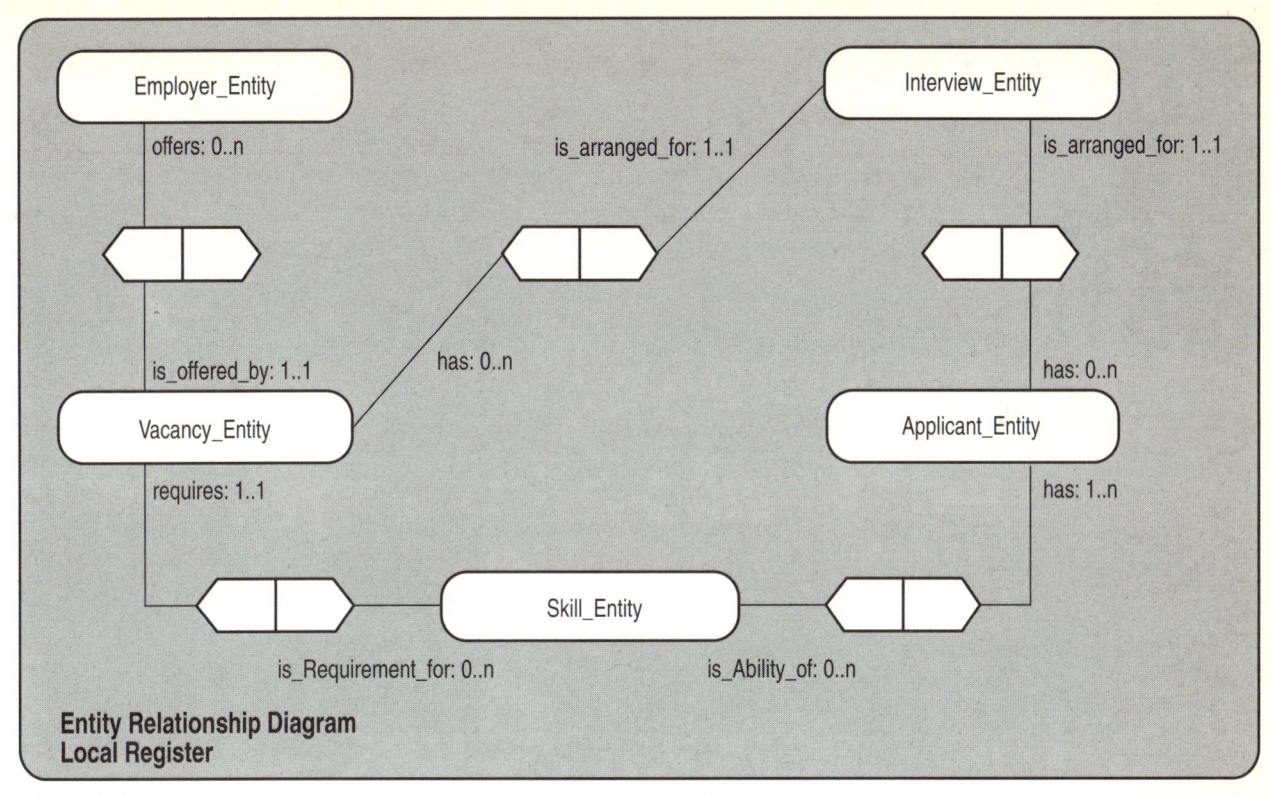

Figure 9.16

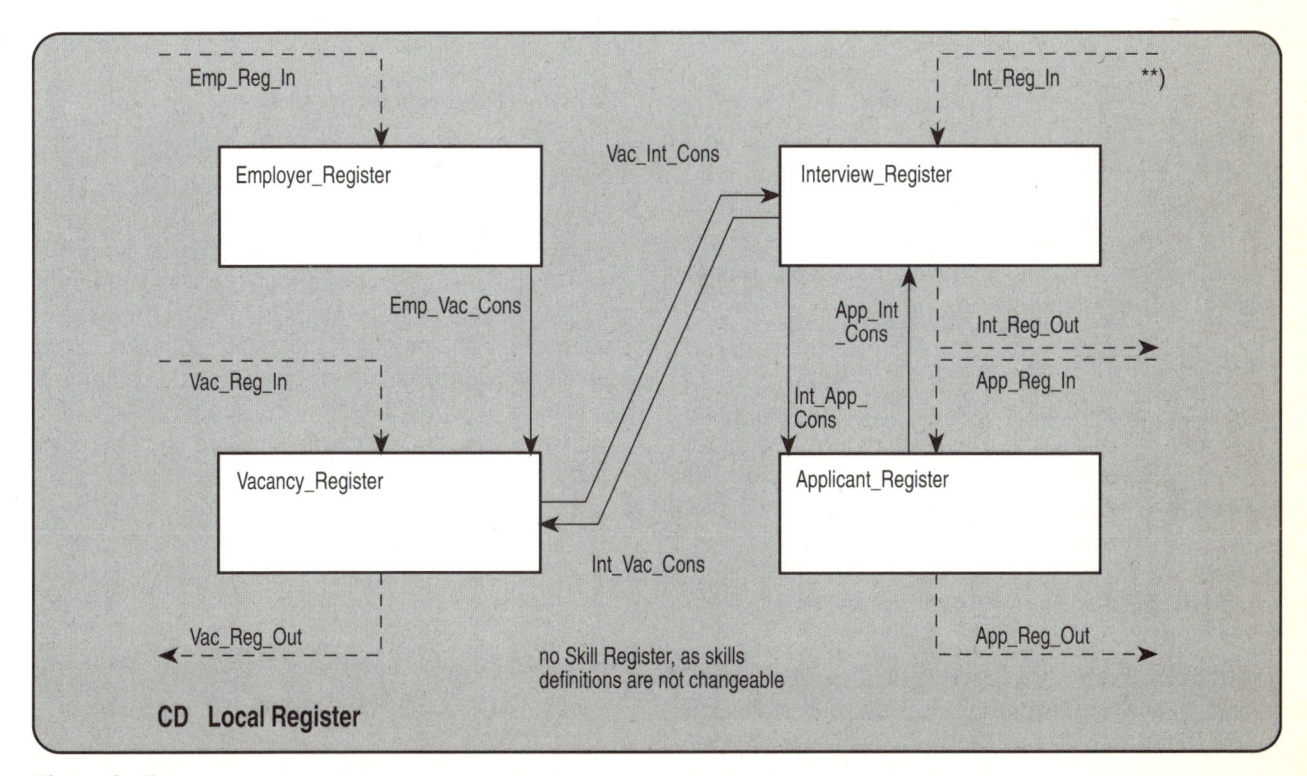

Figure 9.17

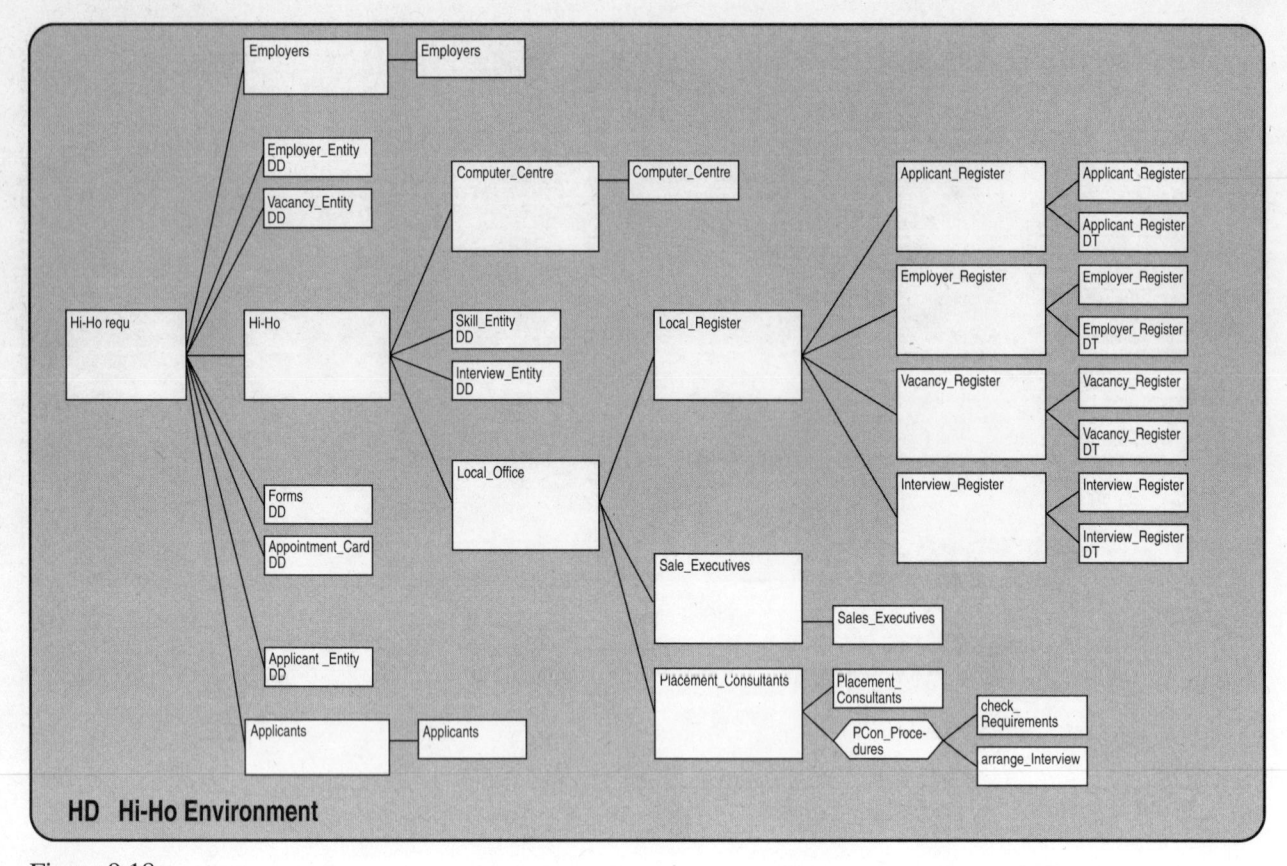

Figure 9.18

Interface tables, hierarchy diagram → intermediate diagrams

The example shows the intermediate diagram that was produced for the return of an interview result to the Hi-Ho company (Fig. 9.19). There is a channel named „return _ Result" in the interface table defining the inputs from the employers.

Intermediate diagrams → process diagrams

The intermediate diagram contains two processing paths for the interview register. These were assembled together with corresponding paths from the other intermediate diagrams to produce the process diagram for the interview register (Fig. 9.20).

Intermediate diagrams → interface tables, communication diagrams

Pairs of communication symbols describing communication between the same objects were collected and the communication lines necessary to establish the connection were introduced into the communication diagrams. The names of the communication symbols were inserted as channel names into the appropriate interface tables, as in figure 9.21.

Transformations between models

The transformation from current system to required system in the case study was trivial. The structure of the system did not change so the whole model was copied. In fact the model of the current system was not developed in full detail, but as soon as the structure and processing were identified the GRAPES team switched to producing the model of the required system. As no new inputs were defined, the modelling of the required system consisted of the intermediate diagrams and the transformations from intermediate diagrams to process diagrams and interface tables.

9.4 Examples of the GRAPES Transformations 113

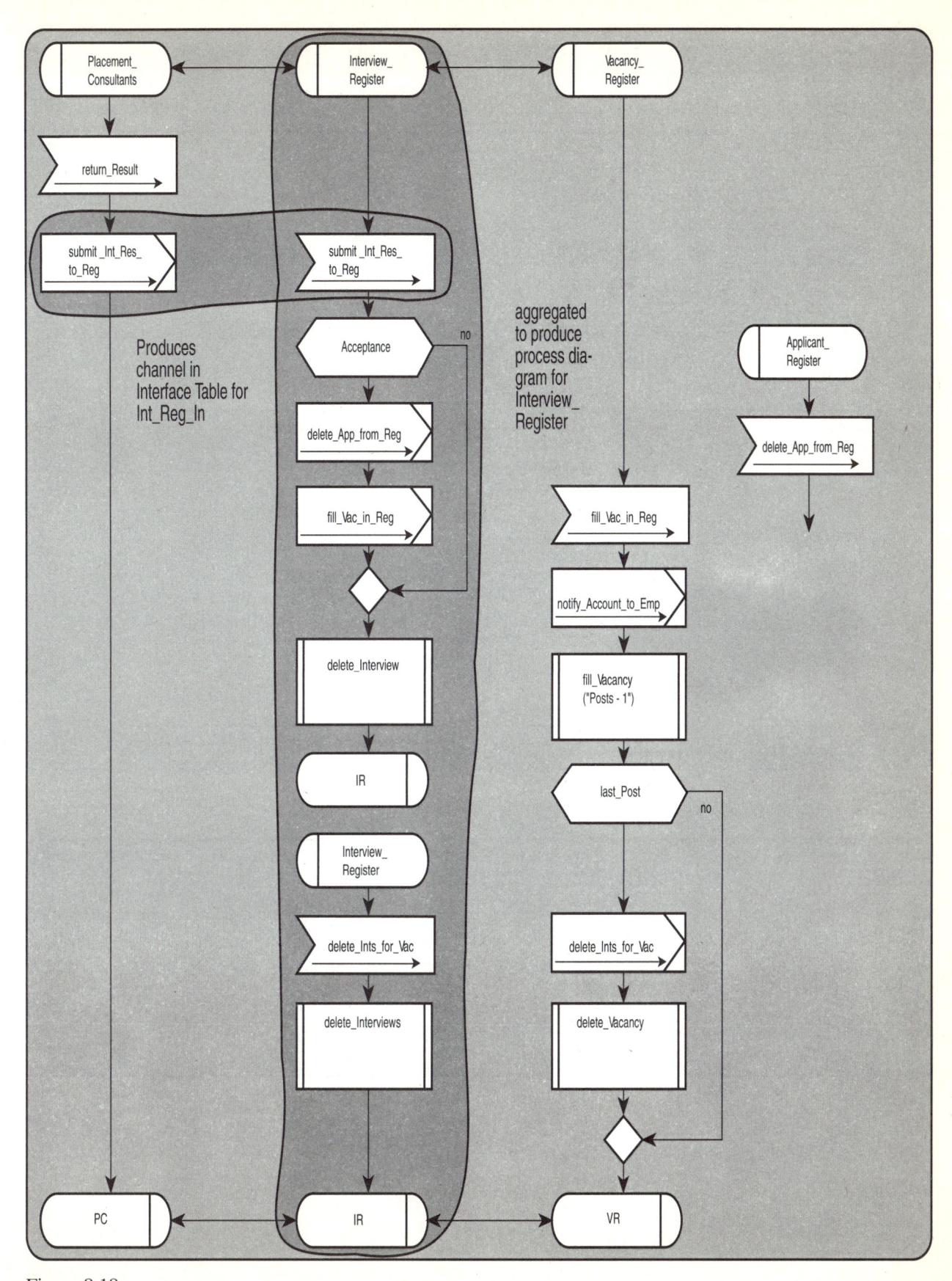

Figure 9.19

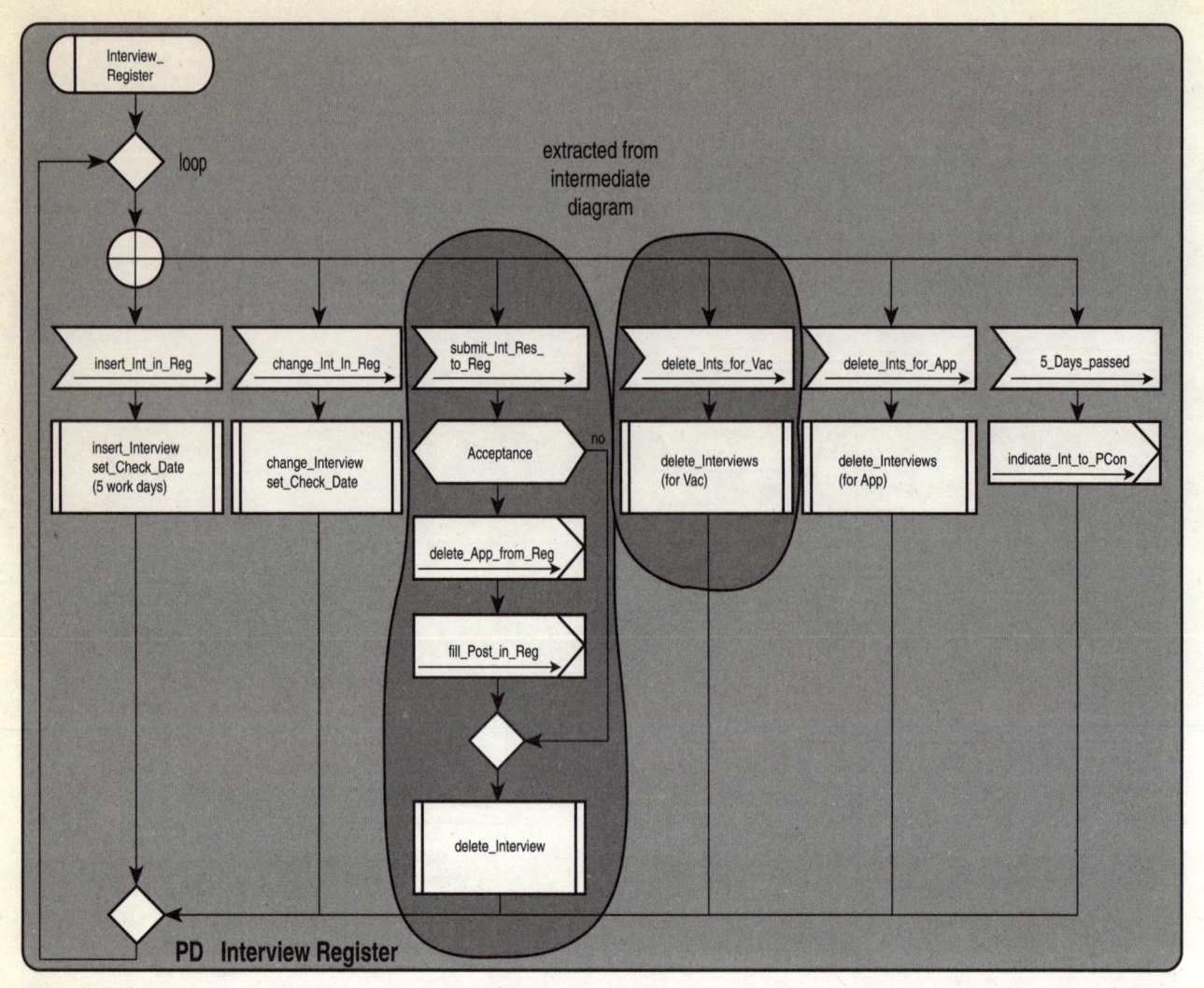

Figure 9.20

Name:	Data Type:	Description:
insert_Int_in_Reg	Interview_Entity	PC_access →
change_Int_in_Reg	Int_Rescheduling	PC_access → SE_access →
submit_Int_Res_to_Reg	Appointment_Card	PC_access →

IT Int_Reg_In

Figure 9.21

9.5 Comparison of the GRAPES and SSADM Transformations

SSADM and GRAPES support similar transformations in the production of the required system data flow models and communication diagrams.

GRAPES uses similar transformations to those of SSADM in building E-R diagrams, for example TNF analysis.

As was remarked in chapter 8, in the case study work, the transformation direction of the GRAPES equivalent of SSADM entity life histories and effect correspondence diagrams is exactly opposite to that of the corresponding SSADM end-products.

In GRAPES, the transformations between the different models are guided by loose rules and generally accepted procedures. Additionally, GRAPES provides automatic consistency checks.

SSADM has more rules to guide transformations and these rules are formulated more strictly than in GRAPES.

The transformation of the final model towards implementation supports some automatic generation of parts of the code and database description. On the database side SSADM seems to do a bit more, while GRAPES seems to be a little stronger in generating code.

Ideally one would like both methods to have even more rules for transformations, and for transformations to be supported by appropriate tools. This is especially true for the transformations between the life cycle phases. While the support for transformations within each phase is currently supported to some extent by GRAPES and SSADM with only minor differences, the support for transformations between the phases is clearly weak in both methods.

10 Conclusions

The two approaches have much in common. Their areas of application and their intended implementations are, for all practical purposes, identical. The SSADM architecture is realizable in GRAPES, and GRAPES is heading in a direction which formalizes this more.

GRAPES is missing two constructs whose presence would otherwise serve to unify the capabilities of the two approaches in the process modelling area. SSADM is missing a capability for reusability of process definition at the conceptual scenario level.

SSADM can profit by adopting some of the formality of GRAPES. GRAPES could benefit from the SSADM knowledge acquisition techniques.

In this chapter we first of all answer the specific questions posed in chapter 1.

Finally we show that there are significant opportunities for harmonizing the two approaches.

10.1 Answers to Specific Questions

Comparison of case studies

How understandable are the results?

This is difficult to answer in direct terms because, of course, each team was more familiar with its own method. (The GRAPES team preferred the smaller CDs to the larger DFDs; Mr Robinson preferred the smaller number of levels in the SSADM DFDs.)

However, on the basis of the number of facts which were represented directly, some knowledge, represented to a degree in both sets of documentation, but not directly in both, can be more easily traced in the SSADM representation.

It takes effort, for example, to trace the effects of post acceptance through the GRAPES documentation. But if adequate tools were provided for the deduction of indirect facts there would be little to choose between the two methods. However, those tools would need to be sophisticated, for example to deduce entity life history structures from state-indicator information.

How easy is it to make changes?

This is a function of at least two variables: direct representation and non-redundancy. Towards the end of the Hi-Ho SSADM development a change had to be made when it was discovered that interviews could not be deleted on refusal of offer or rejection of applicant – because otherwise a new interview might be arranged for a vacancy an applicant had already rejected or for which he had been rejected. It was easy to find what to change. But it was rather more laborious to make the change than had been expected because the actual deletion of failed interviews now could occur on one of several different events. Another dispiriting realization was that because the number of states had changed in one of the life histories several otherwise unaffected update process models had to be changed to include checks for the new states; this ought to have been unnecessary because the new states were equivalent to existing ones, but SSADM does not recognize equivalent states.

In the version of the case study developed using super-events and optimized states in SSADM it was quite easy to make the change.

In the GRAPES model the same effect could be achieved by deleting the procedure call "delete_Interview" in the process diagram "PD Interview Register".

How complete are the results?

Both approaches claim some sort of completeness. It is clear, however, that completeness in the sense that the specifications guarantee to capture every relevant fact cannot be proven. GRAPES completeness is a syntactic completeness – a check that there are no loose ends in the same sense as a compiler guarantees there are no loose ends – for example, that any data used in a process has actually been defined somewhere. SSADM provides some informal checks of this kind of completeness. But SSADM is also aiming at an attempt to enrich the knowledge acquisition process so that it is more likely to capture relevant facts, for example to carry forward semantic implications found in the data models into processing. GRAPES tries to achieve similar results by external (DOMINO) techniques for example in quality assurance.

See also the remarks about fairness in the later section "What are the limits of this study?".

10.1 Answers to Specific Questions

How consistent are the results?

In the SSADM solution we have no guarantee that, for example, data names are used consistently in update process models. We do have a number of guarantees: e.g. that the entity data contents are consistent with the data flow data contents. We know that some data items stored in the entities never appear on any data flow; but we know why this is. In general we feel that the SSADM solution has a high degree of internal consistency.

The GRAPES models have a fully checked internal consistency.

These remarks about consistency apply to the tool-based end-products. We have no guarantee that in transferring the diagrams and definitions to this report that they consistently reflect what is in the tool-based descriptions because they have been transferred piecemeal and updated piecemeal as the tool-based end products have been reworked.

What tool support was needed?

Tools were essential in both the SSADM and GRAPES developments:

- Rapid production of the diagrams
- Consistency checking
- Managing redundancy that the system knew about (e.g. effect correspondence diagrams and entity life histories)
- Generating end-products (e.g. TNF model, update process models)

It is inconceivable that the study could have produced the results it did in the time it did without the level of tool support used.

What further effort would be needed for implementation?

The SSADM dialogue designs are very rudimentary and included for illustrative purposes only. Full design and prototyping of the dialogues is probably the lengthiest activity remaining. It is impossible to give an estimate without knowing about a specific implementation environment.

Database design and optimization would require little more than one or two man-days' effort.

The update process models need to be supplemented with some code to handle event failure (checkpoint, rollback, displaying error messages).

In a friendly environment (e.g. Application Master), very little else remains to be done beyond coding. Test site experience shows the end-products are largely pre-debugged.

For GRAPES a reasonable estimation of the effort needed for implementation is impossible without knowing the implementation and production environments.

At the point reached in the case study development only a logical model of the required system had been produced. From this logical model it would take 2 - 4 man-days' effort to produce an animated prototype.

General comparison

Do the methods suggest the same view of the world?

In communication structure – yes. In static structure – yes. In dynamic structure – to a degree, but SSADM focuses on individual entity behaviours over time, GRAPES focuses on the behaviours of entity sets, and doesn't really see the time sequences of behaviour at the entity level.

Does any difference in world-view have a useful outcome?

Yes. SSADM has richer knowledge acquisition tools as a result of its entity-instance focus.

Are both methods targetting the same set of systems to be described?

Yes. Both methods are aimed at information processing and process control systems. In addition, GRAPES is aiming at the specification of distributed systems; SSADM is currently examining extensions designed to support distributed systems.

Where are the areas of application of the methods?

The main area of application of both methods is the analysis and design of administrative data processing systems. Other areas, e.g. process control or simulation can be tackled too. Both methods have similar intended implementation direction. Both SSADM and GRAPES are used for the specification of applications implemented in COBOL and 4GLs. GRAPES is also aimed at C because Siemens Nixdorf provides a C compiler.

Is the same knowledge about a system captured?

GRAPES and SSADM cover similar sets of facts about a system. Because SSADM can operate at the process array level it can represent dynamic facts more directly than GRAPES. In principle, the GRAPES facts could be transformed into corresponding SSADM facts, but the amount of effort in effecting this transformation manually is such that it may be reasonable to doubt whether people would carry this out in a reliable manner, and thus whether they really have the same "knowledge" in this area. When GRAPES has been extended to include the process array construct the fact representation in this area, and hence the "knowledge", will be much more directly comparable.

The other significant differences in the knowledge are that:

- SSADM has the notion of a success unit whereas GRAPES does not;
- GRAPES has some knowledge of the same process being replicated at several locations whereas SSADM does not;
- SSADM doesn't "know" that some of its knowledge is redundant in the sense that the same processing may be replicated in several entity life histories and effect correspondence diagrams.

What skill levels are necessary to use the method?

SSADM has more techniques so the analyst requires more skills in using a larger tool kit. However the GRAPES designer has to use his intuition to a higher degree than an SSADM analyst, and so needs more experience. In practice, the same thoughts have to be formulated by analysts using either method if equivalent outcomes are to be achieved.

What training is necessary to use the methods?

GRAPES: about 5 days
SSADM: 10-15 days

What tool support is given, and what is required?

A number of CASE tools support SSADM with varying degrees of rigour, automation of diagram production, and consistency checking. GRAPES is supported by the Siemens Nixdorf DOMINO SE environment tools. The amount of indirect fact representation discovered during this study suggests that both methods could benefit from tools that are smarter in their ability to derive facts and present alternative world-views to the systems developers.

How comprehensive are the consistency checks?

GRAPES provides full consistency checks in the manner of a strongly-typed programming language. Because (with the exception of its diagram formation rules) SSADM is based on an informal syntax its checks (other than that diagrams are well-formed) are based on cross-relating designed-in redundancies in its end-products (e.g. that keys in TNF relations correspond to relationships in the LDM).

The focus of GRAPES on formal consistent models with executable description of their dynamic behaviour allows the generation of animation prototypes and in consequence an easier validation of the system's model with the customer.

What support is there for software reuse?

GRAPES formally identifies reusable subroutines. But the need for them depends on human intuition; they would be recognized by inspection of GRAPES definitions. This provides reuse at a programming level rather than an analysis level.

For reuse at analysis level GRAPES offers the concept of object types.

SSADM addresses the identification of reusable code in stage 6 (physical design). This is slightly later than GRAPES does, but SSADM gains the same benefit as GRAPES at coding time.

Earlier in the analysis, update process models and enquiry process models are reusable in several functions. But SSADM currently does not address further reusability at the analysis level – inside an update or enquiry process model. As remarked elsewhere, introducing the concept of a "super event" could provide such analysis-level reusability inside update process models and their antecedent diagrams.

What support is there for structuring?

SSADM supports everything GRAPES does with one exception: in GRAPES it is possible to define types of process and then instantiate them. GRAPES can use this ability to represent distributable functionality, and instantiate that functionality at named locations. In this way it provides some support for Zachman row 3 (distribution) in a way that SSADM does not.

GRAPES does not have support for process arrays (implicit in SSADM). This is what prevents GRAPES currently from being able to operate at the level of the life history of an individual entity. GRAPES also does not provide backtracking / escape constructs.

Where do the methods have strengths and weaknesses?

SSADM is stronger in knowledge acquisition and can state directly more kinds of facts. SSADM is weaker in formal definition and in the syntactic cross-checking ability that goes with increased formality. For example, it would not be possible to build an interpreter which could execute all SSADM descriptions as they stand today and simulate the final system.

GRAPES is more formal. It would be possible to build a GRAPES interpreter as described above. But a GRAPES system, whilst it would be syntactically complete, is less likely to be semantically complete than an SSADM system.

In which areas would both systems benefit from a joint add-on development?

Both systems are heading in an object-oriented direction. Entity life history analysis is clearly a quasi-object-oriented approach; it could be made more clearly object-oriented and simplified by further import of these ideas. GRAPES is committed to developing its object concepts further (process arrays, inheritance). Currently GRAPES is being used together with C++ and CooL in pilot projects. A common approach is clearly possible.

Both organizations have interests in GEMINI. SSADM knowledge acquisition is demonstrably more sophisticated than GRAPES. But there are several knowledge acquisition techniques not included in SSADM. Joint research would profit both organizations.

Both organizations are interested in EuroMethod. Developing some common interchange standards would benefit both organizations in this context.

What are the limits of this study?

The study is limited as follows:
- The early stages of SSADM and GRAPES were not addressed by the case study.
- No prototyping was possible.
- No carrying forward to implementation was possible.
- The comparison is unfair to SSADM in the sense that the GRAPES team had the benefit of knowledge that was gathered using SSADM. The comparison was unfair to GRAPES in the sense that the case study had been worked on over a period of several years by Model Systems, who had the benefit of much greater familiarity.

10.2 Opportunities for Harmonization of GRAPES and SSADM

Life cycle

Consider the following facts:
- DOMINO / GRAPES has only four different milestones in the analysis & design phase.
- SSADM has a much more detailed step structure aimed at knowledge acquisition through the production of intermediate products.
- The same progression from current to required system is visible in both SSADM and DOMINO / GRAPES.
- The application areas of both approaches are almost identical.
- The world-views of the two systems have much in common and are growing closer.

From these facts, it seems reasonable that a harmonized life cycle for the two products could be produced:
- The DOMINO milestones could act as "collection points" for SSADM intermediate products.
- Subject to some prescription of products currently not absolutely required in GRAPES (e.g. an E-R model), it is easy to imagine a composite life cycle in which a project could be developed in GRAPES up to some GRAPES milestone and then further developed in SSADM (or the other way round).

Architecture

GRAPES would need to prescribe a 3-schema approach. If future GRAPES developments follow the paradigm adopted by the GRAPES Hi-Ho team, GRAPES will, in any event, produce 3-schema objects; all that would remain is some formal recognition of which schema the objects belonged to.

Facts represented and syntax

Data flow model / communication diagrams

These documents are more or less equivalent.

Logical data model / E-R diagram

These diagrams are very close. GRAPES supports attributed relationships. If either GRAPES removed this support or SSADM introduced more disciplined key-choice rules (and so became more likely to represent

GRAPES attributed relationships as compound key entities) the diagrams would be almost equivalent. The only other significant difference is that GRAPES supports unnormalized entity data.

Entity life histories etc. / process diagrams

The most significant differences occur here. Harmonization would require the following actions:

- GRAPES needs a process-array mechanism. (Likely to happen soon.)
- GRAPES needs some kind of backtracking construct.
- The production of entity-like objects would need to be prescribed in GRAPES.
- SSADM would need to beef-up its approach to common processing (probably with "super-events")
- SSADM should adopt a formal operation and data syntax.
- SSADM should investigate the use of explicit "timer" objects.

Knowledge collection

GRAPES relies on the quality assurance process to ensure that the analyst has discovered all of the relevant knowledge about the system. A method which ensured more knowledge collection during development would:

- Minimize the chance of the quality assurer's not discovering missing / incorrect facts
- Prevent some costly rework of end-products to take account of knowledge which had been discovered during quality assurance.

SSADM has much more detailed knowledge collection than GRAPES in the area of systems dynamics. Much of this knowledge acquisition could be adopted by GRAPES.

Both approaches would benefit from explicit research to discover more knowledge collection rules. There would be significant synergy between this work and GEMINI.

Tool platform & information interchange

Data flow diagrams to communication diagrams and vice versa

In principle, this should be possible now.

Logical data models to entity relationship diagrams and vice versa

This should be possible now if attributed relationships and unnormalized entities are not involved.

Entity life histories etc. to process diagrams and vice versa

Process diagrams enhanced by object arrays and backtracking are more or less syntactically equivalent to entity life histories and update & enquiry process models. Entity life histories have a strong correspondence to process diagrams which implement a GRAPES decomposition based on entity-like objects. The only things in Jackson diagrams not generatable from the GRAPES process diagrams are the intermediate node names; these could be generatable if GRAPES includes loop and selection names.

Note that interchange of processes requires not just syntactic equivalence, but also semantic equivalence; in fact what is required is a common meta-model.

Annex 1

The Hi-Ho Recruitment Case Study

Introduction

Hi-Ho is an employment agency specializing in professional and management placements.

There are several local offices around the country, all operating independently but with common standards for processing and data. Each keeps a register of recent local vacancies and local job applicants, and tries to match them.

Employers with vacancies in more than one locality are required to register separately at each local office. Employers registered at more than one local office are treated as different employers.

Placement consultants can look up recent vacancies and applicants in the local registers but the company has a computer centre at its head-office which provides overnight turnaround on the full set of vacancies and applicants whose details are kept in the system.

The current Hi-Ho system

Employers

Having agreed terms and conditions, employers notify Hi-Ho of vacancies they wish to fill. There may be one or more posts for a given vacancy.

Vacancies

Only vacancies for the standard Hi-Ho skill categories are accepted. When a vacancy notification arrives, it is assigned to a single primary skill area.

Sales executives enter vacancy details onto forms for submission to the computer system. Overnight, vacancy details are matched with possible applicants and a list of matches (even nil) is sent to the Hi-Ho office where the vacancy is registered.

Potential applicants are selected (the number depending on the number of posts for a vacancy) and contacted by their placement consultant. If they are still available for employment and are interested in the vacancy, then an interview is arranged for them with the employer (see "Interviews").

If a vacancy has not been filled within six months, the employer is contacted and asked whether it is still open.

Applicants

Each applicant fills in a registration form giving his personal details and skills and submits it to his local Hi-Ho office.

Only applicants in the standard Hi-Ho skill categories, as listed on the registration form, are accepted for registration.

Each applicant is assigned to a placement consultant who checks his registration form before submitting it to the computer system. Overnight, the applicant's details are matched with possible vacancies and a list of matches (even nil) is sent to the applicant's placement consultant.

If suitable local vacancies are available, up to about six are selected, a check is made with the employers to ensure that they are still open, and the applicant is asked to attend the Hi-Ho office to discuss them (see "Interviews").

If there are no suitable vacancies the applicant is asked if he wants to change his requirements. If he does not, then he is informed that he will be contacted when suitable vacancies become available.

A review date is set for the applicant. If he has not been placed before this, the applicant is contacted to ask if he has found a job elsewhere, or if he wishes to come into the office to review his requirements.

Interviews (between employer and applicant)

Applicants attend their local Hi-Ho office to discuss vacancies with their placement consultant. The relevant employers are contacted by telephone while the applicant is present, the interviews are arranged and the applicant is given an appointment card for each.

A second copy of each appointment card is sent to the computer system and the employers are sent a written confirmation of each interview appointment.

After the interview the applicant leaves the first copy of the appointment card with the employer, who should return it to Hi-Ho with one of three entries completed:

- "acceptance" (post offered and accepted)
- "refusal" (post offered and refused)
- "rejection" (post not offered to this applicant)

Some employers do not return appointment cards punctually. If, five working days after the date of an interview, the appointment card has not been returned then the employer is contacted and asked for the result.

If the result is "acceptance", "refusal" or "rejection", the placement consultant fills in a duplicate appointment card and submits it to the computer system.

Requirements for the new Hi-Ho system

The current system doesn't work very well. In practice the local registers are of limited use. So much matching has to be done on the head-office computer that the placement consultants routinely use it instead of using the local registers first. This means that applicants are typically not seen by a placement consultant until a day or so after they register.

There is currently a booming job recruitment market. Applicants who aren't found jobs quickly move on to other recruitment agencies. There is a significant fall-off, from the numbers of applicants registering, to the numbers of applicants who actually attend for a subsequent meeting with their placement consultant to review the potential vacancies selected for them by the head-office computer. Hi-Ho management believe that it is desirable to be able to arrange an applicant's first batch of interviews during his attendance at the local office for the purposes of registration.

Similarly, sales executives feel that they are losing business because of their inability to give some indication to employers, when they first contact Hi-Ho, about the likely numbers of qualified applicants.

Sample forms in use in the current system

Applicant details

Hi-Ho Recruitment – New Applicant's Personal Details

Applicant id :..........:

Surname :.....................:   Title :................:

Forenames :.................................:

Home :.................................:

Address :.................................:

Postcode :...........:   Telephone :....................:

Date of Birth :.../.../.....:   Marital Status :..............:

Advert Reference :............:

SKILLS   Enter   1 for more than five years practical experience
                  2 for some practical experience
                  3 for theoretical knowledge only

General Management	:.:	Production Management	:.:
Financial Management	:.:	Marketing	:.:
Purchasing	:.:	Materials Control	:.:
Project Management	:.:	Resource Planning	:.:
Sales Management	:.:	Sales	:.:
Accounting	:.:	Audit	:.:
Quality Assurance	:.:	Management Services	:.:
Systems Analysis	:.:	DP Management	:.:

Minimum annual salary required   :............:

Willing to move to another area ?   yes :.:   no :.:

Office use only:

Hi-Ho Office :.....:   Placement Consultant :..........:

Registration Date :.../.../.....:

Employer Interview Status Report

Hi-Ho Office	12	CHISWICK		September 1986

Employer	Dodds and Partners

Vacancy	101	Sales Representative	No of Posts	2

Applicant Id	Name	Interview Date	Result
12432178654	Jones A	3/9/86	Not Offered
12752654321	Peters B	6/9/86	Offered/Accepted
13245367543	Brown J	29/9/86	Not yet known

Vacancy	103	Sales Manager	No of Posts	1

Applicant Id	Name	Interview Date	Result
14653287623	Williams K	9/9/86	Offered/Refused

Employer	J W Cartwright and Sons

Vacancy	321	Account Manager	No of Posts	0

Applicant Id	Name	Interview Date	Result
16543873215	Bell P	17/9/86	Offered/Accepted

Applicants Suitable for Named Vacancies – 01/07/86

Employer No	Employer Name	Vacancy No	Vacancy Name	Suitable Applicant	
10/14325	Jones Bros	154	Account Clerk	16543873215	Bell P
				29141678988	Freleigh I P
				23456789012	Owen D
10/15432	Smith and Partners	192	Financial Director	16543873215	Bell P
				29141678988	Freleigh I P
				23456789012	Owen D
12/02491	Dodds and Partners	101	Sales Representative	12432178654	Jones A
				12752654321	Peters B
				13245367543	Brown J
				14653287623	Williams K
12/02491	Dodds and Partners	103	Sales Manager	12432178654	Jones A
				12752654321	Peters B
				13245367543	Brown J
				14653287623	Williams K
13/19713	British Grot	003	Invoice Clerk	16543873215	Bell P
				29141678988	Freleigh I P
				23456789012	Owen D

Vacancies Suitable for Named Applicant – 01/07/88

Applicant No	Applicant Name	Skill Area	Suitable Vacancy		
39143558921	R Jimlad	Accounting	10/14325	Jones Bros	154 Account Clerk
			10/15432	Smith and Ptners	192 Financial Director
			13/19713	British Grot	003 Invoice Cler
		Systems Analysis	22/21347	Loamshire HA	001 Project Leader
			23/32456	Blacklung Tobacco	009 Systems Analyst
39275643213	C O Jones	Accounting	10/14325	Jones Bros	154 Account Clerk
			10/15432	Smith and Ptners	192 Financial Director
			13/19713	British Grot	003 Invoice Clerk
		Audit	10/14325	Jones Bros	192 Audit Manager
			17/17865	U R Bent	001 Security Manager
			13/19713	British Grot	002 EDP Auditor
			23/32456	Blacklung Tobacco	008 Audit Clerk

Sample forms in use in the current system

Annex 2

Hi-Ho SSADM Development

Data flow diagrams

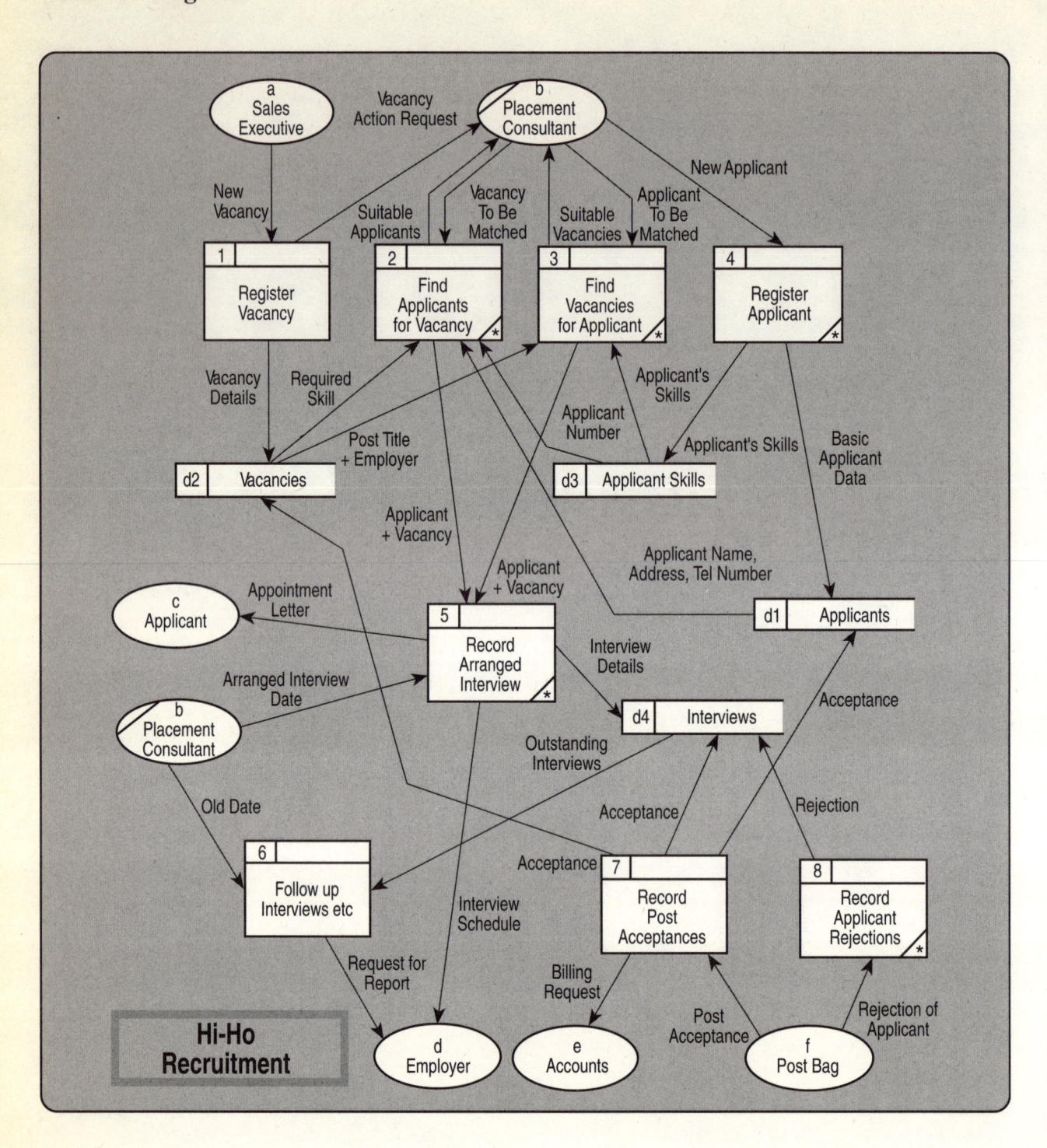

Data flow diagrams

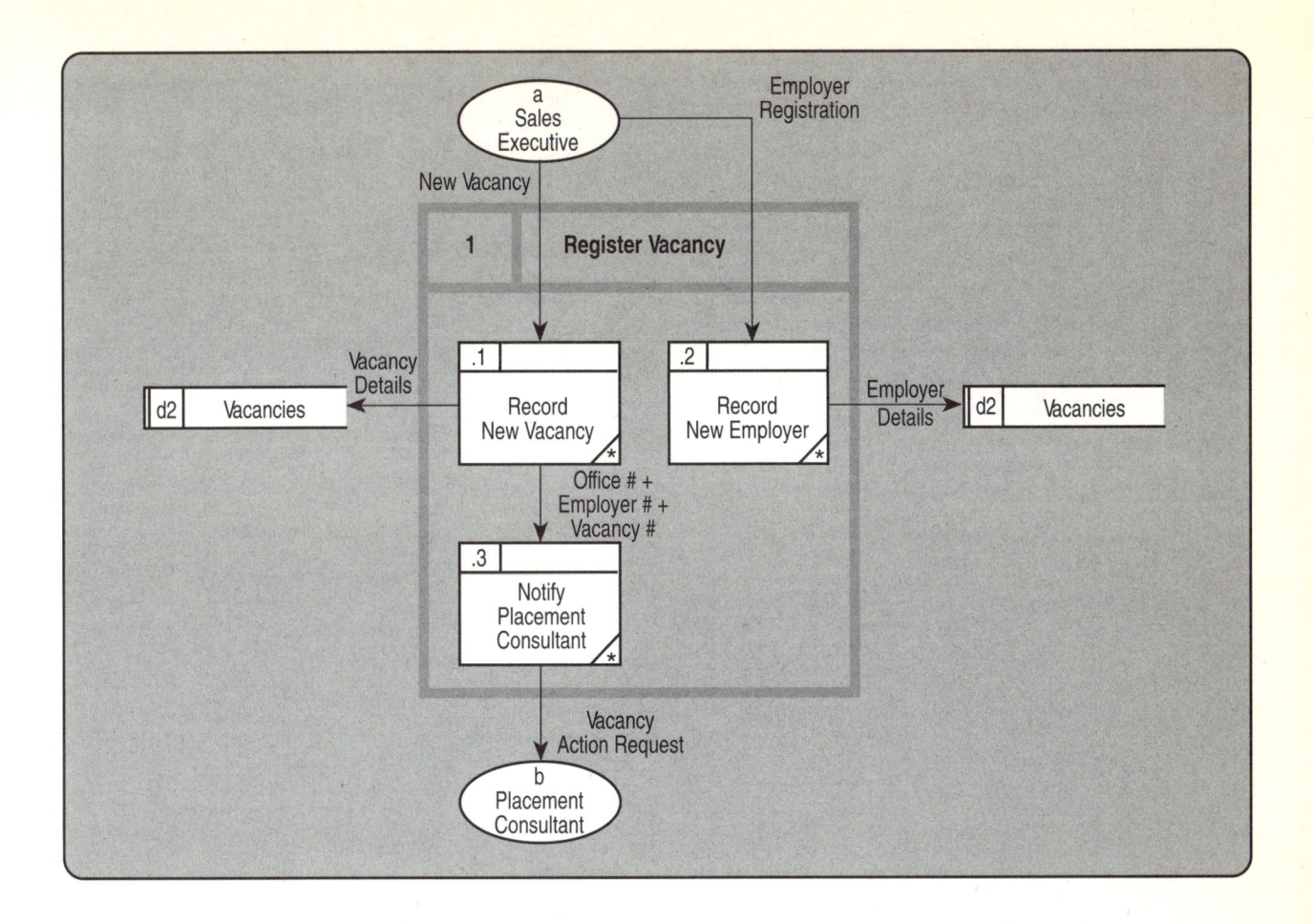

```
                    a
                  Sales          Employer
                Executive       Registration
          New Vacancy

    ┌─────────────────────────────────────────────┐
    │  1         Register Vacancy                 │
    │   ┌──────────┐                              │
    │   │          │                              │
    │                                             │
    │         ┌─.1──────┐      ┌─.2──────┐        │
    │  Vacancy│         │      │         │Employer│
    │  Details│ Record  │      │ Record  │Details │
    │ ◄───────│New Vacancy│   │New Employer├──────►│
    │ d2 Vacancies│   * │      │       * │  d2 Vacancies
    │         └────┬────┘      └─────────┘        │
    │              │ Office # +                   │
    │              │ Employer # +                 │
    │              │ Vacancy #                    │
    │         ┌─.3─▼────┐                         │
    │         │ Notify  │                         │
    │         │Placement│                         │
    │         │Consultant│                        │
    │         │      *  │                         │
    │         └────┬────┘                         │
    └──────────────┼─────────────────────────────┘
                   │ Vacancy
                   │ Action Request
                   ▼
                   b
               Placement
               Consultant
```

Annex 2 Hi-Ho SSADM Development

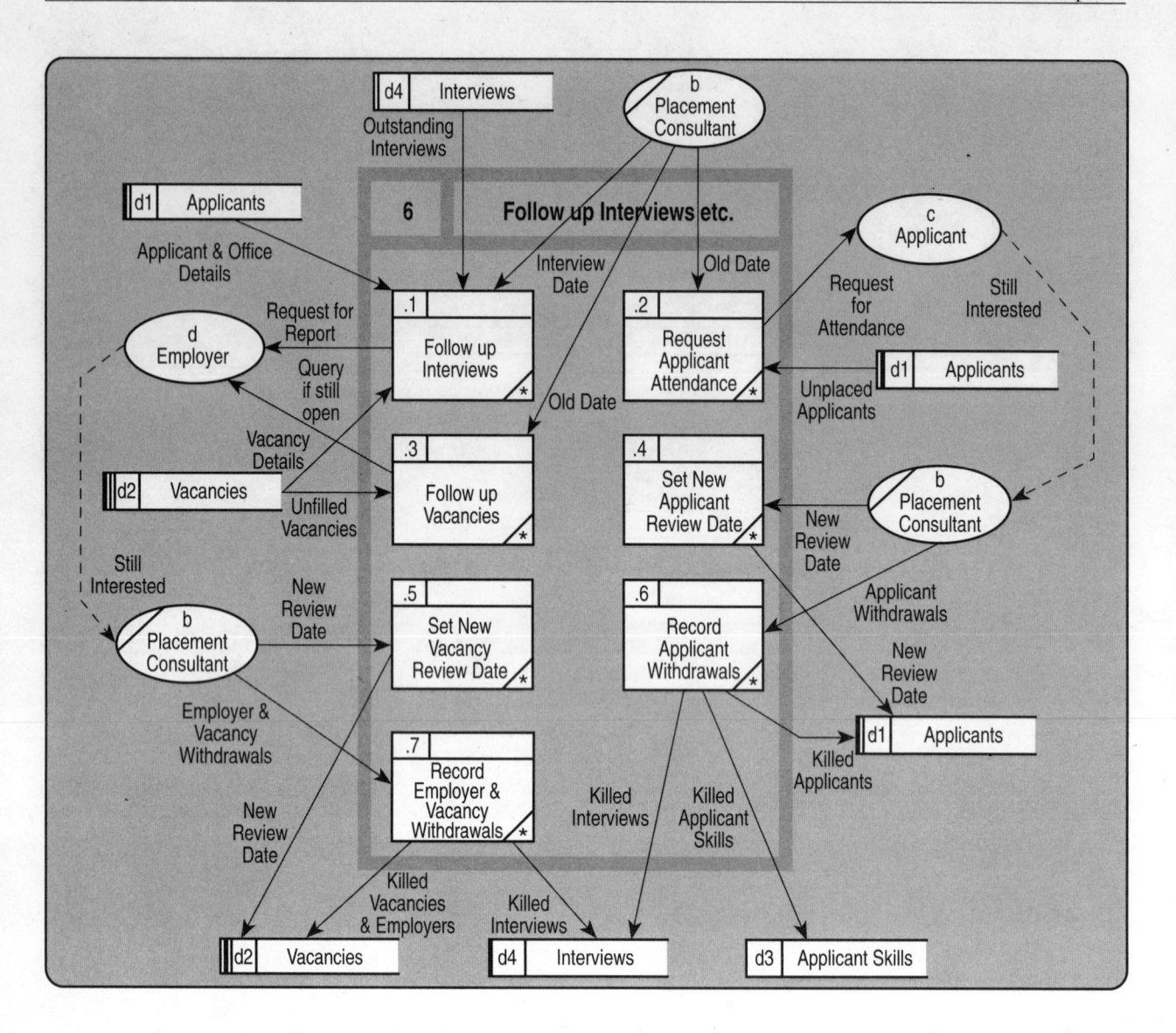

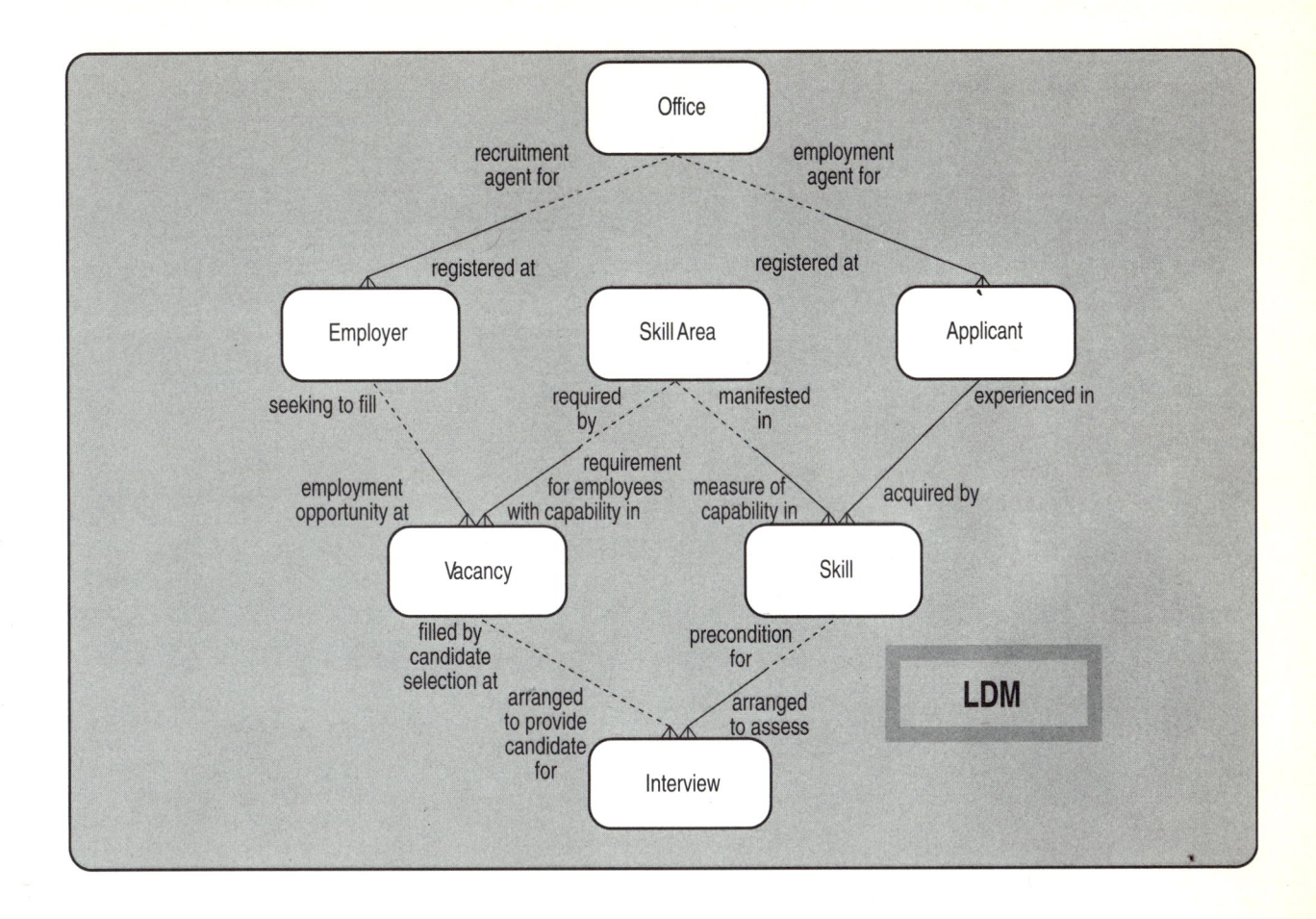

Logical data model

Diagram: Logical Data Model (LDM)

Entities and relationships:

- **Office**
  - recruitment agent for → **Employer**
  - employment agent for → **Applicant**
- **Employer** registered at **Office**
- **Applicant** registered at **Office**
- **Employer** seeking to fill / employment opportunity at → **Vacancy**
- **Skill Area** required by / requirement for employees with capability in → **Vacancy**
- **Skill Area** manifested in / measure of capability in → **Skill**
- **Applicant** experienced in / acquired by → **Skill**
- **Vacancy** filled by candidate selection at / arranged to provide candidate for → **Interview**
- **Skill** precondition for / arranged to assess → **Interview**

Logical data model – data store cross-references

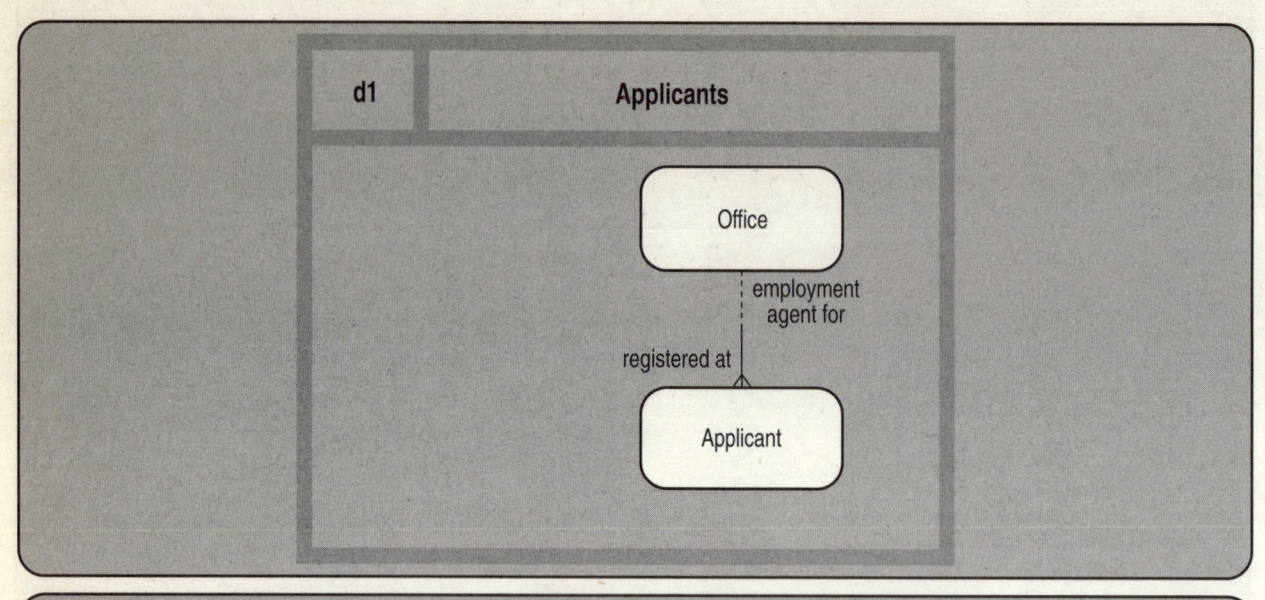

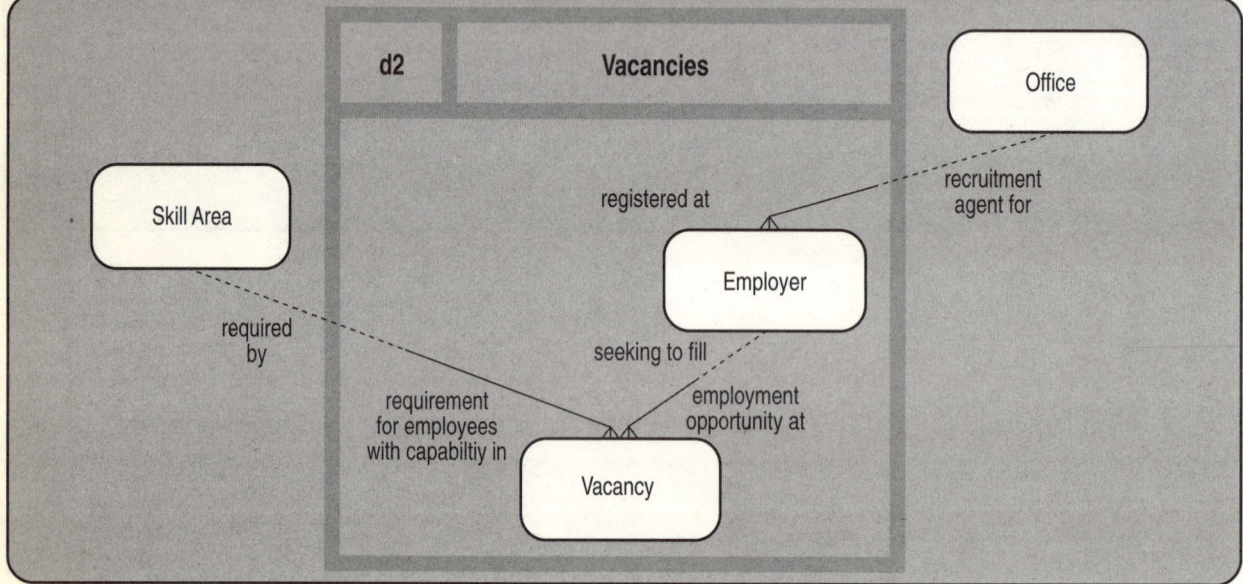

d1 Applicants

- Office
  - employment agent for
  - registered at
- Applicant

d2 Vacancies

- Skill Area
  - required by
  - requirement for employees with capabiltiy in
- Employer
  - registered at
  - recruitment agent for (Office)
  - seeking to fill
  - employment opportunity at
- Vacancy
- Office

Logical data model – data store cross-references

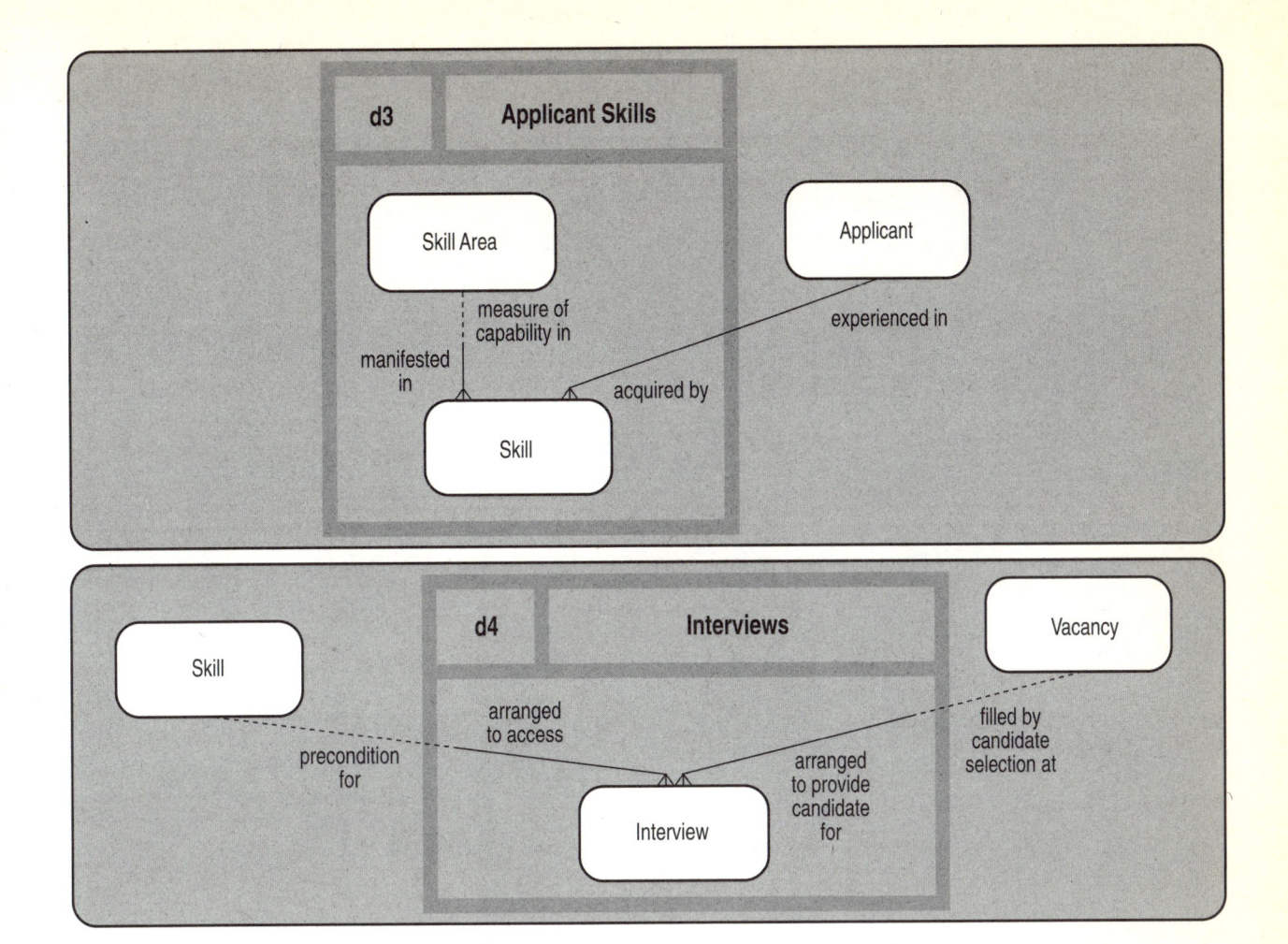

d3 Applicant Skills

- Skill Area — measure of capability in — Skill
- Skill Area — manifested in — Skill
- Applicant — experienced in — Skill
- Skill — acquired by — Applicant

d4 Interviews

- Skill — precondition for — Interview (arranged to access)
- Interview — arranged to provide candidate for — Vacancy
- Vacancy — filled by candidate selection at — Interview

Enquiry access paths

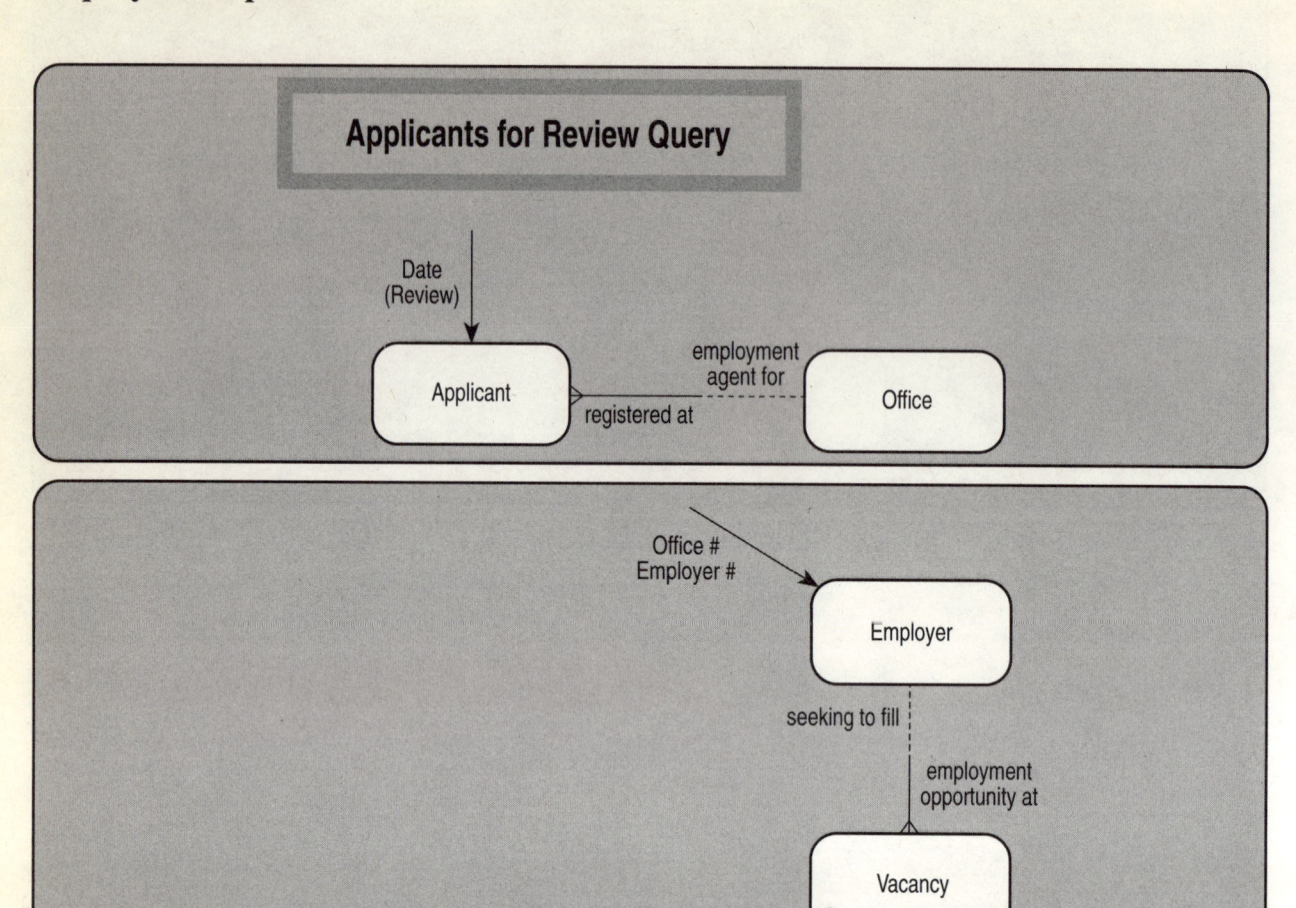

Applicants for Review Query

Date (Review) → Applicant — registered at — Office (employment agent for)

Employer's Interviews Query

Office #, Employer # → Employer

Employer — seeking to fill — Vacancy (employment opportunity at)

Vacancy — filled by candidate selection at / arranged to provide candidate for — Interview

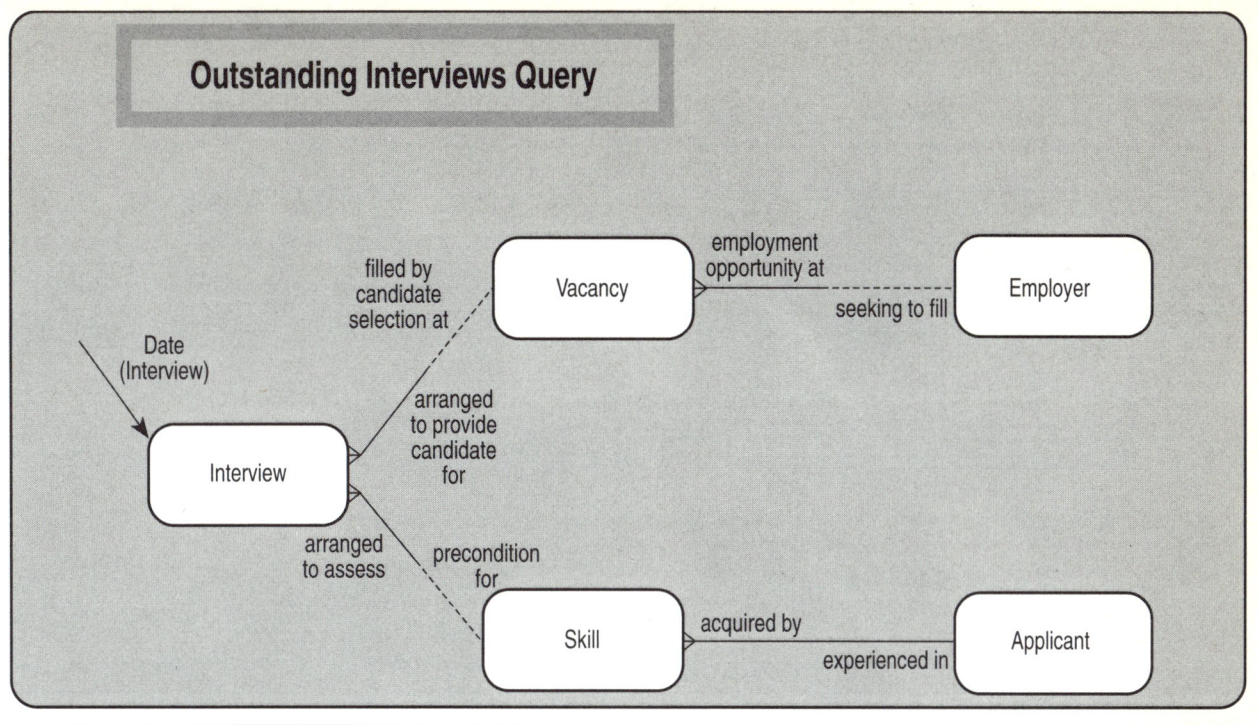

Enquiry access paths

Outstanding Interviews Query

- Date (Interview) → Interview
- Interview — filled by candidate selection at — Vacancy
- Interview — arranged to provide candidate for — Vacancy
- Vacancy — employment opportunity at / seeking to fill — Employer
- Interview — arranged to assess — Skill
- Skill — precondition for — Vacancy
- Skill — acquired by / experienced in — Applicant

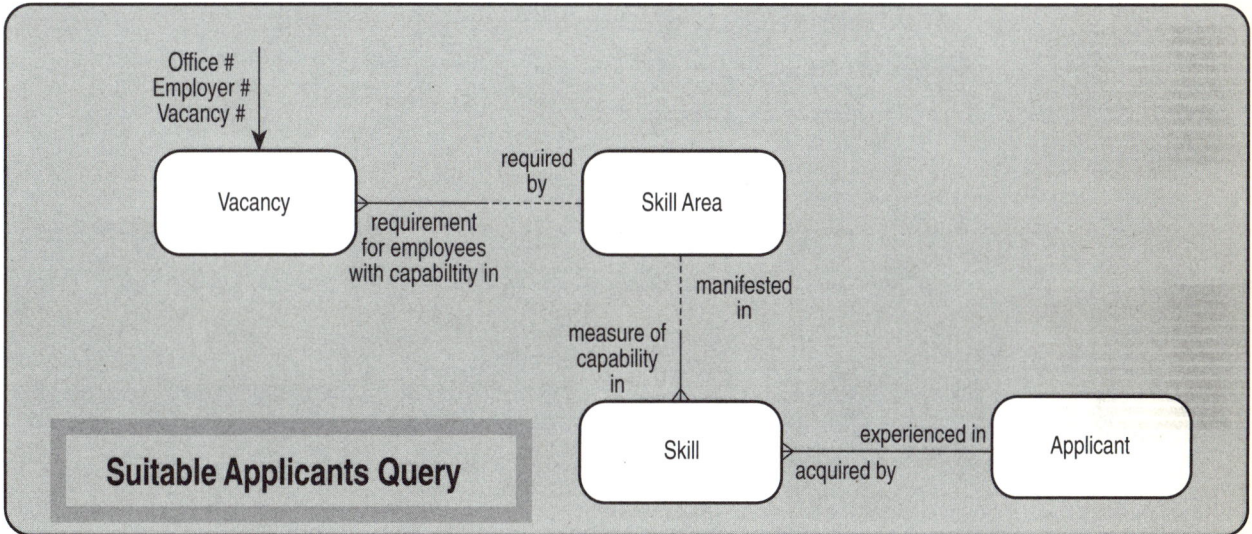

Suitable Applicants Query

- Office #, Employer #, Vacancy # → Vacancy
- Vacancy — required by / requirement for employees with capabiltity in — Skill Area
- Skill Area — manifested in / measure of capability in — Skill
- Skill — experienced in / acquired by — Applicant

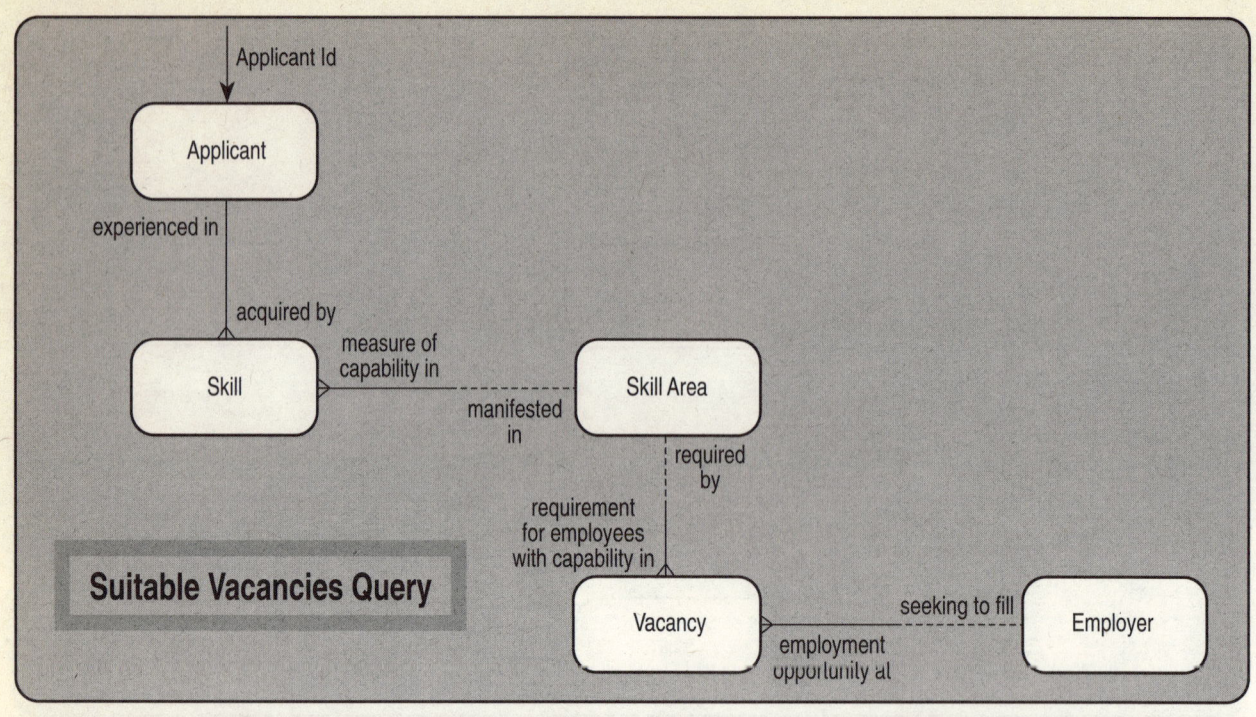

Suitable Vacancies Query

- Applicant Id → **Applicant**
- Applicant — experienced in / acquired by — **Skill**
- Skill — measure of capability in / manifested in — **Skill Area**
- Skill Area — required by / requirement for employees with capability in — **Vacancy**
- Vacancy — seeking to fill / employment opportunity at — **Employer**

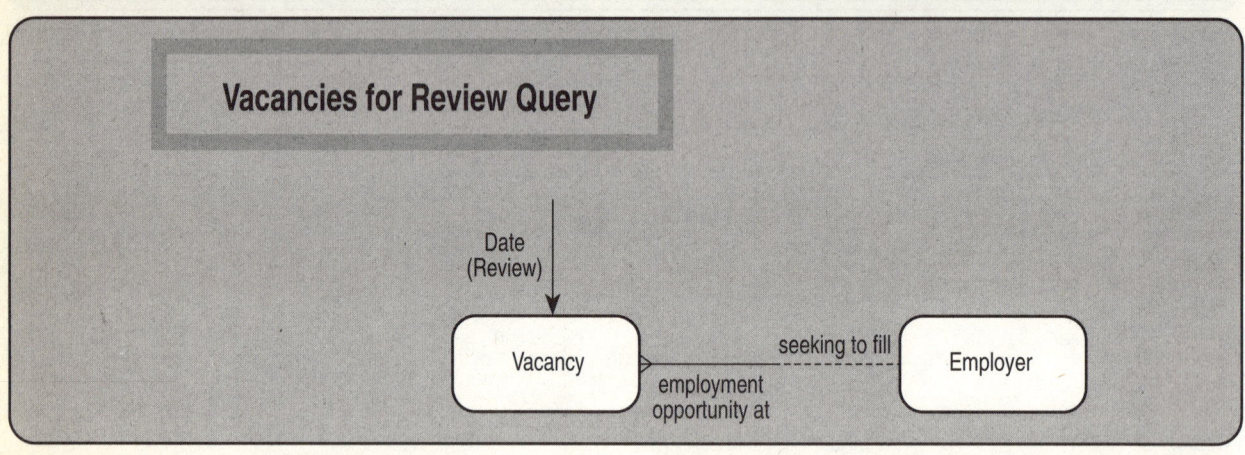

Vacancies for Review Query

- Date (Review) → **Vacancy**
- Vacancy — seeking to fill / employment opportunity at — **Employer**

LDM enhanced with RDA results

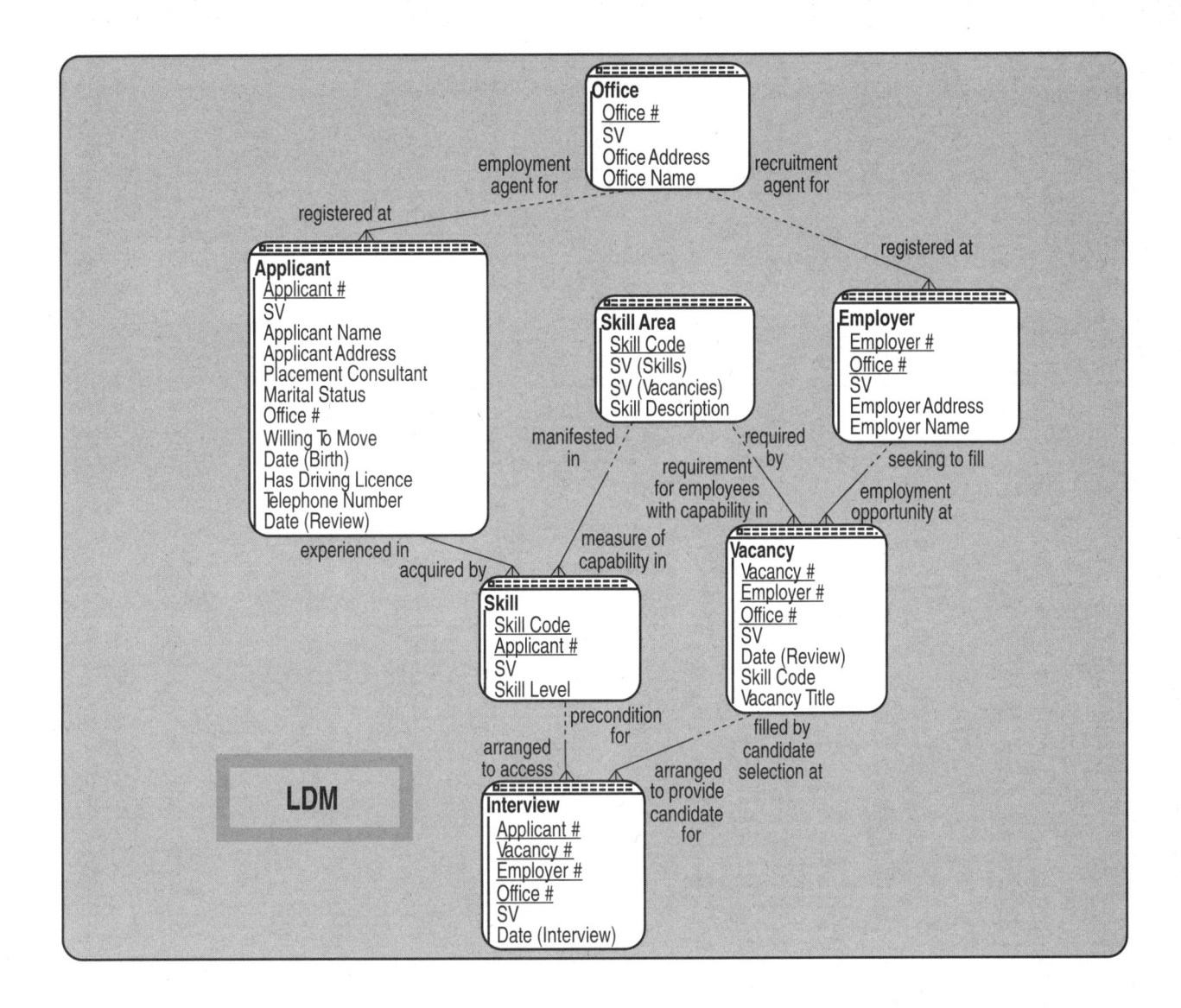

Office
- Office #
- SV
- Office Address
- Office Name

employment agent for — registered at — **Applicant**
- Applicant #
- SV
- Applicant Name
- Applicant Address
- Placement Consultant
- Marital Status
- Office #
- Willing To Move
- Date (Birth)
- Has Driving Licence
- Telephone Number
- Date (Review)

recruitment agent for — registered at — **Employer**
- Employer #
- Office #
- SV
- Employer Address
- Employer Name

Skill Area
- Skill Code
- SV (Skills)
- SV (Vacancies)
- Skill Description

manifested in / required by

requirement for employees with capability in / seeking to fill / employment opportunity at

Vacancy
- Vacancy #
- Employer #
- Office #
- SV
- Date (Review)
- Skill Code
- Vacancy Title

experienced in / acquired by / measure of capability in

Skill
- Skill Code
- Applicant #
- SV
- Skill Level

precondition for / arranged to access / arranged to provide candidate for / filled by candidate selection at

Interview
- Applicant #
- Vacancy #
- Employer #
- Office #
- SV
- Date (Interview)

LDM

Results of ELH analysis, 1st pass

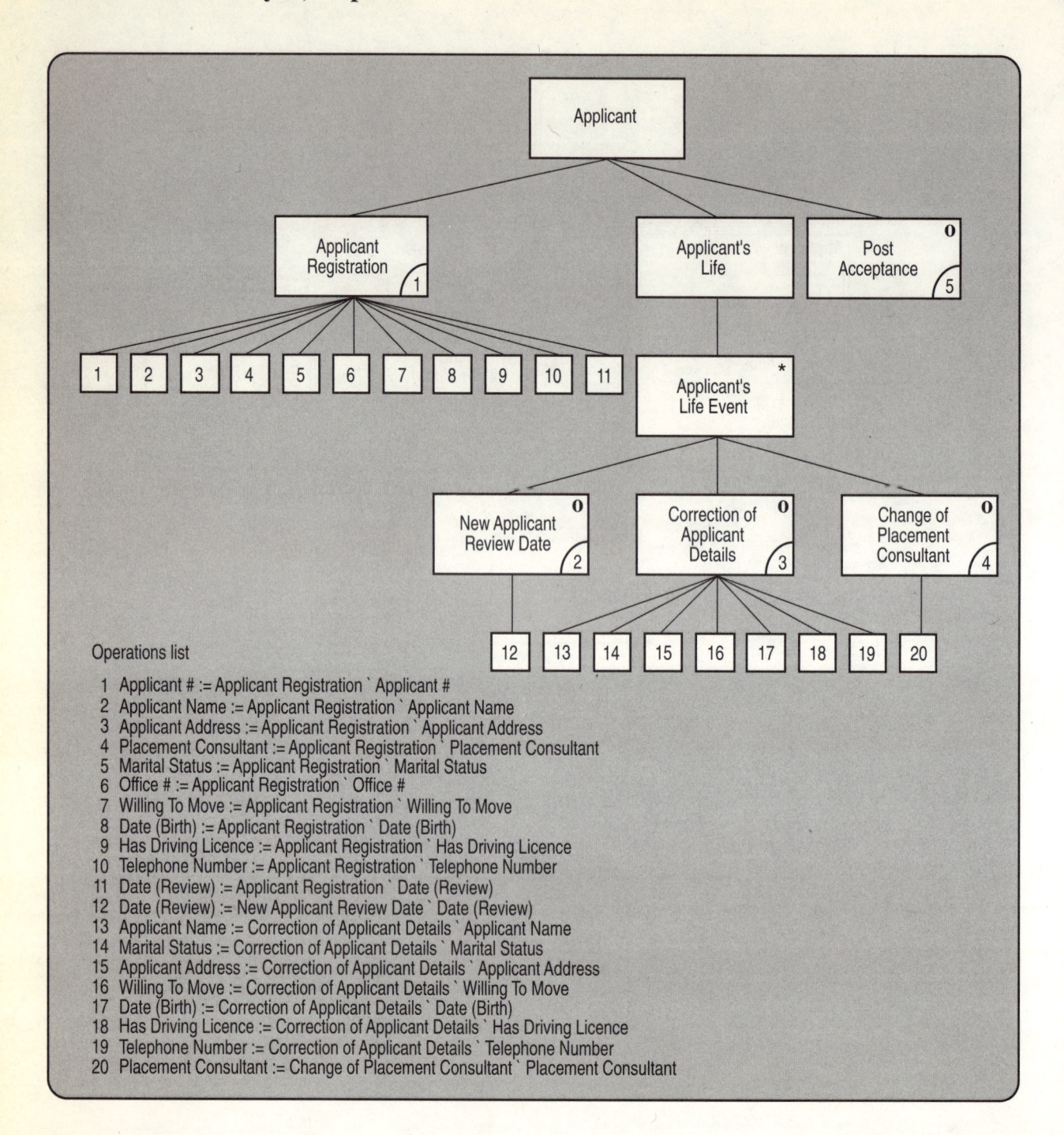

Operations list

1. Applicant # := Applicant Registration ` Applicant #
2. Applicant Name := Applicant Registration ` Applicant Name
3. Applicant Address := Applicant Registration ` Applicant Address
4. Placement Consultant := Applicant Registration ` Placement Consultant
5. Marital Status := Applicant Registration ` Marital Status
6. Office # := Applicant Registration ` Office #
7. Willing To Move := Applicant Registration ` Willing To Move
8. Date (Birth) := Applicant Registration ` Date (Birth)
9. Has Driving Licence := Applicant Registration ` Has Driving Licence
10. Telephone Number := Applicant Registration ` Telephone Number
11. Date (Review) := Applicant Registration ` Date (Review)
12. Date (Review) := New Applicant Review Date ` Date (Review)
13. Applicant Name := Correction of Applicant Details ` Applicant Name
14. Marital Status := Correction of Applicant Details ` Marital Status
15. Applicant Address := Correction of Applicant Details ` Applicant Address
16. Willing To Move := Correction of Applicant Details ` Willing To Move
17. Date (Birth) := Correction of Applicant Details ` Date (Birth)
18. Has Driving Licence := Correction of Applicant Details ` Has Driving Licence
19. Telephone Number := Correction of Applicant Details ` Telephone Number
20. Placement Consultant := Change of Placement Consultant ` Placement Consultant

Results of ELH analysis, 1st pass

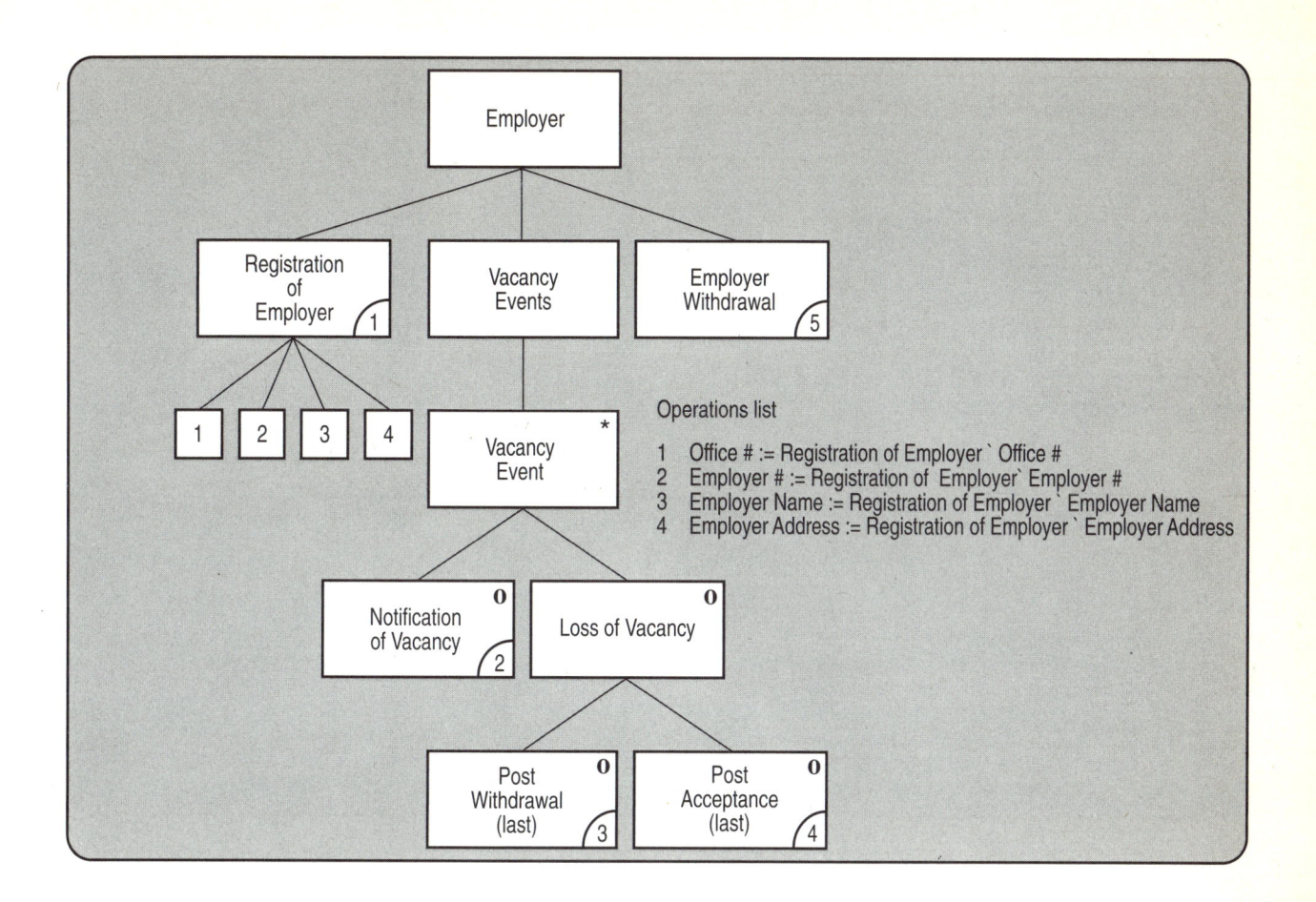

```
                    ┌──────────┐
                    │ Employer │
                    └────┬─────┘
         ┌───────────────┼───────────────┐
   ┌──────────┐    ┌──────────┐    ┌──────────┐
   │Registration│   │ Vacancy  │    │ Employer │
   │    of     │   │  Events  │    │Withdrawal│
   │ Employer 1│   └────┬─────┘    └────────5 │
   └──┬──┬──┬──┘        │
      │  │  │  │        │
    ┌─┬─┬─┬─┐      ┌─────────┐ *
    │1│2│3│4│      │ Vacancy │
    └─┴─┴─┴─┘      │  Event  │
                   └────┬────┘
                ┌───────┴────────┐
           ┌─────────┐ 0    ┌─────────┐ 0
           │Notification│   │ Loss of │
           │ of Vacancy │   │ Vacancy │
           └──────────2┘   └────┬────┘
                        ┌───────┴───────┐
                   ┌─────────┐ 0   ┌─────────┐ 0
                   │  Post   │     │  Post   │
                   │Withdrawal│    │Acceptance│
                   │ (last) 3│     │ (last) 4│
                   └─────────┘     └─────────┘
```

Operations list

1 Office # := Registration of Employer ` Office #
2 Employer # := Registration of Employer ` Employer #
3 Employer Name := Registration of Employer ` Employer Name
4 Employer Address := Registration of Employer ` Employer Address

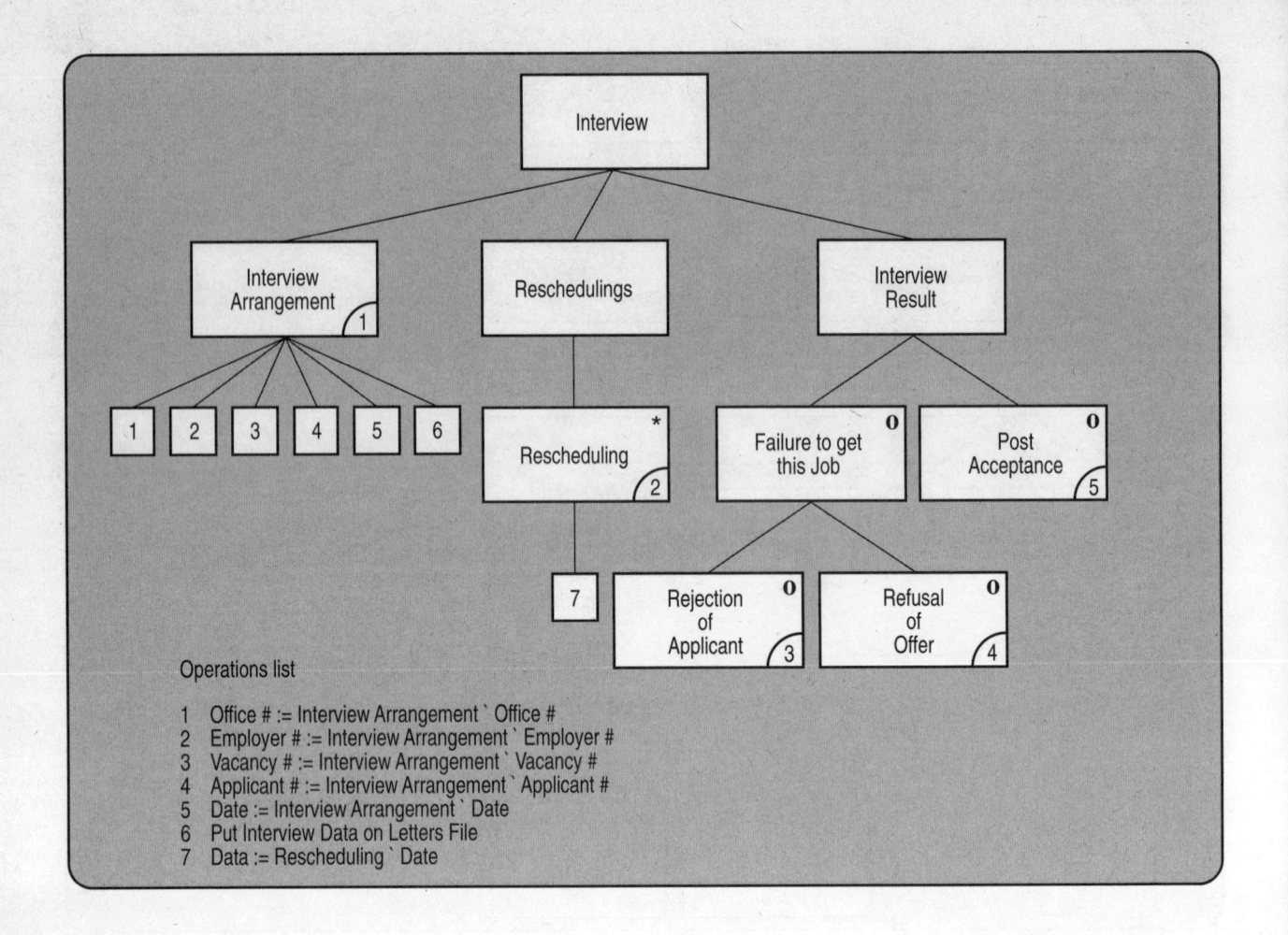

```
Interview
├── Interview Arrangement (1)
│   ├── 1
│   ├── 2
│   ├── 3
│   ├── 4
│   ├── 5
│   └── 6
├── Reschedulings
│   └── Rescheduling * (2)
│       └── 7
└── Interview Result
    ├── Failure to get this Job °
    │   ├── Rejection of Applicant ° (3)
    │   └── Refusal of Offer ° (4)
    └── Post Acceptance ° (5)
```

Operations list

1 Office # := Interview Arrangement ` Office #
2 Employer # := Interview Arrangement ` Employer #
3 Vacancy # := Interview Arrangement ` Vacancy #
4 Applicant # := Interview Arrangement ` Applicant #
5 Date := Interview Arrangement ` Date
6 Put Interview Data on Letters File
7 Data := Rescheduling ` Date

Results of ELH analysis, 1st pass

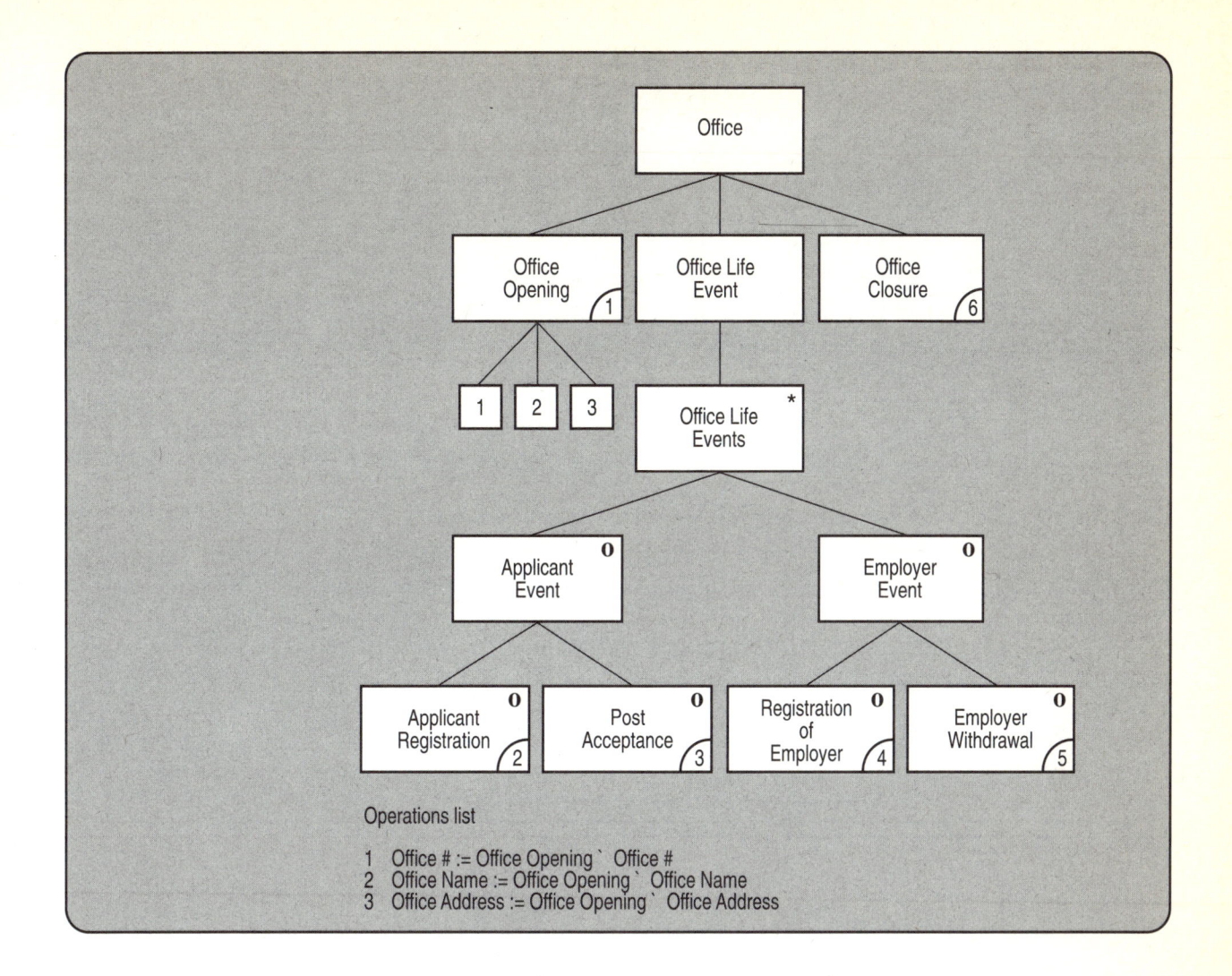

```
                              Office
                                │
         ┌────────────────────┬─┴──────────────────┐
      Office              Office Life            Office
      Opening  (1)        Event                  Closure (6)
         │                   │
      ┌──┼──┐             Office Life *
      1  2  3             Events
                             │
                    ┌────────┴────────┐
                 Applicant (o)     Employer (o)
                 Event             Event
                    │                 │
              ┌─────┴─────┐      ┌────┴─────┐
           Applicant(o)  Post(o) Registration(o) Employer(o)
           Registration  Acceptance  of          Withdrawal
              (2)        (3)     Employer (4)    (5)
```

Operations list

1   Office # := Office Opening ` Office #
2   Office Name := Office Opening ` Office Name
3   Office Address := Office Opening ` Office Address

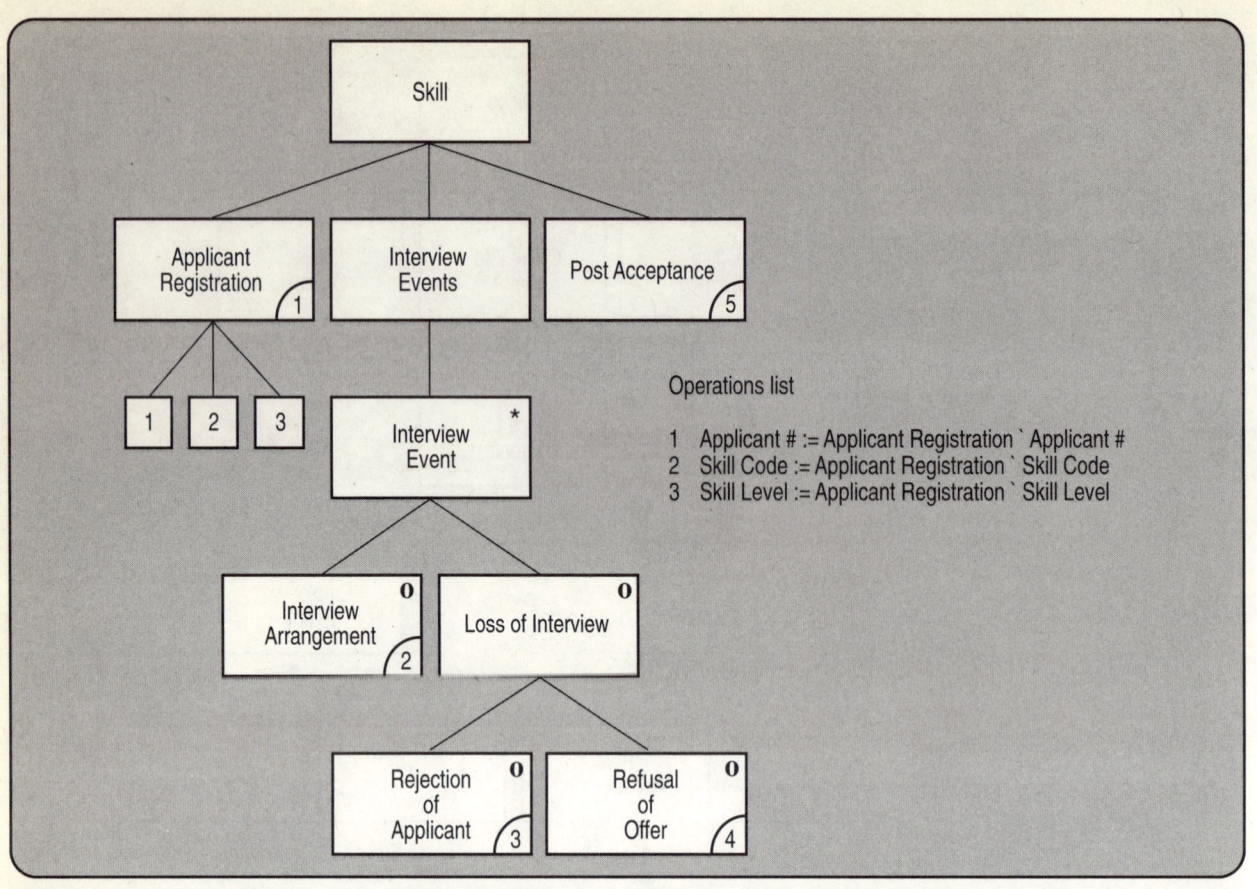

Skill (ELH)

```
                    Skill
        ┌─────────────┼─────────────┐
  Applicant      Interview      Post Acceptance
  Registration   Events              (5)
      (1)            │
   ┌───┼───┐         │
  [1] [2] [3]   Interview Event *
                    ┌────┴────┐
              Interview    Loss of
              Arrangement  Interview
                  (2)         │
                         ┌────┴────┐
                     Rejection  Refusal
                     of         of
                     Applicant  Offer
                       (3)       (4)
```

Operations list

1 Applicant # := Applicant Registration ` Applicant #
2 Skill Code := Applicant Registration ` Skill Code
3 Skill Level := Applicant Registration ` Skill Level

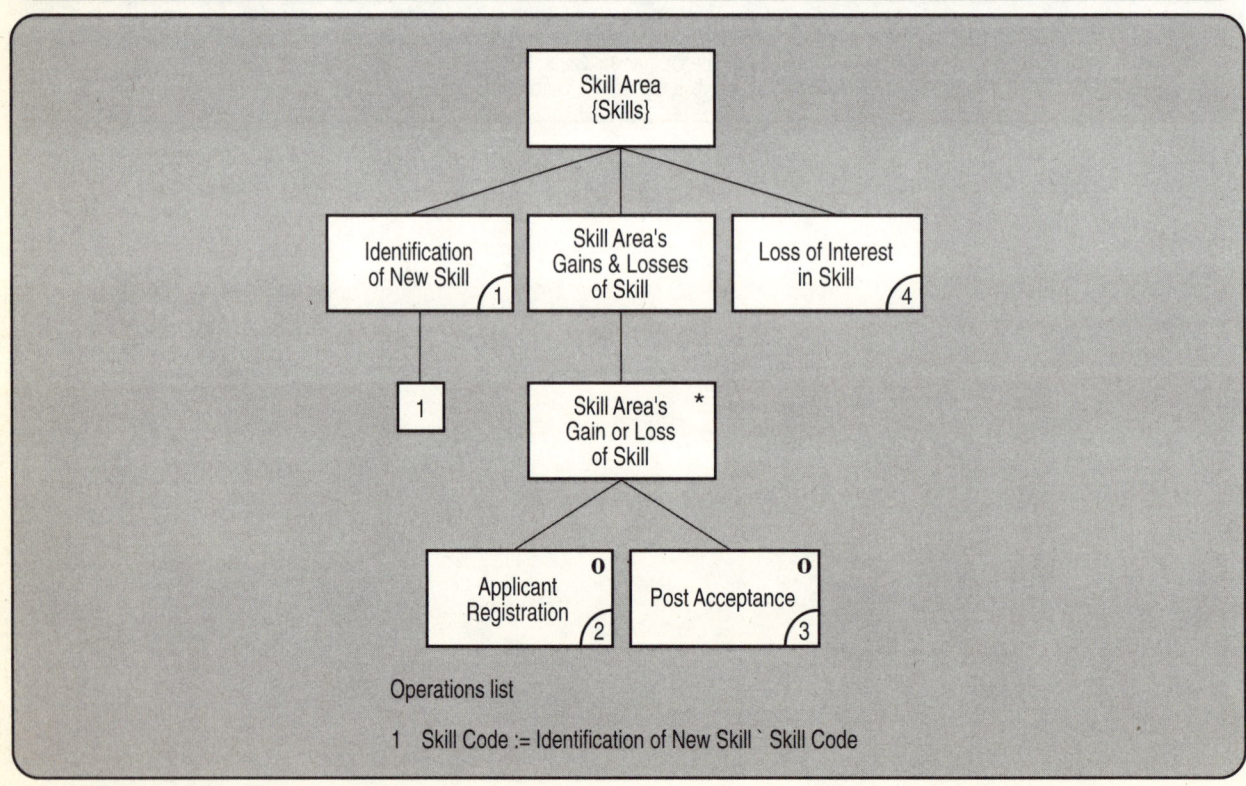

Skill Area {Skills} (ELH)

```
                Skill Area
                {Skills}
        ┌───────────┼────────────┐
  Identification  Skill Area's   Loss of Interest
  of New Skill    Gains & Losses in Skill
      (1)         of Skill          (4)
       │
      [1]      Skill Area's *
               Gain or Loss
               of Skill
                 ┌────┴────┐
             Applicant   Post Acceptance
             Registration    (3)
                 (2)
```

Operations list

1 Skill Code := Identification of New Skill ` Skill Code

Results of ELH analysis, 1st pass

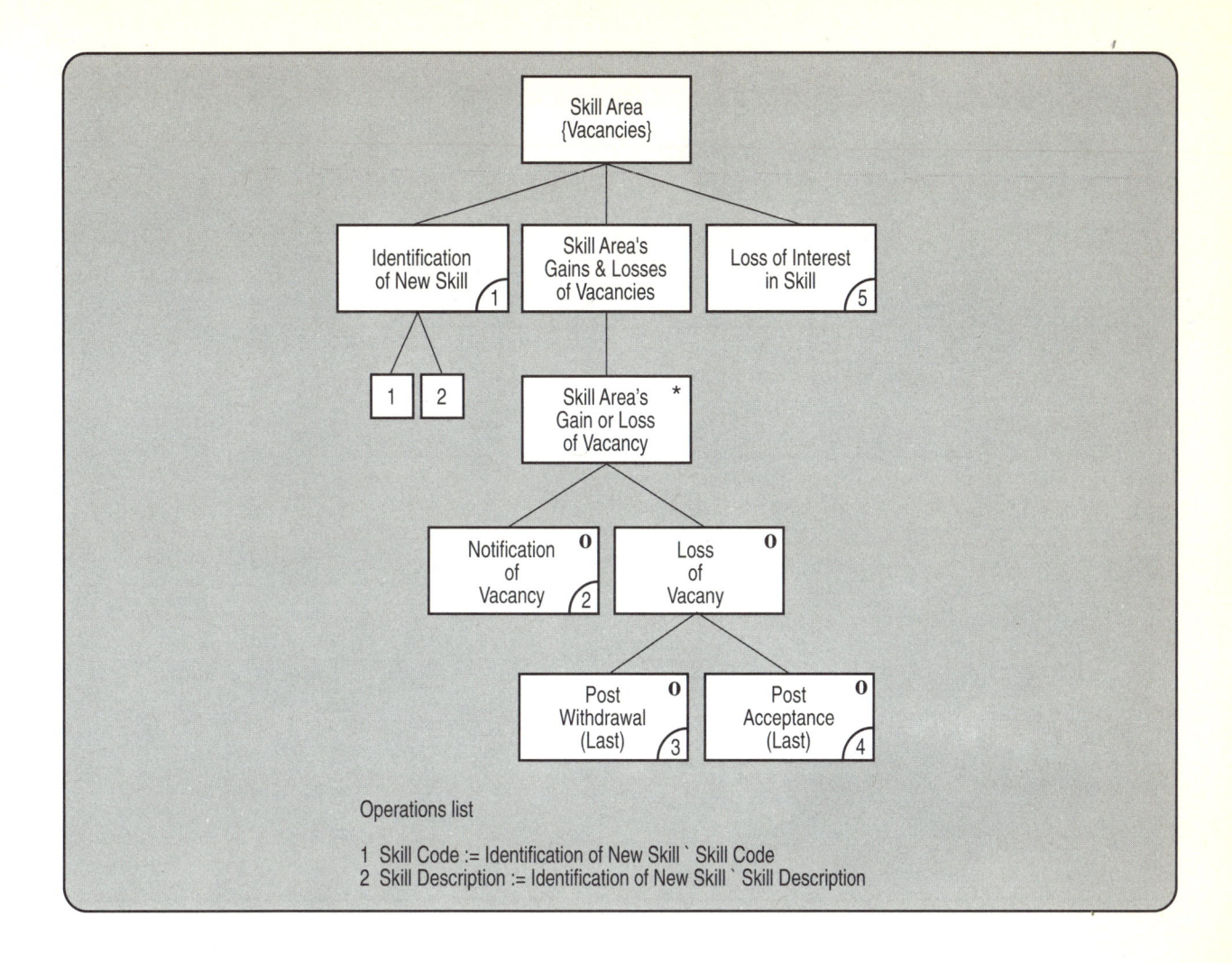

```
                    ┌─────────────┐
                    │  Skill Area │
                    │ {Vacancies} │
                    └──────┬──────┘
          ┌────────────────┼────────────────┐
┌─────────────────┐ ┌─────────────────┐ ┌─────────────────┐
│ Identification  │ │  Skill Area's   │ │ Loss of Interest│
│  of New Skill  1│ │ Gains & Losses  │ │    in Skill    5│
└────────┬────────┘ │  of Vacancies   │ └─────────────────┘
      ┌──┴──┐       └────────┬────────┘
     ┌─┐ ┌─┐         ┌───────────────┐ *
     │1│ │2│         │ Skill Area's  │
     └─┘ └─┘         │ Gain or Loss  │
                     │  of Vacancy   │
                     └───────┬───────┘
                    ┌────────┴────────┐
           ┌────────────────┐o ┌────────────────┐o
           │  Notification  │  │     Loss       │
           │      of        │  │      of        │
           │    Vacancy    2│  │    Vacany      │
           └────────────────┘  └────────┬───────┘
                               ┌────────┴────────┐
                      ┌────────────────┐o ┌────────────────┐o
                      │     Post       │  │     Post       │
                      │   Withdrawal   │  │   Acceptance   │
                      │    (Last)     3│  │    (Last)     4│
                      └────────────────┘  └────────────────┘
```

Operations list

1 Skill Code := Identification of New Skill ` Skill Code
2 Skill Description := Identification of New Skill ` Skill Description

Vacancy

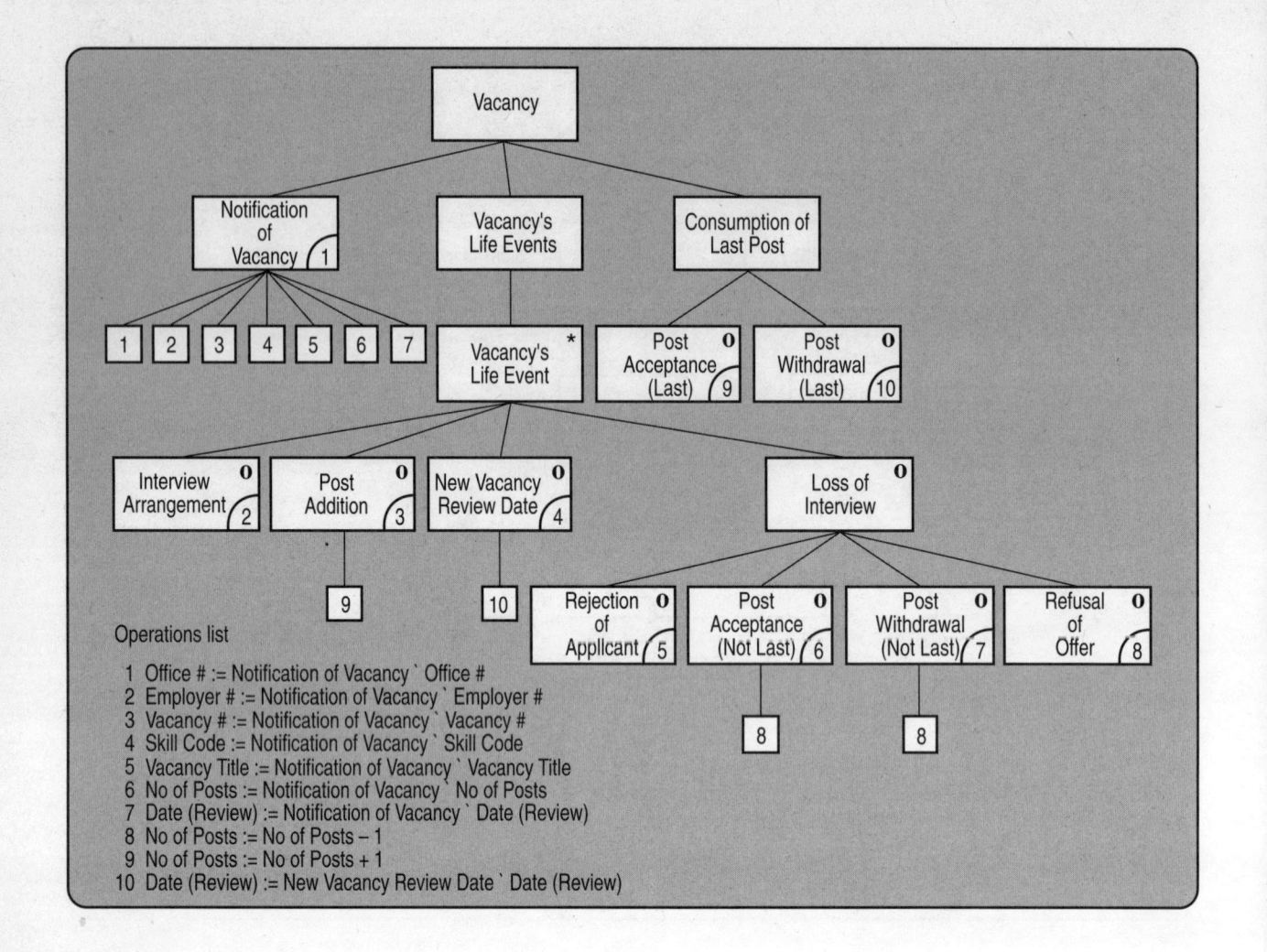

- **Notification of Vacancy** (1)
  - 1, 2, 3, 4, 5, 6, 7
- **Vacancy's Life Events**
  - **Vacancy's Life Event** *
    - Interview Arrangement (2) — O
    - Post Addition (3) — O → 9
    - New Vacancy Review Date (4) — O → 10
    - Loss of Interview — O
      - Rejection of Applicant (5) — O
      - Post Acceptance (Not Last) (6) — O → 8
      - Post Withdrawal (Not Last) (7) — O → 8
      - Refusal of Offer (8) — O
- **Consumption of Last Post**
  - Post Acceptance (Last) (9) — O
  - Post Withdrawal (Last) (10) — O

Operations list

1. Office # := Notification of Vacancy ` Office #
2. Employer # := Notification of Vacancy ` Employer #
3. Vacancy # := Notification of Vacancy ` Vacancy #
4. Skill Code := Notification of Vacancy ` Skill Code
5. Vacancy Title := Notification of Vacancy ` Vacancy Title
6. No of Posts := Notification of Vacancy ` No of Posts
7. Date (Review) := Notification of Vacancy ` Date (Review)
8. No of Posts := No of Posts − 1
9. No of Posts := No of Posts + 1
10. Date (Review) := New Vacancy Review Date ` Date (Review)

Further DFD & LDM documentation

Process Defs 1

Arrange Interviews for Applicants	Register new Applicants, find Vacancies for them, and arrange Interviews.
Arrange Interviews for Vacancies	Act on requests from Sales Executives to find suitable Applicants for a new Vacancy and arrange Interviews.
Do Follow-ups	Follow up on Interviews, stale Vacancies & unplaced Applicants.
Find Applicants For Vacancy	Find all Applicants skilled in the Skill Area required by the Vacancy.
Find Vacancies For Applicant	Find all Vacancies requiring the Skill Areas recorded in the Applicant's Skills.
Follow Up Interviews	Retrieve details of Interviews more than one week old which have still not had any Post Acceptance, Rejection of Applicant or Refusal of Offer and request result from Employer.
Follow Up Vacancies	Retrieve Vacancies which have not been filled and are past their Review Date and request information from Employer about Vacancy status.
Hi-Ho Functions	The Required Hi-Ho System.
Maintain Administrative Data	Maintain Information about Offices and Skills.
Maintain Applicant Data	Record changes to Applicant Details.
Maintain Employers & Vacancies	Maintain details about Employers and Vacancies.
Notify Placement Consultant	Request suitable applicant search by Placement Consultant.
Record Applicant Rejections	Record Interview as having taken place but failed to result in placement.
Record Applicant Withdrawals	Record Applicants withdrawing from Hi-Ho, and kill their Applicant Skills and any outstanding Interviews.
Record Arranged Interview	Record the date of an Interview set up for a particular Applicant and a particular Vacancy.
Record Employer & Vacancy Withdrawals	Record Employers & Vacancies withdrawing from Hi-Ho, and kill any outstanding Interviews.
Record Interview Results	Record results of Interviews from information returned through the postal system by the Employers.
Record Loss of Interest in Skills	Delete obsolete skills.

Record New Employer	Record details about new Employer.
Record New Skills	Record codes and descriptions of newly recognised skills.
Record New Vacancy	Record details about new Vacancy.
Record Offer Refusals	Record refusals of offers by Applicant.
Record Office Closures	Delete closed Hi-Ho Offices.
Record Office Openings	Create records of new Hi-Ho Offices.
Record Post Acceptances	Record Post Acceptances and take whatever consequential action is necessary.
Record Post Addition	Record the fact that an extra post has been added to a Vacancy.
Record Post Withdrawal	Record the fact that a post has been withdrawn from a Vacancy.

Process Defs 2

Register Applicant	Record a new Applicant and classify and record the Applicant's Skills.
Request Applicant Attendance	Retrieve Applicants who have not been placed and are past their Review Date, and request attendance at Hi-Ho Office for consultation with Placement Consultant.
Reschedule Interviews	Record postponement (or, in exceptional cases advancement) of Interviews.
Send Letters	Formally notify Applicants and Employers about Interviews.
Set New Applicant Review Date	Record new review date.
Set New Vacancy Review Date	Record new review date.
Sort Interviews by Applicant and Print	Create Applicant Appointment letters.
Sort Interviews by Employer and Print	Create Employer Interview schedules.

Ext Ent Defs 1

Accounts	That part of Hi-Ho which is concerned with billing Employers and accepting payments.
Applicant	A person seeking a job.
Employer	The person or organisation offering employment.
Hi-Ho Management	That part of Hi-Ho concerned with policy about where the business is done and what kinds of recruitment it covers.
Placement Consultant	An employee of Hi-Ho who is responsible for matching Applicants to Vacancies and arranging Interviews with Employers.
Post Bag	The public postal system.
Sales Executive	An employee of Hi-Ho who persuades Employers to buy Hi-Ho's recruitment services.

Data Store Defs 1

Applicant Skills	Records of Applicant's Skills.
Applicants	Records of the Applicants registered with a particular Hi-Ho office.
Interviews	Records of arranged Interviews.
Vacancies	Records of Employers and their current Vacancies.

Entity Defs 1

Applicant	A person, registered with Hi-Ho, who is seeking a job.	200000	250000	300000	250000
Date	A calendar date. About three month's worth of days are of interest to the system at any point in time.	90	90	90	90
Employer	An employer registered with Hi-Ho for the purposes of recruitment to fill vacancies.	15000	15000	15000	15000
Interview	Interview arranged for Applicant with prospective Employer.	1500000	1700000	1800000	1800000
Office	A Hi-Ho branch office.	150	150	150	150
Skill	An actual skill possessed by an Applicant.	400000	600000	700000	600000
Skill Area	A kind of skill, coded so that requirements of vacancies can be matched against suitable applicants.	16	16	16	16
Vacancy	A set of available posts of the same type offered by an Employer at a given point in time. For example, a Vacancy for 3 programmers. Each post acceptance reduces the number of available posts by one.	40000	45000	50000	45000

Relationship Defs 1

Applicant	Skill	acquired by	experienced in	Mandatory	2	2.4	3.5	2.4
Employer	Vacancy	employment opportunity at	seeking to fill	Mandatory	2.7	3	3.3	3
Office	Applicant	registered at	employment agent for	Mandatory	1333	1667	2000	1667
Office	Employer	registered at	recruitment agent for	Mandatory	100	100	100	100
Skill	Interview	arranged to assess	precondition for	Mandatory	2	3	4.5	3
Skill Area	Skill	measure of capability in	manifested in	Mandatory	25000	37500	43750	37500
Skill Area	Vacancy	requirement for employees with capability in	required by	Mandatory	2500	2812	3125	2812
Vacancy	Interview	arranged to provide candidate for	filled by candidate selection at	Mandatory	37.5	37.8	45	40

Data Item Defs 1

Applicant #	Identifier of Applicant.	9(11)
Applicant Adress	Postal address of Applicant.	X(120)
Applicant Name	Applicant's Name.	X(40)
Date	Day, Month & Year.	9(6)
Employer #	Identifier of an Employer placing vacancies with Hi-Ho.	Z(8)9
Employer Adress	Postal address of Employer.	X(120)
Employer Name	Name of Employer registering jobs with Hi-Ho.	X(50)
Has Driving Licence	Flag with values: 0 - No Licence 1 - Has Licence.	B
Marital Status	Flag indicating marital status: 1 - Single 2 - Married 3 - Divorced 4 - Widowed	9
No of Posts	Number of posts currently unfilled on a Vacancy.	99
Office #	Identifier of Hi-Ho local office.	9(6)
Office Adress	Address of Hi-Ho office.	X(120)
Office Name	Local name of Hi-Ho office.	X(40)
Placement Consultant	Identifier of the placement consultant dealing with the applicant.	9(6)

Further DFD & LDM documentation

Skill Code	Standard code used to classify a particular skill for the purpose of matching applicants and vacancies in Hi-Ho's business.	9(6)
Skill Description	Textual explanation of meaning of Skill Code.	X(100)
Skill Level	Measure of level of skill achieved: 1. Less than 1 year's experience. 2. 1 - 5 years' experience. 3. > 5 years' experience.	9
SV	State Indicator, showing point entity has currently reached in its life history.	XX
Telephone Number	Telephone number.	9(9)
Vacancy #	Serial number within Employer # of Vacancy registered by Employer.	9(5)
Vacancy Title	Employer's description of type of work involved in a Vacancy.	X(100)
Willing To Move	Flag with values: 0 - Not Willing To Move 1 - Willing to Move.	B

Entity life histories after 2nd pass

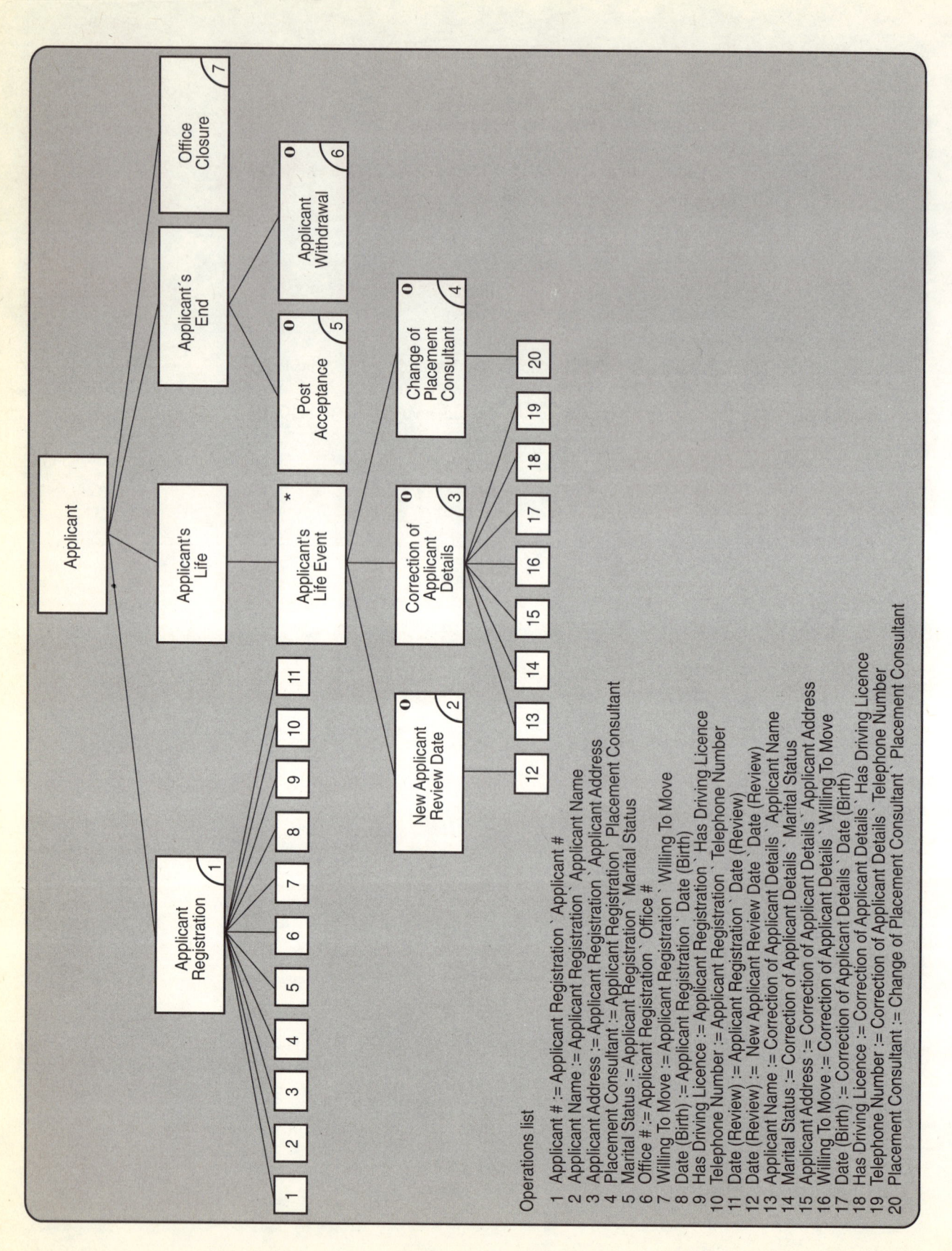

Operations list

1. Applicant # := Applicant Registration ` Applicant #
2. Applicant Name := Applicant Registration ` Applicant Name
3. Applicant Address := Applicant Registration ` Applicant Address
4. Placement Consultant := Applicant Registration ` Placement Consultant
5. Marital Status := Applicant Registration ` Marital Status
6. Office # := Applicant Registration ` Office #
7. Willing To Move := Applicant Registration ` Willing To Move
8. Date (Birth) := Applicant Registration ` Date (Birth)
9. Has Driving Licence := Applicant Registration ` Has Driving Licence
10. Telephone Number := Applicant Registration ` Telephone Number
11. Date (Review) := Applicant Registration ` Date (Review)
12. Date (Review) := New Applicant Review Date ` Date (Review)
13. Applicant Name := Correction of Applicant Details ` Applicant Name
14. Marital Status := Correction of Applicant Details ` Marital Status
15. Applicant Address := Correction of Applicant Details ` Applicant Address
16. Willing To Move := Correction of Applicant Details ` Willing To Move
17. Date (Birth) := Correction of Applicant Details ` Date (Birth)
18. Has Driving Licence := Correction of Applicant Details ` Has Driving Licence
19. Telephone Number := Correction of Applicant Details ` Telephone Number
20. Placement Consultant := Change of Placement Consultant ` Placement Consultant

Entity life histories after 2nd pass

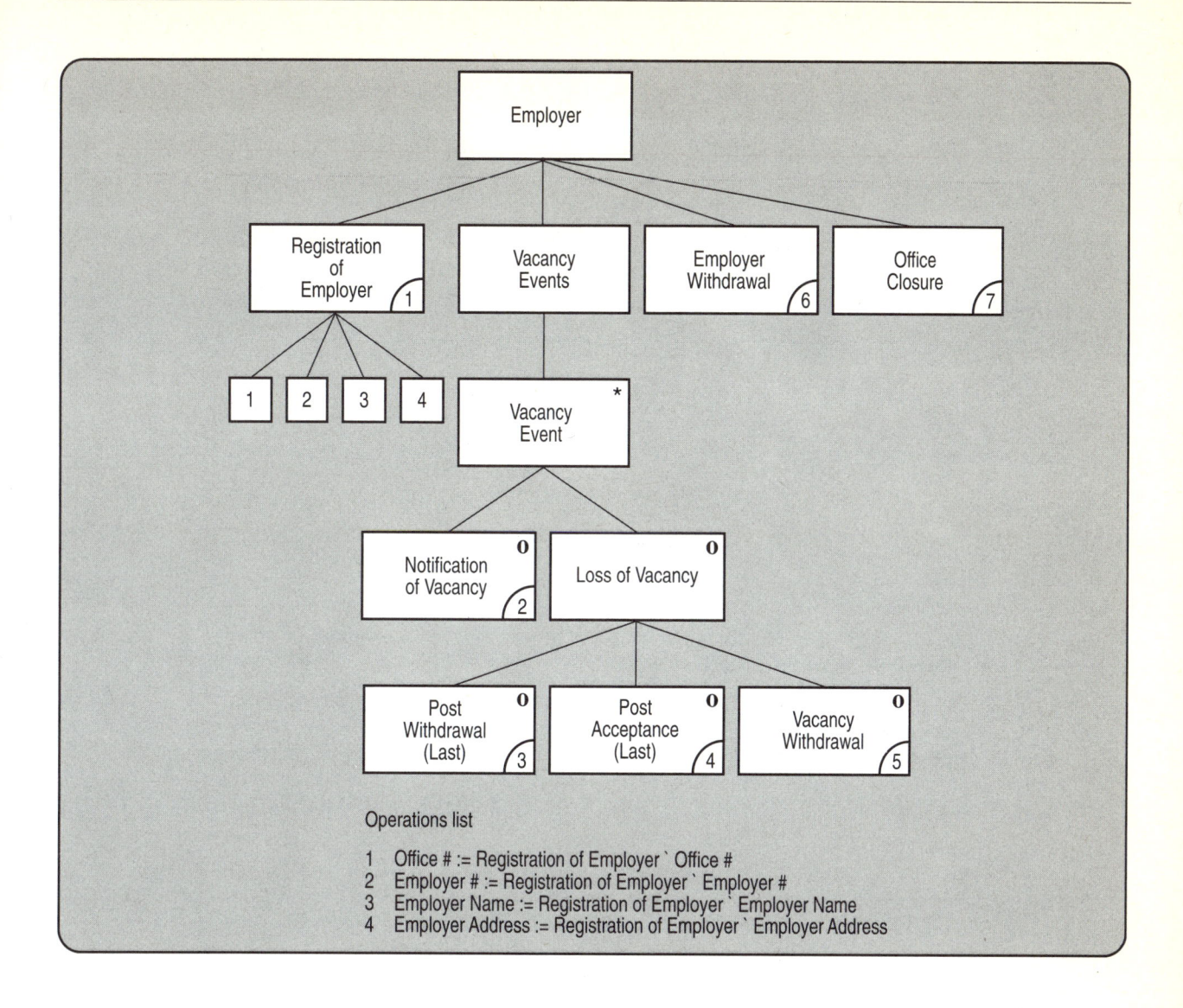

```
                        ┌──────────┐
                        │ Employer │
                        └────┬─────┘
         ┌──────────────┬────┴──────┬──────────────┐
   ┌─────────────┐ ┌──────────┐ ┌──────────┐ ┌──────────┐
   │Registration │ │ Vacancy  │ │ Employer │ │  Office  │
   │     of      │ │  Events  │ │Withdrawal│ │ Closure  │
   │  Employer  1│ │          │ │         6│ │         7│
   └──┬───┬───┬──┘ └────┬─────┘ └──────────┘ └──────────┘
    ┌─┐ ┌─┐ ┌─┐ ┌─┐      │
    │1│ │2│ │3│ │4│  ┌───┴────┐ *
    └─┘ └─┘ └─┘ └─┘  │Vacancy │
                    │ Event  │
                    └───┬────┘
               ┌────────┴────────┐
         ┌───────────┐0    ┌───────────┐0
         │Notification│    │Loss of    │
         │of Vacancy  │    │Vacancy    │
         │          2│    │           │
         └───────────┘    └─────┬─────┘
              ┌─────────────────┼─────────────────┐
        ┌──────────┐0    ┌──────────┐0    ┌──────────┐0
        │  Post    │    │  Post    │    │ Vacancy  │
        │Withdrawal│    │Acceptance│    │Withdrawal│
        │ (Last)  3│    │ (Last)  4│    │         5│
        └──────────┘    └──────────┘    └──────────┘
```

Operations list

1   Office # := Registration of Employer ` Office #
2   Employer # := Registration of Employer ` Employer #
3   Employer Name := Registration of Employer ` Employer Name
4   Employer Address := Registration of Employer ` Employer Address

154   Annex 2 Hi-Ho SSADM Development

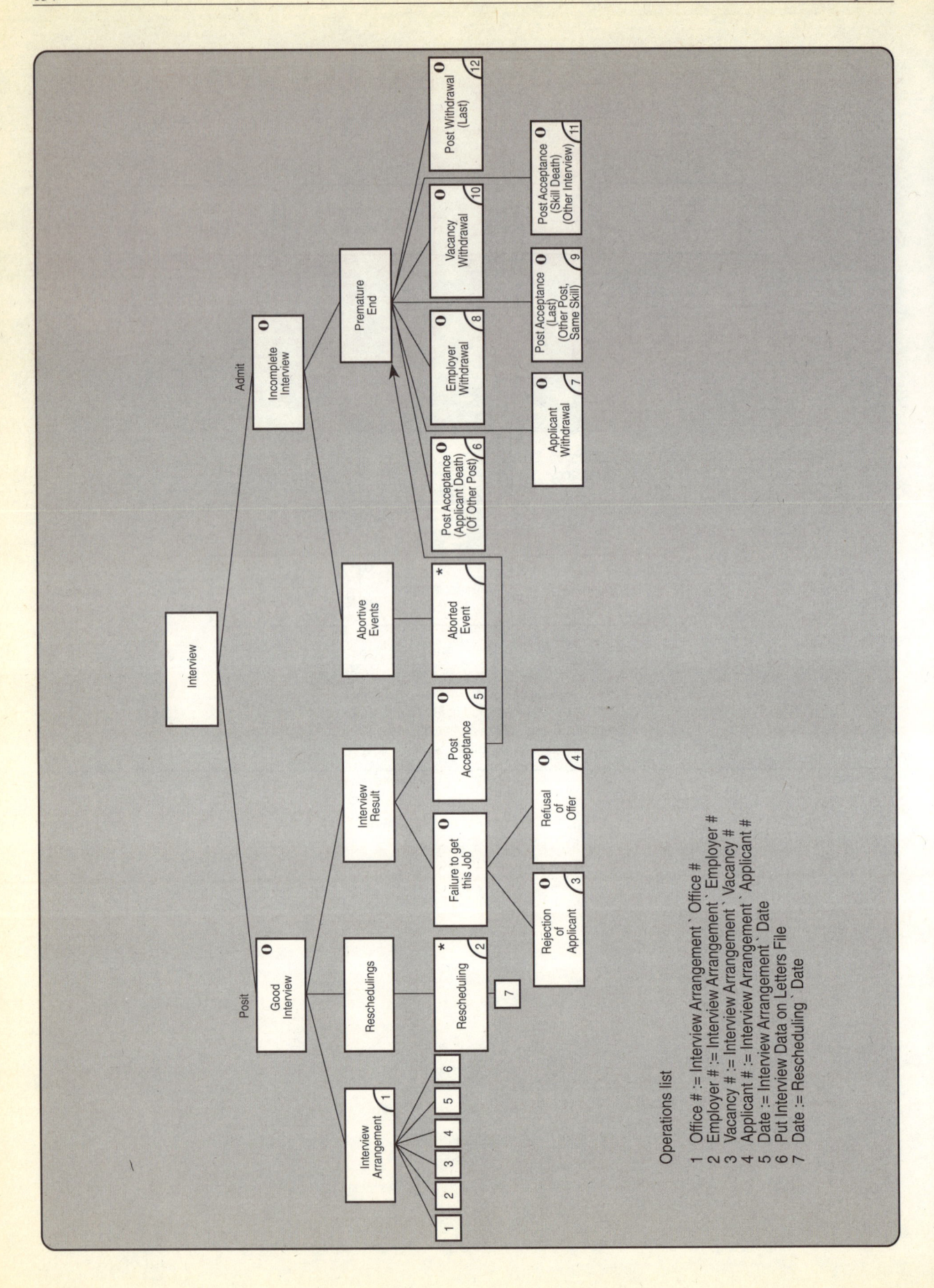

```
                                    Interview
                          ┌────────────┴────────────┐
                       Post                       Admit
                        │                           │
                  Good Interview              Incomplete
                     │                         Interview
         ┌───────────┼───────────┐                 │
    Reschedulings  Interview  Premature End
         │          Result          │
      Rescheduling*    │     ┌──────┼──────┬──────────┬──────────┐
         │       ┌─────┴─────┐  Post    Applicant  Employer  Vacancy   Post Withdrawal
        [7]   Post      Failure  Acceptance Withdrawal Withdrawal Withdrawal    (Last)
              Acceptance  to get  (Applicant    O         O          O           O 12
                 O 5     this Job  Death)      7          8         10
                         O       (Of Other Post)
                      ┌──┴──┐     O 6                Post Acceptance
                  Rejection Refusal                  (Last)        Post Acceptance
                  of       of Offer                  (Other Post,  (Skill Death)
                  Applicant O 4                      Same Skill)   (Other Interview)
                  O 3                                O 9           O 11
```

```
Interview Arrangement  O 1
   │
  [1][2][3][4][5][6]
```

```
Aborted Event *
```

Operations list

1   Office # := Interview Arrangement ` Office #
2   Employer # := Interview Arrangement ` Employer #
3   Vacancy # := Interview Arrangement ` Vacancy #
4   Applicant # := Interview Arrangement ` Applicant #
5   Date := Interview Arrangement ` Date
6   Put Interview Data on Letters File
7   Date := Rescheduling ` Date

Entity life histories after 2nd pass

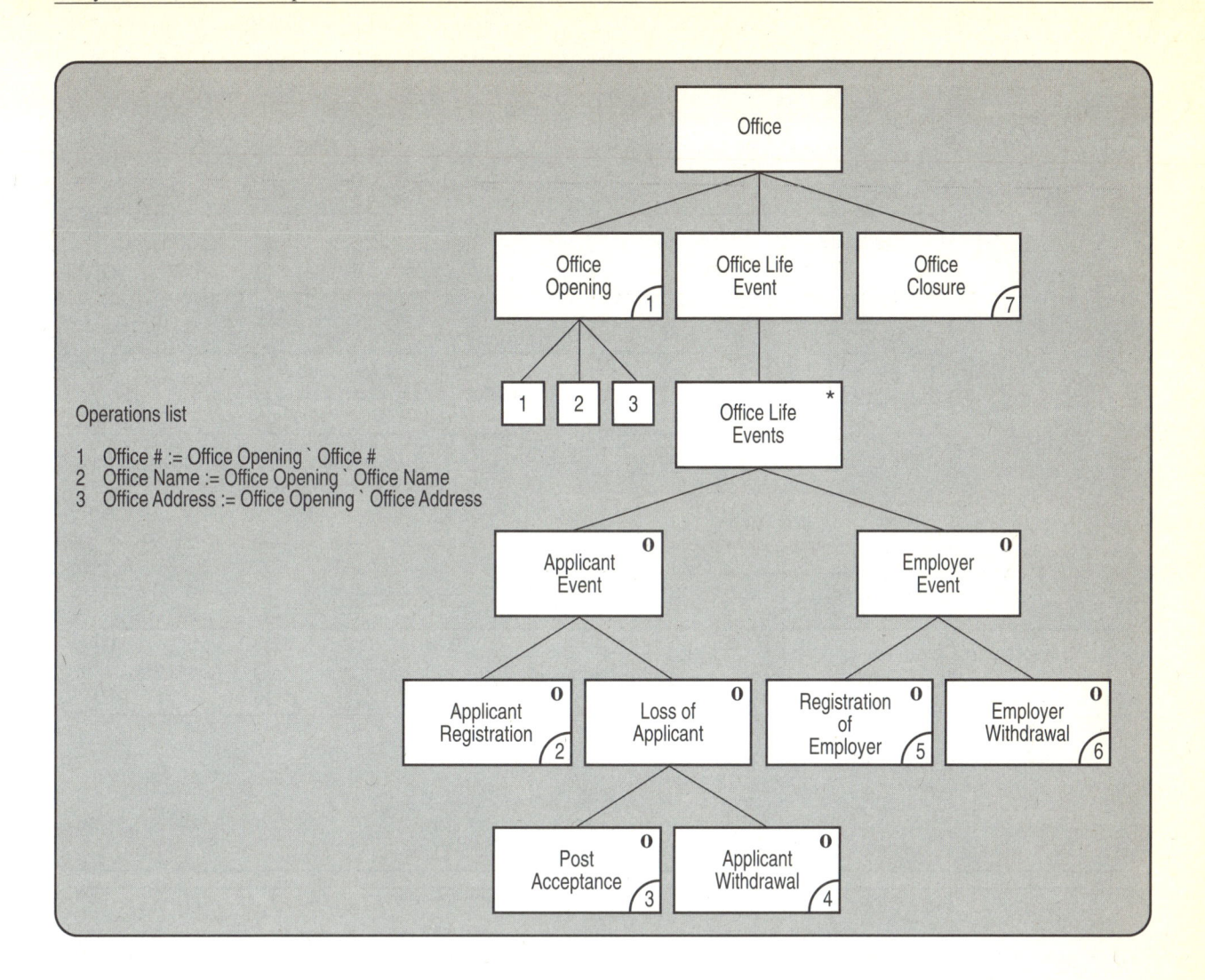

```
                            ┌──────────┐
                            │  Office  │
                            └──────────┘
                   ┌─────────────┼─────────────┐
            ┌──────────┐  ┌──────────┐  ┌──────────┐
            │  Office  │  │Office Life│ │  Office  │
            │ Opening 1│  │  Event   │  │ Closure 7│
            └──────────┘  └──────────┘  └──────────┘
              ┌──┼──┐           │
             [1][2][3]          │
                         ┌──────────┐*
                         │Office Life│
                         │  Events  │
                         └──────────┘
                    ┌──────────┴──────────┐
              ┌──────────┐o         ┌──────────┐o
              │ Applicant│          │ Employer │
              │  Event   │          │  Event   │
              └──────────┘          └──────────┘
              ┌─────┴─────┐         ┌─────┴─────┐
        ┌─────────┐o ┌─────────┐o ┌──────────┐o ┌─────────┐o
        │Applicant│  │ Loss of │  │Registration│ │Employer │
        │Registr. 2│ │Applicant│  │    of    5│ │Withdr.  6│
        └─────────┘  └─────────┘  │ Employer │  └─────────┘
                    ┌─────┴─────┐ └──────────┘
              ┌─────────┐o ┌─────────┐o
              │  Post   │  │Applicant│
              │Acceptance3│ │Withdraw.4│
              └─────────┘  └─────────┘
```

Operations list

1 Office # := Office Opening ` Office #
2 Office Name := Office Opening ` Office Name
3 Office Address := Office Opening ` Office Address

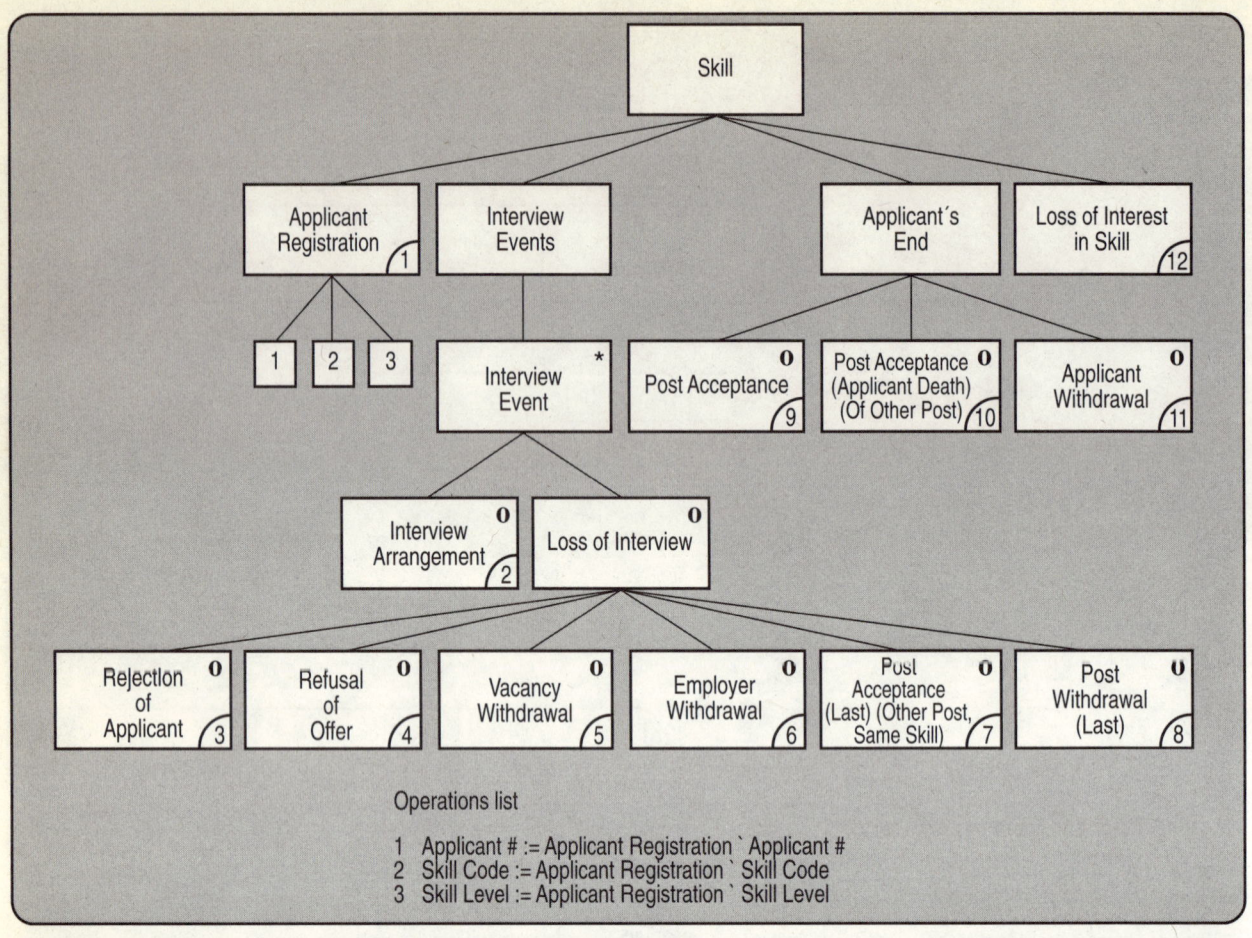

Annex 2 Hi-Ho SSADM Development

Skill

- Skill
  - Applicant Registration (1)
    - 1, 2, 3
  - Interview Events
    - Interview Event *
      - Interview Arrangement (2) o
      - Loss of Interview o
        - Rejection of Applicant (3) o
        - Refusal of Offer (4) o
        - Vacancy Withdrawal (5) o
        - Employer Withdrawal (6) o
        - Post Acceptance (Last) (Other Post, Same Skill) (7) o
        - Post Withdrawal (Last) (8) o
  - Post Acceptance (9) o
  - Applicant's End
    - Post Acceptance (Applicant Death) (Of Other Post) (10) o
    - Applicant Withdrawal (11) o
  - Loss of Interest in Skill (12) o

Operations list

1. Applicant # := Applicant Registration ` Applicant #
2. Skill Code := Applicant Registration ` Skill Code
3. Skill Level := Applicant Registration ` Skill Level

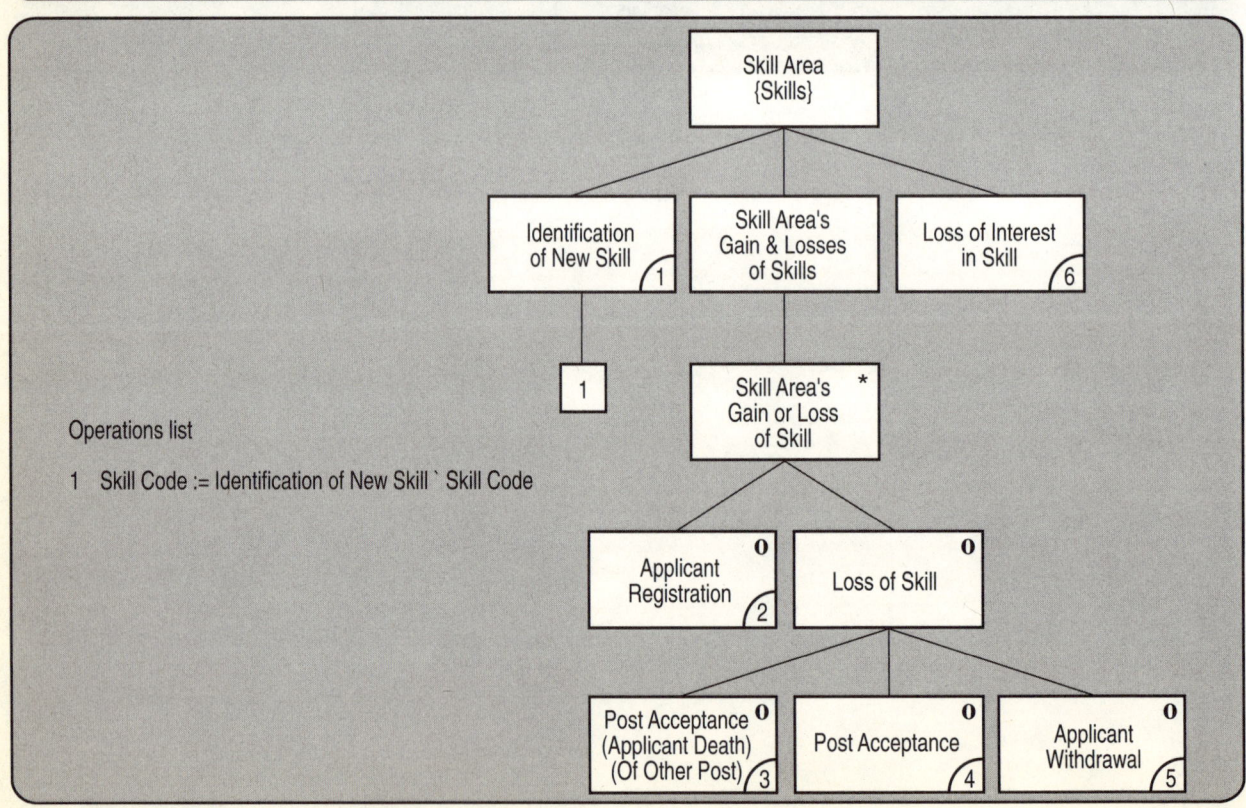

Skill Area {Skills}

- Skill Area {Skills}
  - Identification of New Skill (1)
    - 1
  - Skill Area's Gain & Losses of Skills
    - Skill Area's Gain or Loss of Skill *
      - Applicant Registration (2) o
      - Loss of Skill o
        - Post Acceptance (Applicant Death) (Of Other Post) (3) o
        - Post Acceptance (4) o
        - Applicant Withdrawal (5) o
  - Loss of Interest in Skill (6) o

Operations list

1. Skill Code := Identification of New Skill ` Skill Code

Entity life histories after 2nd pass

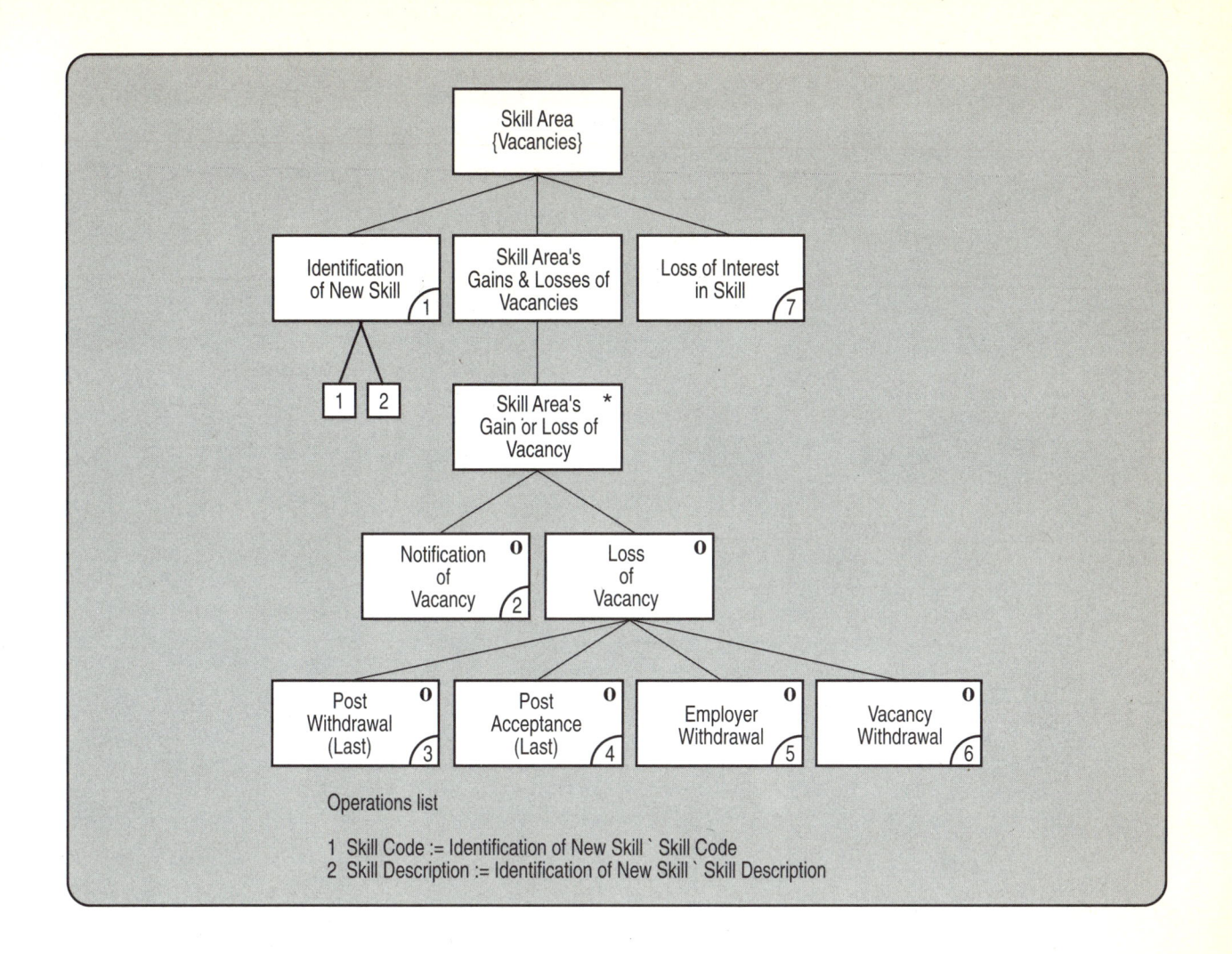

```
                    ┌─────────────┐
                    │ Skill Area  │
                    │ {Vacancies} │
                    └──────┬──────┘
           ┌───────────────┼───────────────┐
    ┌──────┴──────┐ ┌──────┴──────┐ ┌──────┴──────┐
    │Identification│ │ Skill Area's│ │Loss of Interest│
    │ of New Skill│ │Gains & Losses│ │  in Skill   │
    │           1 │ │of Vacancies │ │          7  │
    └──────┬──────┘ └──────┬──────┘ └─────────────┘
        ┌──┴──┐            │
      ┌─┴─┐ ┌─┴─┐    ┌─────┴──────┐
      │ 1 │ │ 2 │    │Skill Area's * │
      └───┘ └───┘    │Gain or Loss of│
                    │  Vacancy    │
                    └──────┬──────┘
                    ┌──────┴──────┐
              ┌─────┴────┐  ┌─────┴────┐
              │Notification│  │  Loss   │
              │    of    o│  │   of   o│
              │ Vacancy  2│  │Vacancy  │
              └───────────┘  └────┬────┘
              ┌───────────┬───────┴───┬───────────┐
         ┌────┴────┐ ┌────┴────┐ ┌────┴────┐ ┌────┴────┐
         │  Post  o│ │  Post  o│ │Employer o│ │Vacancy o│
         │Withdrawal│ │Acceptance│ │Withdrawal│ │Withdrawal│
         │ (Last) 3│ │ (Last) 4│ │        5│ │        6│
         └─────────┘ └─────────┘ └─────────┘ └─────────┘
```

Operations list

1 Skill Code := Identification of New Skill ` Skill Code
2 Skill Description := Identification of New Skill ` Skill Description

Annex 2 Hi-Ho SSADM Development

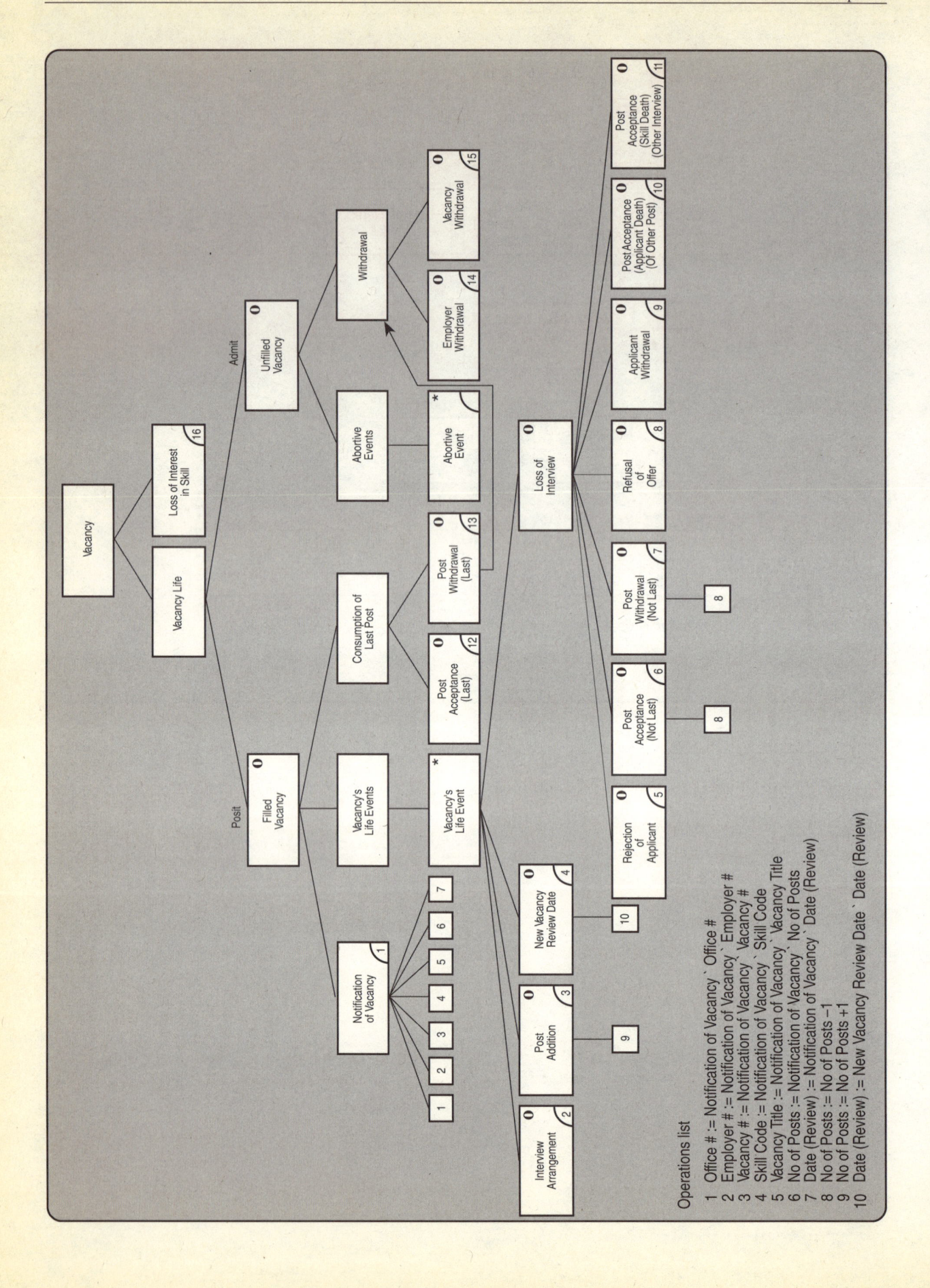

Vacancy
├── Vacancy Life
│   ├── Posit — Filled Vacancy (o)
│   │   └── Vacancy's Life Events
│   │   └── Vacancy's Life Event (*)
│   │   ├── Notification of Vacancy (1)
│   │   │   ├── 1
│   │   │   ├── 2
│   │   │   ├── 3
│   │   │   ├── 4
│   │   │   ├── 5
│   │   │   ├── 6
│   │   │   └── 7
│   │   ├── Interview Arrangement (2) (o)
│   │   ├── Post Addition (3) (o)
│   │   │   └── 9
│   │   ├── New Vacancy Review Date (4) (o)
│   │   │   └── 10
│   │   ├── Rejection of Applicant (5) (o)
│   │   └── Consumption of Last Post
│   │   ├── Post Acceptance (Last) (12) (o)
│   │   └── Post Withdrawal (Last) (13) (o)
│   │   └── Abortive Events
│   │   └── Abortive Event (*)
│   └── Admit — Unfilled Vacancy (o)
│   └── Withdrawal
│   ├── Employer Withdrawal (14) (o)
│   └── Vacancy Withdrawal (15) (o)
└── Loss of Interest in Skill (16) (o)

Loss of Interview
├── Post Acceptance (Not Last) (6) (o)
│   └── 8
├── Post Withdrawal (Not Last) (7) (o)
│   └── 8
├── Refusal of Offer (8) (o)
├── Applicant Withdrawal (9) (o)
├── Post Acceptance (Applicant Death) (Of Other Post) (10) (o)
└── Post Acceptance (Skill Death) (Other Interview) (11) (o)

Operations list

1   Office # := Notification of Vacancy ` Office #
2   Employer # := Notification of Vacancy ` Employer #
3   Vacancy # := Notification of Vacancy ` Vacancy #
4   Skill Code := Notification of Vacancy ` Skill Code
5   Vacancy Title := Notification of Vacancy ` Vacancy Title
6   No of Posts := Notification of Vacancy ` No of Posts
7   Date (Review) := Notification of Vacancy ` Date (Review)
8   No of Posts := No of Posts −1
9   No of Posts := No of Posts +1
10 Date (Review) := New Vacancy Review Date ` Date (Review)

ECDs after 2nd pass

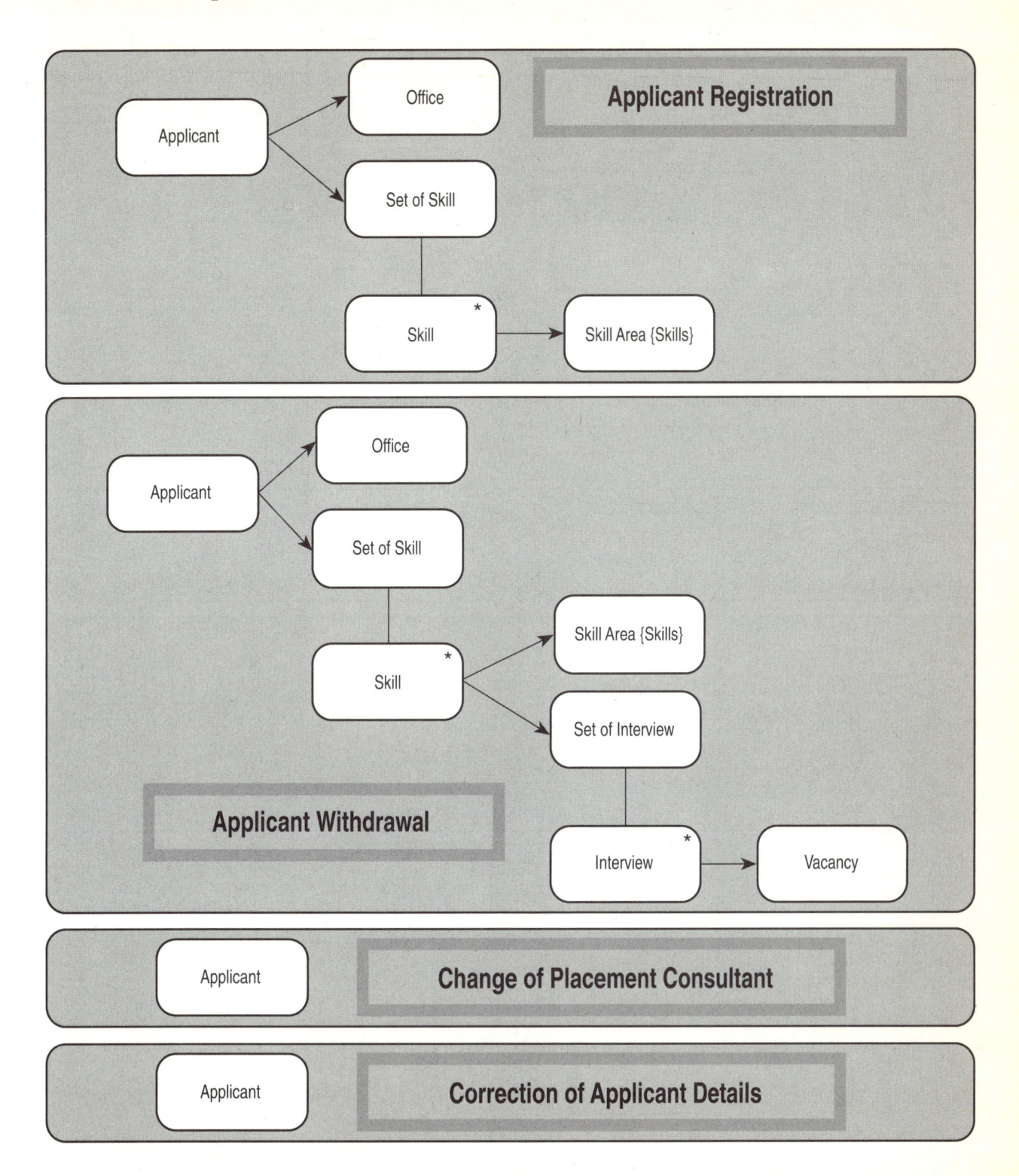

Applicant Registration

Applicant → Office
Applicant → Set of Skill
Set of Skill — Skill *
Skill → Skill Area {Skills}

Applicant Withdrawal

Applicant → Office
Applicant → Set of Skill
Set of Skill — Skill *
Skill → Skill Area {Skills}
Skill → Set of Interview
Set of Interview — Interview *
Interview → Vacancy

Change of Placement Consultant

Applicant

Correction of Applicant Details

Applicant

Employer Withdrawal

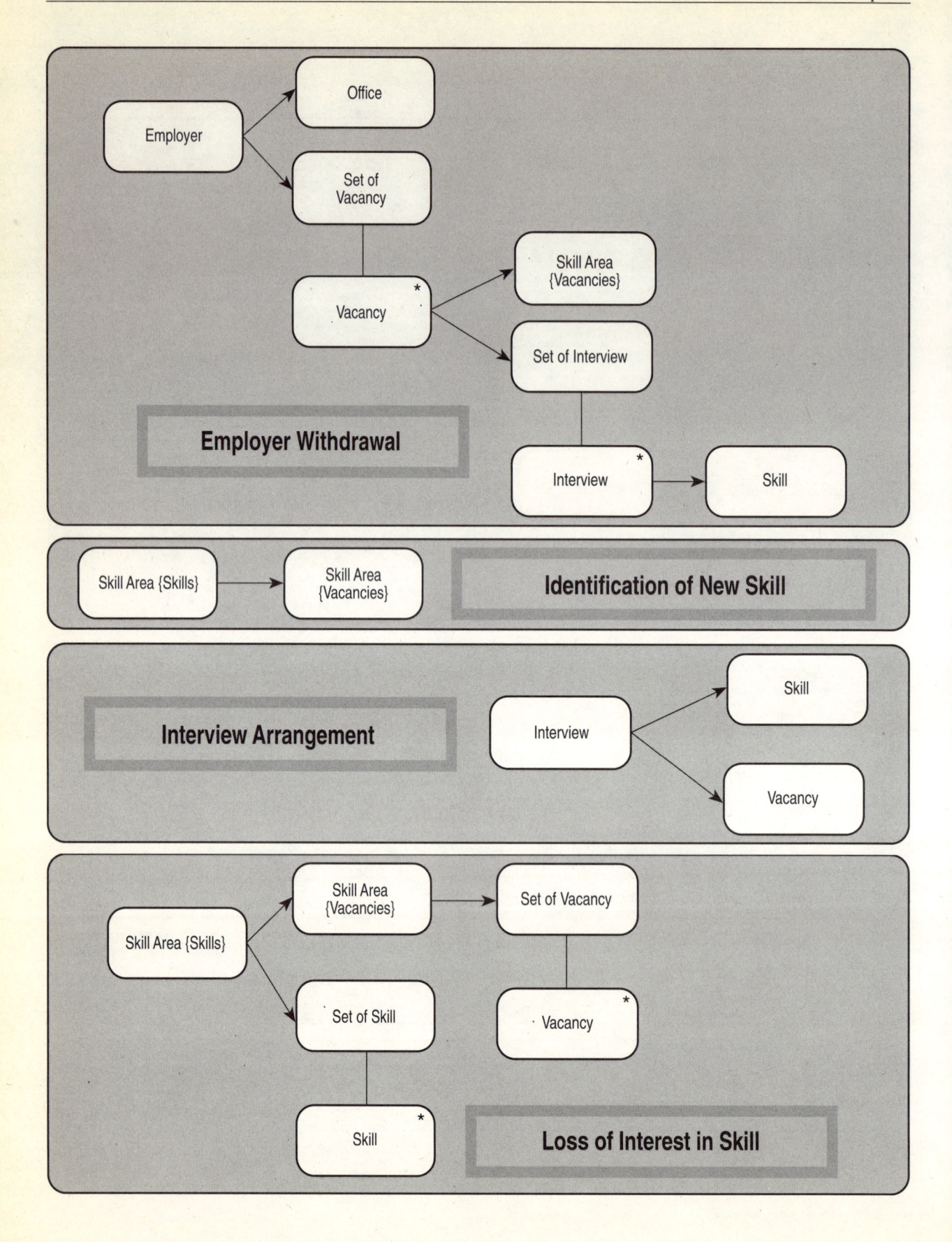

Employer → Office
Employer → Set of Vacancy → Vacancy *
Vacancy → Skill Area {Vacancies}
Vacancy → Set of Interview → Interview * → Skill

Identification of New Skill

Skill Area {Skills} → Skill Area {Vacancies}

Interview Arrangement

Interview → Skill
Interview → Vacancy

Loss of Interest in Skill

Skill Area {Skills} → Skill Area {Vacancies} → Set of Vacancy → Vacancy *
Skill Area {Skills} → Set of Skill → Skill *

ECDs after 2nd pass　　　　　　　　　　　　　　　　　　　　　　　　　　　　　　　　　　　161

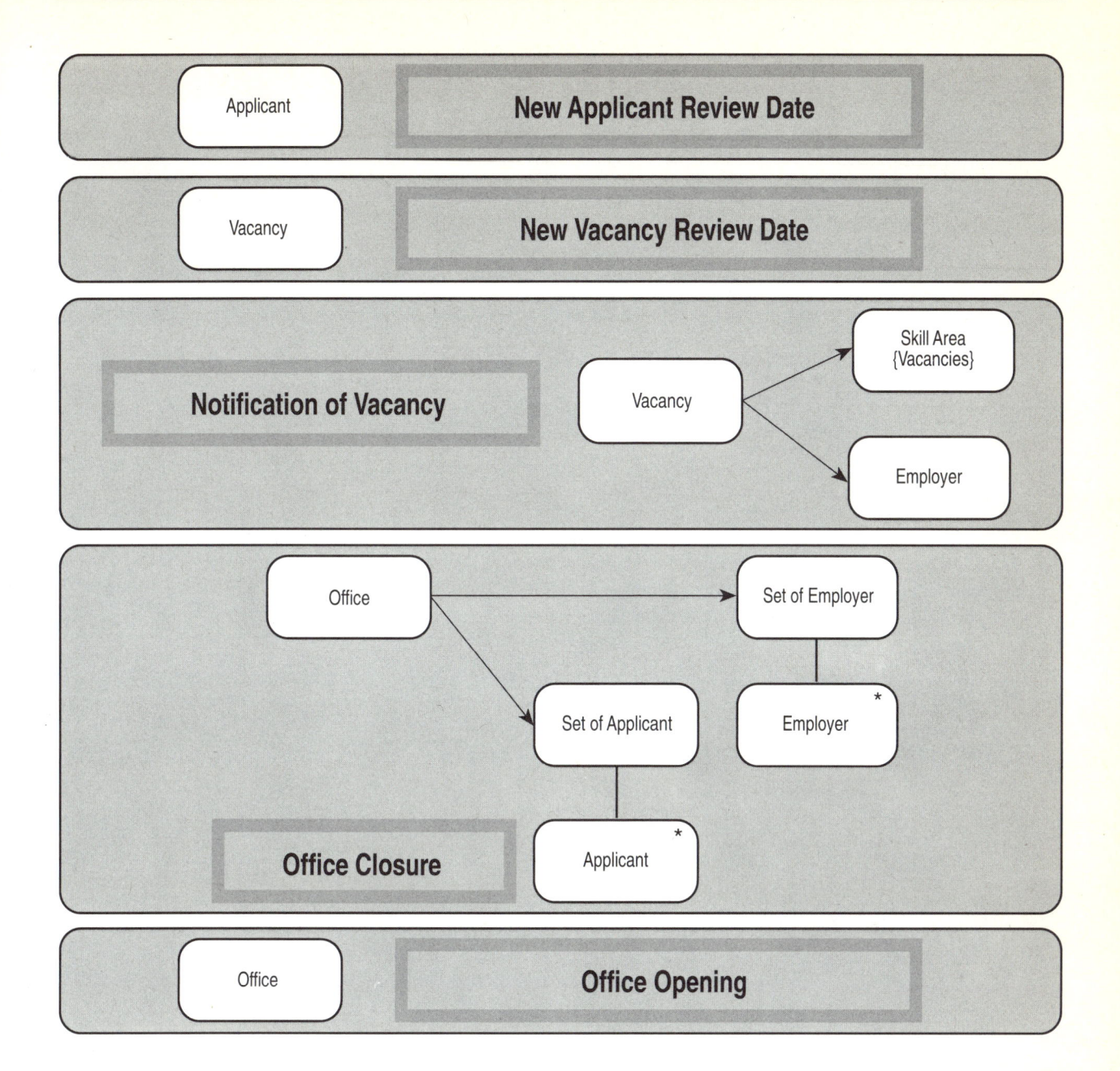

New Applicant Review Date
- Applicant

New Vacancy Review Date
- Vacancy

Notification of Vacancy
- Vacancy
  - Skill Area {Vacancies}
  - Employer

Office Closure
- Office
  - Set of Employer
    - Employer *
  - Set of Applicant
    - Applicant *

Office Opening
- Office

Post Acceptance

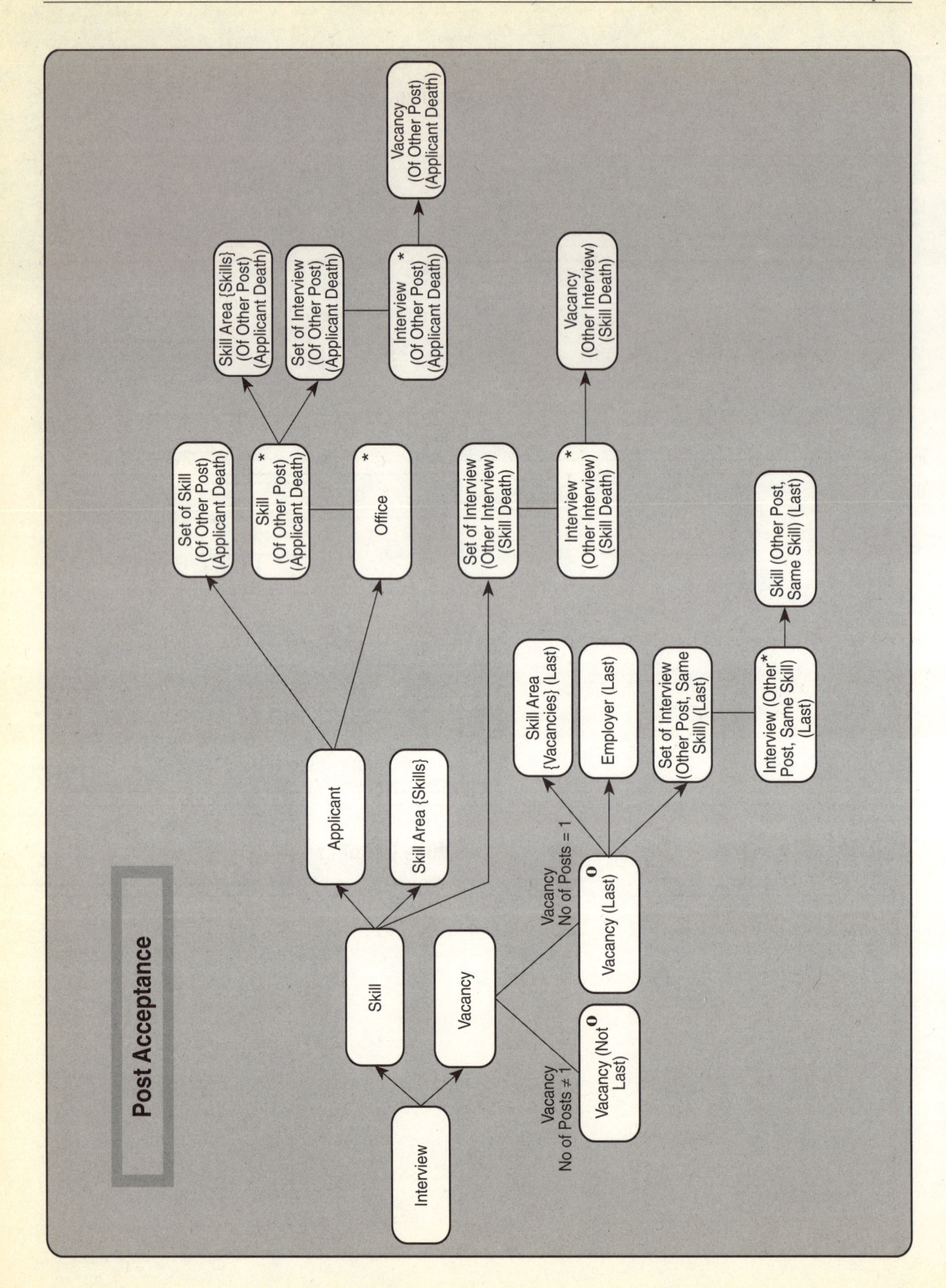

ECDs after 2nd pass

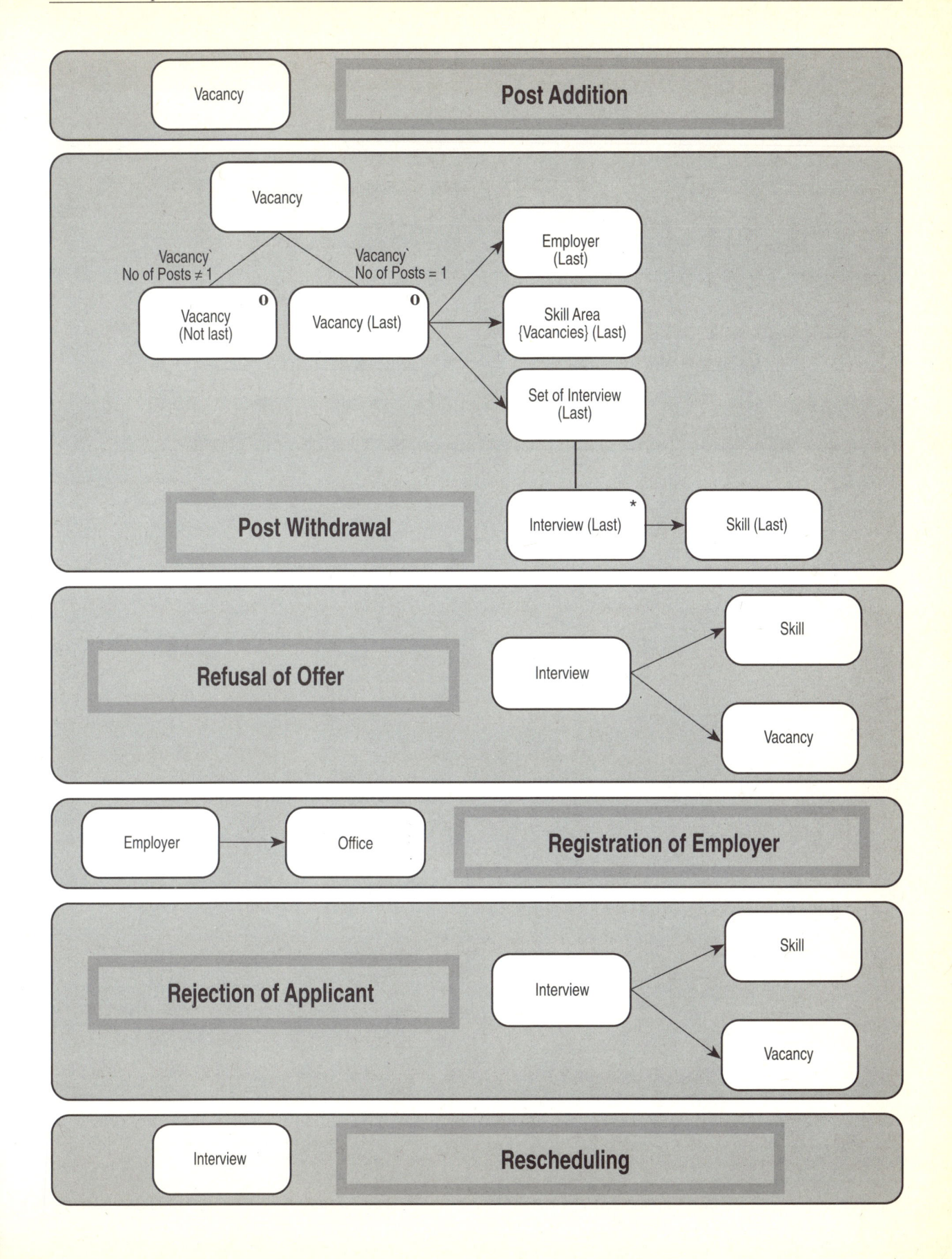

Post Addition

- Vacancy

Post Withdrawal

- Vacancy
  - Vacancy` No of Posts ≠ 1 → Vacancy (Not last) °
  - Vacancy` No of Posts = 1 → Vacancy (Last) °
    - → Employer (Last)
    - → Skill Area {Vacancies} (Last)
    - → Set of Interview (Last)
      - — Interview (Last) * → Skill (Last)

Refusal of Offer

- Interview
  - → Skill
  - → Vacancy

Registration of Employer

- Employer → Office

Rejection of Applicant

- Interview
  - → Skill
  - → Vacancy

Rescheduling

- Interview

Vacancy Withdrawal

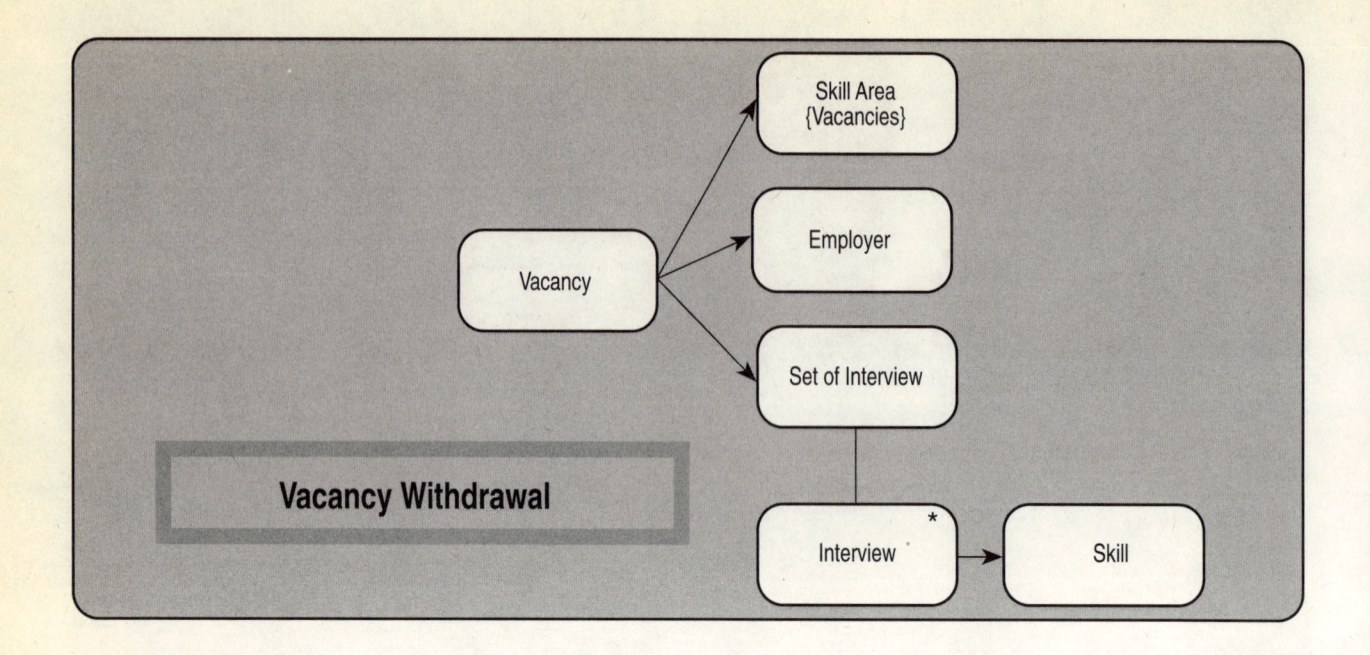

- Vacancy
  - Skill Area {Vacancies}
  - Employer
  - Set of Interview
    - Interview *
      - Skill

After specification of deletion strategy

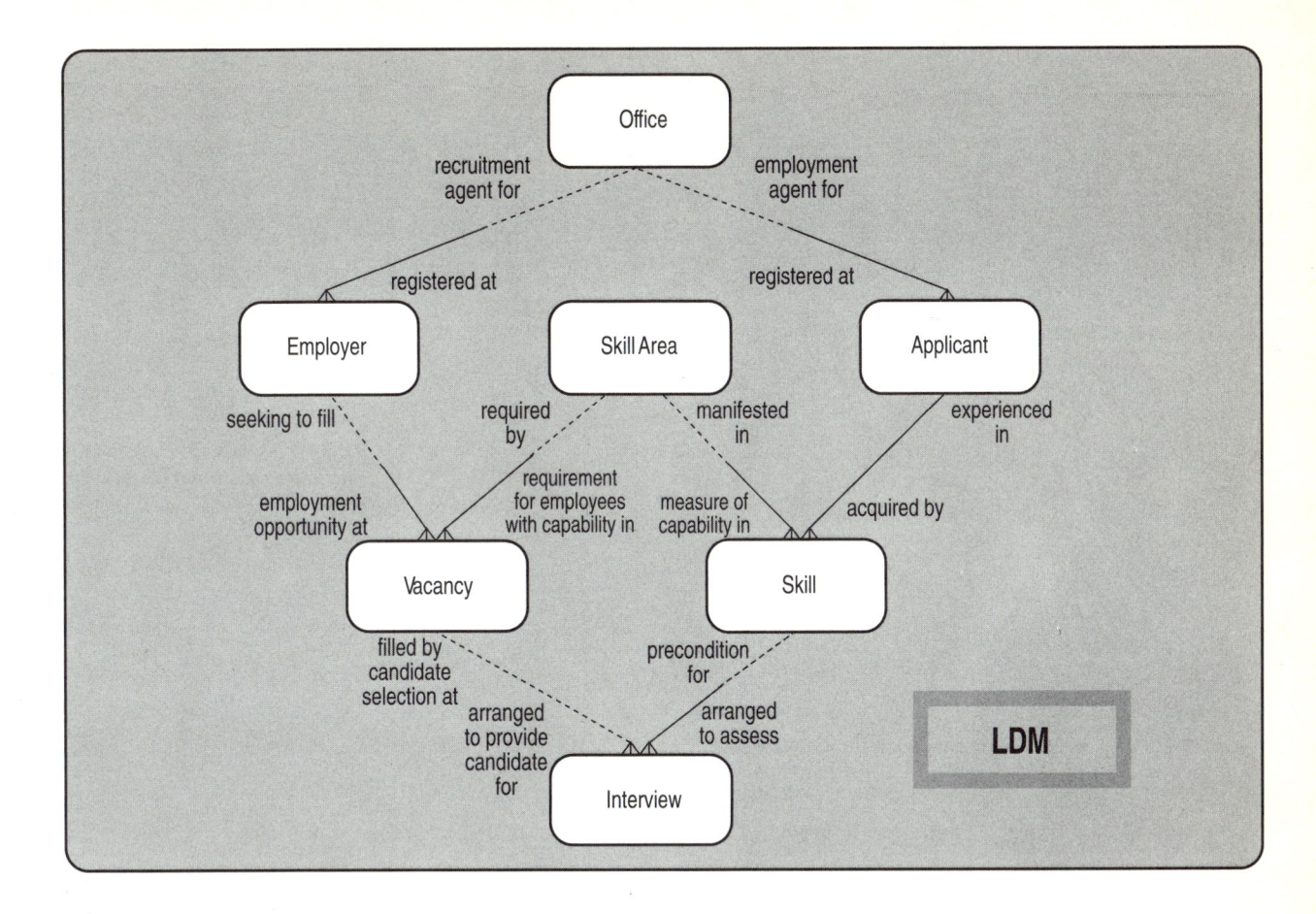

Annex 2 Hi-Ho SSADM Development

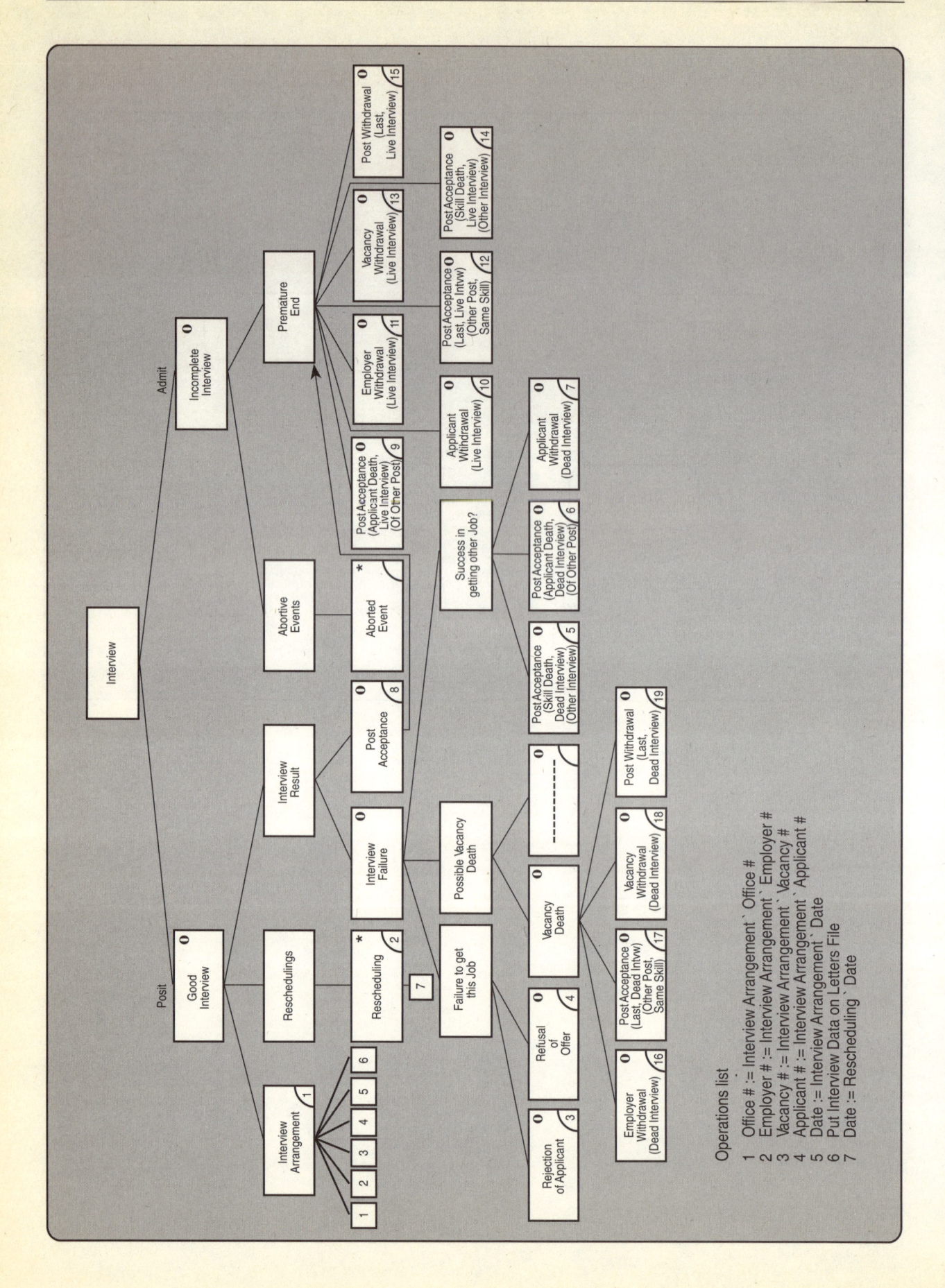

Interview structure diagram:

- Interview
  - Posit: Good Interview
    - Reschedulings
      - Rescheduling *
        - Interview Arrangement (1) [ops: 1,2,3,4,5,6]
        - Rescheduling (2) [op: 7]
    - Interview Result
      - Post Acceptance (8)
      - Interview Failure
        - Possible Vacancy Death
          - Failure to get this Job
            - Refusal of Offer (4)
            - Rejection of Applicant (3)
            - Success in getting other Job?
              - Post Acceptance (Applicant Death, Live Interview) (Of Other Post) (9)
              - Post Acceptance (Applicant Death, Dead Interview) (Of Other Post) (6)
              - Applicant Withdrawal (Live Interview) (10)
              - Applicant Withdrawal (Dead Interview) (7)
          - Post Acceptance (Skill Death, Dead Interview) (Other Interview) (5)
          - Vacancy Death
            - Employer Withdrawal (Dead Interview) (16)
            - Post Acceptance (Last, Dead Intvw) (Other Post, Same Skill) (17)
            - Vacancy Withdrawal (Dead Interview) (18)
            - Post Withdrawal (Last, Dead Interview) (19)
    - Abortive Events
      - Aborted Event *
        - Premature End
          - Employer Withdrawal (Live Interview) (11)
          - Post Acceptance (Last, Live Intvw) (Other Post, Same Skill) (12)
          - Vacancy Withdrawal (Live Interview) (13)
          - Post Acceptance (Skill Death, Live Interview) (Other Interview) (14)
          - Post Withdrawal (Last, Live Interview) (15)
  - Admit: Incomplete Interview

Operations list

1 Office # := Interview Arrangement ` Office #
2 Employer # := Interview Arrangement ` Employer #
3 Vacancy # := Interview Arrangement ` Vacancy #
4 Applicant # := Interview Arrangement ` Applicant #
5 Date := Interview Arrangement ` Date
6 Put Interview Data on Letters File
7 Date := Rescheduling ` Date

After specification of deletion strategy

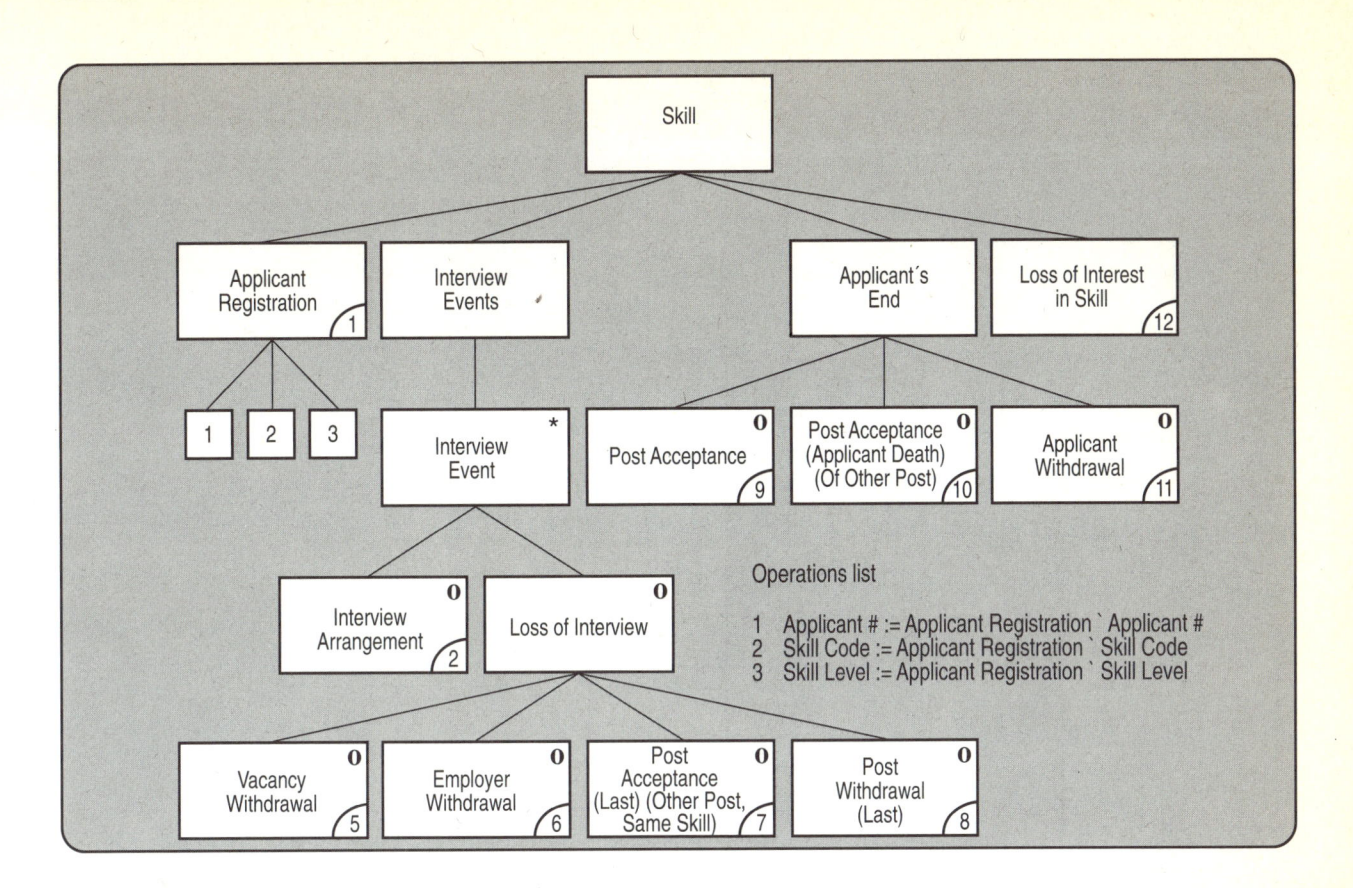

```
                              ┌─────────┐
                              │  Skill  │
                              └────┬────┘
         ┌────────────────┬───────┴────────┬──────────────────┐
   ┌───────────┐   ┌───────────┐     ┌───────────┐     ┌──────────────┐
   │ Applicant │   │ Interview │     │Applicant's│     │Loss of Interest│
   │Registration│  │  Events   │     │    End    │     │   in Skill   │
   └─────┬─────┘ 1 └─────┬─────┘     └─────┬─────┘     └──────────┬───┘12
     ┌───┼───┐           │                 │
   ┌─┐ ┌─┐ ┌─┐     ┌───────────┐*   ┌───────────┐0  ┌────────────┐0  ┌───────────┐0
   │1│ │2│ │3│     │ Interview │    │   Post    │   │Post Acceptance│ │ Applicant │
   └─┘ └─┘ └─┘     │   Event   │    │ Acceptance│   │(Applicant Death)│ │Withdrawal │
                   └─────┬─────┘    └───────────┘ 9 │(Of Other Post)│10└───────────┘11
              ┌──────────┴────────┐
        ┌───────────┐0       ┌───────────┐0
        │ Interview │        │  Loss of  │         Operations list
        │Arrangement│        │ Interview │
        └───────────┘2       └─────┬─────┘         1  Applicant # := Applicant Registration ` Applicant #
                    ┌──────┬──────┴──┬──────┐     2  Skill Code := Applicant Registration ` Skill Code
              ┌──────┐0 ┌──────┐0 ┌──────┐0 ┌──────┐0  3  Skill Level := Applicant Registration ` Skill Level
              │Vacancy│ │Employer│ │ Post │  │ Post │
              │Withdr.│ │Withdr. │ │Accept.│ │Withdr.│
              └──────┘5 └──────┘6 │(Last)│7 │(Last)│8
                                  │(Other│
                                  │Post, │
                                  │Same  │
                                  │Skill)│
                                  └──────┘
```

Annex 2 Hi-Ho SSADM Development

[Entity Life History diagram for Vacancy]

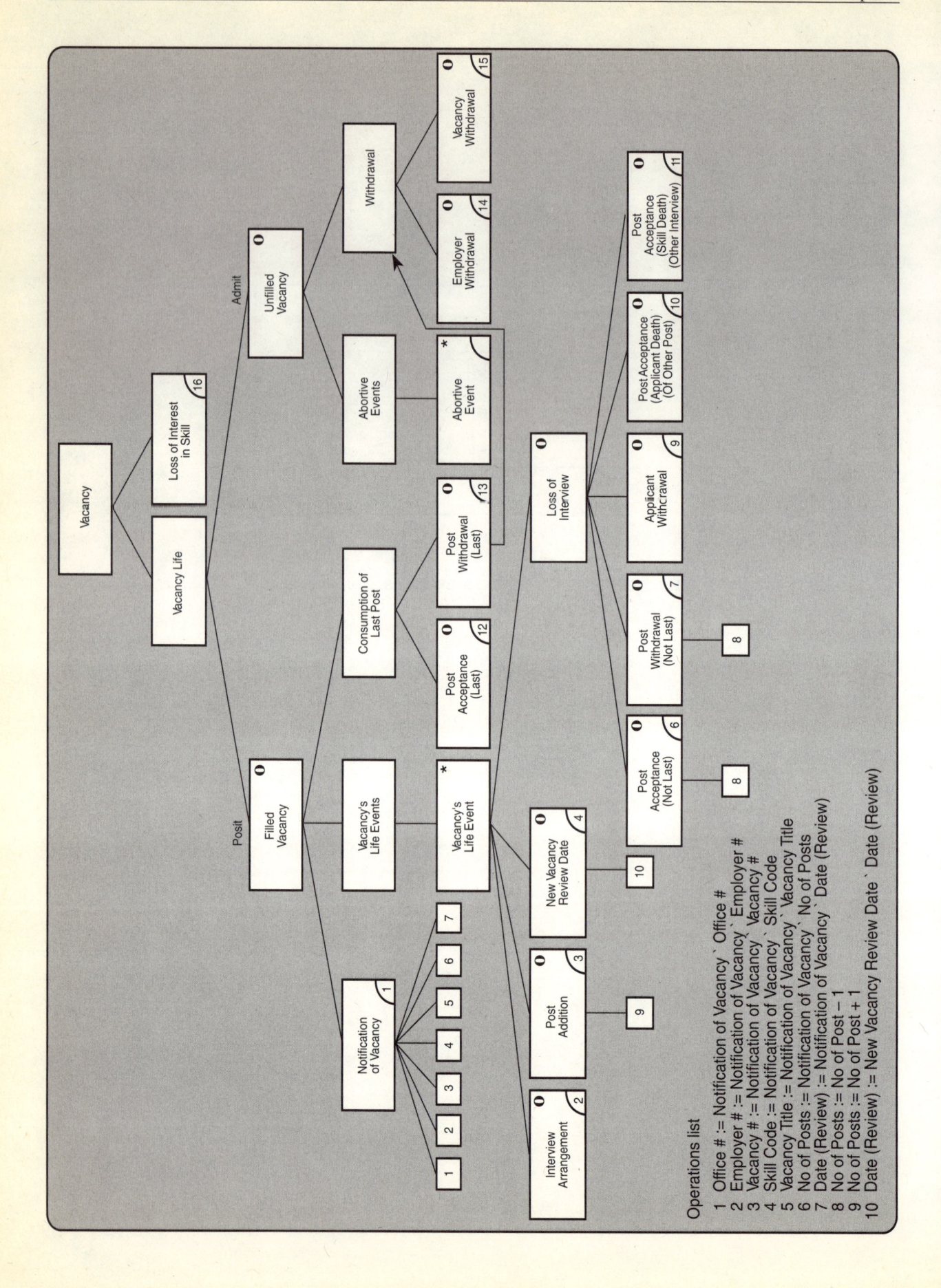

Operations list

1. Office # := Notification of Vacancy ` Office #
2. Employer # := Notification of Vacancy ` Employer #
3. Vacancy # := Notification of Vacancy ` Vacancy #
4. Skill Code := Notification of Vacancy ` Skill Code
5. Vacancy Title := Notification of Vacancy ` Vacancy Title
6. No of Posts := Notification of Vacancy ` No of Posts
7. Date (Review) := Notification of Vacancy ` Date (Review)
8. No of Posts := No of Post − 1
9. No of Posts := No of Post + 1
10. Date (Review) := New Vacancy Review Date ` Date (Review)

After specification of deletion strategy

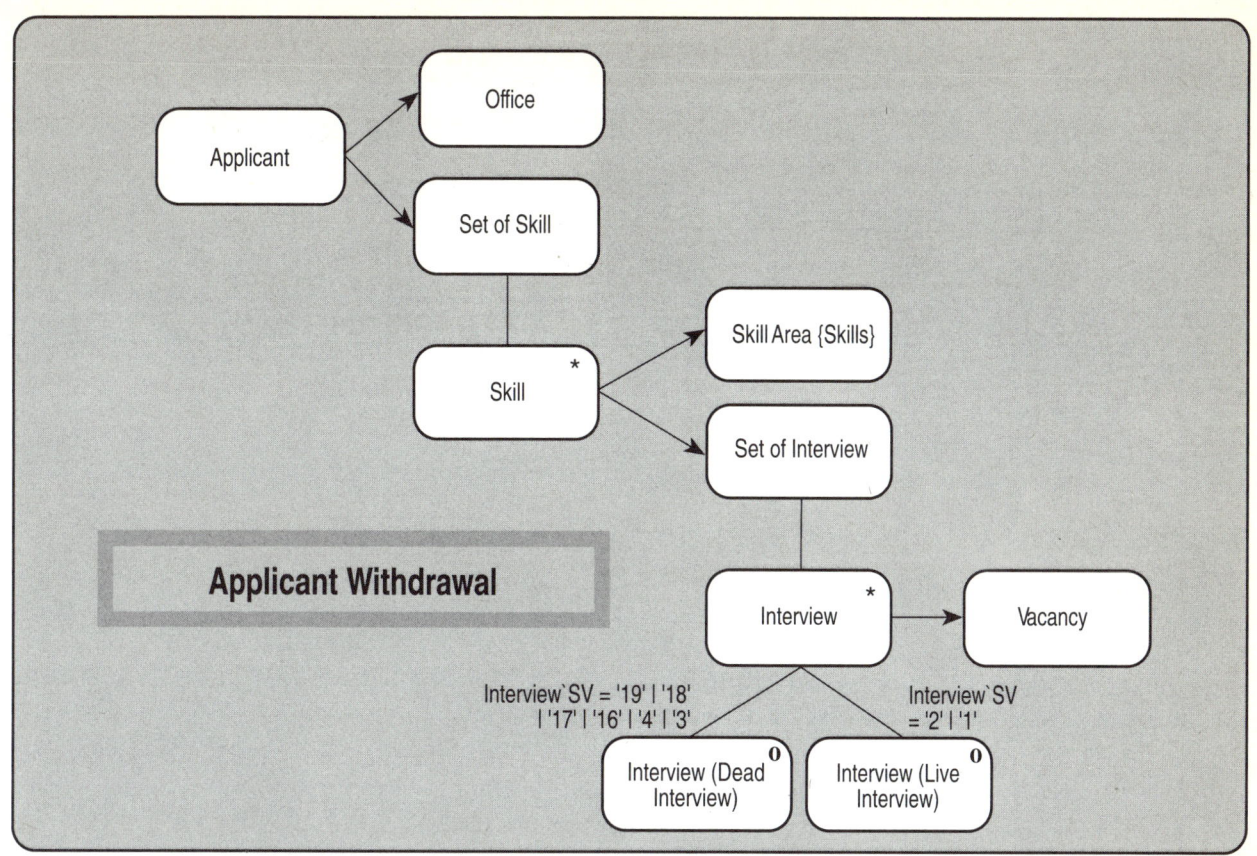

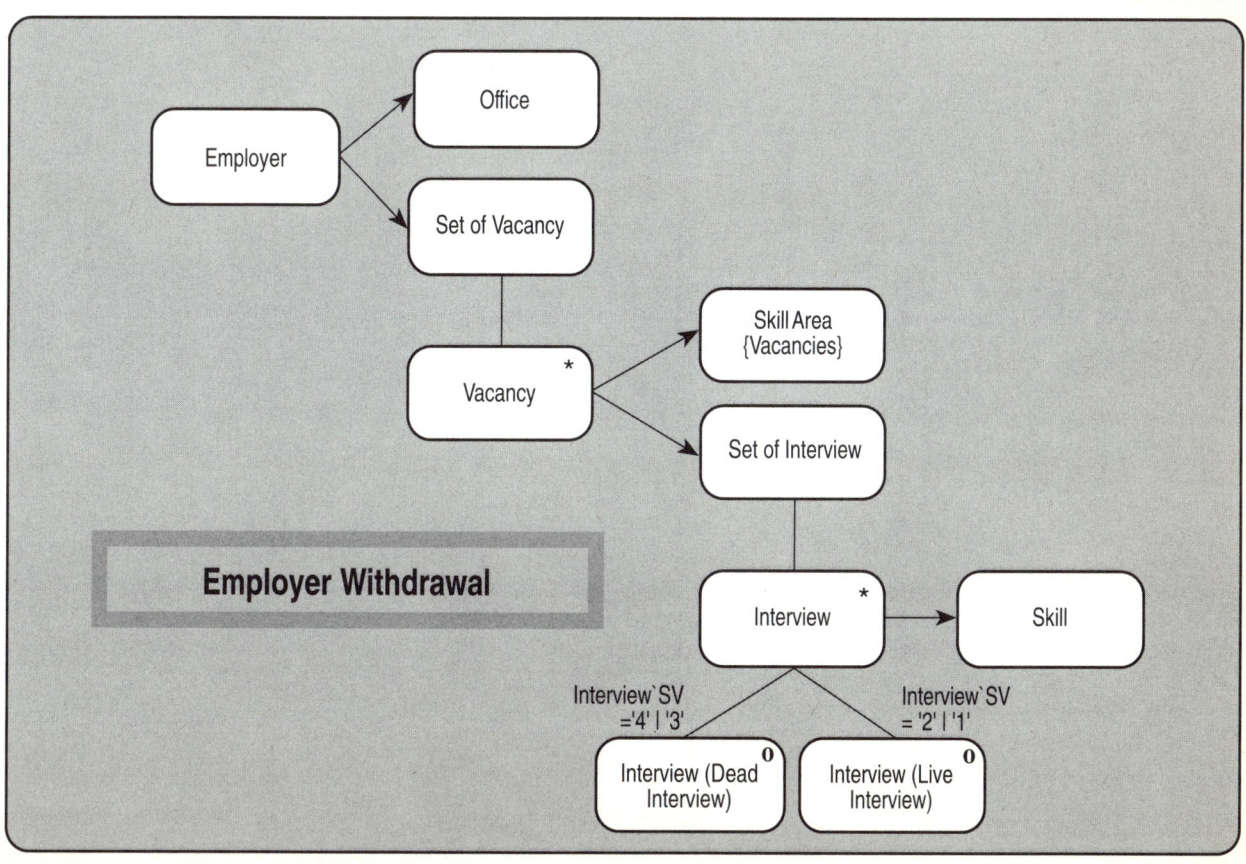

Applicant Withdrawal

- Applicant → Office
- Applicant → Set of Skill
- Set of Skill — Skill *
- Skill → Skill Area {Skills}
- Skill → Set of Interview
- Set of Interview — Interview *
- Interview → Vacancy
- Interview`SV = '19' | '18' | '17' | '16' | '4' | '3' → Interview (Dead Interview) o
- Interview`SV = '2' | '1' → Interview (Live Interview) o

Employer Withdrawal

- Employer → Office
- Employer → Set of Vacancy
- Set of Vacancy — Vacancy *
- Vacancy → Skill Area {Vacancies}
- Vacancy → Set of Interview
- Set of Interview — Interview *
- Interview → Skill
- Interview`SV ='4' | '3' → Interview (Dead Interview) o
- Interview`SV = '2' | '1' → Interview (Live Interview) o

170 — Annex 2 Hi-Ho SSADM Development

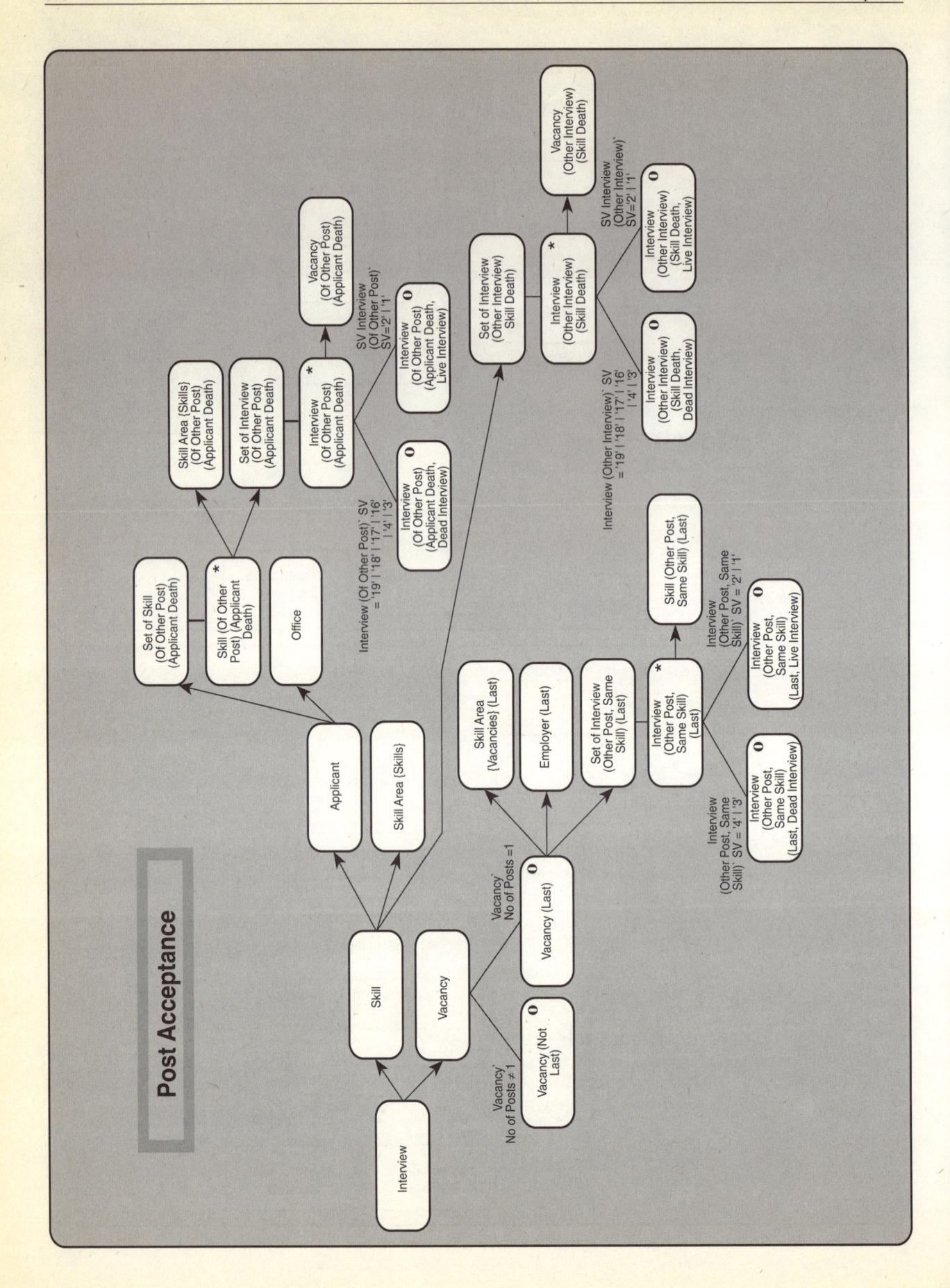

After specification of deletion strategy 171

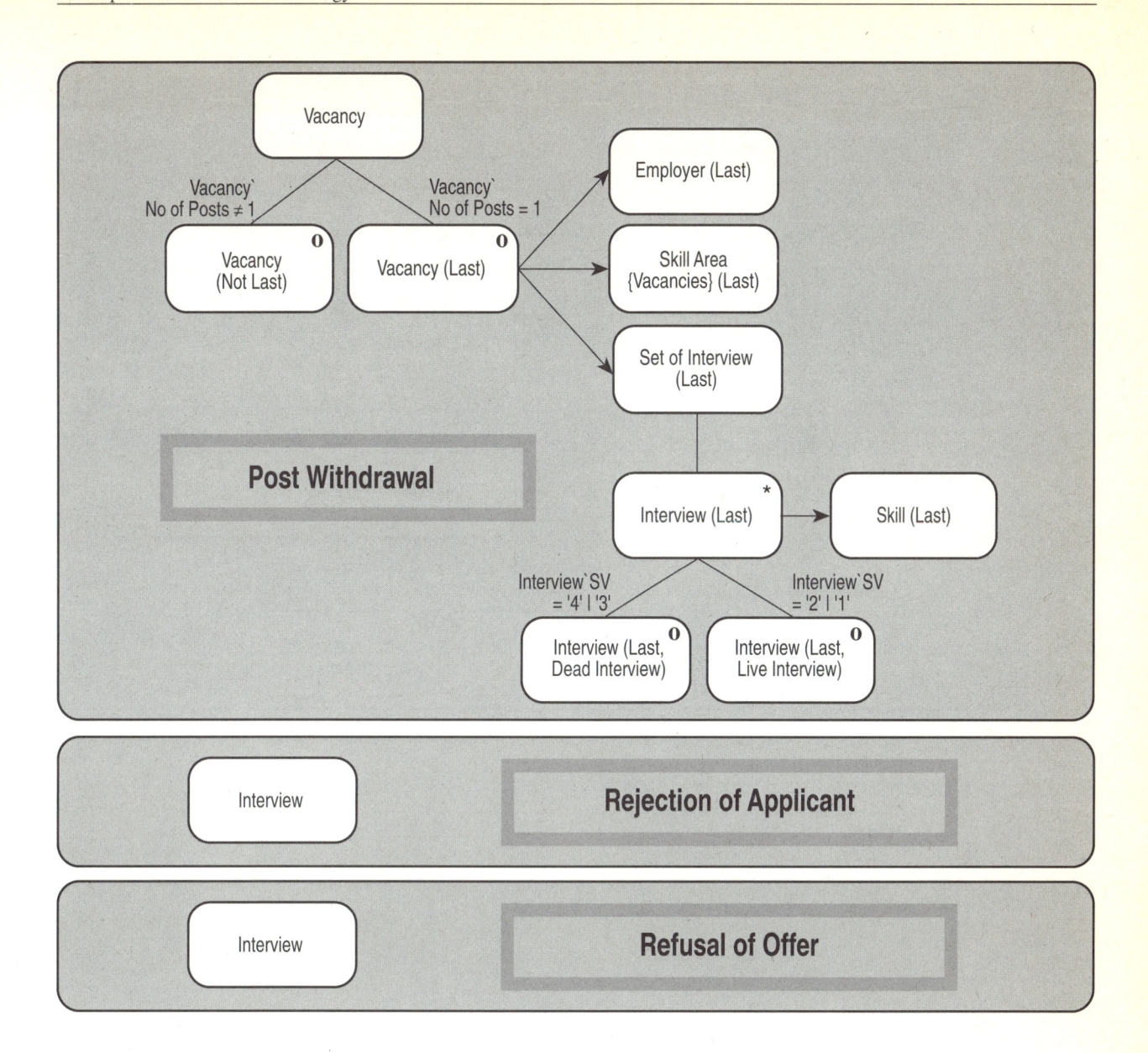

Post Withdrawal

- Vacancy
  - Vacancy`No of Posts ≠ 1 → Vacancy (Not Last) º
  - Vacancy`No of Posts = 1 → Vacancy (Last) º
    - Employer (Last)
    - Skill Area {Vacancies} (Last)
    - Set of Interview (Last)
      - Interview (Last) * → Skill (Last)
        - Interview`SV = '4' | '3' → Interview (Last, Dead Interview) º
        - Interview`SV = '2' | '1' → Interview (Last, Live Interview) º

Rejection of Applicant

- Interview

Refusal of Offer

- Interview

Vacancy Withdrawal

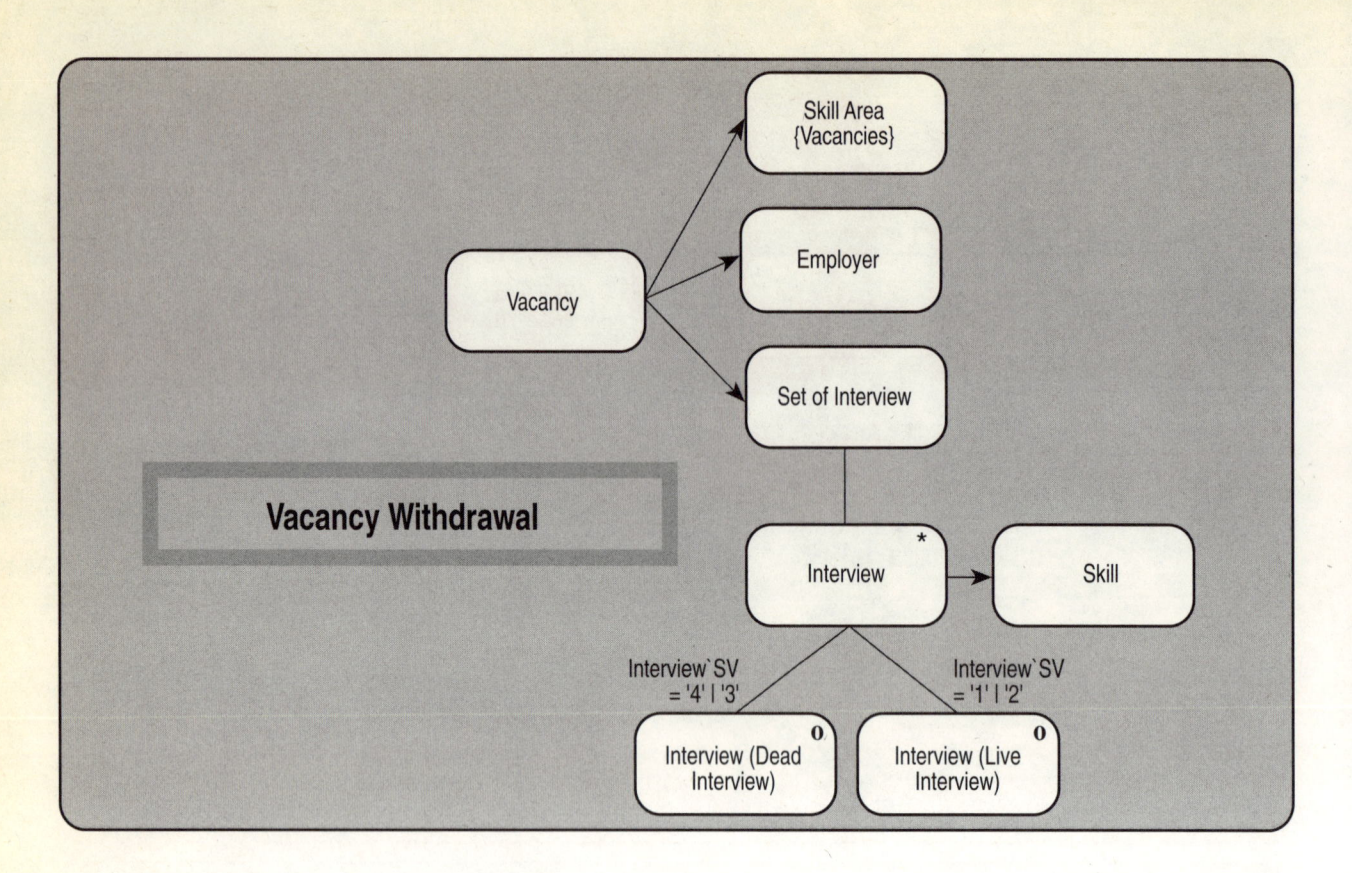

- Vacancy
  - Skill Area {Vacancies}
  - Employer
  - Set of Interview
    - Interview *
      - Skill
      - Interview (Dead Interview) ° [Interview`SV = '4' | '3']
      - Interview (Live Interview) ° [Interview`SV = '1' | '2']

After specification of deletion strategy 173

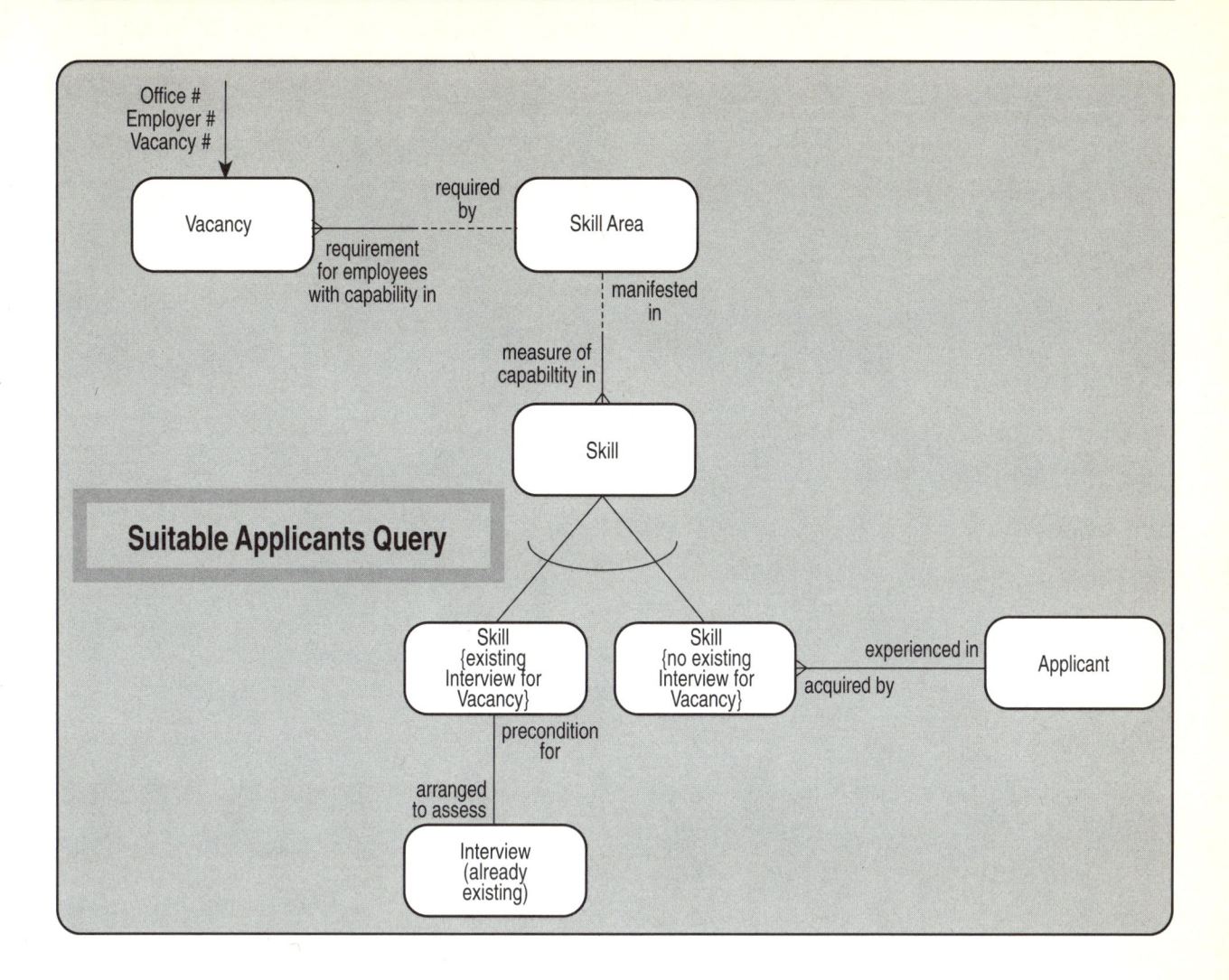

Suitable Applicants Query

- Vacancy (Office #, Employer #, Vacancy #) —required by / requirement for employees with capability in— Skill Area
- Skill Area —manifested in / measure of capabiltity in— Skill
- Skill subtypes: Skill {existing Interview for Vacancy}, Skill {no existing Interview for Vacancy}
- Skill {no existing Interview for Vacancy} —experienced in / acquired by— Applicant
- Skill {existing Interview for Vacancy} —precondition for / arranged to assess— Interview (already existing)

Suitable Vacancies Query

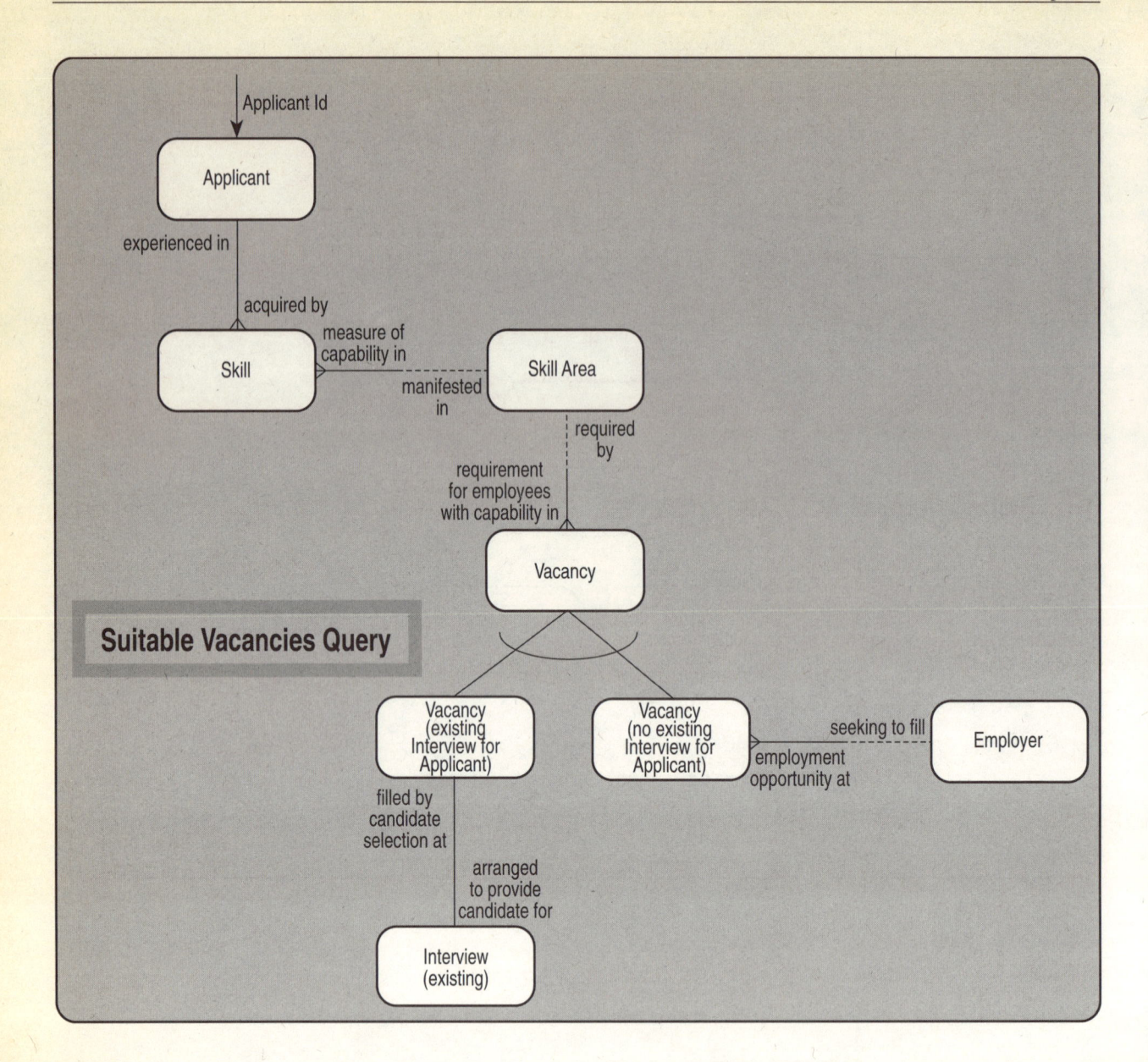

Update process models

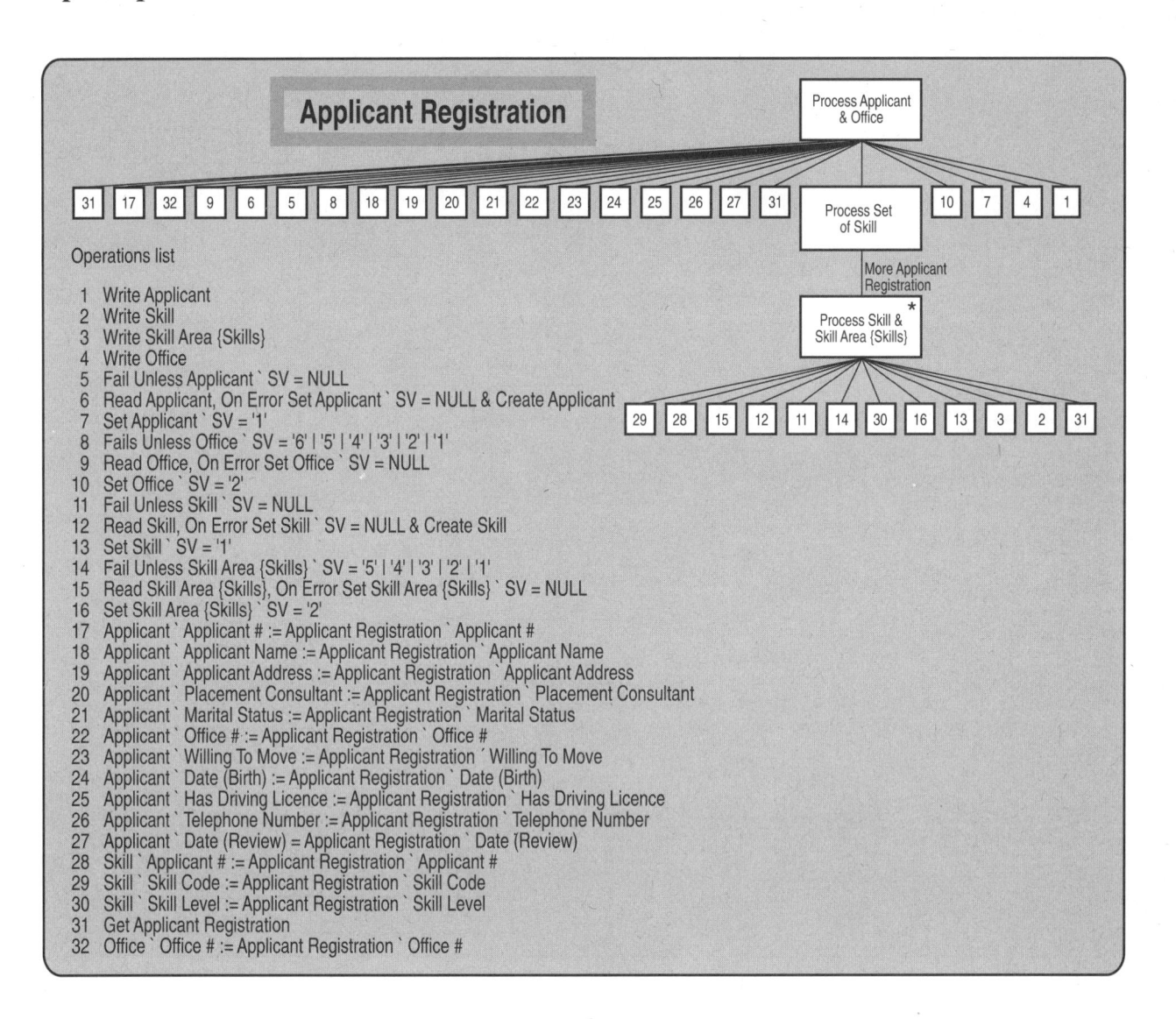

Applicant Registration

Process Applicant & Office
Process Set of Skill
More Applicant Registration
Process Skill & Skill Area {Skills} *

Operations list

1. Write Applicant
2. Write Skill
3. Write Skill Area {Skills}
4. Write Office
5. Fail Unless Applicant ` SV = NULL
6. Read Applicant, On Error Set Applicant ` SV = NULL & Create Applicant
7. Set Applicant ` SV = '1'
8. Fails Unless Office ` SV = '6' | '5' | '4' | '3' | '2' | '1'
9. Read Office, On Error Set Office ` SV = NULL
10. Set Office ` SV = '2'
11. Fail Unless Skill ` SV = NULL
12. Read Skill, On Error Set Skill ` SV = NULL & Create Skill
13. Set Skill ` SV = '1'
14. Fail Unless Skill Area {Skills} ` SV = '5' | '4' | '3' | '2' | '1'
15. Read Skill Area {Skills}, On Error Set Skill Area {Skills} ` SV = NULL
16. Set Skill Area {Skills} ` SV = '2'
17. Applicant ` Applicant # := Applicant Registration ` Applicant #
18. Applicant ` Applicant Name := Applicant Registration ` Applicant Name
19. Applicant ` Applicant Address := Applicant Registration ` Applicant Address
20. Applicant ` Placement Consultant := Applicant Registration ` Placement Consultant
21. Applicant ` Marital Status := Applicant Registration ` Marital Status
22. Applicant ` Office # := Applicant Registration ` Office #
23. Applicant ` Willing To Move := Applicant Registration ´ Willing To Move
24. Applicant ` Date (Birth) := Applicant Registration ` Date (Birth)
25. Applicant ` Has Driving Licence := Applicant Registration ` Has Driving Licence
26. Applicant ` Telephone Number := Applicant Registration ` Telephone Number
27. Applicant ` Date (Review) = Applicant Registration ` Date (Review)
28. Skill ` Applicant # := Applicant Registration ` Applicant #
29. Skill ` Skill Code := Applicant Registration ` Skill Code
30. Skill ` Skill Level := Applicant Registration ` Skill Level
31. Get Applicant Registration
32. Office ` Office # := Applicant Registration ` Office #

Applicant Withdrawal

Structure chart:

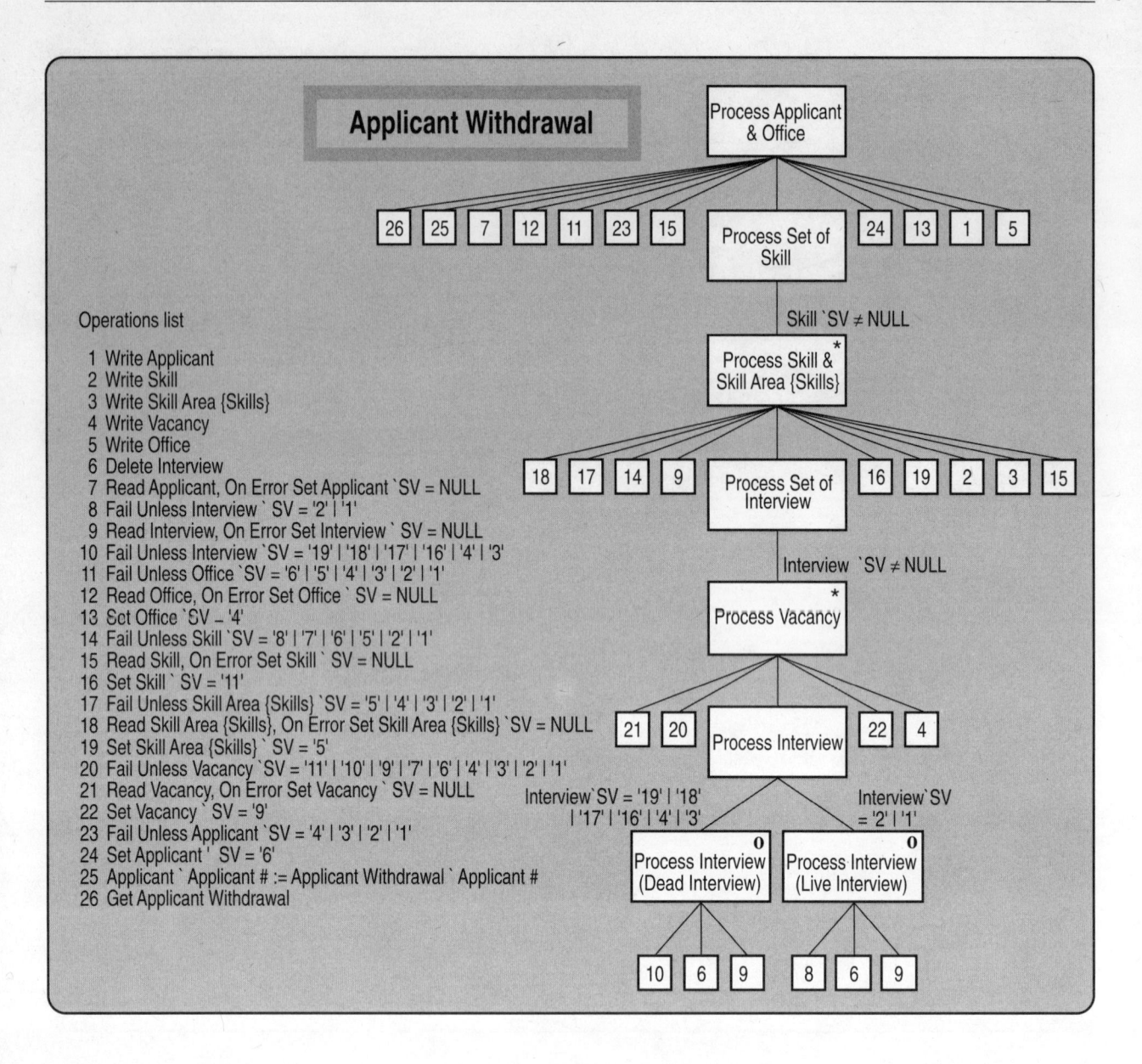

- **Process Applicant & Office** (root)
  - Operations: 26, 25, 7, 12, 11, 23, 15, 24, 13, 1, 5
  - **Process Set of Skill**
    - Skill `SV ≠ NULL
    - **Process Skill & Skill Area {Skills}** *
      - Operations: 18, 17, 14, 9, 16, 19, 2, 3, 15
      - **Process Set of Interview**
        - Interview `SV ≠ NULL
        - **Process Vacancy** *
          - Operations: 21, 20, 22, 4
          - **Process Interview**
            - **Process Interview (Dead Interview)** o
              - Interview `SV = '19' | '18' | '17' | '16' | '4' | '3'
              - Operations: 10, 6, 9
            - **Process Interview (Live Interview)** o
              - Interview `SV = '2' | '1'
              - Operations: 8, 6, 9

Operations list

1. Write Applicant
2. Write Skill
3. Write Skill Area {Skills}
4. Write Vacancy
5. Write Office
6. Delete Interview
7. Read Applicant, On Error Set Applicant `SV = NULL
8. Fail Unless Interview ` SV = '2' | '1'
9. Read Interview, On Error Set Interview ` SV = NULL
10. Fail Unless Interview `SV = '19' | '18' | '17' | '16' | '4' | '3'
11. Fail Unless Office `SV = '6' | '5' | '4' | '3' | '2' | '1'
12. Read Office, On Error Set Office ` SV = NULL
13. Set Office `SV = '4'
14. Fail Unless Skill `SV = '8' | '7' | '6' | '5' | '2' | '1'
15. Read Skill, On Error Set Skill ` SV = NULL
16. Set Skill ` SV = '11'
17. Fail Unless Skill Area {Skills} `SV = '5' | '4' | '3' | '2' | '1'
18. Read Skill Area {Skills}, On Error Set Skill Area {Skills} `SV = NULL
19. Set Skill Area {Skills} ` SV = '5'
20. Fail Unless Vacancy `SV = '11' | '10' | '9' | '7' | '6' | '4' | '3' | '2' | '1'
21. Read Vacancy, On Error Set Vacancy ` SV = NULL
22. Set Vacancy ` SV = '9'
23. Fail Unless Applicant `SV = '4' | '3' | '2' | '1'
24. Set Applicant ' SV = '6'
25. Applicant ` Applicant # := Applicant Withdrawal ` Applicant #
26. Get Applicant Withdrawal

Change of Placement Consultant

```
                    Process Applicant
                    /   /   |   |   \   \
                   7   6   2   4   3   5   1
```

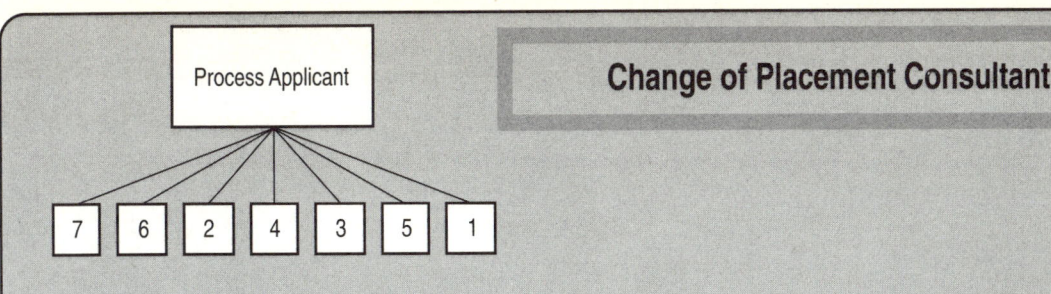

Operations list

1. Write Applicant
2. Read Applicant, On Error Set Applicant ` SV = NULL
3. Applicant ` Placement Consultant := Change of Placement Consultant ` Placement Consultant
4. Fail Unless Applicant ` SV = '4' | '3' | '2' | '1'
5. Set Applicant ` SV = '4'
6. Applicant ` Applicant # := Change of Placement Consultant ` Applicant #
7. Get Change of Placement Consultant

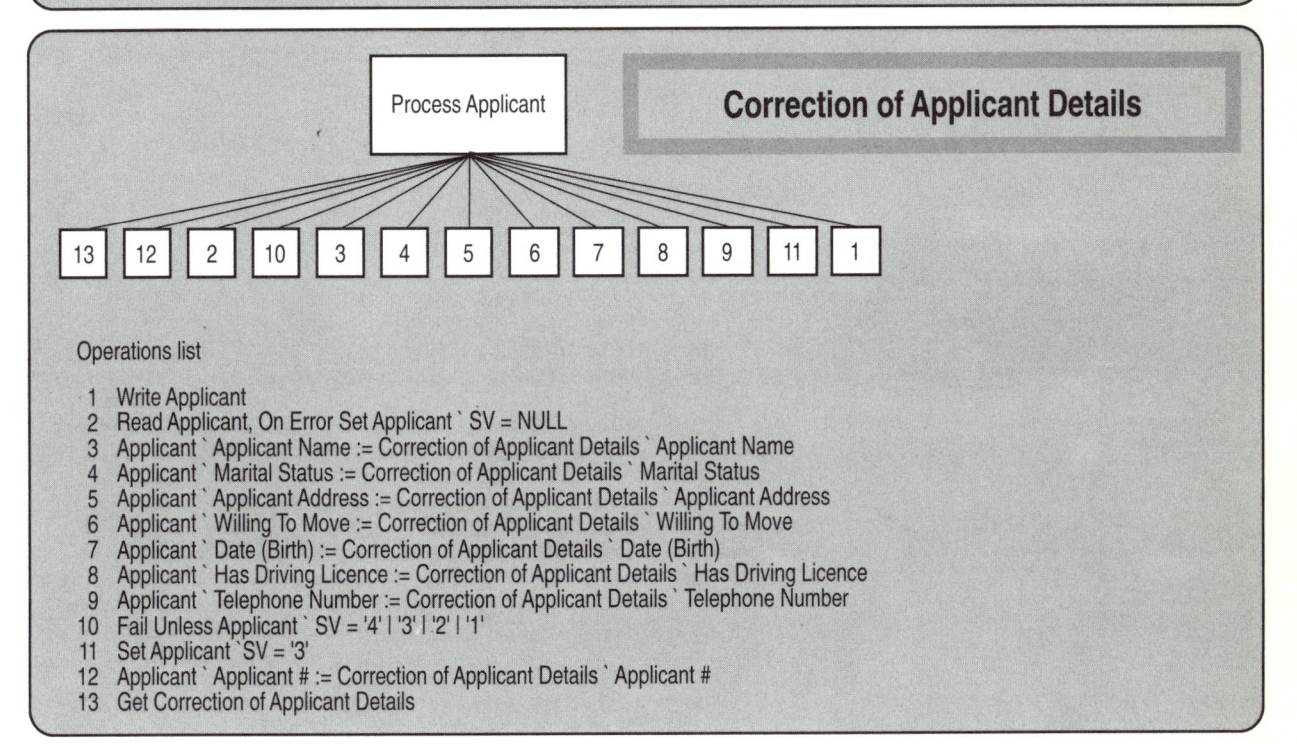

Correction of Applicant Details

```
                         Process Applicant
        /    /    /    /    /    |    \    \    \    \    \    \
       13  12   2   10   3   4   5   6   7   8   9   11   1
```

Operations list

1. Write Applicant
2. Read Applicant, On Error Set Applicant ` SV = NULL
3. Applicant ` Applicant Name := Correction of Applicant Details ` Applicant Name
4. Applicant ` Marital Status := Correction of Applicant Details ` Marital Status
5. Applicant ` Applicant Address := Correction of Applicant Details ` Applicant Address
6. Applicant ` Willing To Move := Correction of Applicant Details ` Willing To Move
7. Applicant ` Date (Birth) := Correction of Applicant Details ` Date (Birth)
8. Applicant ` Has Driving Licence := Correction of Applicant Details ` Has Driving Licence
9. Applicant ` Telephone Number := Correction of Applicant Details ` Telephone Number
10. Fail Unless Applicant ` SV = '4' | '3' | '2' | '1'
11. Set Applicant ` SV = '3'
12. Applicant ` Applicant # := Correction of Applicant Details ` Applicant #
13. Get Correction of Applicant Details

Employer Withdrawal

```
                              Process Employer
                                  & Office
     ┌────┬────┬────┬────┬────┬────┬────┬────┬────────┬────┬────┬────┬────┐
    [29] [28] [27] [9] [16] [15] [8] [25]  Process   [10] [17] [1] [5]
                                           Set of
                                           Vacancy
                                              │
                                    Vacancy `SV ≠ NULL
                                              │
                                       Process Vacancy *
                                         & Skill Area
                                         {Vacancies}
                                              │
              ┌────┬────┬────┬────┬──────────┬────┬────┬────┬────┐
             [22] [21] [24] [12]  Process   [26] [23] [4] [3] [25]
                                  Set of
                                  Interview
                                      │
                            Interview `SV ≠ NULL
                                      │
                                Process Skill *
                                      │
                     ┌────┬────┬──────────────┬────┬────┐
                    [19] [18]  Process        [20] [2]
                               Interview
                                  │
                    Interview `SV         Interview `SV
                    ='4' | '3'            ='2' | '1'
                         │                     │
                  Process Interview °   Process Interview °
                  (Dead Interview)      (Live Interview)
                         │                     │
                  ┌───┬───┬───┬───┐     ┌───┬───┬───┐
                 [13][14][7][12]       [11][6][12]
```

Operations list

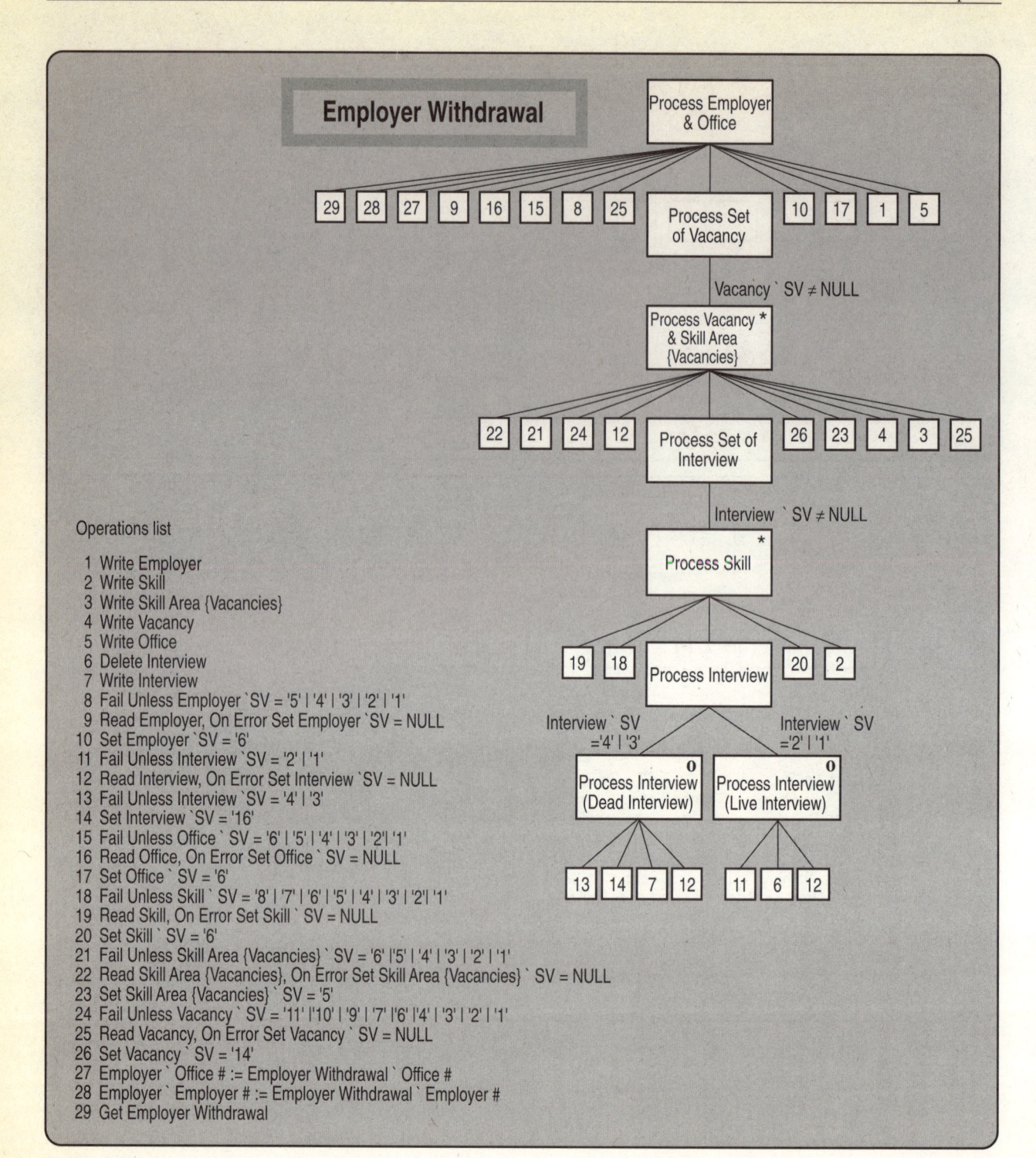

1 Write Employer
2 Write Skill
3 Write Skill Area {Vacancies}
4 Write Vacancy
5 Write Office
6 Delete Interview
7 Write Interview
8 Fail Unless Employer `SV = '5' | '4' | '3' | '2' | '1'
9 Read Employer, On Error Set Employer `SV = NULL
10 Set Employer `SV = '6'
11 Fail Unless Interview `SV = '2' | '1'
12 Read Interview, On Error Set Interview `SV = NULL
13 Fail Unless Interview `SV = '4' | '3'
14 Set Interview `SV = '16'
15 Fail Unless Office ` SV = '6' | '5' | '4' | '3' | '2'| '1'
16 Read Office, On Error Set Office ` SV = NULL
17 Set Office ` SV = '6'
18 Fail Unless Skill ` SV = '8' | '7' | '6' | '5' | '4' | '3' | '2'| '1'
19 Read Skill, On Error Set Skill ` SV = NULL
20 Set Skill ` SV = '6'
21 Fail Unless Skill Area {Vacancies} ` SV = '6' |'5' | '4' | '3' | '2' | '1'
22 Read Skill Area {Vacancies}, On Error Set Skill Area {Vacancies} ` SV = NULL
23 Set Skill Area {Vacancies} ` SV = '5'
24 Fail Unless Vacancy ` SV = '11' |'10' | '9' | '7' |'6' |'4' | '3' | '2' | '1'
25 Read Vacancy, On Error Set Vacancy ` SV = NULL
26 Set Vacancy ` SV = '14'
27 Employer ` Office # := Employer Withdrawal ` Office #
28 Employer ` Employer # := Employer Withdrawal ` Employer #
29 Get Employer Withdrawal

Update process models

Identification of New Skill

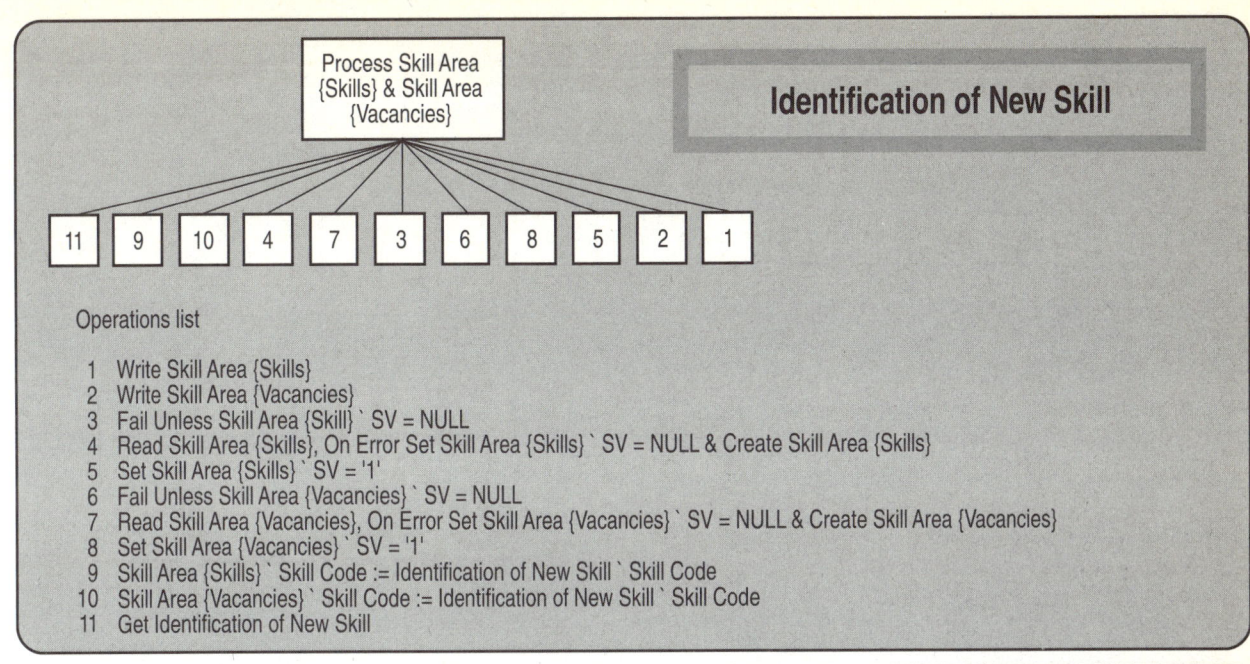

Process Skill Area {Skills} & Skill Area {Vacancies}

| 11 | 9 | 10 | 4 | 7 | 3 | 6 | 8 | 5 | 2 | 1 |

Operations list

1. Write Skill Area {Skills}
2. Write Skill Area {Vacancies}
3. Fail Unless Skill Area {Skill} ` SV = NULL
4. Read Skill Area {Skills}, On Error Set Skill Area {Skills} ` SV = NULL & Create Skill Area {Skills}
5. Set Skill Area {Skills} ` SV = '1'
6. Fail Unless Skill Area {Vacancies} ` SV = NULL
7. Read Skill Area {Vacancies}, On Error Set Skill Area {Vacancies} ` SV = NULL & Create Skill Area {Vacancies}
8. Set Skill Area {Vacancies} ` SV = '1'
9. Skill Area {Skills} ` Skill Code := Identification of New Skill ` Skill Code
10. Skill Area {Vacancies} ` Skill Code := Identification of New Skill ` Skill Code
11. Get Identification of New Skill

Interview Arrangement

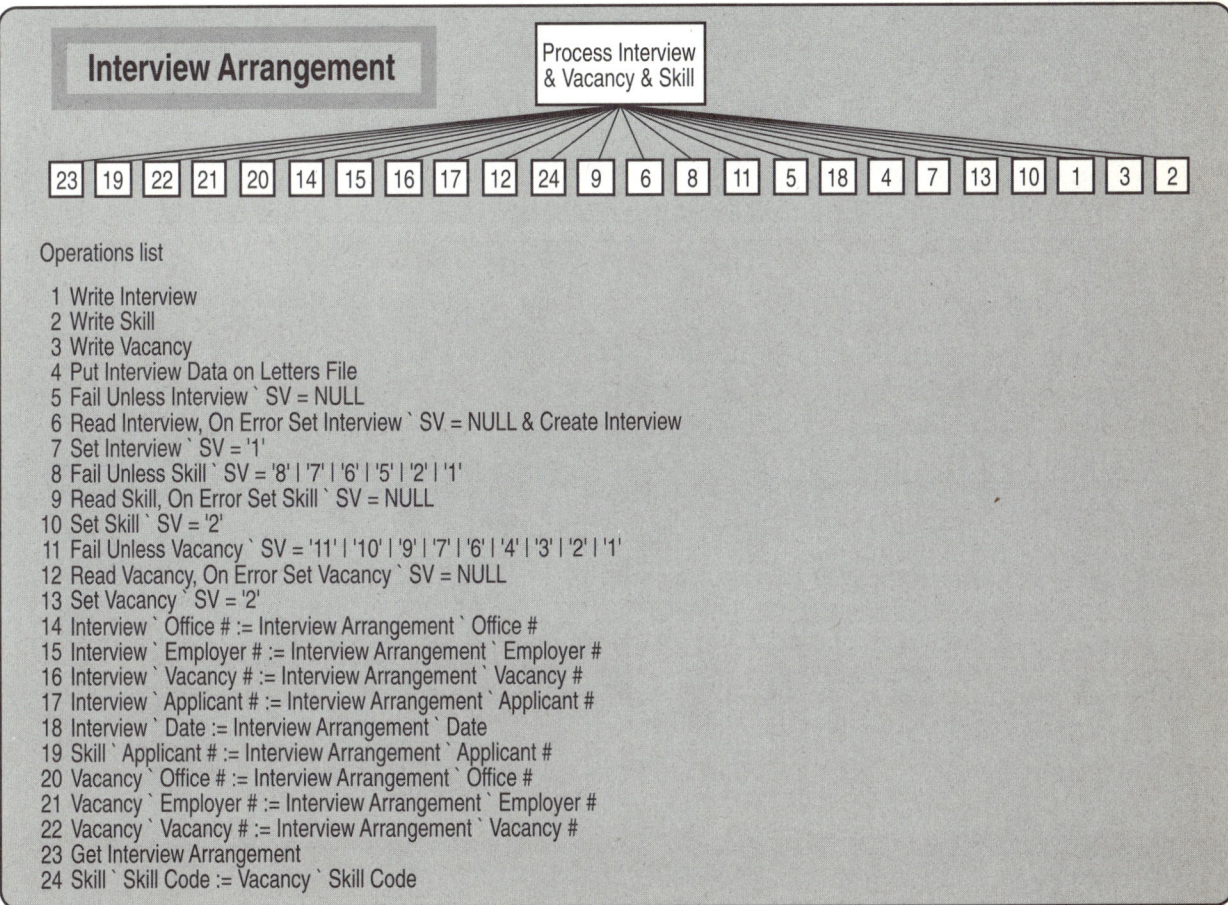

Process Interview & Vacancy & Skill

| 23 | 19 | 22 | 21 | 20 | 14 | 15 | 16 | 17 | 12 | 24 | 9 | 6 | 8 | 11 | 5 | 18 | 4 | 7 | 13 | 10 | 1 | 3 | 2 |

Operations list

1. Write Interview
2. Write Skill
3. Write Vacancy
4. Put Interview Data on Letters File
5. Fail Unless Interview ` SV = NULL
6. Read Interview, On Error Set Interview ` SV = NULL & Create Interview
7. Set Interview ` SV = '1'
8. Fail Unless Skill ` SV = '8' | '7' | '6' | '5' | '2' | '1'
9. Read Skill, On Error Set Skill ` SV = NULL
10. Set Skill ` SV = '2'
11. Fail Unless Vacancy ` SV = '11' | '10' | '9' | '7' | '6' | '4' | '3' | '2' | '1'
12. Read Vacancy, On Error Set Vacancy ` SV = NULL
13. Set Vacancy ` SV = '2'
14. Interview ` Office # := Interview Arrangement ` Office #
15. Interview ` Employer # := Interview Arrangement ` Employer #
16. Interview ` Vacancy # := Interview Arrangement ` Vacancy #
17. Interview ` Applicant # := Interview Arrangement ` Applicant #
18. Interview ` Date := Interview Arrangement ` Date
19. Skill ` Applicant # := Interview Arrangement ` Applicant #
20. Vacancy ` Office # := Interview Arrangement ` Office #
21. Vacancy ` Employer # := Interview Arrangement ` Employer #
22. Vacancy ` Vacancy # := Interview Arrangement ` Vacancy #
23. Get Interview Arrangement
24. Skill ` Skill Code := Vacancy ` Skill Code

Loss of Interest in Skill

Structure:
- Process Skill Area {Skills} & Skill Area {Vacancies}
  - [15] [14] [13] [10] [8] [9] [7] [2] [3] [6] [12]
  - Process Set of Skill
    - Skill ` SV ≠ NULL
    - Process Skill *
      - [5] [1] [6]
  - Process Set of Vacancy
    - Vacancy ` SV ≠ NULL
    - Process Vacancy *
      - [11] [4] [12]

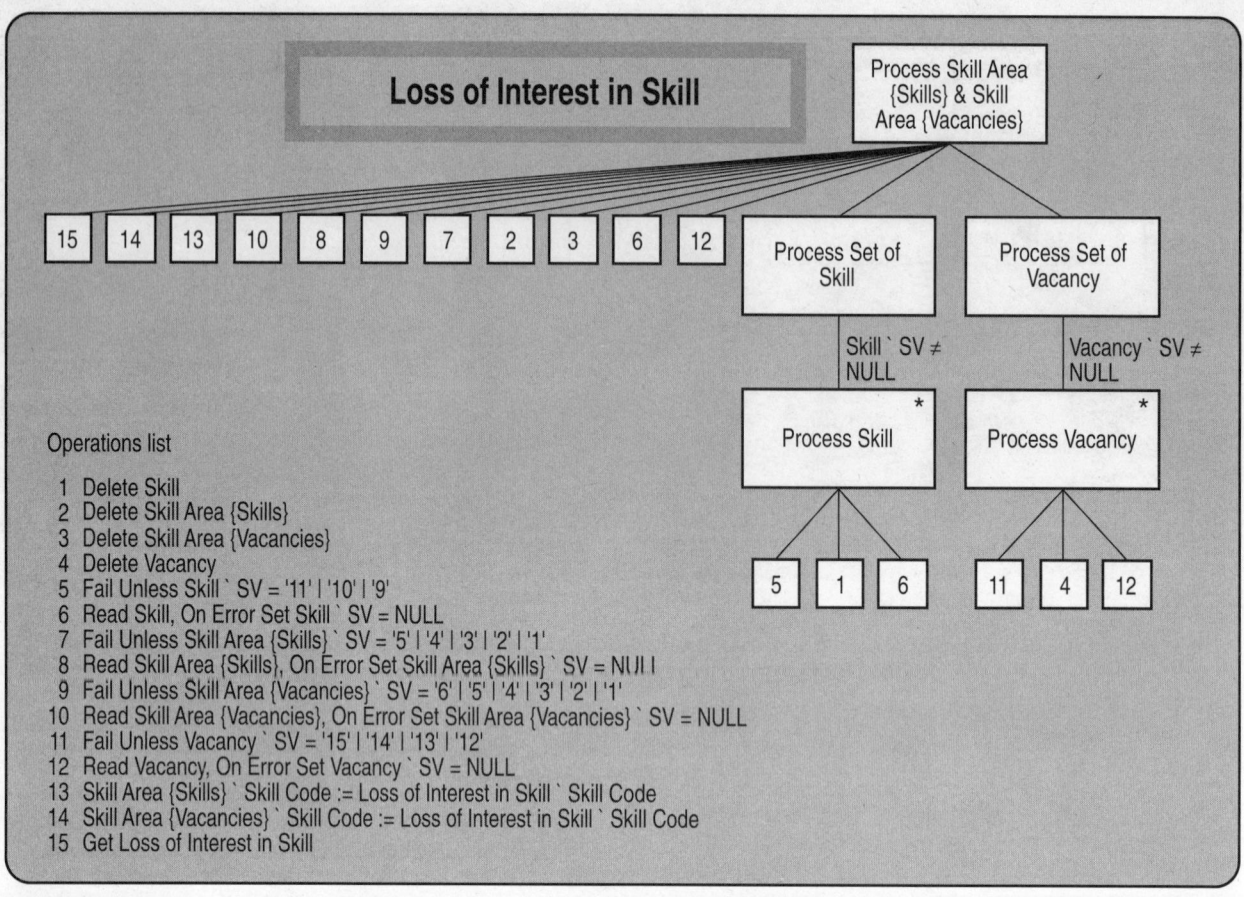

Operations list
1. Delete Skill
2. Delete Skill Area {Skills}
3. Delete Skill Area {Vacancies}
4. Delete Vacancy
5. Fail Unless Skill ` SV = '11' | '10' | '9'
6. Read Skill, On Error Set Skill ` SV = NULL
7. Fail Unless Skill Area {Skills} ` SV = '5' | '4' | '3' | '2' | '1'
8. Read Skill Area {Skills}, On Error Set Skill Area {Skills} ` SV = NULL
9. Fail Unless Skill Area {Vacancies} ` SV = '6' | '5' | '4' | '3' | '2' | '1'
10. Read Skill Area {Vacancies}, On Error Set Skill Area {Vacancies} ` SV = NULL
11. Fail Unless Vacancy ` SV = '15' | '14' | '13' | '12'
12. Read Vacancy, On Error Set Vacancy ` SV = NULL
13. Skill Area {Skills} ` Skill Code := Loss of Interest in Skill ` Skill Code
14. Skill Area {Vacancies} ` Skill Code := Loss of Interest in Skill ` Skill Code
15. Get Loss of Interest in Skill

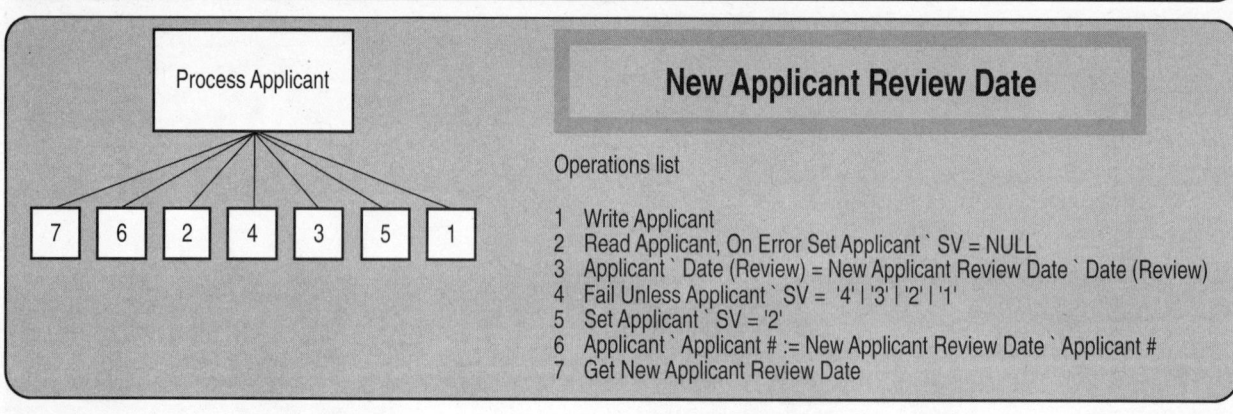

New Applicant Review Date

Structure:
- Process Applicant
  - [7] [6] [2] [4] [3] [5] [1]

Operations list
1. Write Applicant
2. Read Applicant, On Error Set Applicant ` SV = NULL
3. Applicant ` Date (Review) = New Applicant Review Date ` Date (Review)
4. Fail Unless Applicant ` SV = '4' | '3' | '2' | '1'
5. Set Applicant ` SV = '2'
6. Applicant ` Applicant # := New Applicant Review Date ` Applicant #
7. Get New Applicant Review Date

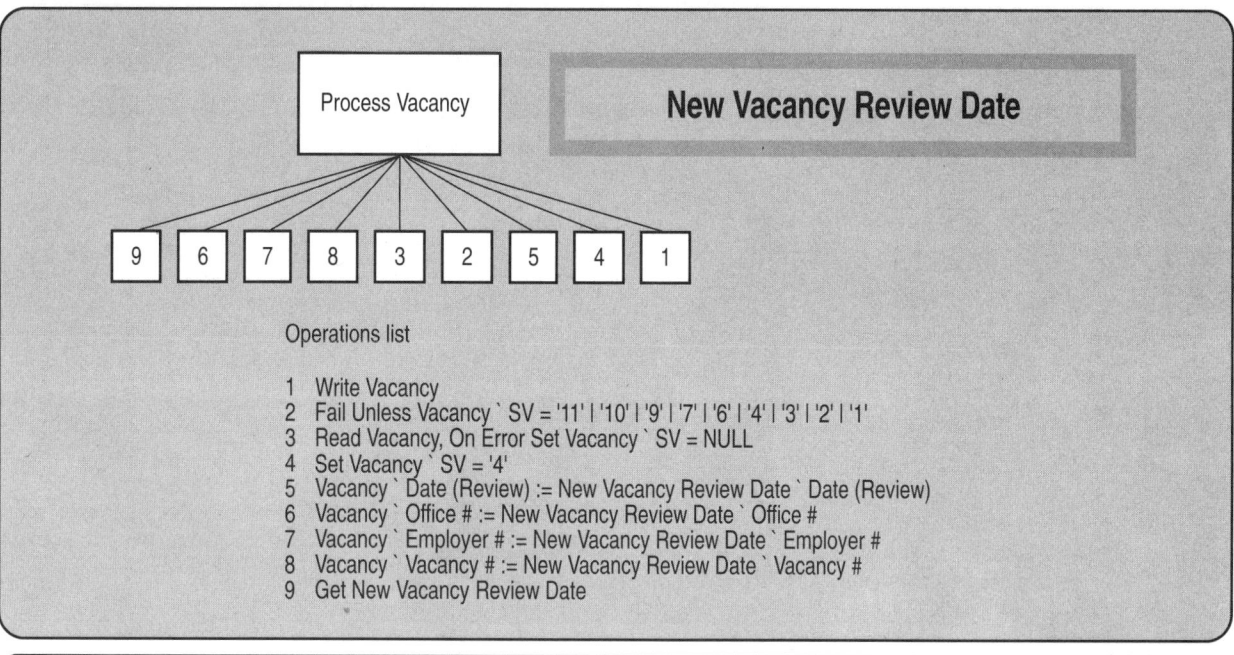

New Vacancy Review Date

Process Vacancy

9 6 7 8 3 2 5 4 1

Operations list

1. Write Vacancy
2. Fail Unless Vacancy ` SV = '11' | '10' | '9' | '7' | '6' | '4' | '3' | '2' | '1'
3. Read Vacancy, On Error Set Vacancy ` SV = NULL
4. Set Vacancy ` SV = '4'
5. Vacancy ` Date (Review) := New Vacancy Review Date ` Date (Review)
6. Vacancy ` Office # := New Vacancy Review Date ` Office #
7. Vacancy ` Employer # := New Vacancy Review Date ` Employer #
8. Vacancy ` Vacancy # := New Vacancy Review Date ` Vacancy #
9. Get New Vacancy Review Date

Notification of Vacancy

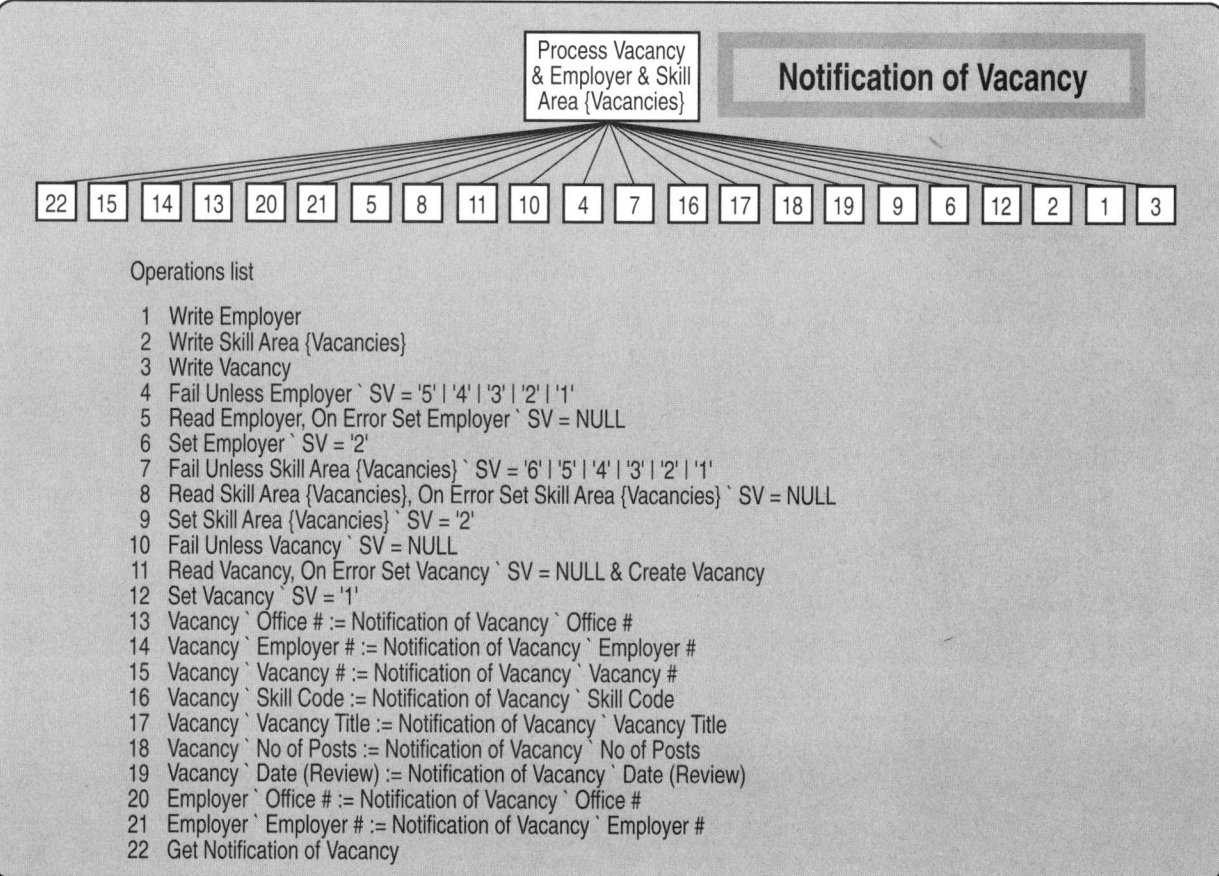

Process Vacancy & Employer & Skill Area {Vacancies}

22 15 14 13 20 21 5 8 11 10 4 7 16 17 18 19 9 6 12 2 1 3

Operations list

1. Write Employer
2. Write Skill Area {Vacancies}
3. Write Vacancy
4. Fail Unless Employer ` SV = '5' | '4' | '3' | '2' | '1'
5. Read Employer, On Error Set Employer ` SV = NULL
6. Set Employer ` SV = '2'
7. Fail Unless Skill Area {Vacancies} ` SV = '6' | '5' | '4' | '3' | '2' | '1'
8. Read Skill Area {Vacancies}, On Error Set Skill Area {Vacancies} ` SV = NULL
9. Set Skill Area {Vacancies} ` SV = '2'
10. Fail Unless Vacancy ` SV = NULL
11. Read Vacancy, On Error Set Vacancy ` SV = NULL & Create Vacancy
12. Set Vacancy ` SV = '1'
13. Vacancy ` Office # := Notification of Vacancy ` Office #
14. Vacancy ` Employer # := Notification of Vacancy ` Employer #
15. Vacancy ` Vacancy # := Notification of Vacancy ` Vacancy #
16. Vacancy ` Skill Code := Notification of Vacancy ` Skill Code
17. Vacancy ` Vacancy Title := Notification of Vacancy ` Vacancy Title
18. Vacancy ` No of Posts := Notification of Vacancy ` No of Posts
19. Vacancy ` Date (Review) := Notification of Vacancy ` Date (Review)
20. Employer ` Office # := Notification of Vacancy ` Office #
21. Employer ` Employer # := Notification of Vacancy ` Employer #
22. Get Notification of Vacancy

Office Closure

```
                        Process Office
       ┌───┬───┬───┬───┬───┬──────┴──────┬─────────────┬───┐
      11  10   8   7   6              Process Set    Process Set   3
                                      of Applicant   of Employer
                                    Applicant`SV ≠   Employer`SV ≠
                                      NULL              NULL
                                         │ *              │ *
                                   Process Applicant  Process Employer
                                    ┌──┬──┐             ┌──┬──┐
                                    9  1  4             5  2  6
```

Operations list

1. Delete Applicant
2. Delete Employer
3. Delete Office
4. Read Applicant, On Error Set Applicant ` SV = NULL
5. Fail Unless Employer ` SV = '6'
6. Read Employer, On Error Set Employer `SV = NULL
7. Fail Unless Office ` SV = '6' | '5' | '4' | '3' | '2' | '1'
8. Read Office, On Error Set Office ` SV = NULL
9. Fail Unless Applicant `SV = '6' | '5'
10. Office ` Office # := Office Closure ` Office #
11. Get Office Closure

Office Opening

```
       Process Office
    ┌──┬──┬──┬──┬──┬──┐
    8  5  3  2  6  7  4  1
```

Operations list

1. Write Office
2. Fail Unless Office ` SV = NULL
3. Read Office, On Error Set Office ` SV = NULL & Create Office
4. Set Office ` SV = '1'
5. Office ` Office # := Office Opening ` Office #
6. Office ` Office Name := Office Opening ` Office Name
7. Office ` Office Address := Office Opening ` Office Address
8. Get Office Opening

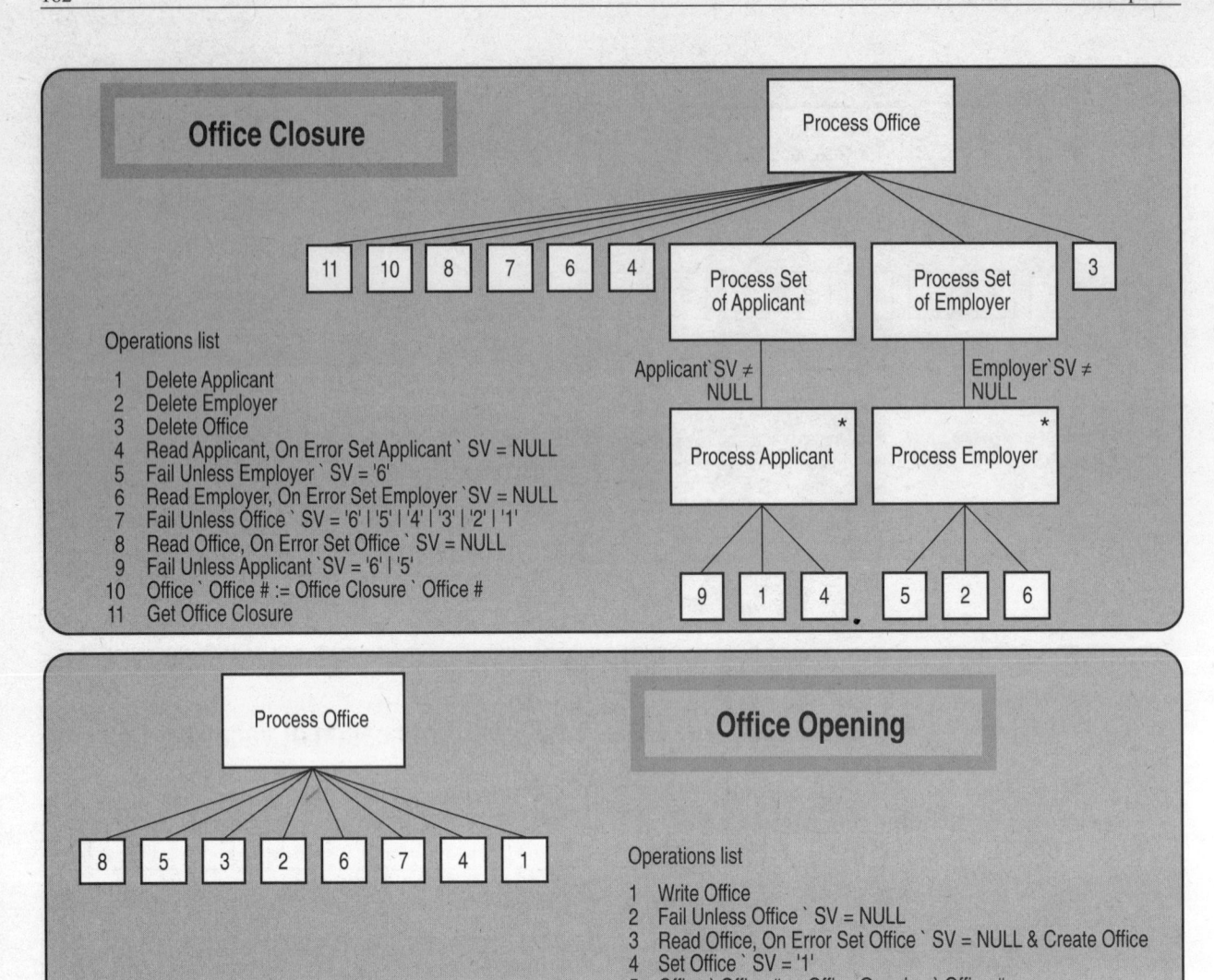

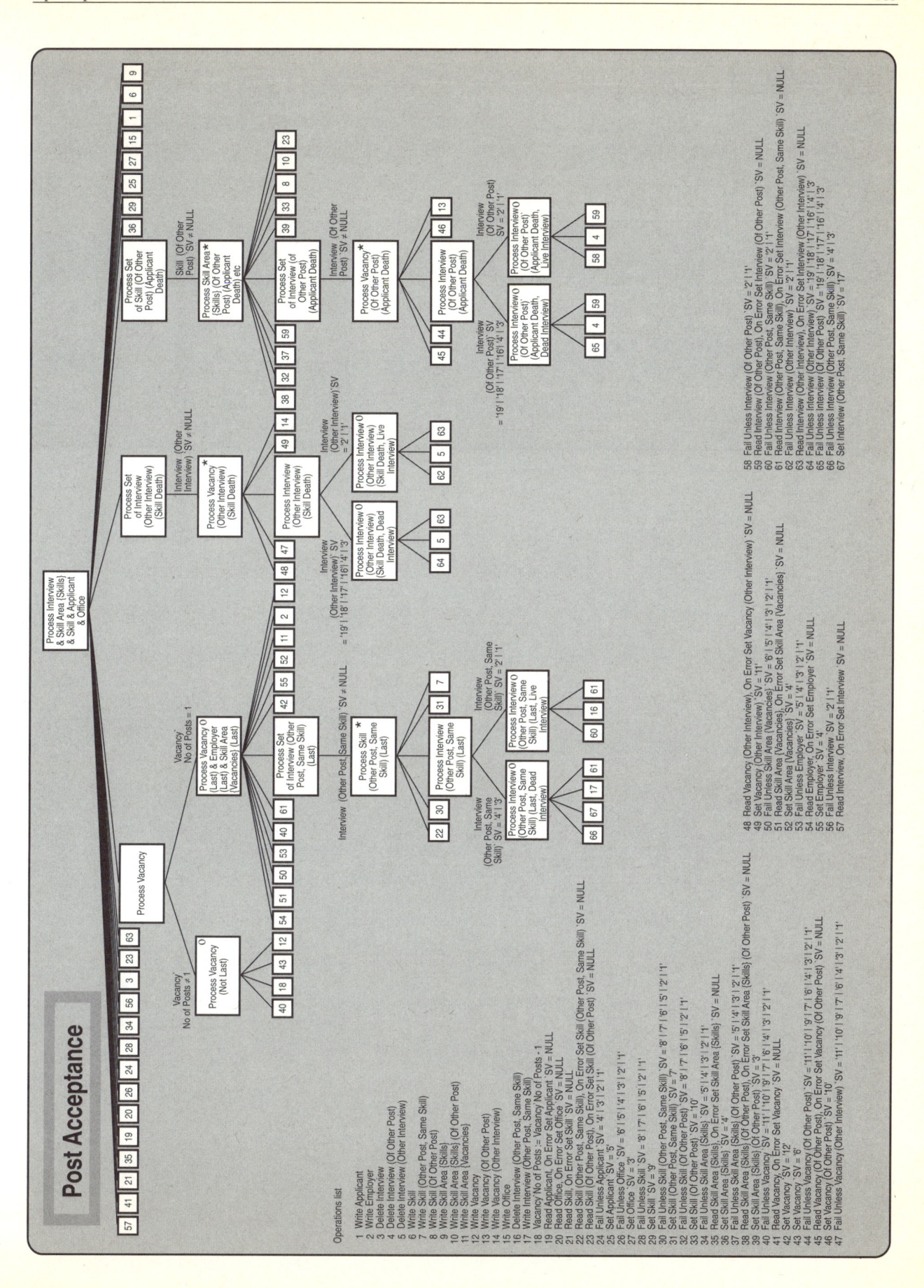

Update process models 183

Post Acceptance

Operations list

1. Write Applicant
2. Write Employer
3. Delete Interview (Of Other Post)
4. Delete Interview (Other Interview)
5. Delete Interview (Other Interview)
6. Write Skill
7. Write Skill (Other Post, Same Skill)
8. Write Skill (Of Other Post)
9. Write Skill Area (Skills)
10. Write Skill Area (Skills) (Of Other Post)
11. Write Skill Area (Vacancies)
12. Write Vacancy
13. Write Vacancy (Of Other Post)
14. Write Vacancy (Other Interview)
15. Write Office
16. Delete Interview (Other Post, Same Skill)
17. Write Interview (Other Post, Same Skill)
18. Vacancy No of Posts = Vacancy No of Posts - 1
19. Write Vacancy (Of Other Post) 'SV' = '10'
20. Read Applicant, On Error Set Applicant 'SV' = NULL
21. Read Skill, On Error Set Office 'SV' = NULL
22. Read Skill, On Error Set Skill 'SV' = NULL
23. Read Skill (Other Post, Same Skill), On Error Set Skill (Other Post, Same Skill) 'SV' = NULL
24. Read Skill (Of Other Post), On Error Set Skill (Of Other Post) 'SV' = NULL
25. Fail Unless Applicant 'SV' = '4' | '3' | '2' | '1'
26. Set Applicant 'SV' = '5'
27. Fail Unless Office 'SV' = '3'
28. Set Office 'SV' = '6' | '5' | '4' | '3' | '2' | '1'
29. Set Skill 'SV' = '9'
30. Fail Unless Skill (Other Post, Same Skill) 'SV' = '8' | '7' | '6' | '5' | '2' | '1'
31. Set Skill (Other Post, Same Skill) 'SV' = '7'
32. Fail Unless Skill (Of Other Post) 'SV' = '8' | '7' | '6' | '5' | '2' | '1'
33. Fail Unless Skill Area (Skills) 'SV' = '5' | '4' | '3' | '2' | '1'
34. Fail Unless Skill Area (Skills) 'SV' = NULL
35. Read Skill Area (Skills), On Error Set Skill Area (Skills) 'SV' = NULL
36. Set Skill Area (Skills) 'SV' = '4'
37. Fail Unless Skill Area (Skills) (Of Other Post) 'SV' = '5' | '4' | '3' | '2' | '1'
38. Read Skill Area (Skills) (Of Other Post), On Error Set Skill Area (Skills) (Of Other Post) 'SV' = NULL
39. Set Skill Area (Skills) (Of Other Post) 'SV' = '3'
40. Fail Unless Vacancy 'SV' = '11' | '10' | '9' | '7' | '6' | '4' | '3' | '2' | '1'
41. Read Vacancy, On Error Set Vacancy 'SV' = NULL
42. Set Vacancy 'SV' = '12'
43. Set Vacancy 'SV' = '8'
44. Fail Unless Vacancy (Of Other Post) 'SV' = '11' | '10' | '9' | '7' | '6' | '4' | '3' | '2' | '1'
45. Read Vacancy (Of Other Post), On Error Set Vacancy (Of Other Post) 'SV' = NULL
46. Set Vacancy (Of Other Post) 'SV' = '10'
47. Fail Unless Vacancy (Other Interview) 'SV' = '11' | '10' | '9' | '7' | '6' | '4' | '3' | '2' | '1'
48. Read Vacancy (Other Interview), On Error Set Vacancy (Other Interview) 'SV' = NULL
49. Set Vacancy (Other Interview) 'SV' = '11'
50. Fail Unless Skill Area (Vacancies), On Error Set Skill Area (Vacancies) 'SV' = '6' | '5' | '4' | '3' | '2' | '1'
51. Read Skill Area (Vacancies), On Error Set Skill Area (Vacancies) 'SV' = NULL
52. Set Skill Area (Vacancies) 'SV' = '4'
53. Fail Unless Employer 'SV' = '5' | '4' | '3' | '2' | '1'
54. Read Employer, On Error Set Employer 'SV' = NULL
55. Set Employer 'SV' = '4'
56. Fail Unless Interview 'SV' = '2' | '1'
57. Read Interview, On Error Set Interview 'SV' = NULL
58. Fail Unless Interview (Of Other Post) 'SV' = '2' | '1'
59. Read Interview (Of Other Post), On Error Set Interview (Of Other Post) 'SV' = NULL
60. Fail Unless Interview (Other Post, Same Skill) 'SV' = '2' | '1'
61. Read Interview (Other Post, Same Skill), On Error Set Interview (Other Post, Same Skill) 'SV' = NULL
62. Fail Unless Interview (Other Interview) 'SV' = '2' | '1'
63. Read Interview (Other Interview), On Error Set Interview (Other Interview) 'SV' = NULL
64. Fail Unless Interview (Other Interview) 'SV' = NULL
65. Set Interview (Other Interview) 'SV' = '19' | '18' | '17' | '16' | '4' | '3'
66. Fail Unless Interview (Other Post, Same Skill) 'SV' = '4' | '3'
67. Set Interview (Other Post, Same Skill) 'SV' = '17'

Post Addition

Process Vacancy

| 9 | 6 | 7 | 8 | 4 | 3 | 2 | 5 | 1 |

Operations list

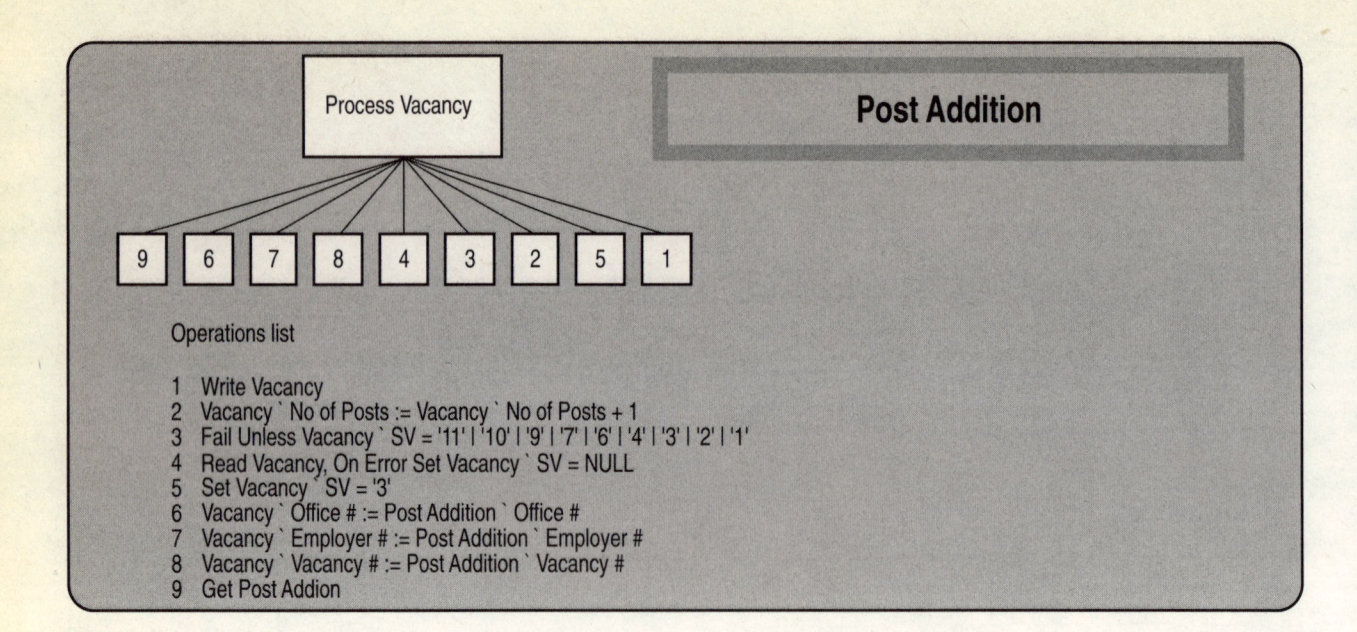

1 Write Vacancy
2 Vacancy ` No of Posts := Vacancy ` No of Posts + 1
3 Fail Unless Vacancy ` SV = '11' | '10' | '9' | '7' | '6' | '4' | '3' | '2' | '1'
4 Read Vacancy, On Error Set Vacancy ` SV = NULL
5 Set Vacancy ` SV = '3'
6 Vacancy ` Office # := Post Addition ` Office #
7 Vacancy ` Employer # := Post Addition ` Employer #
8 Vacancy ` Vacancy # := Post Addition ` Vacancy #
9 Get Post Addion

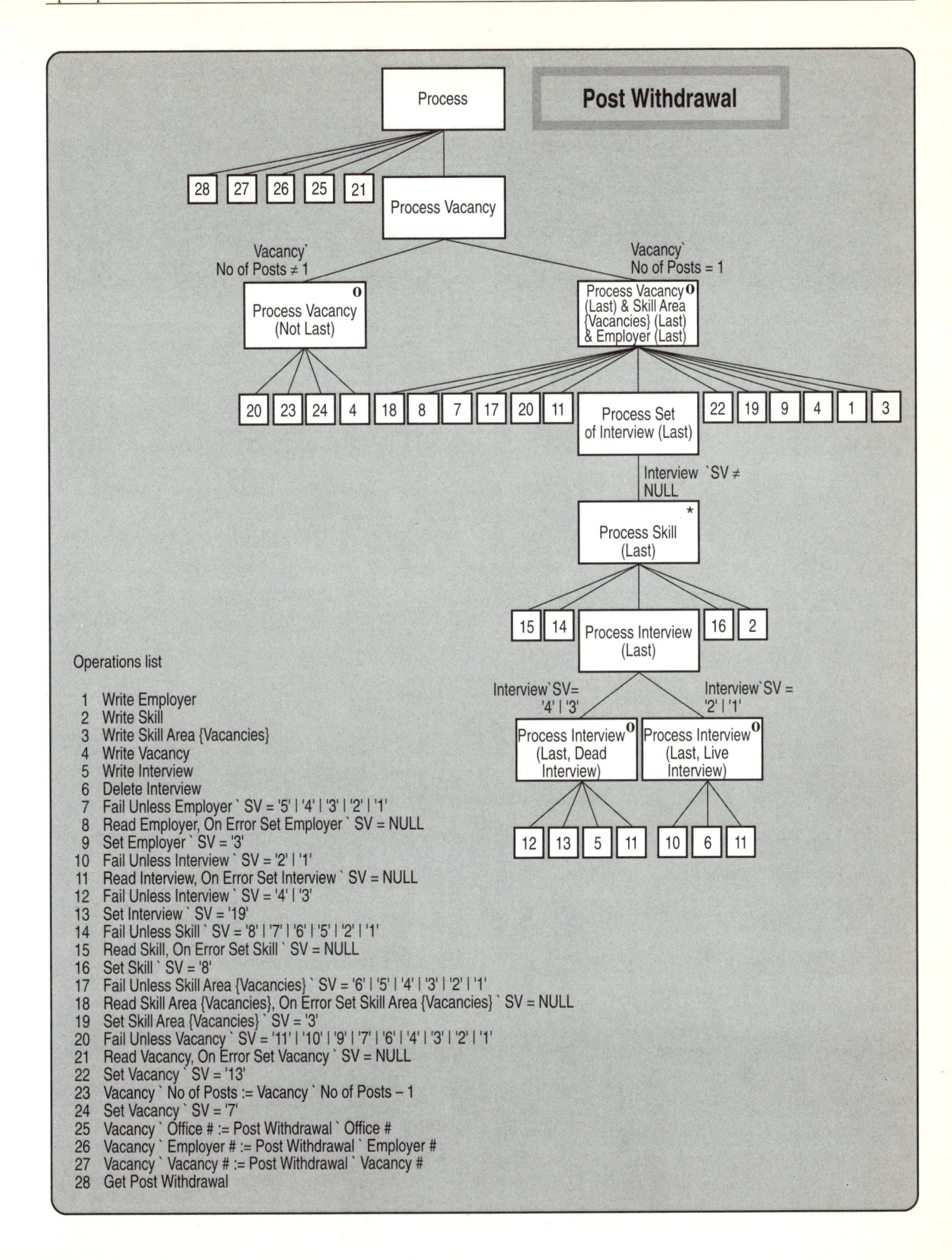

Update process models

Post Withdrawal

Operations list

1. Write Employer
2. Write Skill
3. Write Skill Area {Vacancies}
4. Write Vacancy
5. Write Interview
6. Delete Interview
7. Fail Unless Employer ` SV = '5' | '4' | '3' | '2' | '1'
8. Read Employer, On Error Set Employer ` SV = NULL
9. Set Employer ` SV = '3'
10. Fail Unless Interview ` SV = '2' | '1'
11. Read Interview, On Error Set Interview ` SV = NULL
12. Fail Unless Interview ` SV = '4' | '3'
13. Set Interview ` SV = '19'
14. Fail Unless Skill ` SV = '8' | '7' | '6' | '5' | '2' | '1'
15. Read Skill, On Error Set Skill ` SV = NULL
16. Set Skill ` SV = '8'
17. Fail Unless Skill Area {Vacancies} ` SV = '6' | '5' | '4' | '3' | '2' | '1'
18. Read Skill Area {Vacancies}, On Error Set Skill Area {Vacancies} ` SV = NULL
19. Set Skill Area {Vacancies} ` SV = '3'
20. Fail Unless Vacancy ` SV = '11' | '10' | '9' | '7' | '6' | '4' | '3' | '2' | '1'
21. Read Vacancy, On Error Set Vacancy ` SV = NULL
22. Set Vacancy ` SV = '13'
23. Vacancy ` No of Posts := Vacancy ` No of Posts – 1
24. Set Vacancy ` SV = '7'
25. Vacancy ` Office # := Post Withdrawal ` Office #
26. Vacancy ` Employer # := Post Withdrawal ` Employer #
27. Vacancy ` Vacancy # := Post Withdrawal ` Vacancy #
28. Get Post Withdrawal

Refusal of Offer

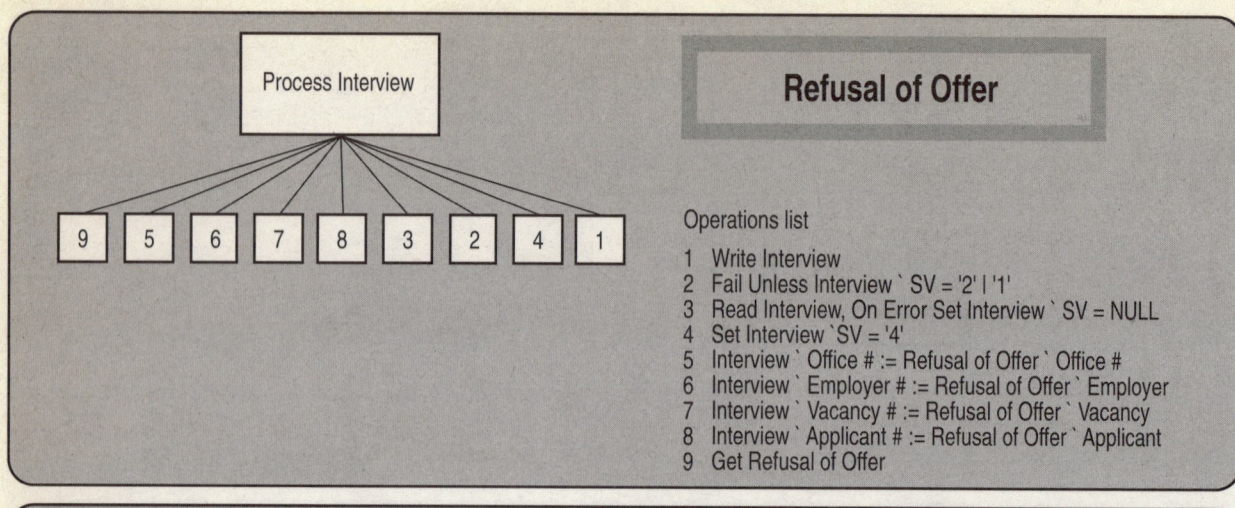

Process Interview
— 9, 5, 6, 7, 8, 3, 2, 4, 1

Operations list

1. Write Interview
2. Fail Unless Interview ` SV = '2' | '1'
3. Read Interview, On Error Set Interview ` SV = NULL
4. Set Interview ` SV = '4'
5. Interview ` Office # := Refusal of Offer ` Office #
6. Interview ` Employer # := Refusal of Offer ` Employer
7. Interview ` Vacancy # := Refusal of Offer ` Vacancy
8. Interview ` Applicant # := Refusal of Offer ` Applicant
9. Get Refusal of Offer

Registration of Employer

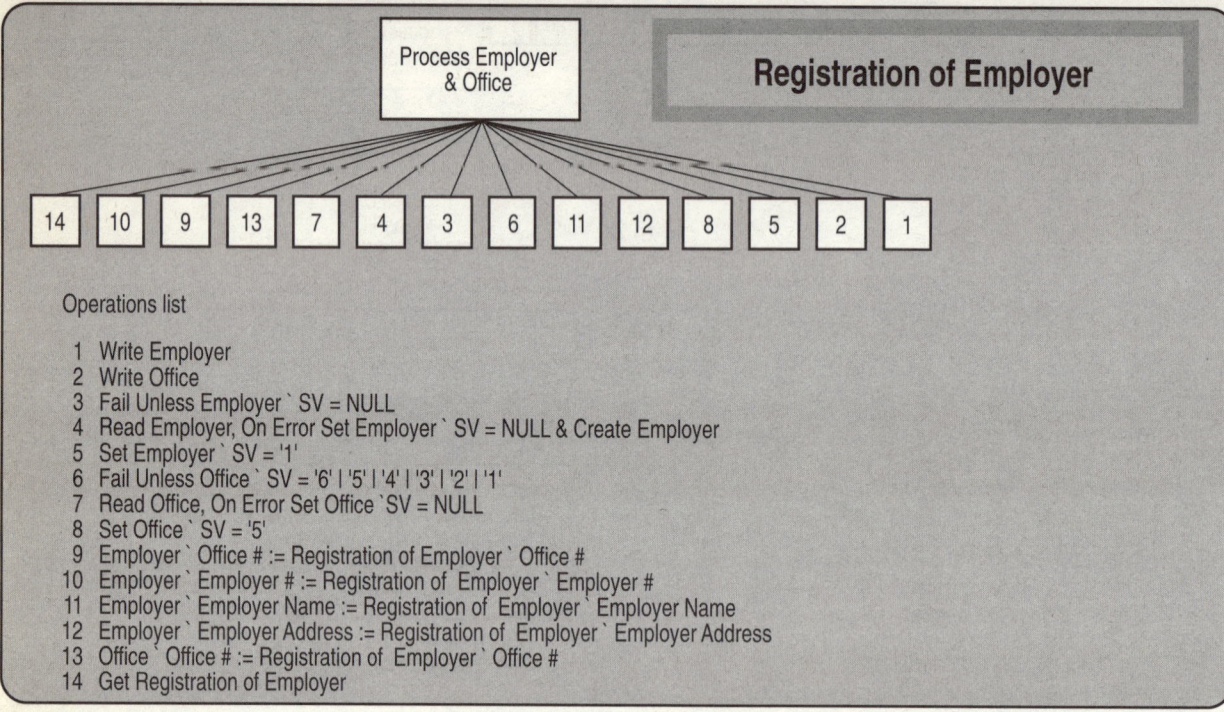

Process Employer & Office
— 14, 10, 9, 13, 7, 4, 3, 6, 11, 12, 8, 5, 2, 1

Operations list

1. Write Employer
2. Write Office
3. Fail Unless Employer ` SV = NULL
4. Read Employer, On Error Set Employer ` SV = NULL & Create Employer
5. Set Employer ` SV = '1'
6. Fail Unless Office ` SV = '6' | '5' | '4' | '3' | '2' | '1'
7. Read Office, On Error Set Office ` SV = NULL
8. Set Office ` SV = '5'
9. Employer ` Office # := Registration of Employer ` Office #
10. Employer ` Employer # := Registration of Employer ` Employer #
11. Employer ` Employer Name := Registration of Employer ` Employer Name
12. Employer ` Employer Address := Registration of Employer ` Employer Address
13. Office ` Office # := Registration of Employer ` Office #
14. Get Registration of Employer

Rejection of Applicant

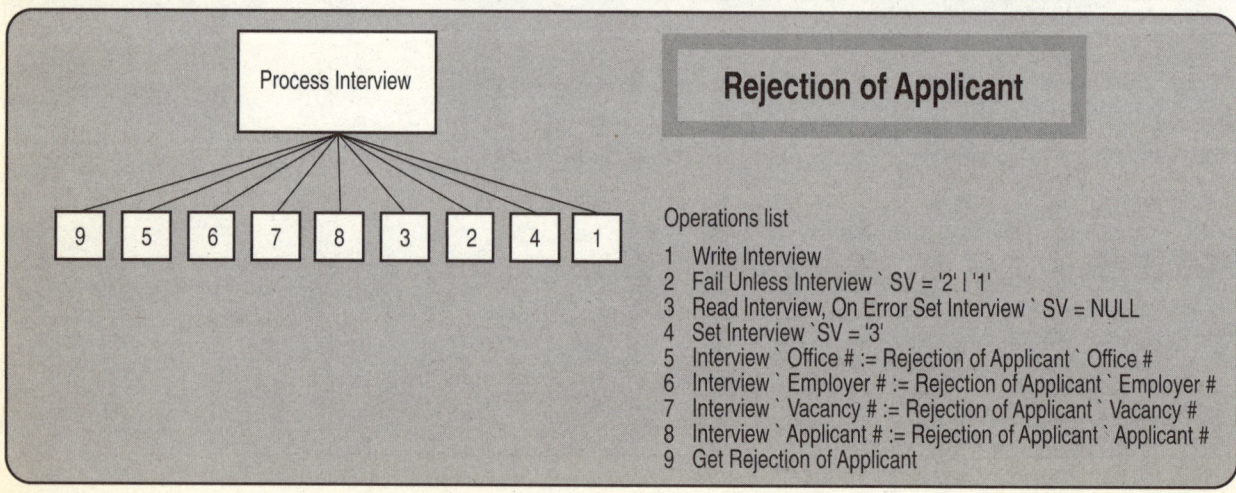

Process Interview
— 9, 5, 6, 7, 8, 3, 2, 4, 1

Operations list

1. Write Interview
2. Fail Unless Interview ` SV = '2' | '1'
3. Read Interview, On Error Set Interview ` SV = NULL
4. Set Interview ` SV = '3'
5. Interview ` Office # := Rejection of Applicant ` Office #
6. Interview ` Employer # := Rejection of Applicant ` Employer #
7. Interview ` Vacancy # := Rejection of Applicant ` Vacancy #
8. Interview ` Applicant # := Rejection of Applicant ` Applicant #
9. Get Rejection of Applicant

Rescheduling

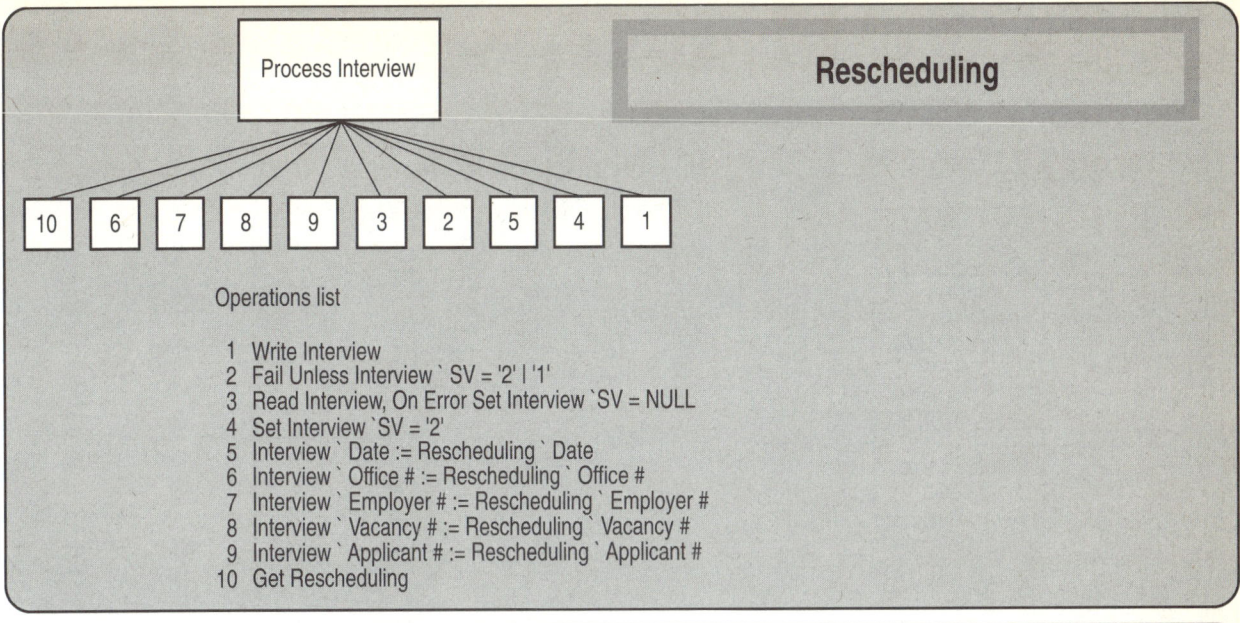

Process Interview
[10] [6] [7] [8] [9] [3] [2] [5] [4] [1]

Operations list

1. Write Interview
2. Fail Unless Interview ` SV = '2' | '1'
3. Read Interview, On Error Set Interview ` SV = NULL
4. Set Interview ` SV = '2'
5. Interview ` Date := Rescheduling ` Date
6. Interview ` Office # := Rescheduling ` Office #
7. Interview ` Employer # := Rescheduling ` Employer #
8. Interview ` Vacancy # := Rescheduling ` Vacancy #
9. Interview ` Applicant # := Rescheduling ` Applicant #
10. Get Rescheduling

Vacancy Withdrawal

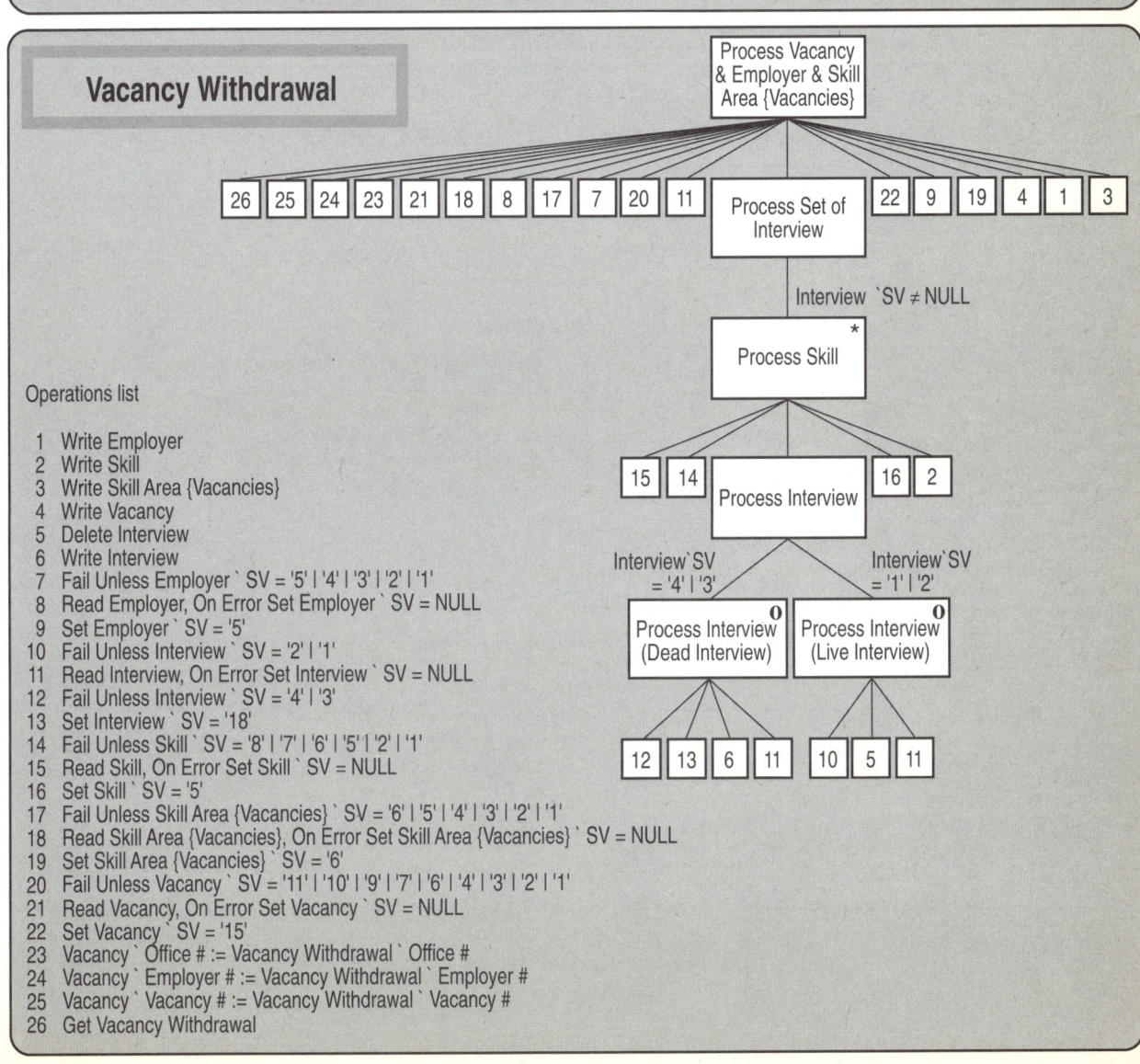

Process Vacancy & Employer & Skill Area {Vacancies}
[26] [25] [24] [23] [21] [18] [8] [17] [7] [20] [11] Process Set of Interview [22] [9] [19] [4] [1] [3]

Interview ` SV ≠ NULL

Process Skill *

[15] [14] Process Interview [16] [2]

Interview`SV = '4' | '3'   Interview`SV = '1' | '2'

Process Interview (Dead Interview)°   Process Interview (Live Interview)°

[12] [13] [6] [11]   [10] [5] [11]

Operations list

1. Write Employer
2. Write Skill
3. Write Skill Area {Vacancies}
4. Write Vacancy
5. Delete Interview
6. Write Interview
7. Fail Unless Employer ` SV = '5' | '4' | '3' | '2' | '1'
8. Read Employer, On Error Set Employer ` SV = NULL
9. Set Employer ` SV = '5'
10. Fail Unless Interview ` SV = '2' | '1'
11. Read Interview, On Error Set Interview ` SV = NULL
12. Fail Unless Interview ` SV = '4' | '3'
13. Set Interview ` SV = '18'
14. Fail Unless Skill ` SV = '8' | '7' | '6' | '5' | '2' | '1'
15. Read Skill, On Error Set Skill ` SV = NULL
16. Set Skill ` SV = '5'
17. Fail Unless Skill Area {Vacancies} ` SV = '6' | '5' | '4' | '3' | '2' | '1'
18. Read Skill Area {Vacancies}, On Error Set Skill Area {Vacancies} ` SV = NULL
19. Set Skill Area {Vacancies} ` SV = '6'
20. Fail Unless Vacancy ` SV = '11' | '10' | '9' | '7' | '6' | '4' | '3' | '2' | '1'
21. Read Vacancy, On Error Set Vacancy ` SV = NULL
22. Set Vacancy ` SV = '15'
23. Vacancy ` Office # := Vacancy Withdrawal ` Office #
24. Vacancy ` Employer # := Vacancy Withdrawal ` Employer #
25. Vacancy ` Vacancy # := Vacancy Withdrawal ` Vacancy #
26. Get Vacancy Withdrawal

Enquiry process models

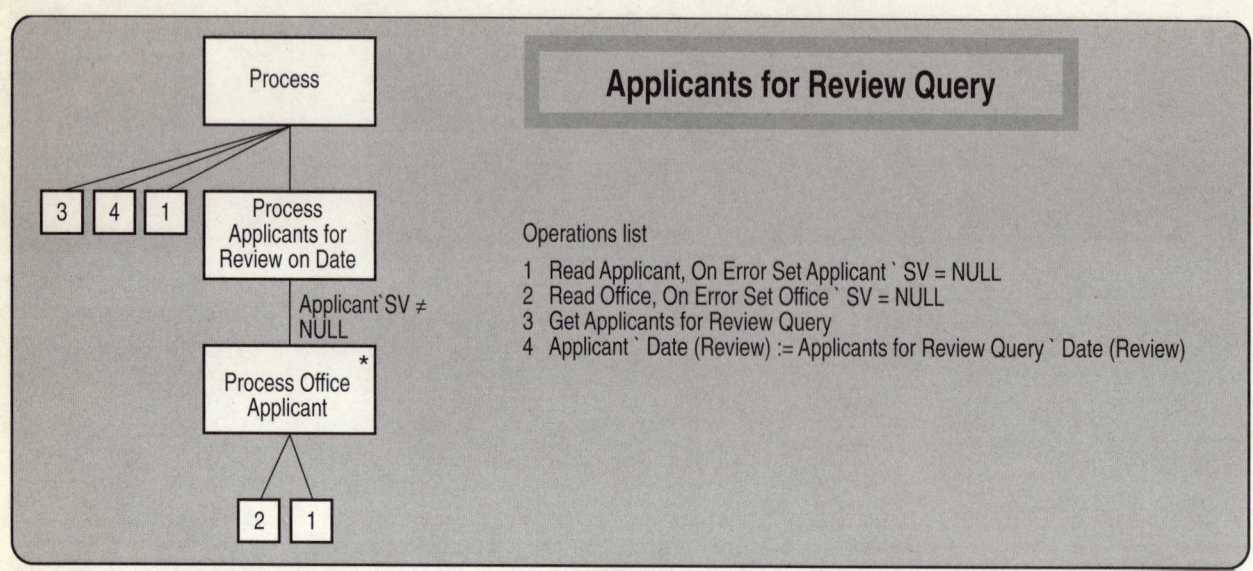

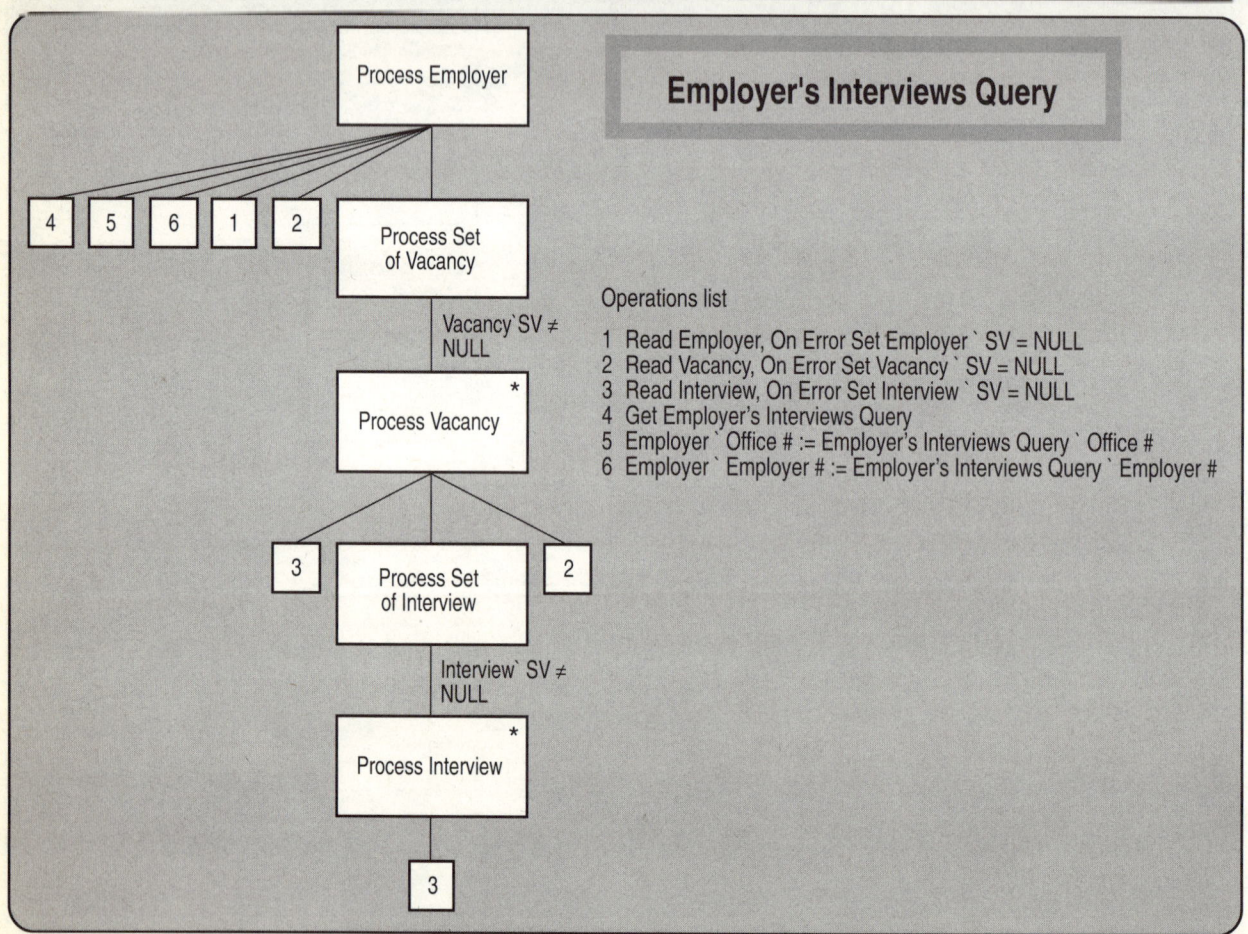

Applicants for Review Query

Operations list
1. Read Applicant, On Error Set Applicant ` SV = NULL
2. Read Office, On Error Set Office ` SV = NULL
3. Get Applicants for Review Query
4. Applicant ` Date (Review) := Applicants for Review Query ` Date (Review)

Employer's Interviews Query

Operations list
1. Read Employer, On Error Set Employer ` SV = NULL
2. Read Vacancy, On Error Set Vacancy ` SV = NULL
3. Read Interview, On Error Set Interview ` SV = NULL
4. Get Employer's Interviews Query
5. Employer ` Office # := Employer's Interviews Query ` Office #
6. Employer ` Employer # := Employer's Interviews Query ` Employer #

Enquiry process models

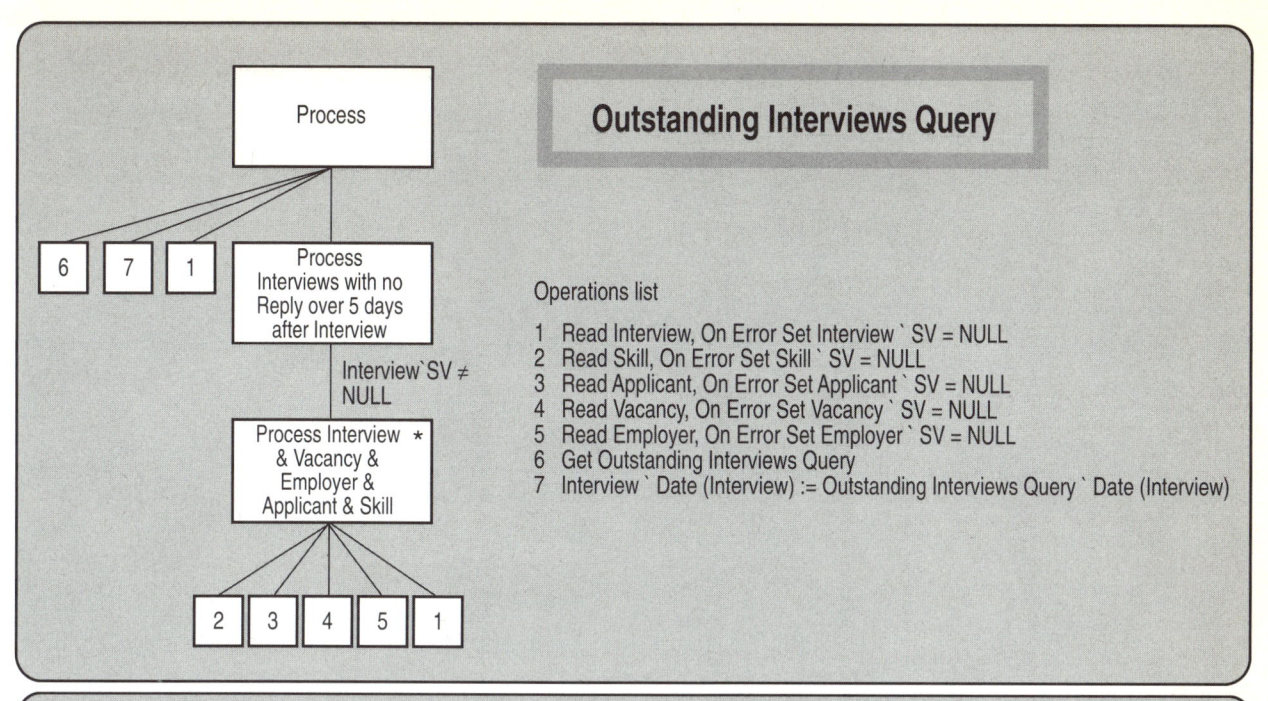

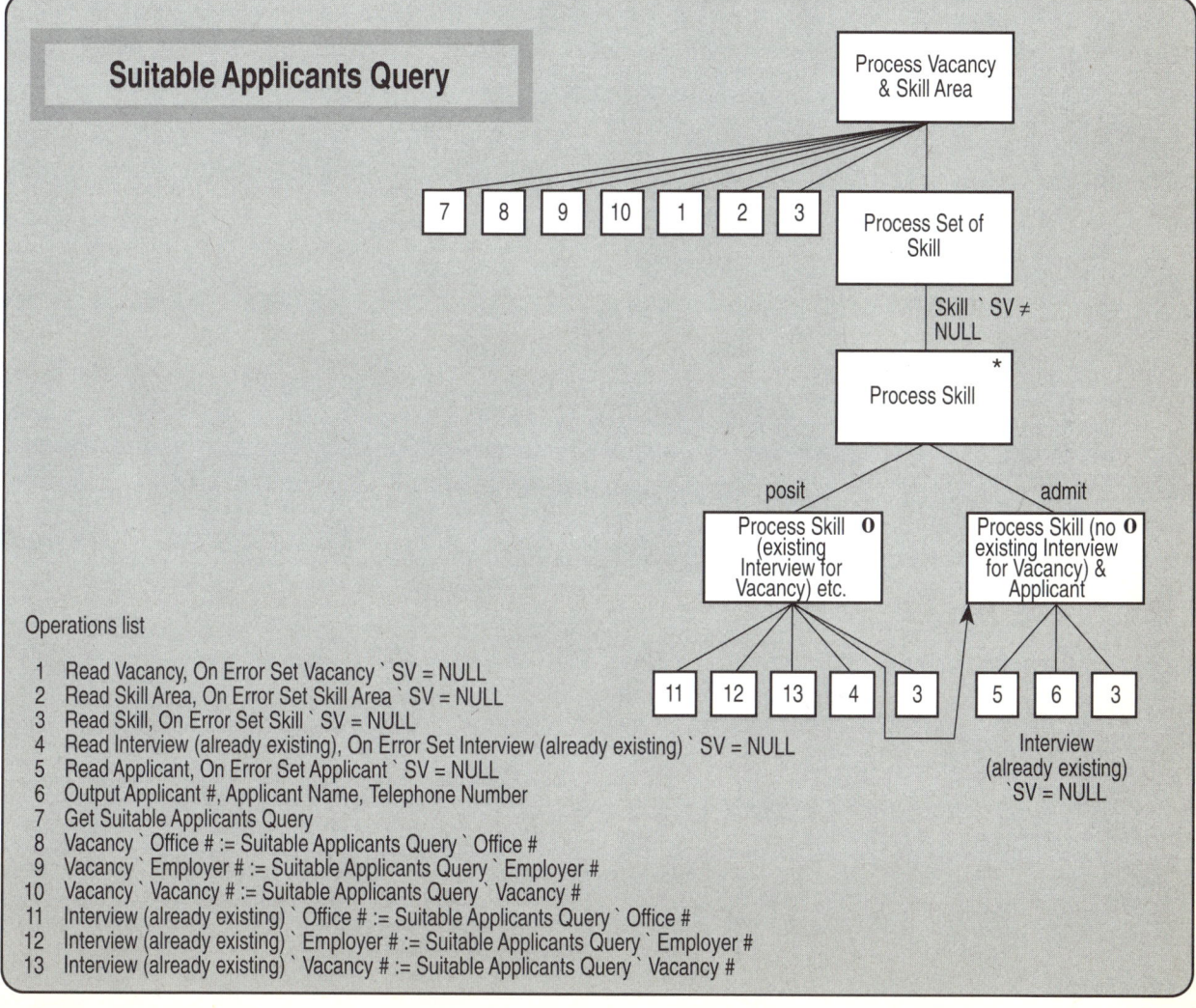

Outstanding Interviews Query

```
            Process
           /  |  \  \
          /   |   \  \
       [6] [7] [1]  Process
                    Interviews with no
                    Reply over 5 days
                    after Interview
                         |
                    Interview`SV ≠
                         NULL
                         |
                    Process Interview  *
                    & Vacancy &
                    Employer &
                    Applicant & Skill
                    /  |  |  |  \
                  [2][3][4][5][1]
```

Operations list

1. Read Interview, On Error Set Interview ` SV = NULL
2. Read Skill, On Error Set Skill ` SV = NULL
3. Read Applicant, On Error Set Applicant ` SV = NULL
4. Read Vacancy, On Error Set Vacancy ` SV = NULL
5. Read Employer, On Error Set Employer ` SV = NULL
6. Get Outstanding Interviews Query
7. Interview ` Date (Interview) := Outstanding Interviews Query ` Date (Interview)

Suitable Applicants Query

```
                                    Process Vacancy
                                    & Skill Area
                                   / / / / | \ \
                                  / / / /  |  \ \
                           [7][8][9][10][1][2][3]  Process Set of
                                                    Skill
                                                      |
                                                 Skill ` SV ≠
                                                    NULL
                                                      |
                                                             *
                                                  Process Skill
                                                   /         \
                                                posit        admit
                                                 /             \
                                    Process Skill  o    Process Skill (no  o
                                    (existing            existing Interview
                                    Interview for        for Vacancy) &
                                    Vacancy) etc.        Applicant
                                    / | | | \           / | \
                                 [11][12][13][4][3]   [5][6][3]
                                                      Interview
                                                      (already existing)
                                                      ` SV = NULL
```

Operations list

1. Read Vacancy, On Error Set Vacancy ` SV = NULL
2. Read Skill Area, On Error Set Skill Area ` SV = NULL
3. Read Skill, On Error Set Skill ` SV = NULL
4. Read Interview (already existing), On Error Set Interview (already existing) ` SV = NULL
5. Read Applicant, On Error Set Applicant ` SV = NULL
6. Output Applicant #, Applicant Name, Telephone Number
7. Get Suitable Applicants Query
8. Vacancy ` Office # := Suitable Applicants Query ` Office #
9. Vacancy ` Employer # := Suitable Applicants Query ` Employer #
10. Vacancy ` Vacancy # := Suitable Applicants Query ` Vacancy #
11. Interview (already existing) ` Office # := Suitable Applicants Query ` Office #
12. Interview (already existing) ` Employer # := Suitable Applicants Query ` Employer #
13. Interview (already existing) ` Vacancy # := Suitable Applicants Query ` Vacancy #

Suitable Vacancies Query

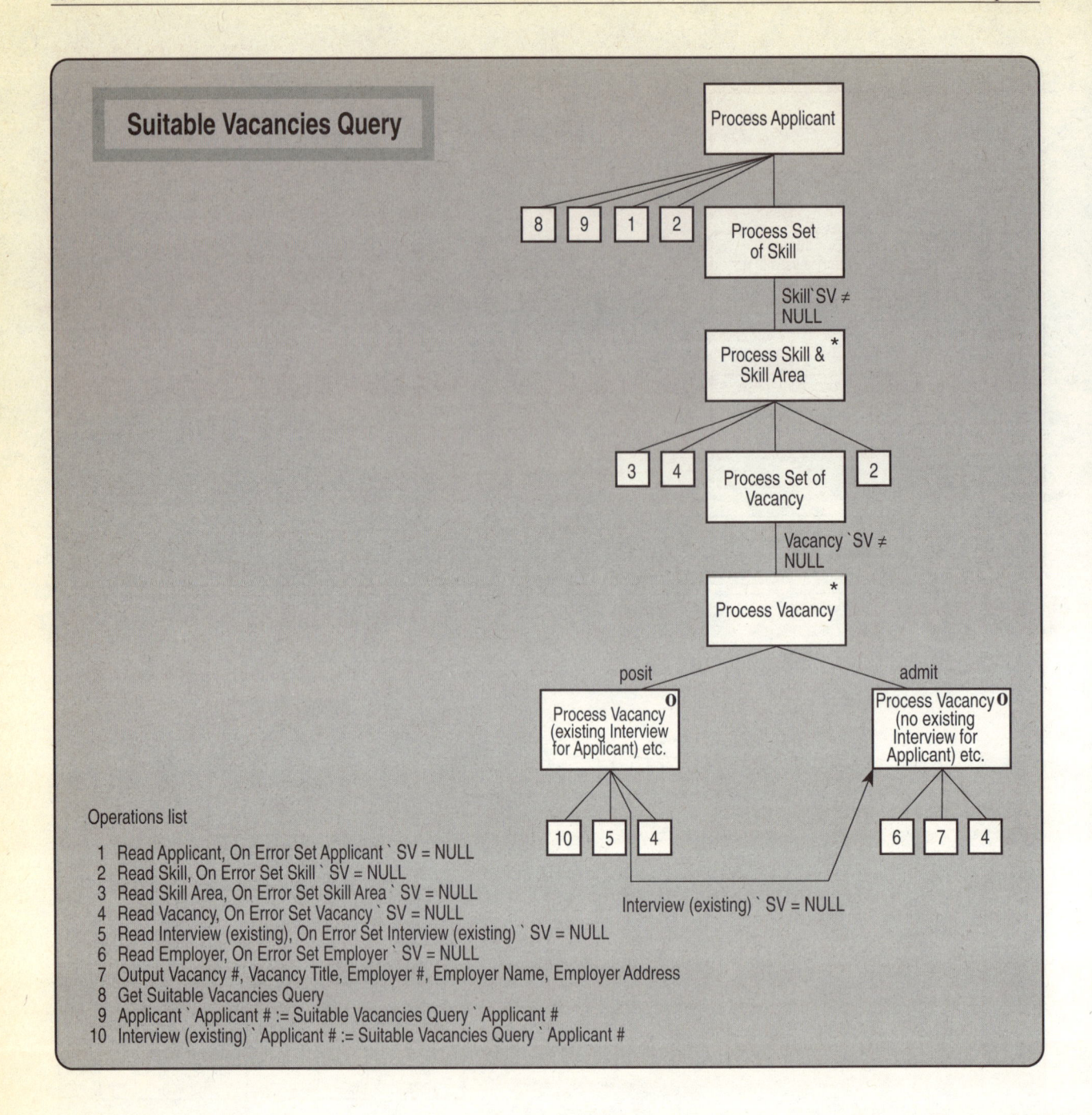

```
                                    Process Applicant
                                   /   |   |   |    \
                                  /    |   |   |     \
                                [8]  [9] [1] [2]   Process Set
                                                    of Skill
                                                       |
                                                  Skill`SV ≠
                                                     NULL
                                                       |
                                                  Process Skill &  *
                                                   Skill Area
                                                  /    |       \
                                                 /     |        \
                                              [3]  [4]  Process Set of  [2]
                                                         Vacancy
                                                            |
                                                      Vacancy`SV ≠
                                                          NULL
                                                            |
                                                      Process Vacancy  *
                                                       /          \
                                                   posit          admit
                                                    /                \
                                        Process Vacancy o      Process Vacancy o
                                        (existing Interview     (no existing
                                         for Applicant) etc.    Interview for
                                                                Applicant) etc.
                                         /    |    \             /    |    \
                                       [10]  [5]  [4]          [6]  [7]  [4]

                                    Interview (existing) ` SV = NULL
```

Operations list

1. Read Applicant, On Error Set Applicant ` SV = NULL
2. Read Skill, On Error Set Skill ` SV = NULL
3. Read Skill Area, On Error Set Skill Area ` SV = NULL
4. Read Vacancy, On Error Set Vacancy ` SV = NULL
5. Read Interview (existing), On Error Set Interview (existing) ` SV = NULL
6. Read Employer, On Error Set Employer ` SV = NULL
7. Output Vacancy #, Vacancy Title, Employer #, Employer Name, Employer Address
8. Get Suitable Vacancies Query
9. Applicant ` Applicant # := Suitable Vacancies Query ` Applicant #
10. Interview (existing) ` Applicant # := Suitable Vacancies Query ` Applicant #

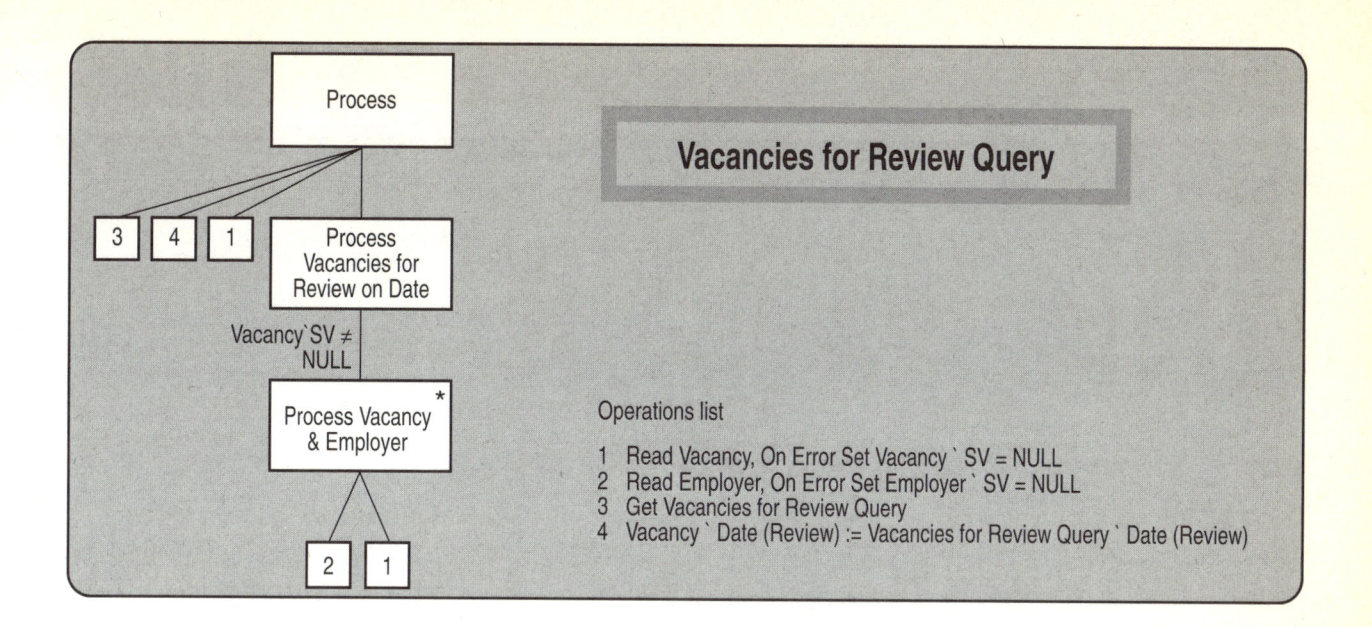

Vacancies for Review Query

Operations list

1. Read Vacancy, On Error Set Vacancy ` SV = NULL
2. Read Employer, On Error Set Employer ` SV = NULL
3. Get Vacancies for Review Query
4. Vacancy ` Date (Review) := Vacancies for Review Query ` Date (Review)

192 Annex 2 Hi-Ho SSADM Development

Function DFD equivalents

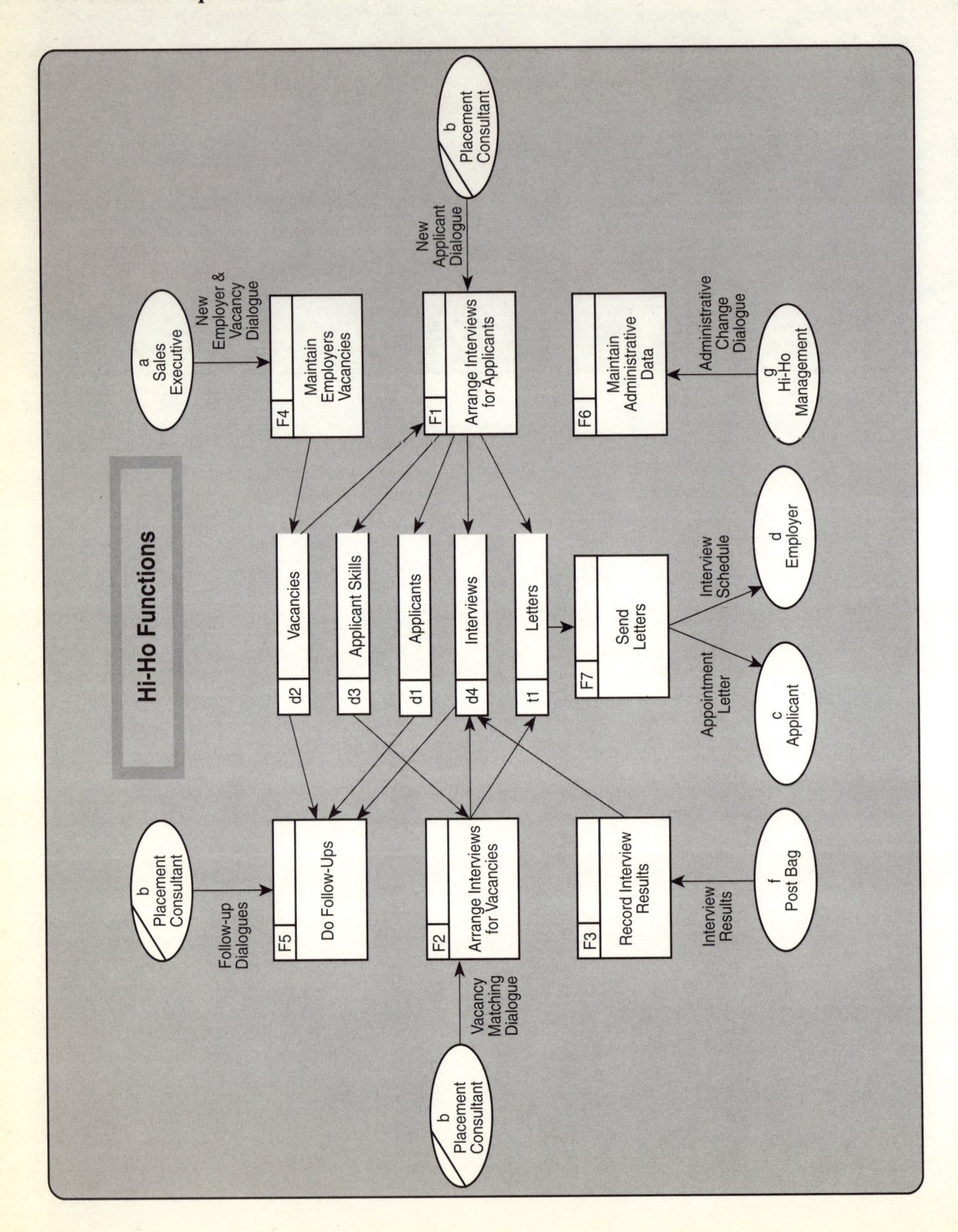

Function DFD equivalents

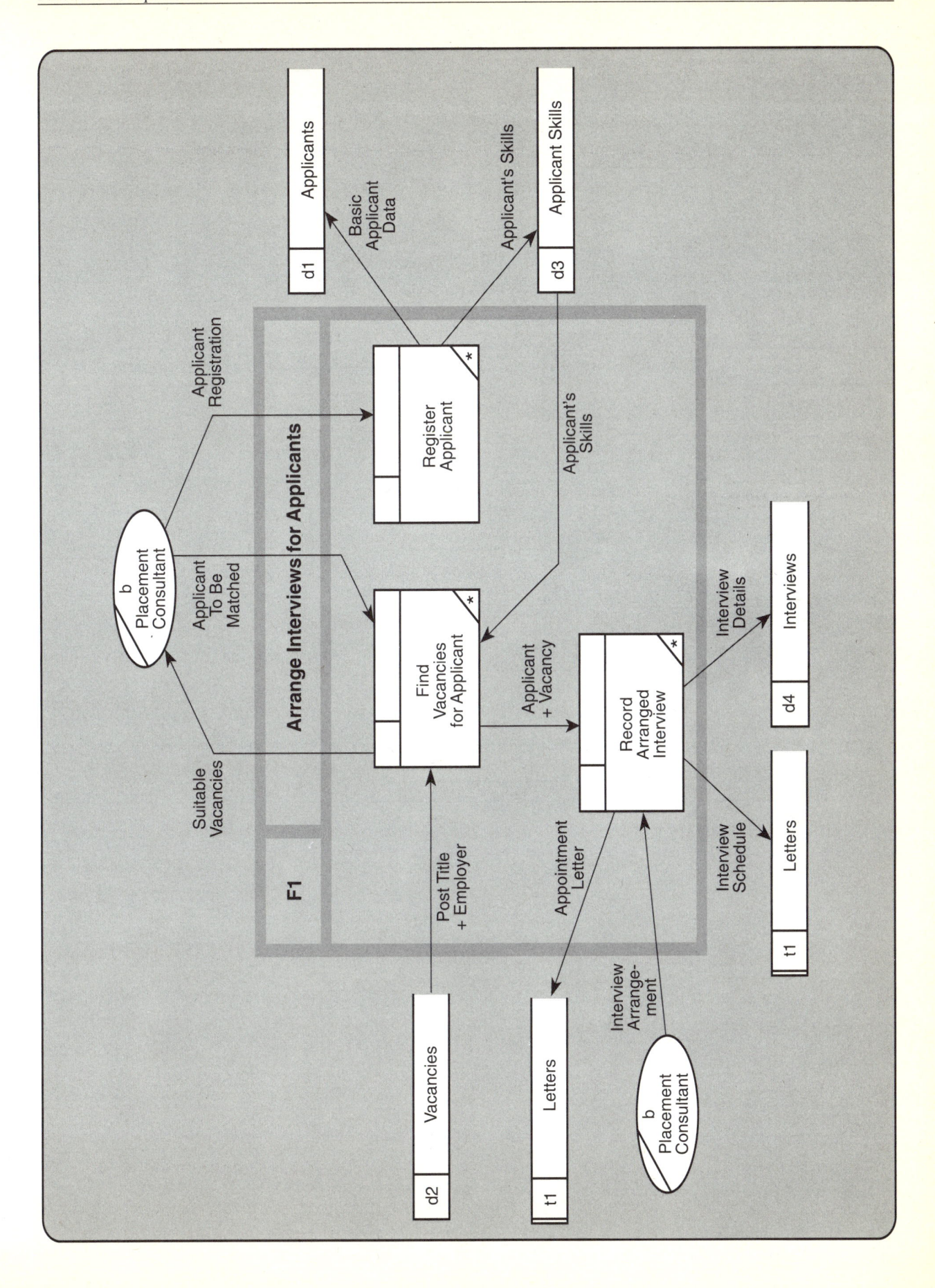

Arrange Interviews for Applicants (F1)

- Placement Consultant (b) → Applicant Registration → Register Applicant
- Register Applicant → Basic Applicant Data → d1 Applicants
- Register Applicant → Applicant's Skills → d3 Applicant Skills
- Placement Consultant (b) → Applicant To Be Matched → Find Vacancies for Applicant
- d3 Applicant Skills → Applicant's Skills → Find Vacancies for Applicant
- d2 Vacancies → Post Title + Employer → Find Vacancies for Applicant
- Find Vacancies for Applicant → Suitable Vacancies → Placement Consultant (b)
- Find Vacancies for Applicant → Applicant + Vacancy → Record Arranged Interview
- Placement Consultant (b) → Interview Arrangement → Record Arranged Interview
- Record Arranged Interview → Interview Details → d4 Interviews
- Record Arranged Interview → Appointment Letter → t1 Letters
- Record Arranged Interview → Interview Schedule → t1 Letters

F2 Arrange Interviews for Vacancies

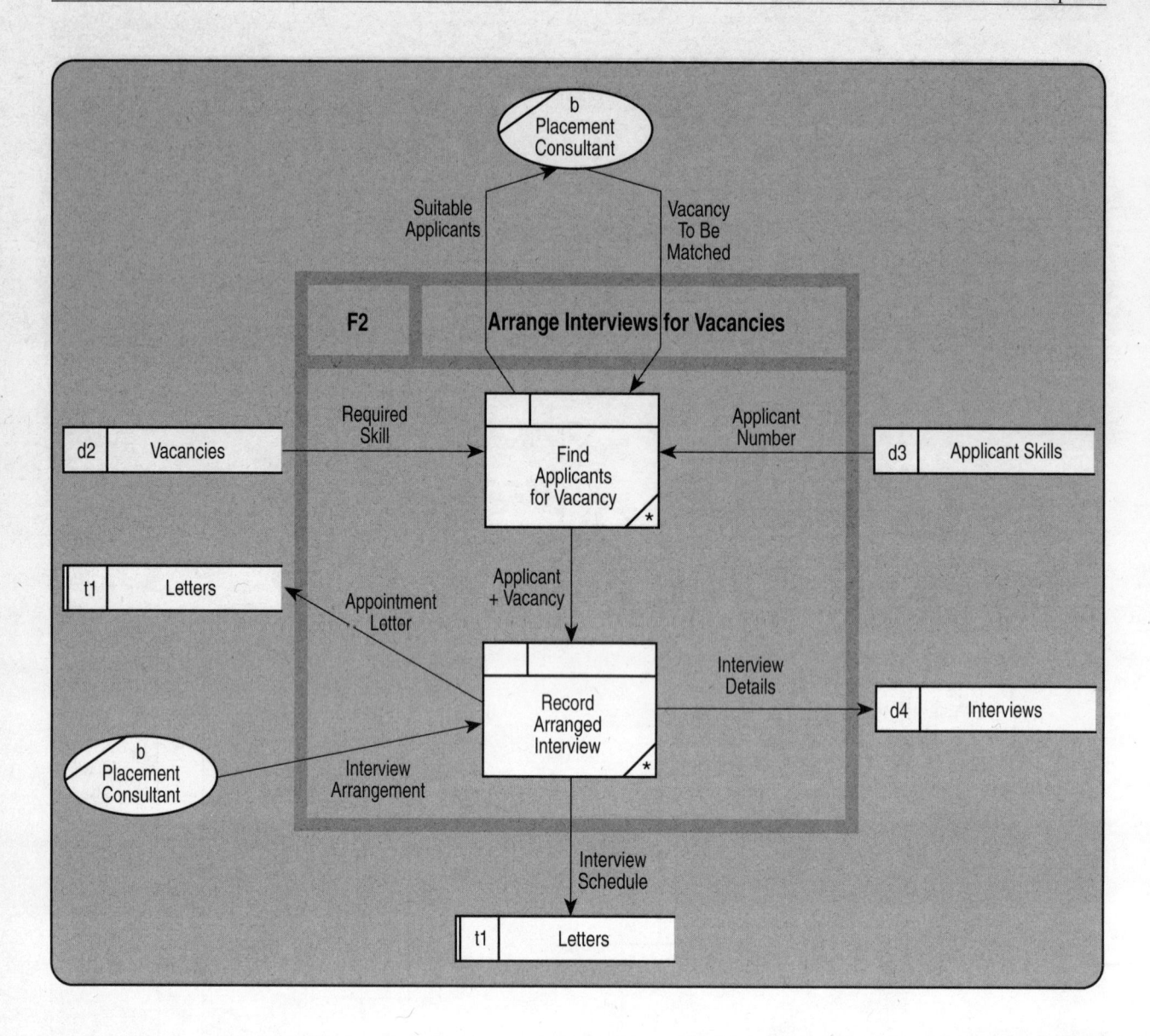

Function DFD equivalents 195

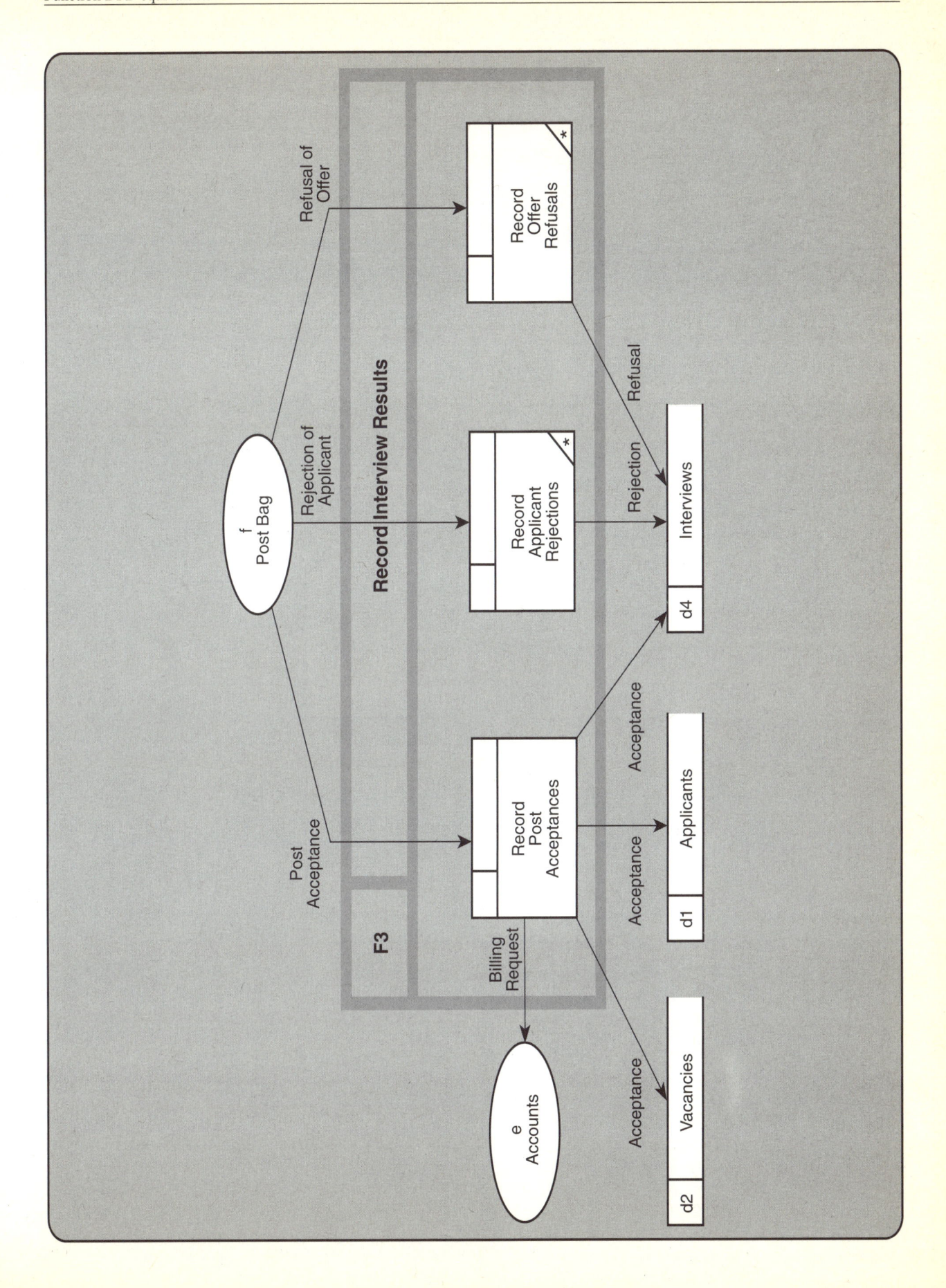

Record Interview Results (F3)

- **Post Bag (f)** → Refusal of Offer → **Record Offer Refusals** *
- Record Offer Refusals → Refusal → **Interviews (d4)**
- Post Bag → Rejection of Applicant → **Record Applicant Rejections** *
- Record Applicant Rejections → Rejection → **Interviews (d4)**
- Post Bag → Post Acceptance → **Record Post Acceptances**
- Record Post Acceptances → Acceptance → **Interviews (d4)**
- Record Post Acceptances → Acceptance → **Applicants (d1)**
- Record Post Acceptances → Acceptance → **Vacancies (d2)**
- Record Post Acceptances → Billing Request → **Accounts (e)**

Annex 2 Hi-Ho SSADM Development

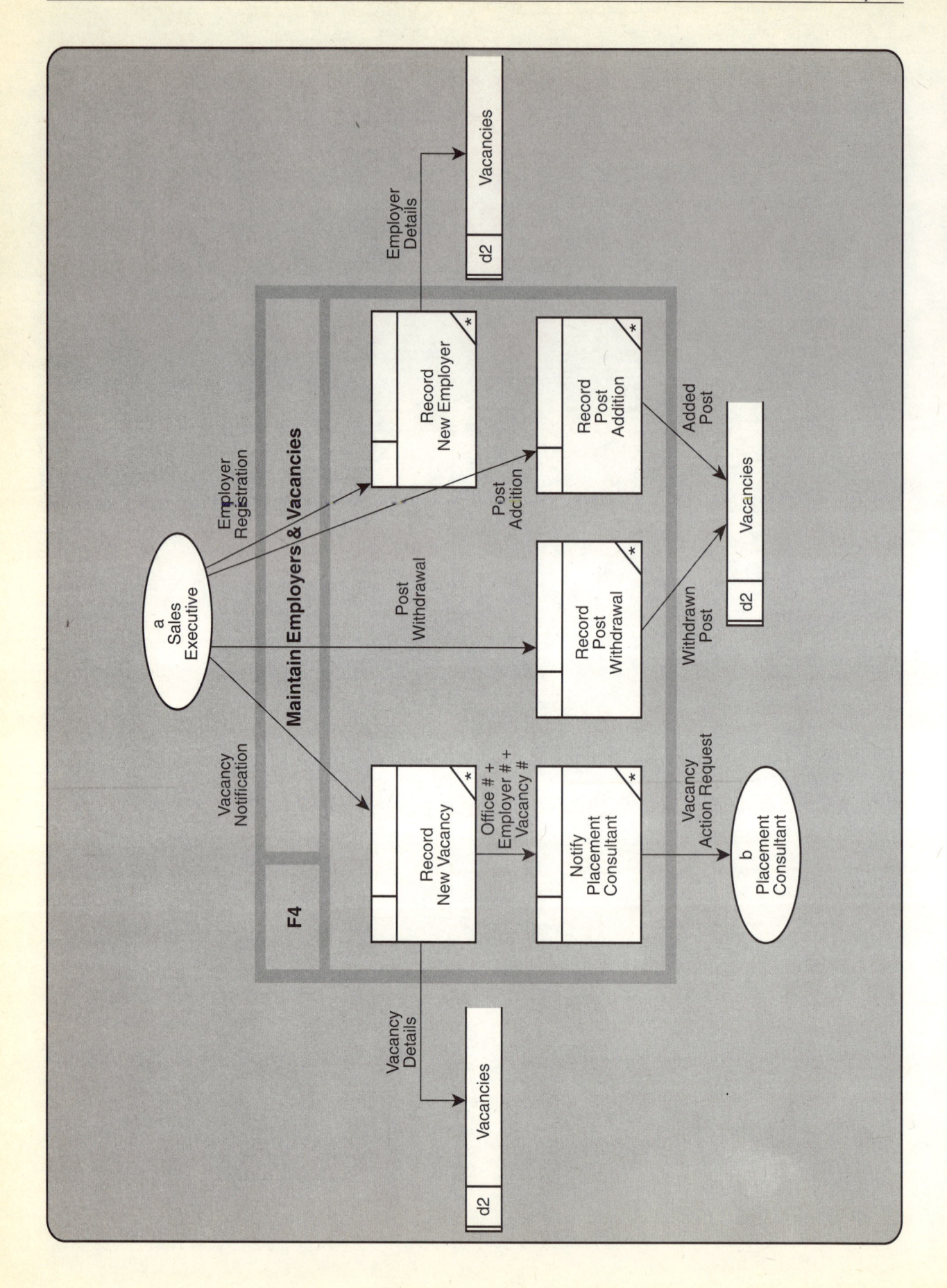

Function DFD equivalents 197

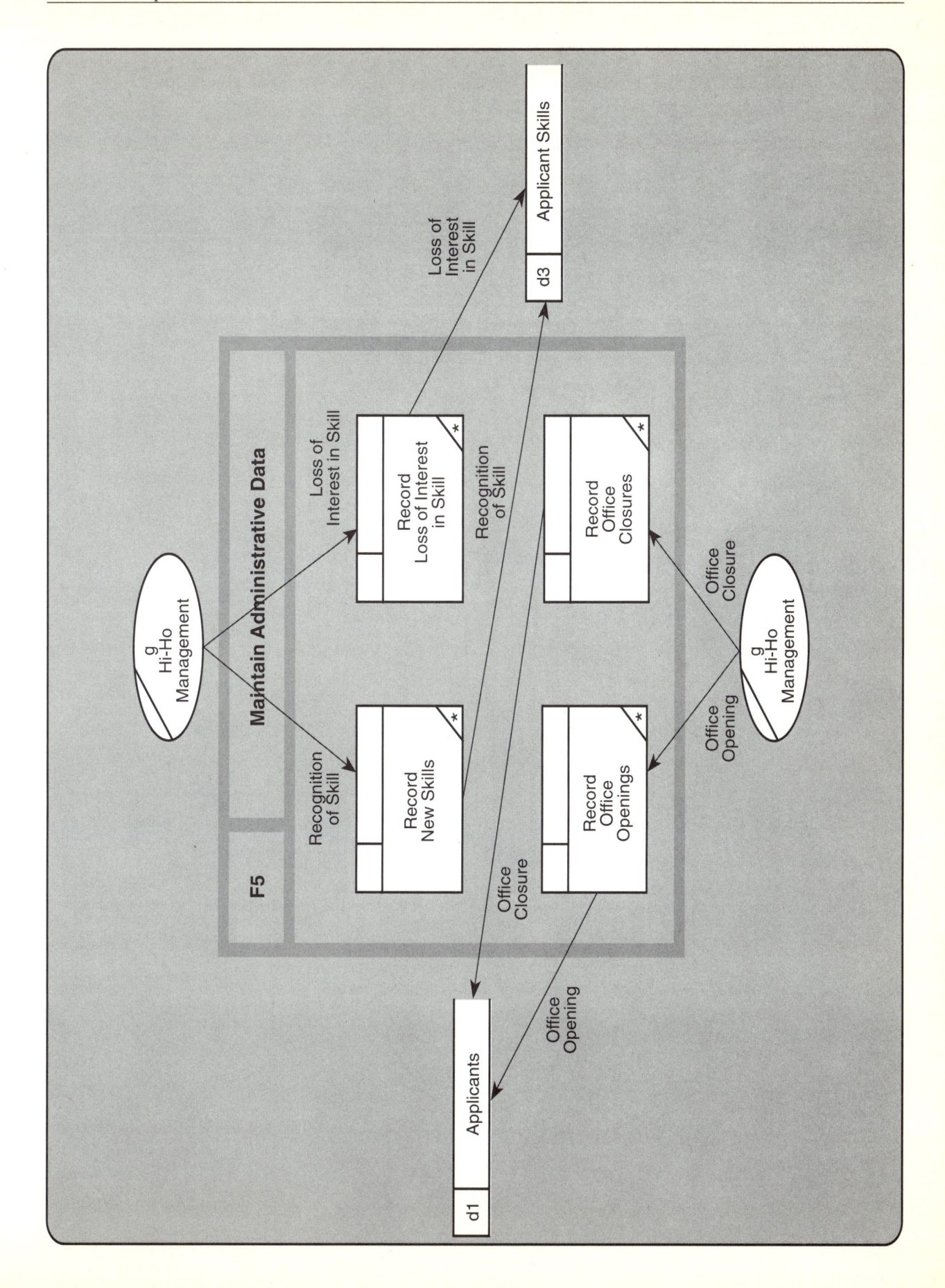

Annex 2 Hi-Ho SSADM Development

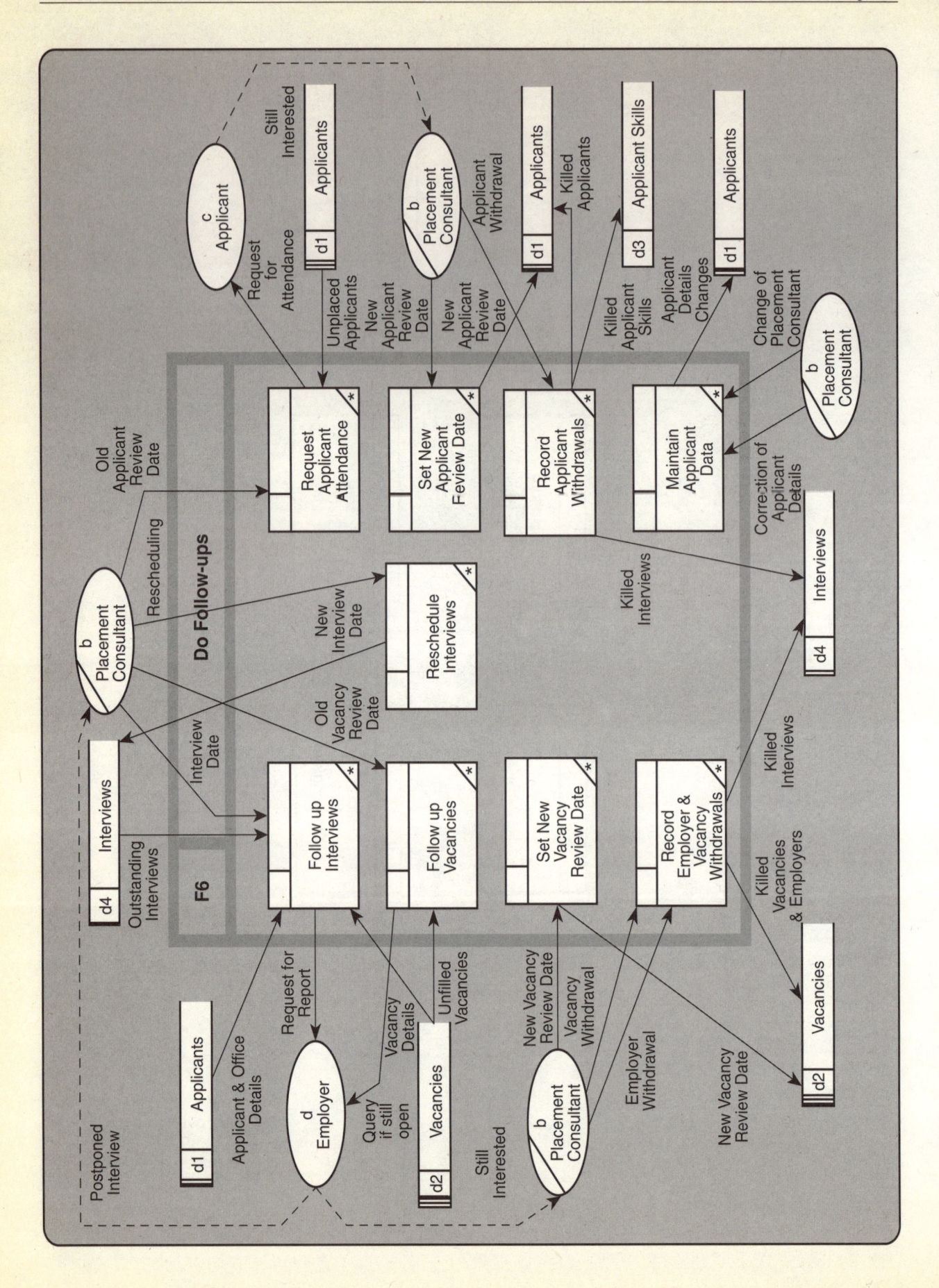

Function DFD equivalents

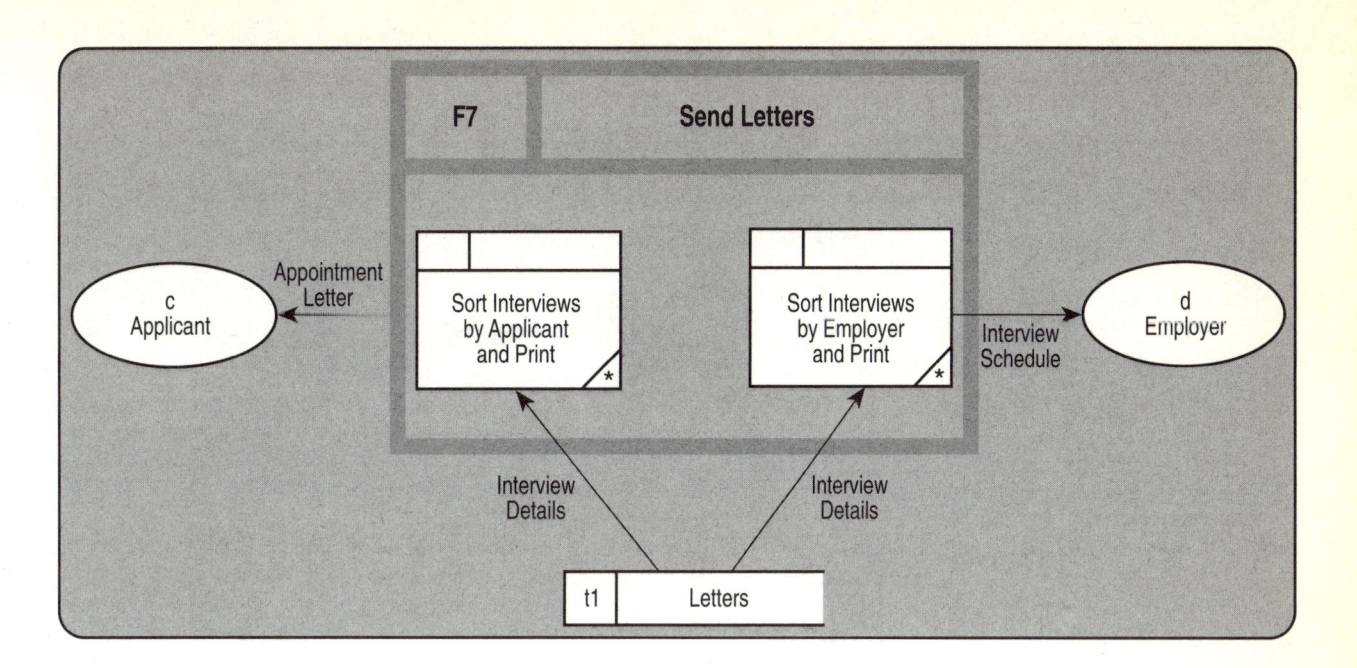

Function input structures

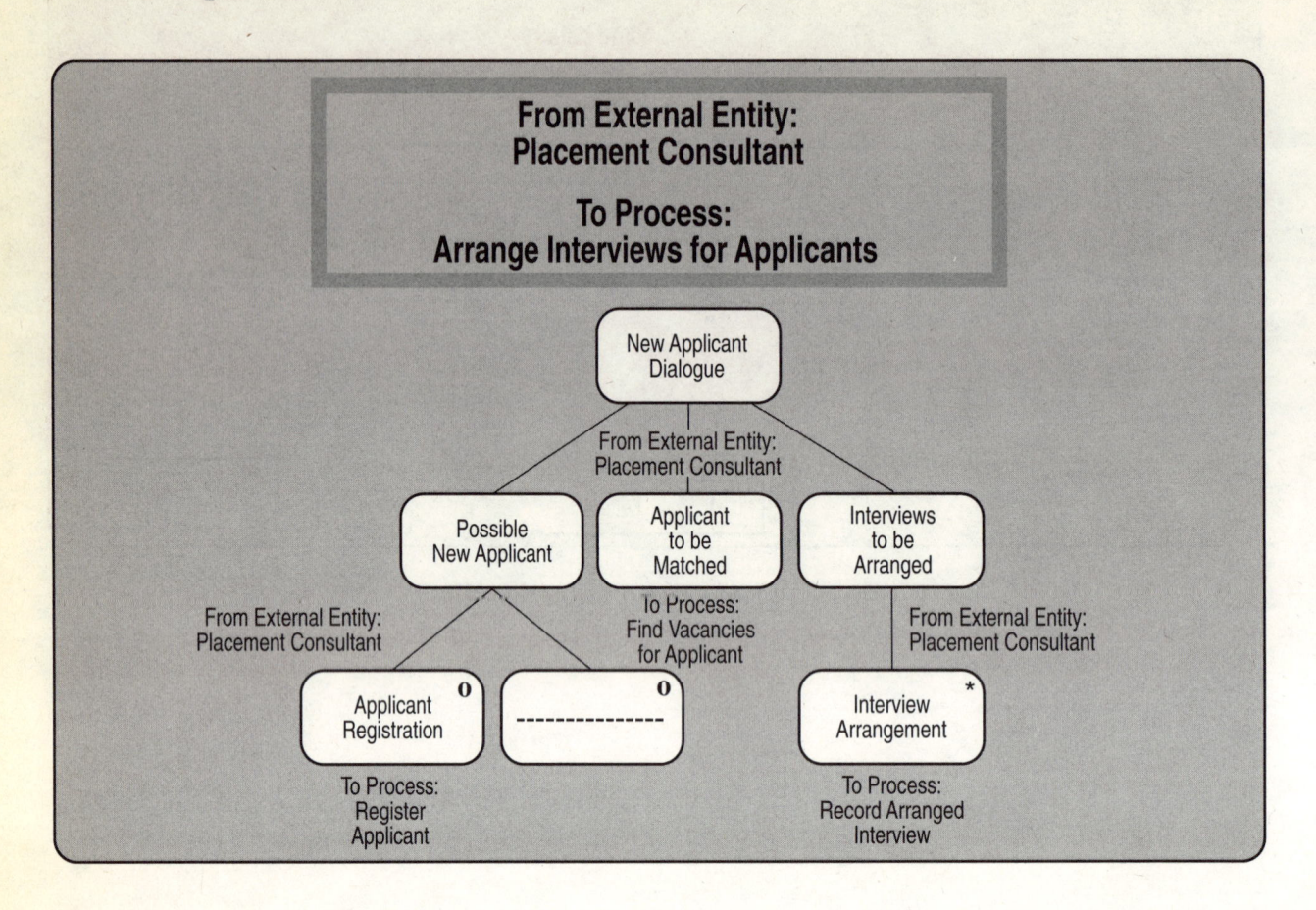

**From External Entity:
Placement Consultant**

**To Process:
Arrange Interviews for Applicants**

- New Applicant Dialogue
  - *From External Entity: Placement Consultant*
    - Possible New Applicant
      - *From External Entity: Placement Consultant*
        - Applicant Registration °
          - *To Process: Register Applicant*
        - ---------------- °
    - Applicant to be Matched
      - *To Process: Find Vacancies for Applicant*
    - Interviews to be Arranged
      - *From External Entity: Placement Consultant*
        - Interview Arrangement *
          - *To Process: Record Arranged Interview*

Function input structures 201

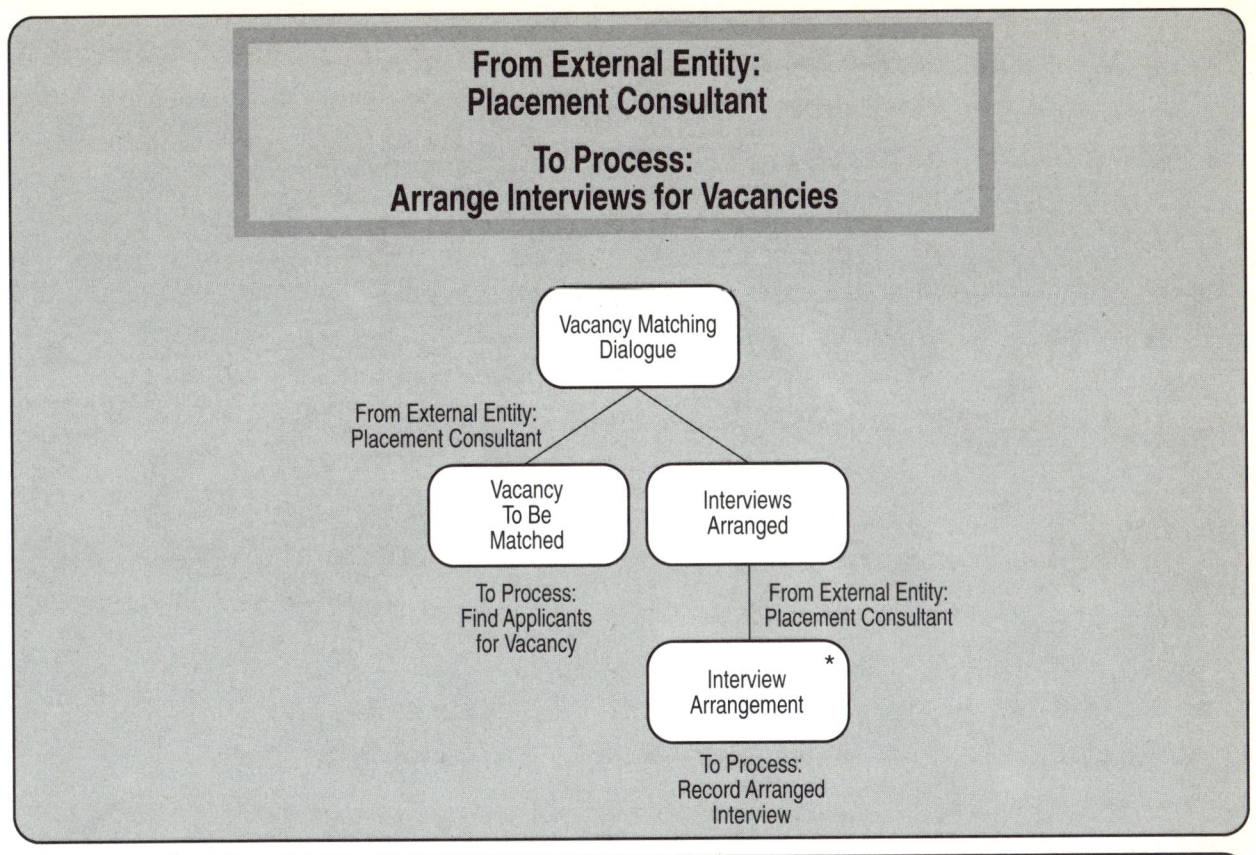

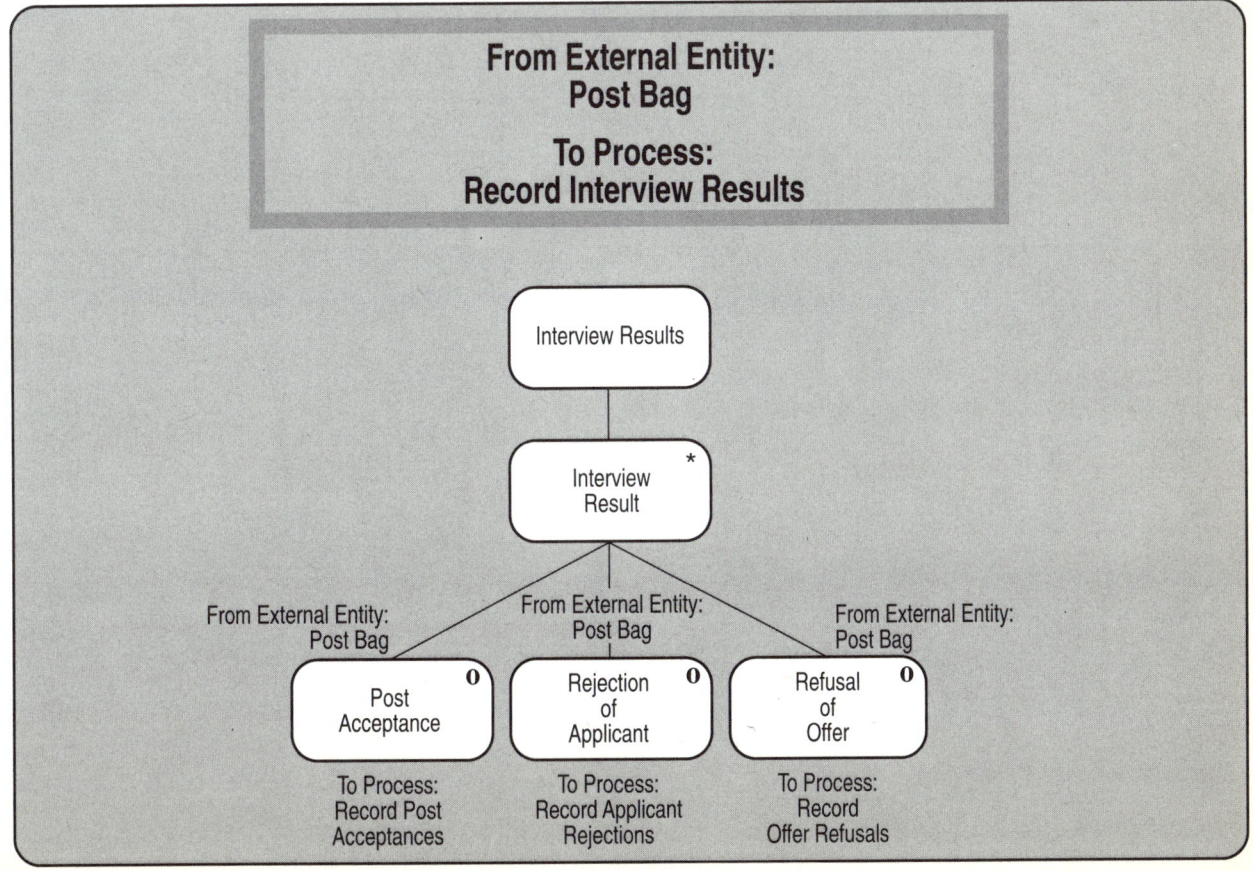

**From External Entity:
Placement Consultant**

**To Process:
Arrange Interviews for Vacancies**

```
                    Vacancy Matching
                        Dialogue
                       /          \
  From External Entity:            \
  Placement Consultant              \
                   /                 \
              Vacancy            Interviews
              To Be              Arranged
              Matched                  \
                                        From External Entity:
         To Process:                    Placement Consultant
         Find Applicants                        \
         for Vacancy                       Interview        *
                                           Arrangement

                                           To Process:
                                           Record Arranged
                                           Interview
```

**From External Entity:
Post Bag**

**To Process:
Record Interview Results**

```
                Interview Results
                        |
                   Interview           *
                   Result
              /         |         \
  From External    From External    From External
  Entity:          Entity:          Entity:
  Post Bag         Post Bag         Post Bag
     |                |                |
   Post  o        Rejection  o      Refusal   o
   Acceptance     of                of
                  Applicant         Offer

   To Process:    To Process:       To Process:
   Record Post    Record Applicant  Record
   Acceptances    Rejections        Offer Refusals
```

**From External Entity:
Sales Executive**

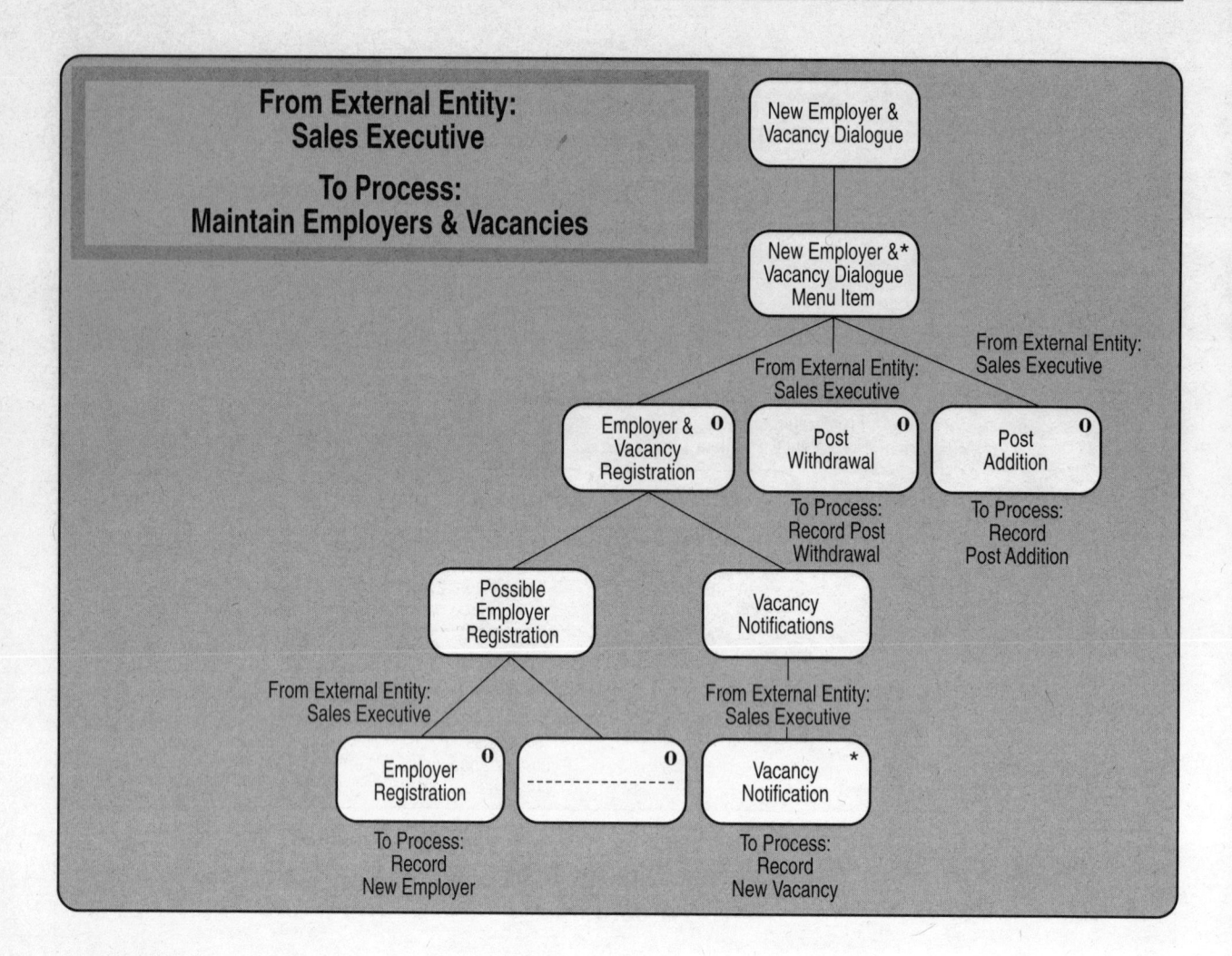

**To Process:
Maintain Employers & Vacancies**

- New Employer & Vacancy Dialogue
  - New Employer & Vacancy Dialogue Menu Item *
    - Employer & Vacancy Registration °
      - Possible Employer Registration
        - Employer Registration °
          - *From External Entity: Sales Executive*
          - *To Process: Record New Employer*
        - ---------------- °
      - Vacancy Notifications
        - Vacancy Notification *
          - *From External Entity: Sales Executive*
          - *To Process: Record New Vacancy*
    - Post Withdrawal °
      - *From External Entity: Sales Executive*
      - *To Process: Record Post Withdrawal*
    - Post Addition °
      - *From External Entity: Sales Executive*
      - *To Process: Record Post Addition*

Function input structures

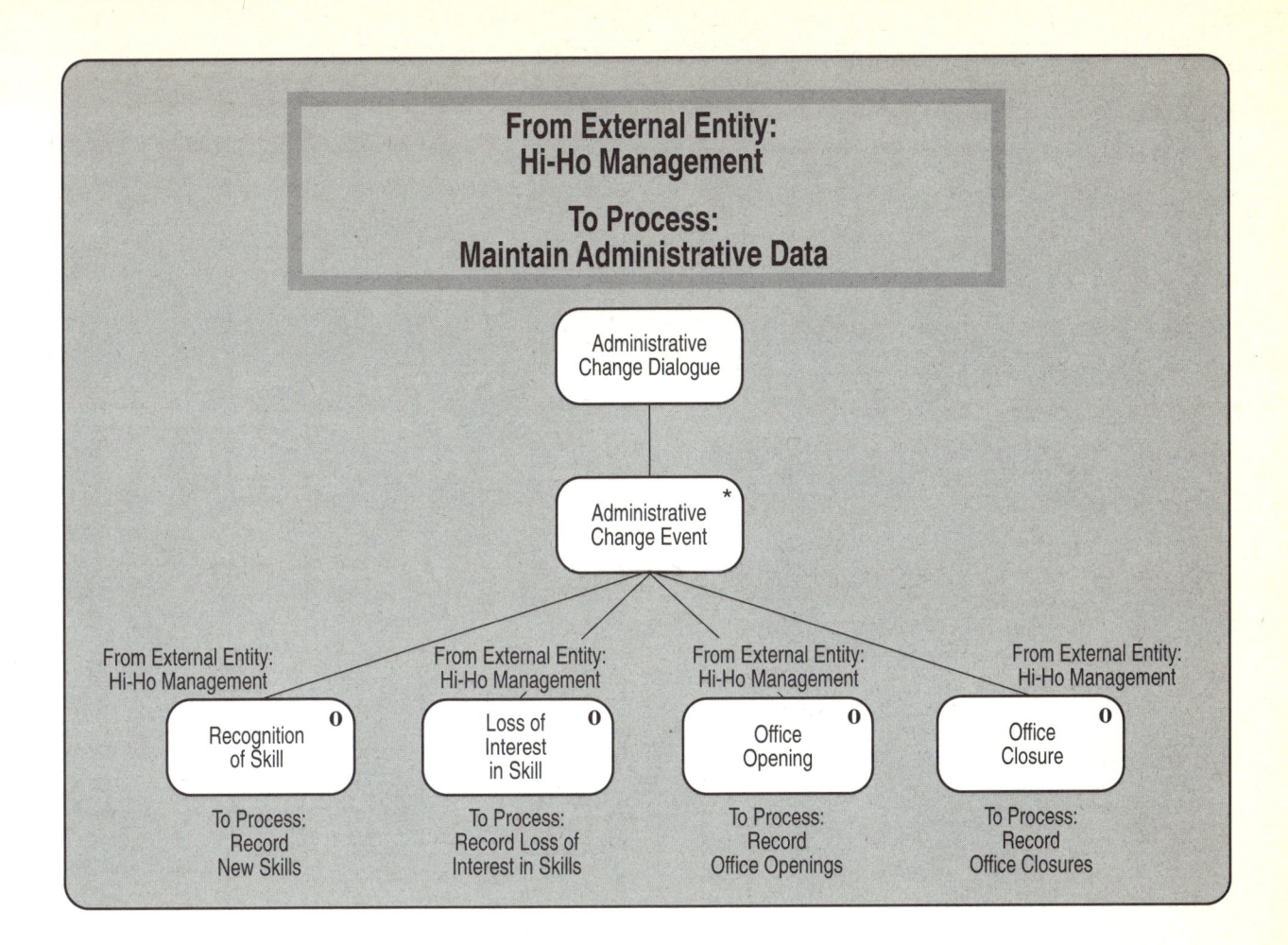

**From External Entity:
Hi-Ho Management**

**To Process:
Maintain Administrative Data**

- Administrative Change Dialogue
  - Administrative Change Event *
    - From External Entity: Hi-Ho Management
      - Recognition of Skill (o)
      - To Process: Record New Skills
    - From External Entity: Hi-Ho Management
      - Loss of Interest in Skill (o)
      - To Process: Record Loss of Interest in Skills
    - From External Entity: Hi-Ho Management
      - Office Opening (o)
      - To Process: Record Office Openings
    - From External Entity: Hi-Ho Management
      - Office Closure (o)
      - To Process: Record Office Closures

Annex 2 Hi-Ho SSADM Development

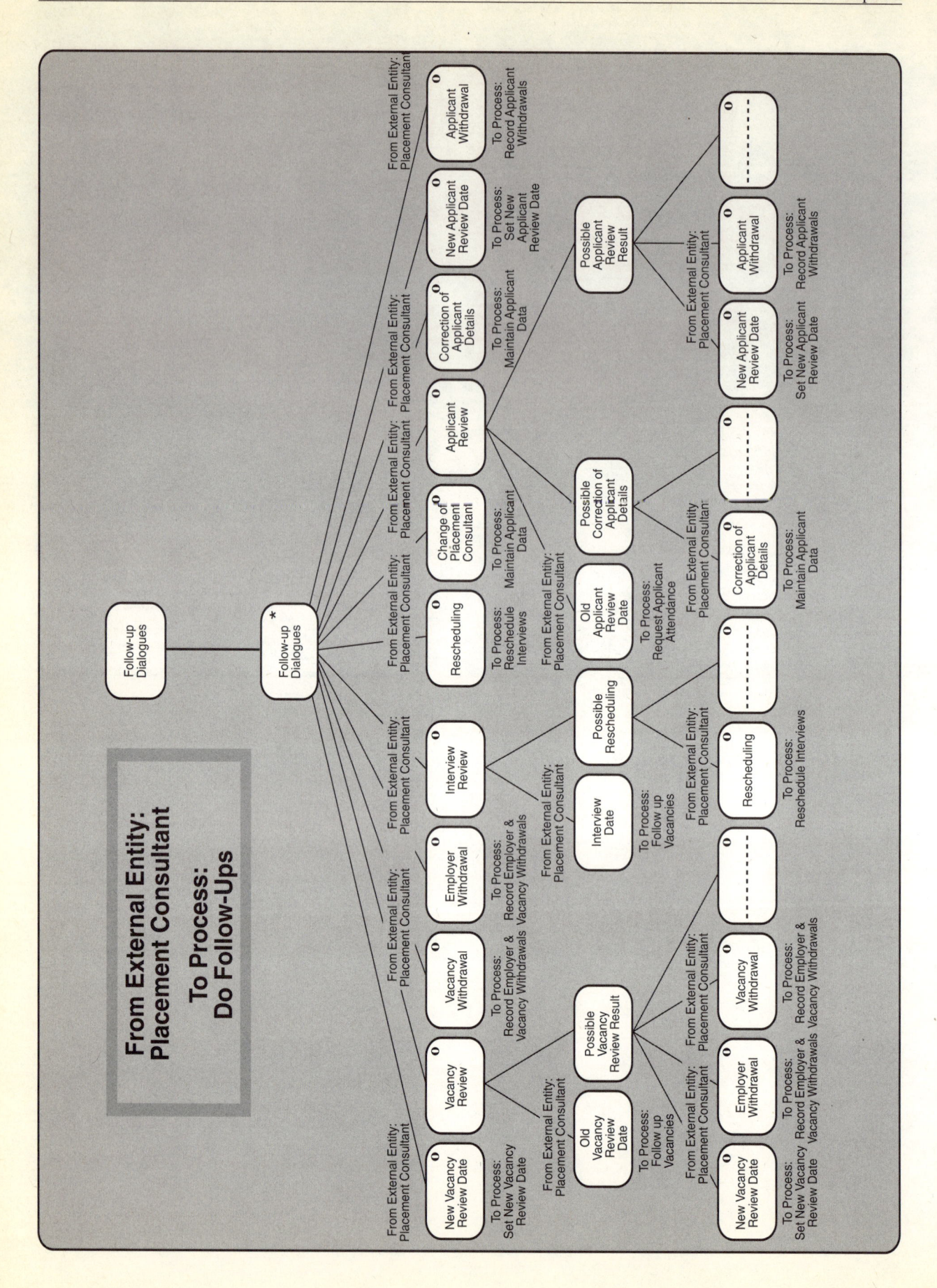

Probable function processing from stage 6

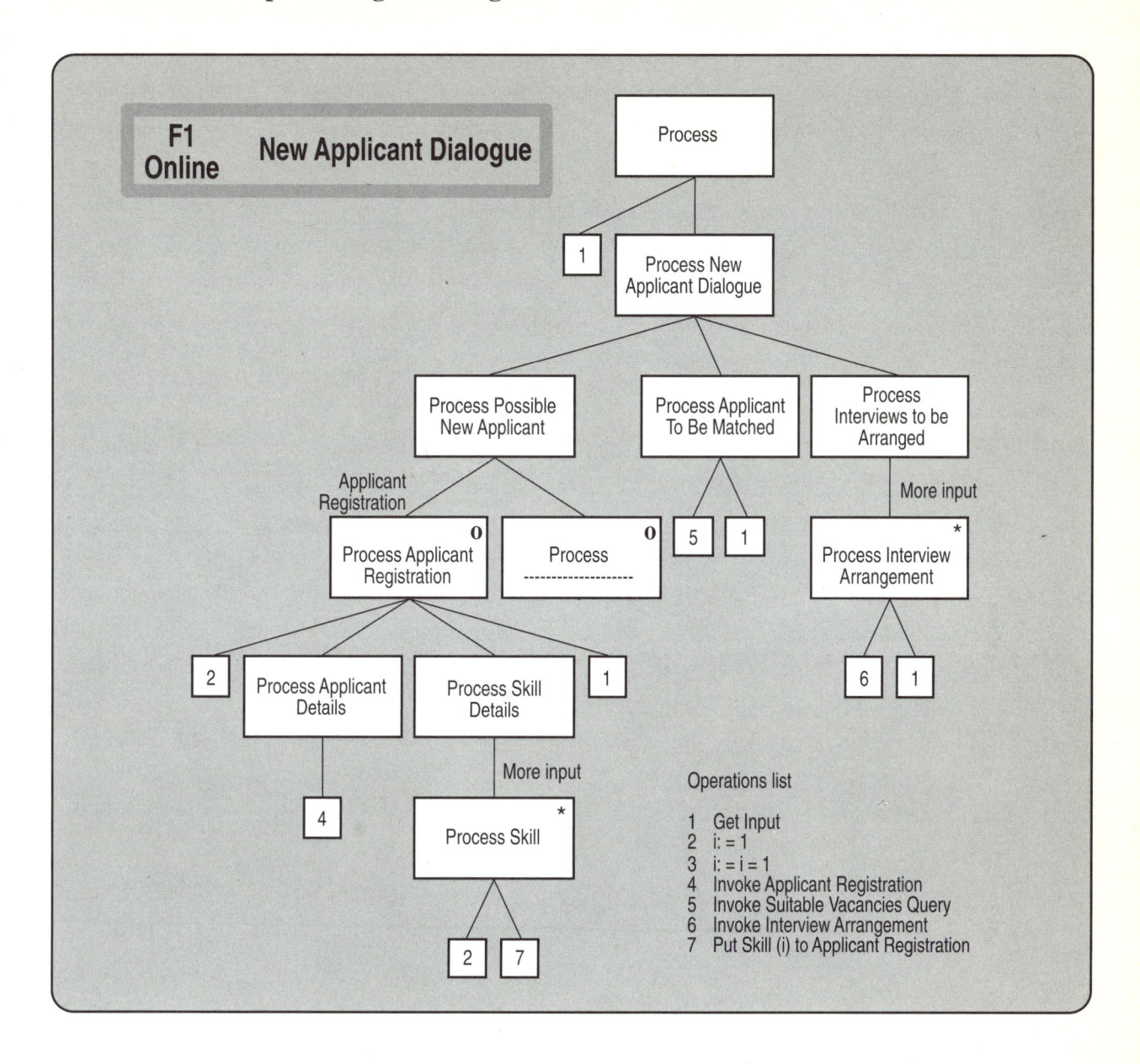

F1 Online — **New Applicant Dialogue**

```
                                    Process
                                       |
                                       1
                                       |
                            Process New
                            Applicant Dialogue
                            /       |        \
                           /        |         \
              Process Possible  Process Applicant  Process
              New Applicant     To Be Matched      Interviews to be
                                                   Arranged
                                                          \
                                                        More input
              Applicant                                     |
              Registration                                  *
                 o            o       5    1        Process Interview
         Process Applicant  Process                 Arrangement
         Registration       ------                    /    \
              /     \                                6      1
             2       1
         Process Applicant  Process Skill
         Details            Details
             |                  |
             4              More input
                                |
                                *
                            Process Skill
                              /    \
                             2      7
```

Operations list

1. Get Input
2. i: = 1
3. i: = i = 1
4. Invoke Applicant Registration
5. Invoke Suitable Vacancies Query
6. Invoke Interview Arrangement
7. Put Skill (i) to Applicant Registration

F2 Online — Vacancy Matching Dialogue

```
                          Process
                            |
                   ┌────────┴────────┐
                   1        Process Vacancy
                            Matching Dialogue
                                  |
                      ┌───────────┴───────────┐
              Process Vacancy          Process Interview
              To Be Matched                Arranged
                    |                         |
                 ┌──┴──┐                   More input
                 3     1                      |
                                     Process Interview  *
                                        Arrangement
                                              |
                                           ┌──┴──┐
                                           2     1
```

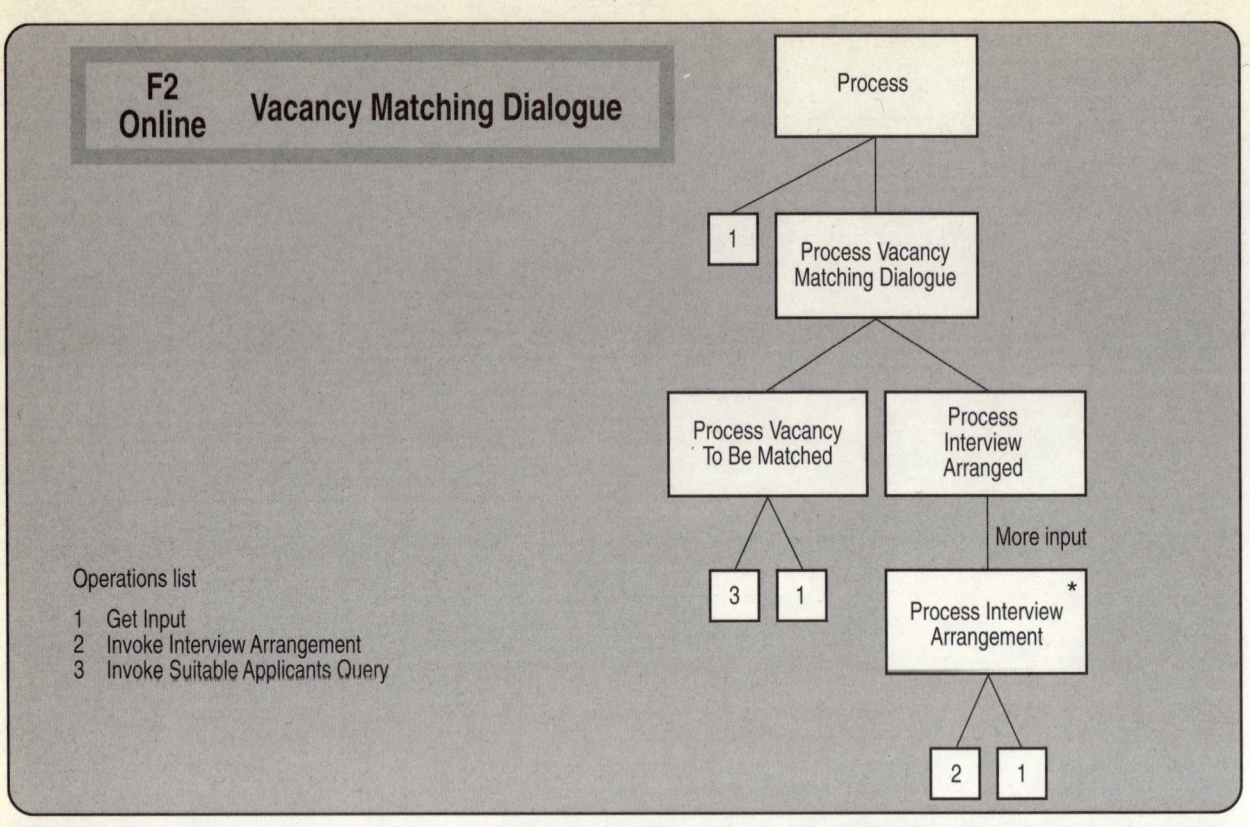

Operations list

1. Get Input
2. Invoke Interview Arrangement
3. Invoke Suitable Applicants Query

F3 Batch — Interview Results

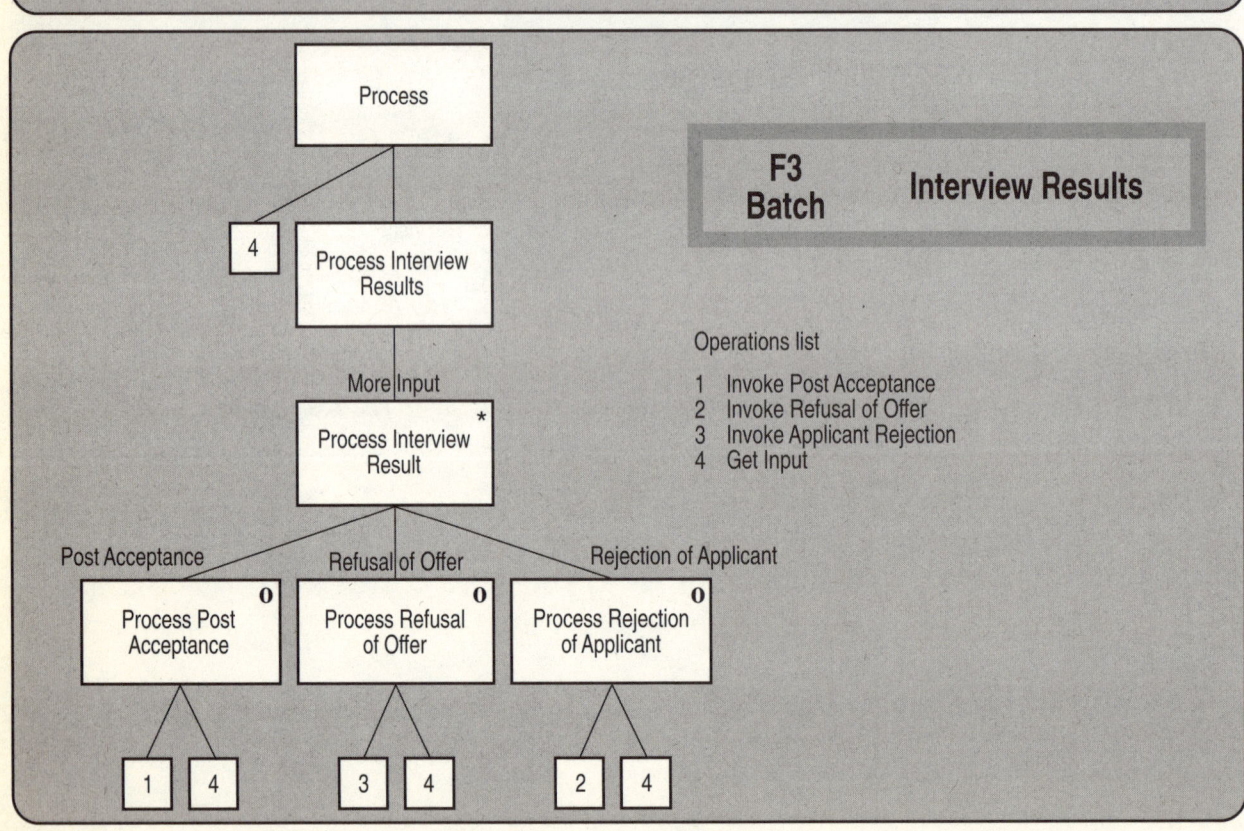

```
                    Process
                       |
              ┌────────┴────────┐
              4        Process Interview
                          Results
                              |
                          More Input
                              |
                       Process Interview  *
                           Result
                              |
        ┌─────────────────────┼─────────────────────┐
   Post Acceptance      Refusal of Offer      Rejection of Applicant
        |                     |                     |
   Process Post       ° Process Refusal     ° Process Rejection   °
   Acceptance              of Offer              of Applicant
        |                     |                     |
     ┌──┴──┐              ┌──┴──┐               ┌──┴──┐
     1     4              3     4               2     4
```

Operations list

1. Invoke Post Acceptance
2. Invoke Refusal of Offer
3. Invoke Applicant Rejection
4. Get Input

Probable function processing from stage 6

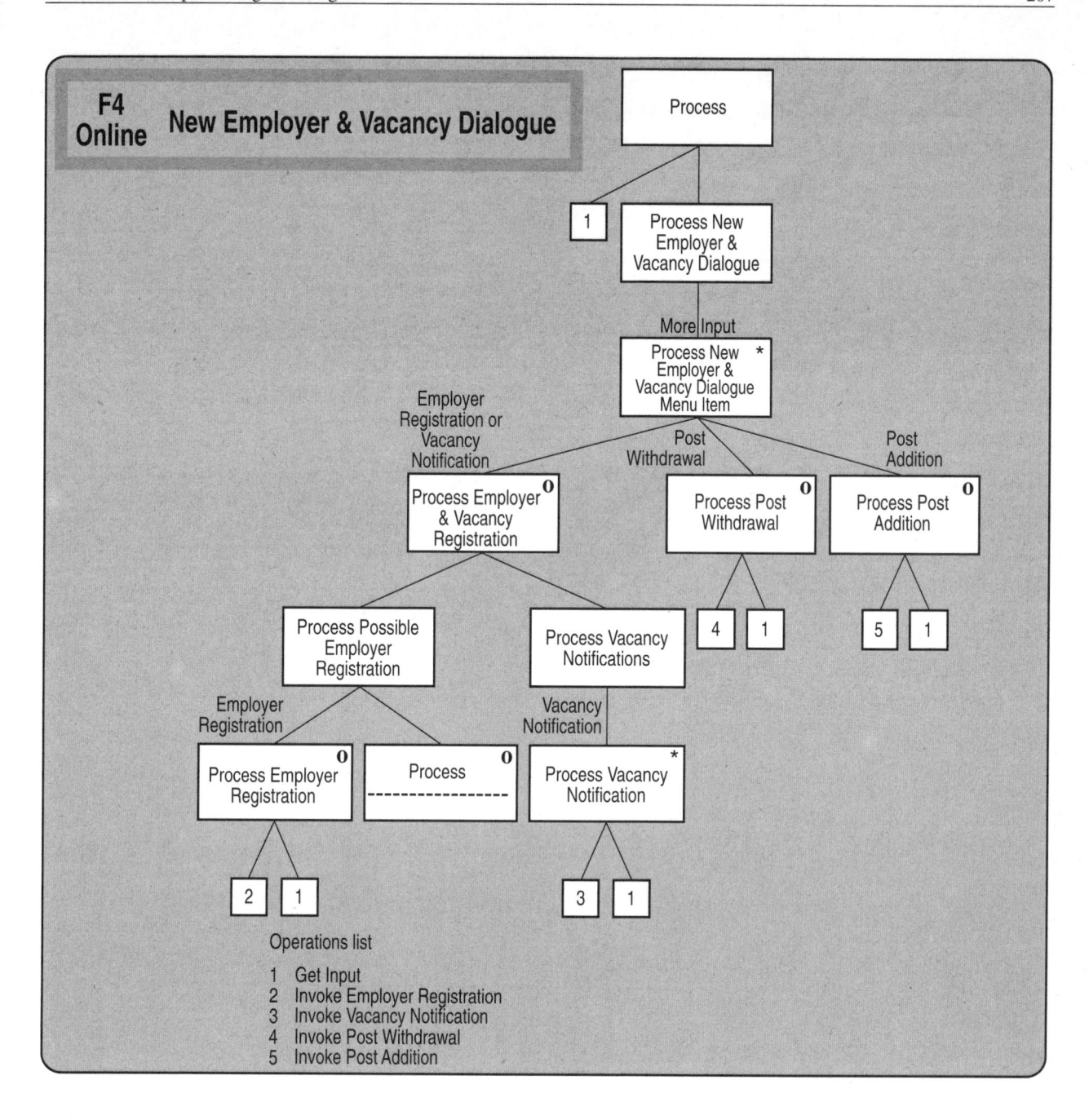

F4 Online — New Employer & Vacancy Dialogue

```
                                    Process
                                       |
                                    1  |
                            Process New Employer &
                              Vacancy Dialogue
                                       |
                                  More Input
                                       |
                            Process New Employer &     *
                              Vacancy Dialogue
                                 Menu Item
        _____|_____
       |                           |                           |
  Employer                    Post                        Post
  Registration or             Withdrawal                  Addition
  Vacancy
  Notification
       |                           |                           |
  Process Employer°          Process Post°             Process Post°
  & Vacancy                  Withdrawal                Addition
  Registration
       |                        |    |                   |    |
       |                        4    1                   5    1
   ____|____
  |         |
  Process Possible      Process Vacancy
  Employer              Notifications
  Registration
       |                        |
  Employer                   Vacancy
  Registration               Notification
   ____|____                    |
  |         |                Process Vacancy     *
  Process   Process°         Notification
  Employer° -------
  Registration
   |    |                     |    |
   2    1                     3    1
```

Operations list

1. Get Input
2. Invoke Employer Registration
3. Invoke Vacancy Notification
4. Invoke Post Withdrawal
5. Invoke Post Addition

F5 Online — Administrative Change Dialogue

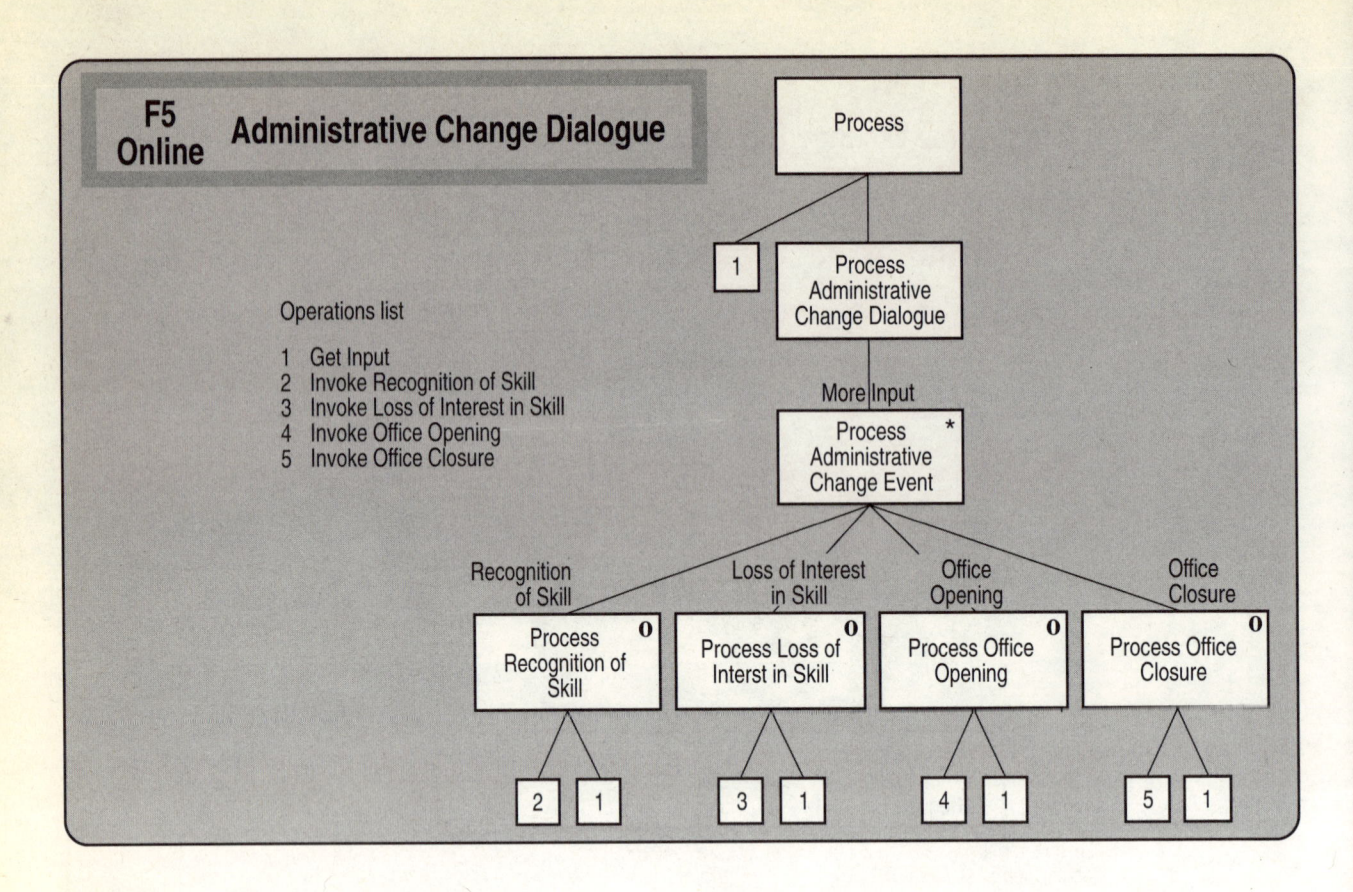

Operations list

1. Get Input
2. Invoke Recognition of Skill
3. Invoke Loss of Interest in Skill
4. Invoke Office Opening
5. Invoke Office Closure

Structure diagram:

- Process
  - Process Administrative Change Dialogue [1]
    - More Input
      - Process Administrative Change Event *
        - Recognition of Skill → Process Recognition of Skill ○ — operations [2, 1]
        - Loss of Interest in Skill → Process Loss of Interst in Skill ○ — operations [3, 1]
        - Office Opening → Process Office Opening ○ — operations [4, 1]
        - Office Closure → Process Office Closure ○ — operations [5, 1]

Probable function processing from stage 6

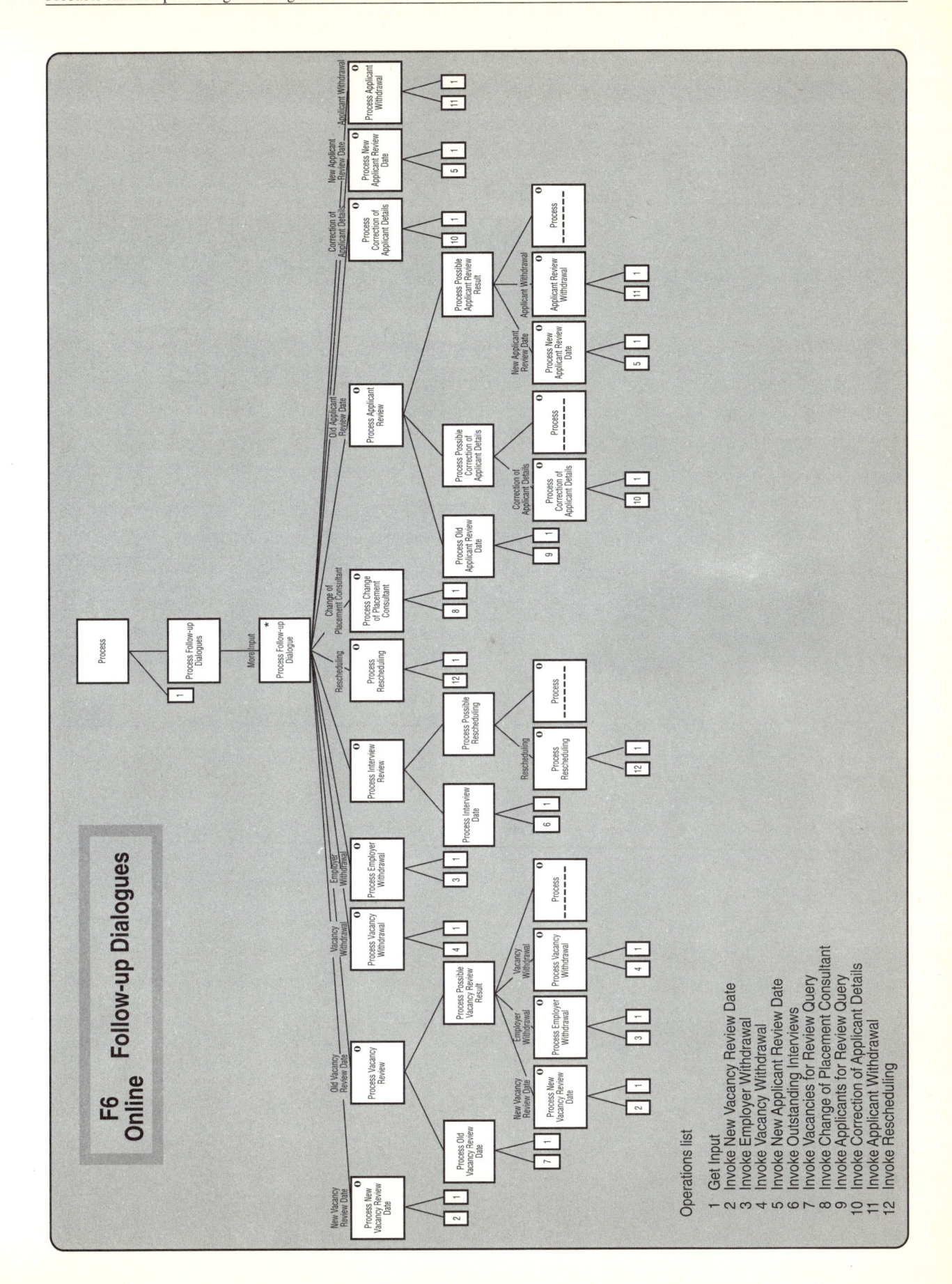

F6 Follow-up Dialogues Online

Operations list

1. Get Input
2. Invoke New Vacancy Review Date
3. Invoke Employer Withdrawal
4. Invoke Vacancy Withdrawal
5. Invoke New Applicant Review Date
6. Invoke Outstanding Interviews
7. Invoke Vacancies for Review Query
8. Invoke Change of Placement Consultant
9. Invoke Applicants for Review Query
10. Invoke Correction of Applicant Details
11. Invoke Applicant Withdrawal
12. Invoke Rescheduling

F7 Batch Send Letters

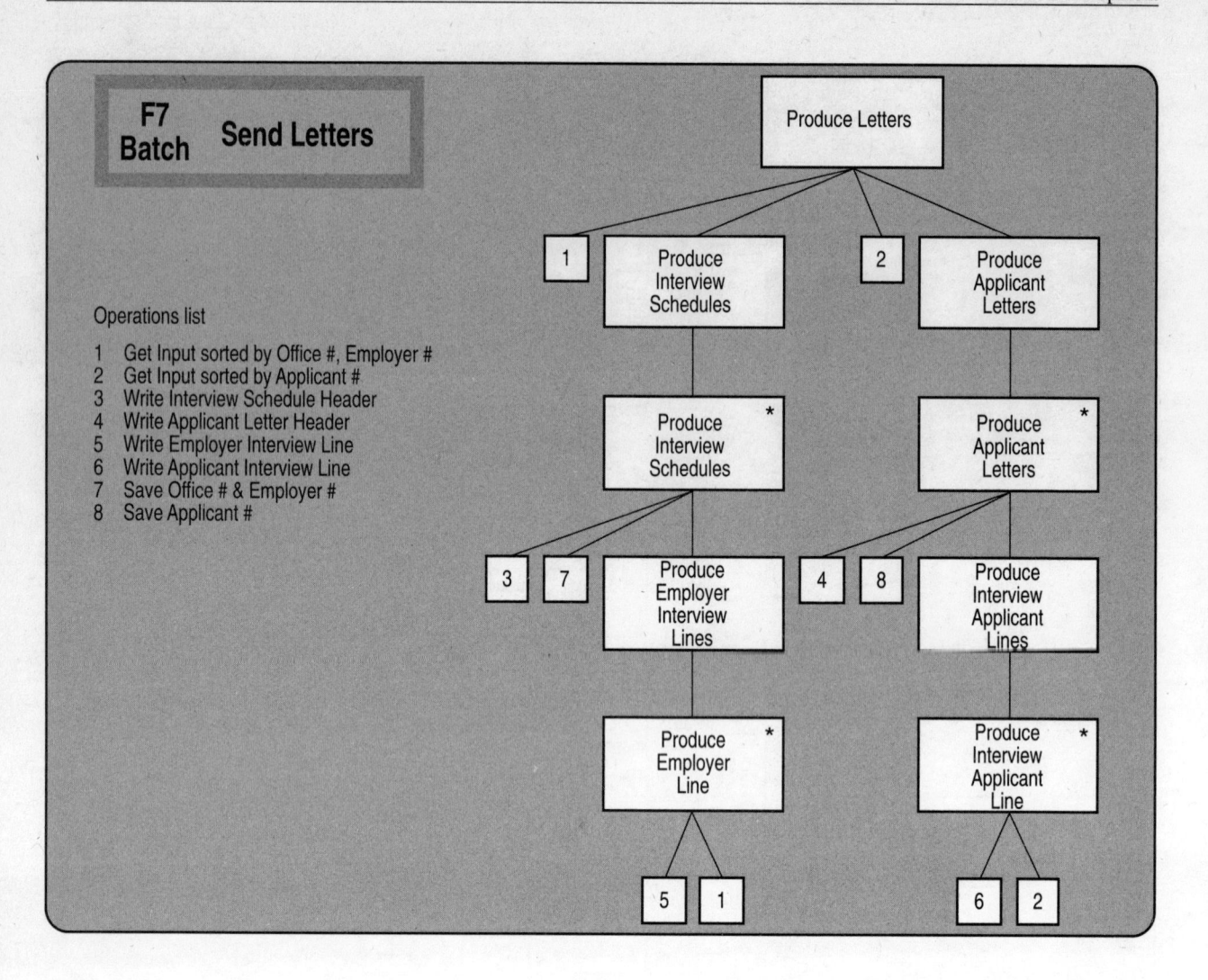

Operations list

1. Get Input sorted by Office #, Employer #
2. Get Input sorted by Applicant #
3. Write Interview Schedule Header
4. Write Applicant Letter Header
5. Write Employer Interview Line
6. Write Applicant Interview Line
7. Save Office # & Employer #
8. Save Applicant #

```
                         Produce Letters
                        /              \
                   1 /                    \ 2
            Produce Interview          Produce Applicant
               Schedules                    Letters
                   |                           |
            Produce Interview *         Produce Applicant *
               Schedules                    Letters
                /      \                    /       \
             3 / 7 \              4 / 8 \
            Produce Employer            Produce Interview
             Interview Lines            Applicant Lines
                   |                           |
            Produce Employer *          Produce Interview *
                 Line                      Applicant Line
                /     \                    /     \
              5       1                   6       2
```

Input/output data

```
                          DataFlow Defs - Extended 1

01 Applicant
   Registration
   Placement              Register Applicant
   Consultant             (Process)
   (Extl Entity)
   02 Applicant #
   02 Applicant Name
   02 Applicant Address
   02 Placement
      Consultant
   02 Office #
   02 Telephone Number
   02 Has Driving
      Licence
   02 Date (Birth)
   02 Willing To Move
   02 Marital Status
   02 Skills
      03 Skill                                            REPEATS
         04 Skill Code
         04 Skill Level
         04 Skill Description
00 0

01 Applicant To Be
   Matched
   Placement              Find Vacancies for
   Consultant             Applicant
   (Extl Entity)          (Process)
   02 Applicant #
00 0

01 Applicant
   Withdrawal
   Placement              Record Applicant
   Consultant             Withdrawals
   (Extl Entity)          (Process)
   02 Applicant #
00 0
```

DataFlow Defs - Extended 2

01 Appointment
   Letter
      Sort Interviews by   Applicant
      Applicant and Print (Extl Entity)
      (Process)
   02 Applicant #
   02 Applicant Name
   02 Applicant Adress
   02 Office #
   02 Interviews
      03 Interview   REPEATS
         04 Employer #
         04 Employer Name
         04 Employer Address
         04 Vacancy #
         04 Vacancy Title
         04 Data (Interview)
00 0

01 Billing Request
      Record Post   Accounts
      Acceptances   (Extl Entity)
      (Process)
   02 Office #
   02 Employer #
   02 Vacancy #
   02 Applicant #
00 0

01 Change of
   Placement
   Consultant
      Placement   Maintain Applicant
      Consultant   Data
      (Extl Entity)   (Process)
   02 Applicant #
   02 Placement
      Consultant
00 0

DataFlow Defs - Extended 3

01 Correction of
   Applicant Details

   Placement   Maintain Applicant
   Consultant   Data
   (Extl Entity)   (Process)

  02 <u>Applicant #</u>
  02 Applicant Name
  02 Applicant Address
  02 Marital Status
  02 Willing To Move
  02 Date (Birth)
  02 Has Driving
     Licence
  02 Telephone Number

00 0

01 Employer
   Registration

   Sales Executive   Record New
   (Extl Entity)   Employer
                          (Process)

  02 <u>Office #</u>
  02 <u>Employer #</u>
  02 Employer Name
  02 Employer Address

00 0

01 Employer
   Withdrawal

   Placement   Record Employer
   Consultant   & Vacancy
   (Extl Entity)   Withdrawals
                          (Process)

  02 <u>Office #</u>
  02 <u>Employer #</u>

00 0

01 Interview
   Arrangement

   Placement   Record Arranged
   Consultant   Interview
   (Extl Entity)   (Process)

  02 <u>Applicant #</u>
  02 <u>Vacancy #</u>
  02 <u>Employer #</u>
  02 <u>Office #</u>
  02 Date (Interview)

00 0

DataFlow Defs - Extended 4

01 Interview Date

   Placement   Follow up
   Consultant   Interviews
   (Extl Entity)   (Process)

   02 <u>Date</u>

00 0

01 Interview Schedule

   Sort Interviews by   Employer
   Employer and Print (Extl Entity)
   (Process)

   02 <u>Office #</u>

   02 <u>Employer #</u>

   02 Employer Name

   02 Employer Address

   02 Vacancies

      03 Vacancy   REPEATS

         04 <u>Vacancy #</u>

         04 Vacancy Title

         04 Interviews

            05 Interview   REPEATS

               06 <u>Applicant #</u>

               06 Applicant Name

               06 Date (Interview)

00 0

01 Loss of Interest in Skill

   Hi-Ho Management Record Loss of
   (Extl Entity)   Interest in Skills
                             (Process)

   02 <u>Skill Code</u>

00 0

01 New Applicant Review Date

   Placement   Set New Applicant
   Consultant   Review Date
   (Extl Entity)   (Process)

   02 <u>Date</u>

00 0

01 New Vacancy Review Date

   Placement   Set New Vacancy
   Consultant   Review Date
   (Extl Entity)   (Process)

   02 <u>Date</u>

00 0

Input/output data

DataFlow Defs - Extended 5

01 Office Closure
    Hi-Ho Management Record Office
    (Extl Entity)   Closures
                                 (Process)

  02 Office #
00 0

01 Office Opening
    Hi-Ho Management Record Office
    (Extl Entity)   Openings
                                 (Process)

  02 Office #
  02 Office Name
  02 Office Adress
00 0

01 Old Applicant
   Review Date
    Placement   Request Applicant
    Consultant   Attendance
    (Extl Entity)   (Process)

  02 Date
00 0

01 Old Vacancy
   Review Date
    Placement   Follow up
    Consultant   Vacancies
    (Extl Entity)   (Process)

  02 Date
00 0

01 Post Acceptance
    Post Bag   Record Post
    (Extl Entity)   Acceptances
                           (Process)

  02 Office #
  02 Employer #
  02 Vacancy #
  02 Applicant #
00 0

01 Post Addition
    Sales Executive   Record Post
    (Extl Entity)   Addition
                           (Process)

  02 Office #
  02 Employer #
  02 Vacancy #
00 0

DataFlow Defs - Extended 6

01 Post -Withdrawal
    Sales Executive   Record Post
    (Extl Entity)   Withdrawal
                        (Process)
  02 <u>Office #</u>
  02 <u>Employer #</u>
  02 <u>Vacancy #</u>
00 0

01 Query if still open
    Follow up   Employer
    Vacancies   (Extl Entity)
    (Process)
  02 <u>Office #</u>
  02 <u>Employer #</u>
  02 Employer Name
  02 <u>Vacancy #</u>
  02 Vacancy Title
  02 Employer Address
00 0

01 Recognition of
   Skill
    Hi-Ho Management Record New Skills
    (Extl Entity)   (Process)
  02 <u>Skill Code</u>
  02 Skill Description
00 0

01 Refusal of Offer
    Post Bag   Record Offer
    (Extl Entity)   Refusals
                        (Process)
  02 <u>Office #</u>
  02 <u>Employer #</u>
  02 <u>Vacancy #</u>
  02 <u>Applicant #</u>
00 0

01 Rejection of
   Applicant
    Post Bag   Record Applicant
    (Extl Entity)   Rejections
                        (Process)
  02 <u>Office #</u>
  02 <u>Employer #</u>
  02 <u>Vacancy #</u>
  02 <u>Applicant #</u>
00 0

DataFlow Defs - Extended 7

01 Request for
   Attendance
   Request Applicant   Applicant
   Attendance   (Extl Entity)
   (Process)
  02 Applicant #
  02 Applicant Address
  02 Applicant Name
  02 Office #
  02 Office Address
  02 Office Name
00 0

01 Request for Report
   Follow up   Employer
   Interviews   (Extl Entity)
   (Process)
  02 Office #
  02 Employer #
  02 Employer Name
  02 Employer Address
  02 Vacancy #
  02 Vacancy Title
  02 Office Name
  02 Office Address
  02 Applicant #
  02 Applicant Name
00 0

01 Rescheduling
   Placement   Reschedule
   Consultant   Interviews
   (Extl Entity)   (Process)
  02 Applicant #
  02 Vacancy #
  02 Employer #
  02 Office #
  02 Date (Interview)
00 0

DataFlow Defs - Extended 8

01 Suitable
   Vacancies
   Find Vacancies for   Placement
   Applicant   Consultant
   (Process)   (Extl Entity)

  02 <u>Applicant #</u>
  02 Applicant Name
  02 Vacancies by
     Skills
    03 Skill   REPEATS
      04 <u>Skill Code</u>
      04 Vacancies for Skills
        05 Vacancy   REPEATS
          06 <u>Employer #</u>
          06 <u>Office #</u>
          06 <u>Vacancy #</u>
          06 Employer Name
          06 Vacancy Title
00 0

01 Vacancy Action
   Request
   Notify Placement   Placement
   Consultant   Consultant
   (Process)   (Extl Entity)

  02 <u>Office #</u>
  02 <u>Employer #</u>
  02 Employer Name
  02 Employer Address
  02 <u>Vacancy #</u>
  02 Vacancy Title
00 0

01 Vacancy
   Notification
   Sales Executive   Record New
   (Extl Entity)   Vacancy
                            (Process)

  02 <u>Office #</u>
  02 <u>Employer #</u>
  02 Employer Name
  02 Employer Address
  02 <u>Vacancy #</u>
  02 Vacancy Title
  02 Skill Code
  02 Date (Review)
00 0

Input/output data 219

DataFlow Defs - Extended 9

01 Vacancy To Be
   Matched
   Placement   Find Applicants for
   Consultant   Vacancy
   (Extl Entity)   (Process)
   02 Office #
   02 Employer #
   02 Vacancy #
00 0

01 Vacancy
   Withdrawal
   Placement   Record Employer
   Consultant   & Vacancy
   (Extl Entity)   Withdrawals
                        (Process)
   02 Office #
   02 Employer #
   02 Vacancy #
00 0

Annex 3

Hi-Ho GRAPES Development

Model of
Required System

Static structure

External relations of the Hi-Ho company

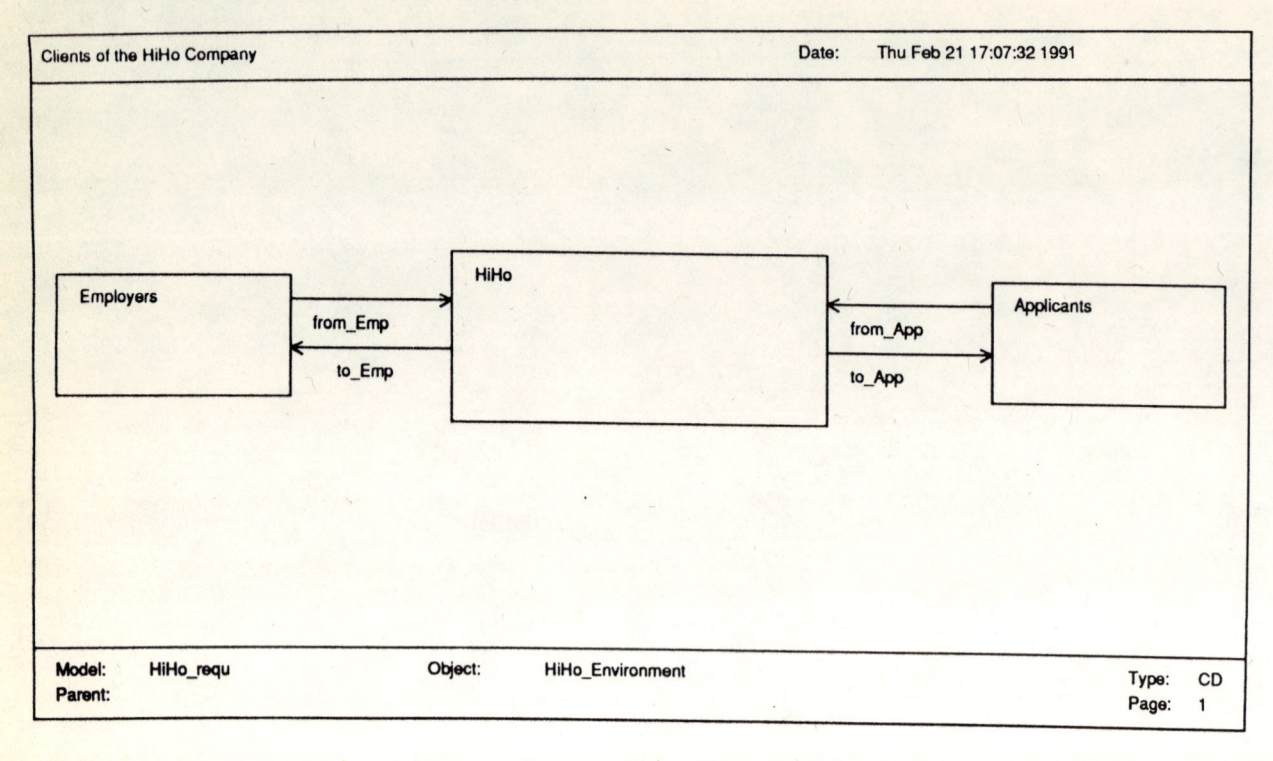

Clients of the HiHo Company — Date: Thu Feb 21 17:07:32 1991

```
  Employers  ──from_Emp──▶  HiHo  ◀──from_App──  Applicants
             ◀──to_Emp───         ───to_App──▶
```

Model: HiHo_requ Object: HiHo_Environment Type: CD
Parent: Page: 1

Structure of the Hi-Ho company

Structure of the HiHo Company — Date: Thu Feb 21 17:07:38 1991

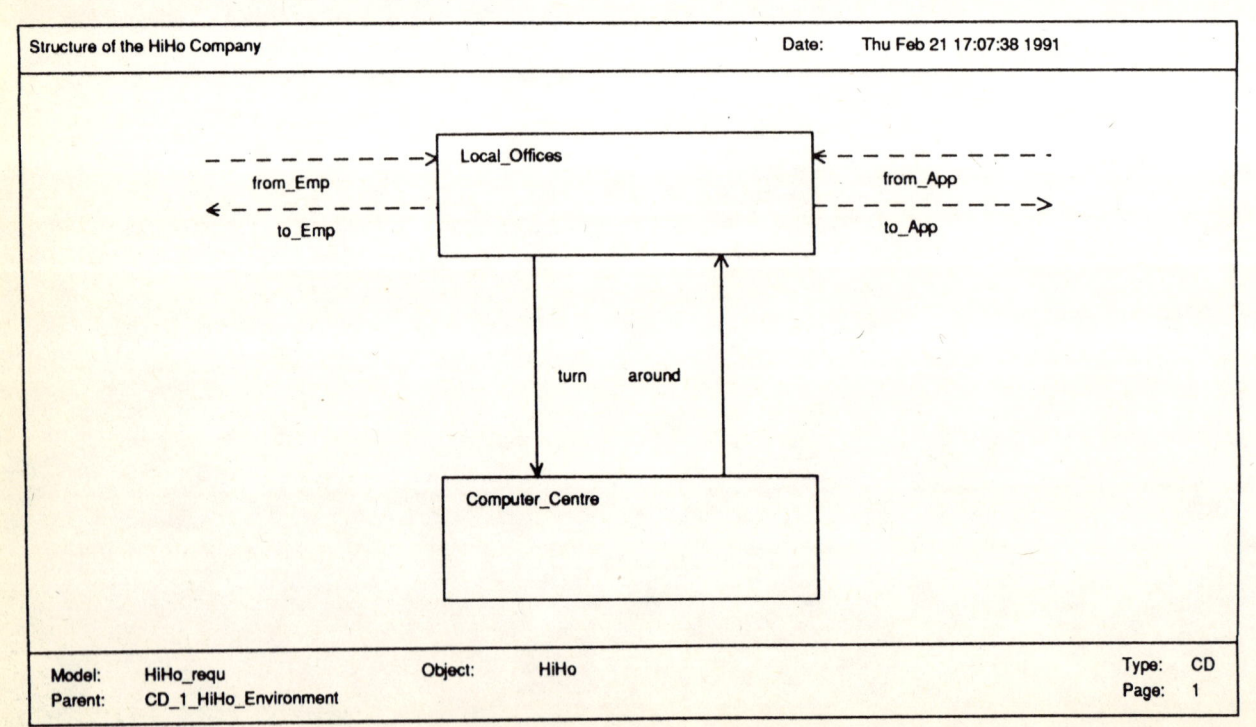

```
  ──from_Emp──▶  Local_Offices  ◀──from_App──
  ◀──to_Emp───                  ───to_App──▶
                      │  ▲
                      │  │
                   turn  around
                      ▼  │
                 Computer_Centre
```

Model: HiHo_requ Object: HiHo Type: CD
Parent: CD_1_HiHo_Environment Page: 1

Static structure 225

Structure of local offices

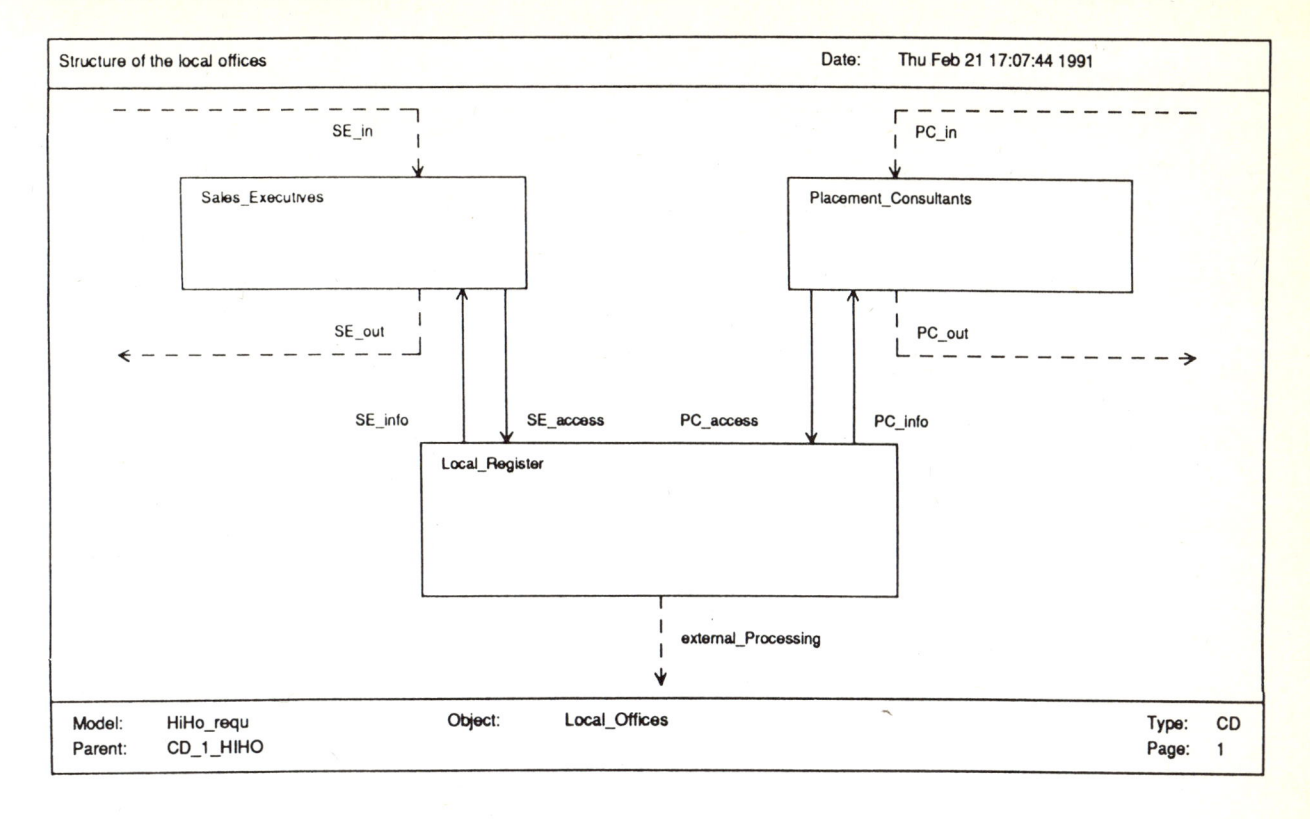

Structure of the local offices — Date: Thu Feb 21 17:07:44 1991

- SE_in → Sales_Executives
- SE_out ←
- PC_in → Placement_Consultants
- PC_out →
- SE_info, SE_access, PC_access, PC_info → Local_Register
- external_Processing ↓

Model: HiHo_requ
Parent: CD_1_HIHO
Object: Local_Offices
Type: CD
Page: 1

Structure of the local register

Structure of the local registers — Date: Fri Aug 2 14:14:57 1991

- Emp_Reg_In → Employer_Register
- Vac_Int_Cons → Interview_Register
- Int_Reg_In ↓
- App_Int_Cons, Int_Reg_Out →
- Vac_Reg_In → Vacancy_Register
- Emp_Vac_Cons
- Int_App_Cons → Applicant_Register
- App_Reg_In
- Int_Vac_Cons
- Vac_Reg_Out ←
- App_Reg_Out →
- Vac_App_Cons
- App_Vac_Cons

Model: HiHo_requ
Parent: CD_1_LOCAL_OFFICES
Object: Local_Register
Type: CD
Page: 1

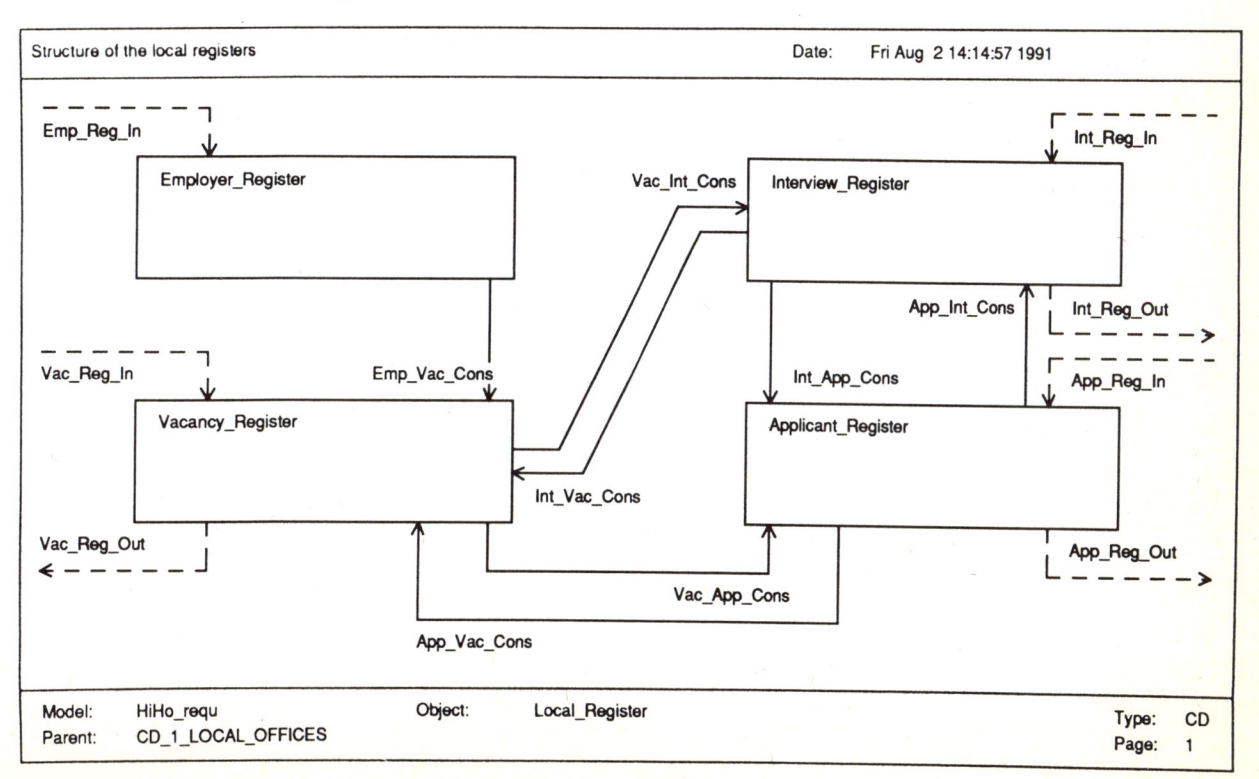

Communication

Input to local offices

Outside view

		Date: Fri Aug 2 13:14:51 1991
Name:	**Data Type:**	**Description:**
request_Conditions	Address	via SE_in to Sales_Executives
register_Employer	Employer_Form	-"-
withdraw_Employer	Emp_Ident	-"-
notify_Vacancy	Vacancy_Form	-"-
withdraw_Vacancy	Vac_Ident	-"-
reschedule_Interview_for_Vac	Int_Rescheduling	-"-
return_Result	Appointment_Card	via PC_in to Placement_Consultants
agree_Int_to_PCon	boolean	-"-
submit_Int_Res_to_PCon	Int_Result	-"-
confirm_Vac_to_PCon	boolean	-"-
confirm_Vac_to_SExe	boolean	via SE_in to Sales_Executives

Model: HiHo_requ Interface: from_Emp Type: IT
Parent: CD_1_HIHO Page: 1

		Date: Fri Aug 2 13:14:56 1991
Name:	**Data Type:**	**Description:**
register_Applicant	Applicant_Form	via PC_in to Placement_Consultants
withdraw_Applicant	App_Ident	-"-
review_Applicant	App_Ident	-"-
arrange_Interview	App_Ident	-"-
reschedule_Interview_for_App	Int_Rescheduling	-"-
submit_Reqs_to_PCon	App_Requ_Notification	-"-

Model: HiHo_requ Interface: from_App Type: IT
Parent: CD_1_HIHO Page: 1

		Date: Fri Aug 2 13:15:00 1991
Name:	**Data Type:**	**Description:**
return_Match_to_PCon	List_of_Matches	via PC_in to Placement_Consultants
return_OM_to_SExe	List_of_Matches	via SE_in to Sales_Executives
return_OM_to_PCon	List_of_Matches	via PC_in to Placement_Consultants

Model: HiHo_requ Interface: around Type: IT
Parent: CD_1_HIHO Page: 1

Inside view

		Date: Fri Aug 2 13:34:59 1991
Name:	**Data Type:**	**Description:**
request_Conditions	Address	from Employer via from_Emp
register_Employer	Employer_Form	_"_
withdraw_Employer	Emp_Ident	_"_
notify_Vacancy	Vacancy_Form	_"_
withdraw_Vacancy	Vac_Ident	_"_
reschedule_Interview_for_Vac	Int_Rescheduling	_"_
confirm_Vac_to_SExe	boolean	_"_
return_OM_to_SExe	List_of_Matches	from Computer_Centre via around

Model: HiHo_requ
Parent: CD_1_LOCAL_OFFICES
Interface: SE_in
Type: IT
Page: 1

		Date: Fri Aug 2 13:34:53 1991
Name:	**Data Type:**	**Description:**
register_Applicant	Applicant_Entity	from Applicant via from_App
review_Applicant	App_Ident	_"_
withdraw_Applicant	App_Ident	_"_
arrange_Interview	App_Ident	_"_
reschedule_Interview_for_App	Int_Rescheduling	_"_
return_Result	Appointment_Card	from Employer via from_Emp
submit_Reqs_to_PCon	App_Requ_Notification	from Applicant via from_App
confirm_Vac_to_PCon	boolean	from Employer via from_Emp
agree_Int_to_PCon	boolean	_"_
submit_Int_Res_to_PCon	Int_Result	_"_
return_Match_to_PCon	List_of_Matches	from Computer_Centre via around
return_OM_to_PCon	List_of_Matches	_"_

Model: HiHo_requ
Parent: CD_1_LOCAL_OFFICES
Interface: PC_in
Type: IT
Page: 1

Output of local offices

Outside view

		Date: Fri Aug 2 13:15:04 1991	
Name:	**Data Type:**	**Description:**	
return_Cond_to_Emp	Condition_Letter	from Sales_Executives via SE_out	
notify_Accout_to_Emp	Invoice	from Vacancy_Register via external_Processing	
check_Int_with_Emp	Int_Date	from Placement_Consultants via PC_out	
confirm_Int_to_Emp	Confirmation_Letter	-"-	
request_Int_Res_from_Emp	Int_Ident	-"-	
check_Vac_with_Emp	Vac_Ident	from Sales_Executives via SE_out	
		and Placement_Consultants via PC_out	
notify_Apps_to_Emp	Vac_Apps_Notification	from Sales Executives via SE_out	
Model: HiHo_requ Parent: CD_1_HIHO	Interface: to_Emp		Type: IT Page: 1

		Date: Fri Aug 2 13:15:08 1991	
Name:	**Data Type:**	**Description:**	
confirm_Rgst_to_App	Confirmation_Letter	from Placement_Consultants via PC_out	
request_Reqs_from_App	signal	-"-	
invite_App_for_Rev	Invitation_Letter	-"-	
invite_App_for_Argmt	Invitation_Letter	-"-	
confirm_Int_to_App	Appointment_Card	-"-	
Model: HiHo_requ Parent: CD_1_HIHO	Interface: to_App		Type: IT Page: 1

		Date: Fri Aug 2 12:59:10 1991	
Name:	**Data Type:**	**Description:**	
submit_Vacs_to_Cntr	Vacancy_Storage	from Vacancy_Register via external_Processing	
submit_Apps_to_Cntr	Applicant_Storage	from Applicant_Register via external Processing	
online_Vac_to_Cntr	Vacancy_Entity	from Vacancy_Register via external_Processing	
online_Vacs_to_Cntr	Vacancy_Storage	-"-	
online_App_to_Cntr	Applicant_Entity	from Applicant_Register via external_Processing	
online_Apps_to_Cntr	Applicant_Storage	-"-	
Model: HiHo_requ Parent: CD_1_HIHO	Interface: turn		Type: IT Page: 1

Inside view

		Date: Fri Aug 2 13:35:11 1991	
Name:	**Data Type:**	**Description:**	
return_Cond_to_Emp	Condition_Letter	via to_Emp to Employer	
check_Vac_with_Emp	Vac_Ident	- " -	
notify_Apps_to_Emp	Vac_Apps_Notification	- " -	
Model: HiHo_requ	Interface: SE_out		Type: IT
Parent: CD_1_LOCAL_OFFICES			Page: 1

		Date: Fri Aug 2 13:35:05 1991	
Name:	**Data Type:**	**Description:**	
invite_App_for_Argmt	Invitation_Letter	via to_App to Applicant	
invite_App_for_Rev	Invitation_Letter	- " -	
confirm_Rgst_to_App	Confirmation_Letter	- " -	
confirm_Int_to_App	Appointment_Card	- " -	
request_Reqs_from_App	signal	- " -	
check_Vac_with_Emp	Vac_Ident	via to_Emp to Employer	
check_Int_with_Emp	Int_Date	- " -	
confirm_Int_to_Emp	Confimation_Letter	- " -	
request_Int_Res_from_Emp	Int_Ident	- " -	
Model: HiHo_requ	Interface: PC_out		Type: IT
Parent: CD_1_LOCAL_OFFICES			Page: 1

		Date: Fri Aug 2 13:35:16 1991	
Name:	**Data Type:**	**Description:**	
submit_Vacs_to_Cntr	Vacancy_Storage	from Vacancy_Register to Computer_Centre	
submit_Apps_to_Cntr	Applicant_Storage	from Applicant_Register to Computer_Centre	
notify_Account_to_Emp	Invoice	from Vacancy_Register to Employer	
online_Vac_to_Cntr	Vacancy_Entity	from Vacancy_Register to Computer_Centre	
online_Vacs_to_Cntr	Vacancy_Storage	- " -	
online_App_to_Cntr	Applicant_Entity	from Applicant_Register to Computer_Centre	
online_Apps_to_Cntr	Applicant_Storage	- " -	
Model: HiHo_requ	Interface: external_Processing		Type: IT
Parent: CD_1_LOCAL_OFFICES			Page: 1

Input of local register

Outside view

Name:	Data Type:	Description:
insert_Emp_in_Reg	Employer_Entity	via Emp_Reg_In to Employer_Register
delete_Emp_from_Reg	Emp_Ident	- " -
insert_Vac_in_Reg	Vacancy_Entity	via Vac_Reg_In to Vacancy_Register
delete_Vac_from_Reg	Vac_Ident	- " -
change_Int_in_Reg	Int_Rescheduling	via Int_Reg_In to Interview_Register
request_Match_for_Vac	Vac_Ident	vie Vac_Reg_In to Vacancy_Register

Date: Fri Aug 2 13:46:47 1991

Model: HiHo_requ
Parent: OD_1_LOCAL_OFFICES
Interface: SE_access
Type: IT
Page: 1

Name:	Data Type:	Description:
insert_App_in_Reg	Applicant_Entity	via App_Reg_In to Applicant_Register
change_App_in_Reg	Applicant_Entity	- " -
delete_App_from_Reg	App_Ident	- " -
insert_Int_in_Reg	Interview_Entity	via Int_Reg_In to Interview_Register
change_Int_in_Reg	Int_Rescheduling	- " -
submit_Int_Res_to_Reg	Appointment_Card	- " -
delete_Vac_from_Reg	Vac_Ident	via Vac_Reg_In to Vacancy_Register
request_Match_for_App	App_Ident	via App_Reg_In to Applicant_Register

Date: Fri Aug 2 13:46:43 1991

Model: HiHo_requ
Parent: CD_1_LOCAL_OFFICES
Interface: PC_access
Type: IT
Page: 1

Inside view

		Date: Fri Aug 2 14:15:08 1991	
Name:	Data Type:	Description:	
insert_Emp_in_Reg	Employer_Entity	from Sales_Executives via SE_access	
delete_Emp_from_Reg	Emp_Ident	-"-	
Model: HiHo_requ Parent: CD_1_LOCAL_REGISTER	Interface: Emp_Reg_In		Type: IT Page: 1

		Date: Fri Aug 2 14:15:03 1991	
Name:	Data Type:	Description:	
insert_App_in_Reg	Applicant_Entity	from Placement_Consultants via PC_access	
change_App_in_Reg	Applicant_Entity	-"-	
delete_App_from_Reg	App_Ident	-"-	
request_Match_for_App	App_Ident	-"-	
Model: HiHo_requ Parent: CD_1_LOCAL_REGISTER	Interface: App_Reg_In		Type: IT Page: 1

		Date: Fri Aug 2 14:15:17 1991	
Name:	Data Type:	Description:	
insert_Int_in_Reg	Interview_Entity	from Placement_Consultants via PC–access	
submit_Int_Res_to_Reg	Appointment_Card	-"-	
change_Int_in_Reg	Int_Rescheduling	from Placement_Consultants via PC_access and Sales_Executives via SE_access	
Model: HiHo_requ Parent: CD_1_LOCAL_REGISTER	Interface: Int_Reg_In		Type: IT Page: 1

		Date: Fri Aug 2 14:15:13 1991	
Name:	Data Type:	Description:	
insert_Vac_in_Reg	Vac_Entity	from Sales_Executives via SE_access	
delete_Vac_from_Reg	Vac_Ident	from Sales_Executives via SE_access and Placement_Consultants via PC–access	
request_Match_for_Vac	Vac_Ident	from Sales_Executives via SE_access	
Model: HiHo_requ Parent: CD_1_LOCAL_REGISTER	Interface: Vac_Reg_In		Type: IT Page: 1

Output of local register

Outside view

		Date: Fri Aug 2 13:46:51 1991
Name:	**Data Type:**	**Description:**
indicate_Vac_to_SExe	Vac_Ident	from Vacancy_Register via Vac_Reg_Out
Model: HiHo_requ Parent: CD_1_LOCAL_OFFICES	Interface: SE_info	Type: IT Page: 1

		Date: Fri Aug 2 13:46:38 1991
Name:	**Data Type:**	**Description:**
indicate_App_to_PCon	App_Ident	from Applicant_Register via App_Reg_Out
indicate_Int_to_PCon	Int_Ident	from Interview_Register via Int_Reg_Out
Model: HiHo_requ Parent: CD_1_LOCAL_OFFICES	Interface: PC_info	Type: IT Page: 1

		Date: Fri Aug 2 13:35:16 1991
Name:	**Data Type:**	**Description:**
submit_Vacs_to_Cntr	Vacancy_Storage	from Vacancy_Register to Computer_Centre
submit_Apps_to_Cntr	Applicant_Storage	from Applicant_Register to Computer_Centre
notify_Account_to_Emp	Invoice	from Vacancy_Register to Employer
online_Vac_to_Cntr	Vacancy_Entity	from Vacancy_Register to Computer_Centre
online_Vacs_to_Cntr	Vacancy_Storage	_"_
online_App_to_Cntr	Applicant_Entity	from Applicant_Register to Computer_Centre
online_Apps_to_Cntr	Applicant_Storage	_"_
Model: HiHo_requ Parent: CD_1_LOCAL_OFFICES	Interface: external_Processing	Type: IT Page: 1

Inside view

Name:	Data Type:	Description:
		Date: Fri Aug 2 14:29:12 1991
indicate_App_to_PCon	App_Ident	via PC_info to Placement_Consultants
submit_Apps_to_Cntr	Applicant_Storage	via extrenal_Processing to Computer_Centre
online_App_to_Cntr	Applicant_Entity	_"_
online_Apps_to_Cntr	Applicant_Storage	_"_

Model: HiHo_requ Interface: App_Reg_Out Type: IT
Parent: CD_1_LOCAL_REGISTER Page: 1

Name:	Data Type:	Description:
		Date: Fri Aug 2 14:29:21 1991
indicate_Int_to_PCon	Int_Ident	via PC_info to Placement_Consultants

Model: HiHo_requ Interface: Int_Reg_Out Type: IT
Parent: CD_1_LOCAL_REGISTER Page: 1

Name:	Data Type:	Description:
		Date: Fri Aug 2 14:29:17 1991
indicate_Vac_to_SExe	Vac_Ident	via SE_info to Sales_executives
notify_Account_to_Emp	Invoice	via external_Processing to Employer
submit_Vacs_to_Cntr	Vacancy_Storage	via external_Processing to Computer_Centre
online_Vac_to_Cntr	Vacancy_Entity	_"_
online_Vacs_to_Cntr	Vacancy_Storage	_"_

Model: HiHo_requ Interface: Vac_Reg_Out Type: IT
Parent: CD_1_LOCAL_REGISTER Page: 1

Internal communication of local register

		Date: Fri Mar 22 13:41:32 1991
Name:	Data Type:	Description:
delete_Ints_for_App	App_Ident	
Model: HiHo_requ Parent: CD_1_LOCAL_REGISTER	Interface: App_Int_Cons	Type: IT Page: 1

		Date: Fri Mar 22 14:01:04 1991
Name:	Data Type:	Description:
delete_Ints_for_Vac	Vac_Ident	
Model: HiHo_requ Parent: CD_1_LOCAL_REGISTER	Interface: Vac_Int_Cons	Type: IT Page: 1

		Date: Fri Mar 22 13:42:35 1991
Name:	Data Type:	Description:
delete_App_from_Reg	App_Ident	
Model: HiHo_requ Parent: CD_1_LOCAL_REGISTER	Interface: Int_App_Cons	Type: IT Page: 1

		Date: Fri Mar 22 13:43:04 1991
Name:	Data Type:	Description:
fill_Post_in_Vac	Vac_Ident	
Model: HiHo_requ Parent: CD_1_LOCAL_REGISTER	Interface: Int_Vac_Cons	Type: IT Page: 1

		Date: Fri Mar 22 13:42:20 1991
Name:	Data Type:	Description:
delete_Vacs_for_Emp	Emp_Ident	
Model: HiHo_requ Parent: CD_1_LOCAL_REGISTER	Interface: Emp_Vac_Cons	Type: IT Page: 1

Communication

		Date:	Fri Mar 22 13:41:56 1991
Name:	Data Type:	Description:	
request_Match_for_Vacs	signal		

Model: HiHo_requ Interface: App_Vac_Cons Type: IT
Parent: CD_1_LOCAL_REGISTER Page: 1

		Date:	Fri Mar 22 14:00:56 1991
Name:	Data Type:	Description:	
request_Match_for_Apps	signal		

Model: HiHo_requ Interface: Vac_App_Cons Type: IT
Parent: CD_1_LOCAL_REGISTER Page: 1

Behaviour

Process objects of local offices

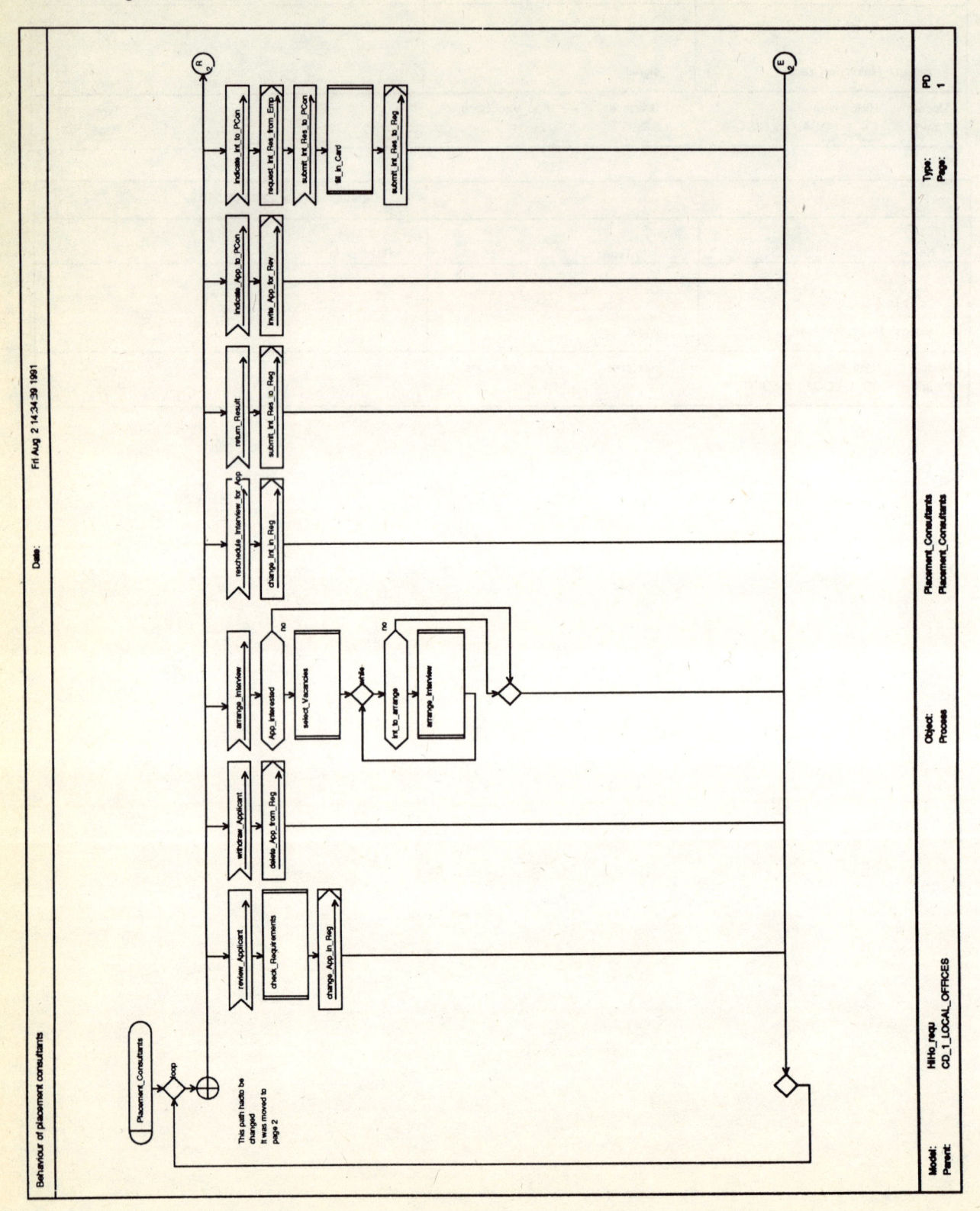

Behaviour

Behaviour of the placement consultants   Date: Thu Mar 21 10:55:21 1991

Model: HiHo_requ   Object: Placement_Consultants
Parent: CD_1_LOCAL_OFFICES   Process: Placement_Consultants
Type: PD
Page: 2

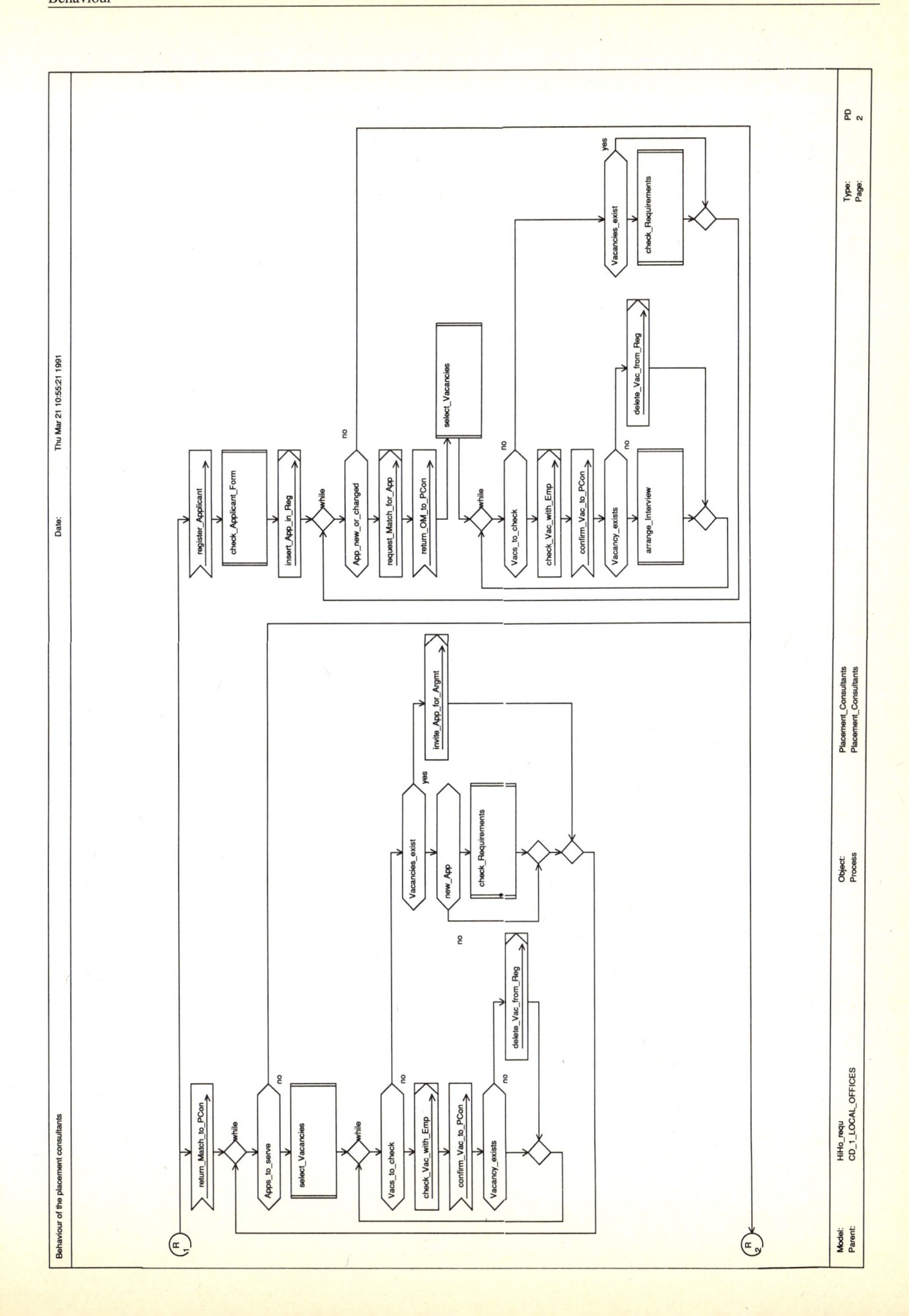

Annex 3 Hi-Ho GRAPES Development

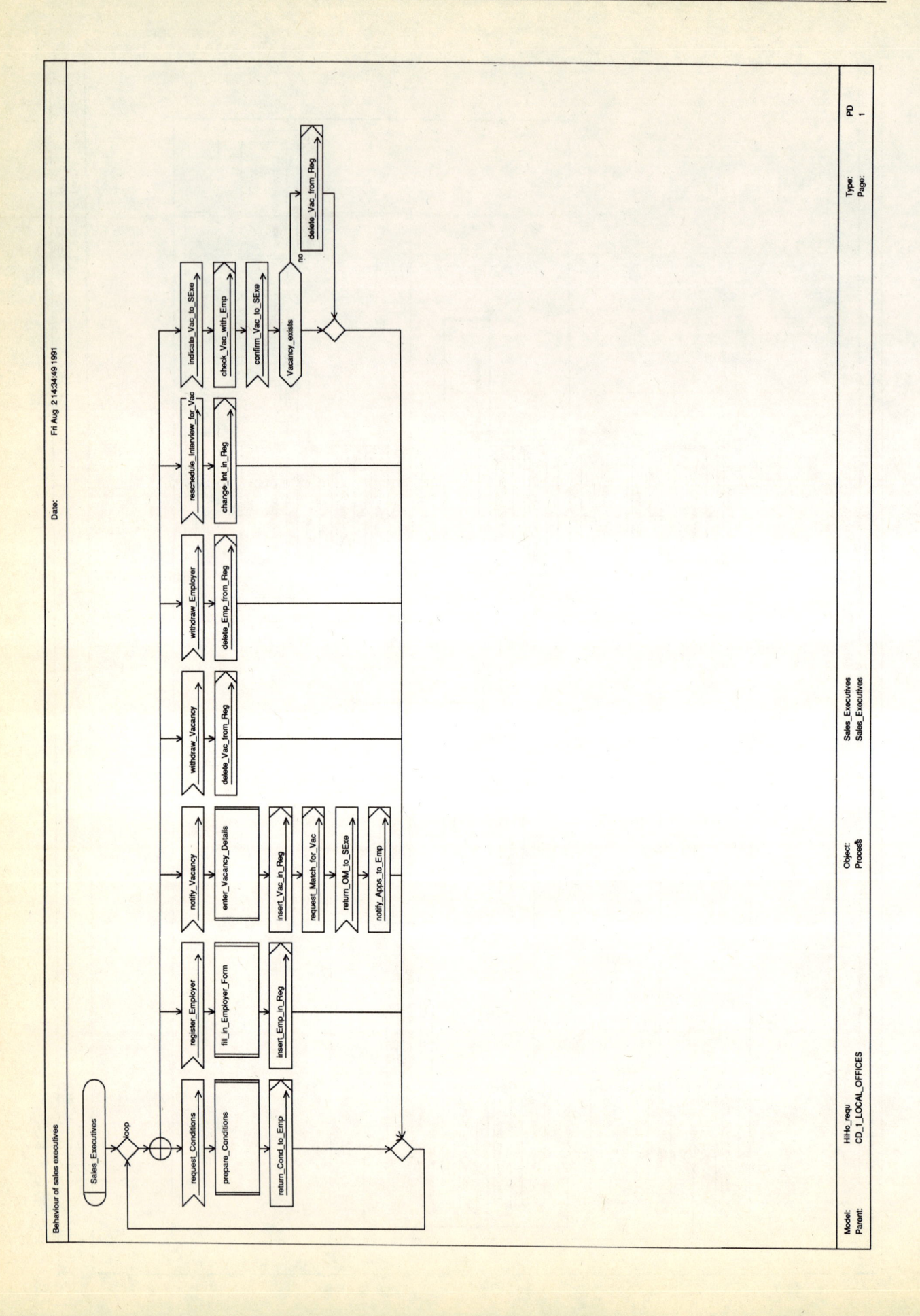

Process objects in local register

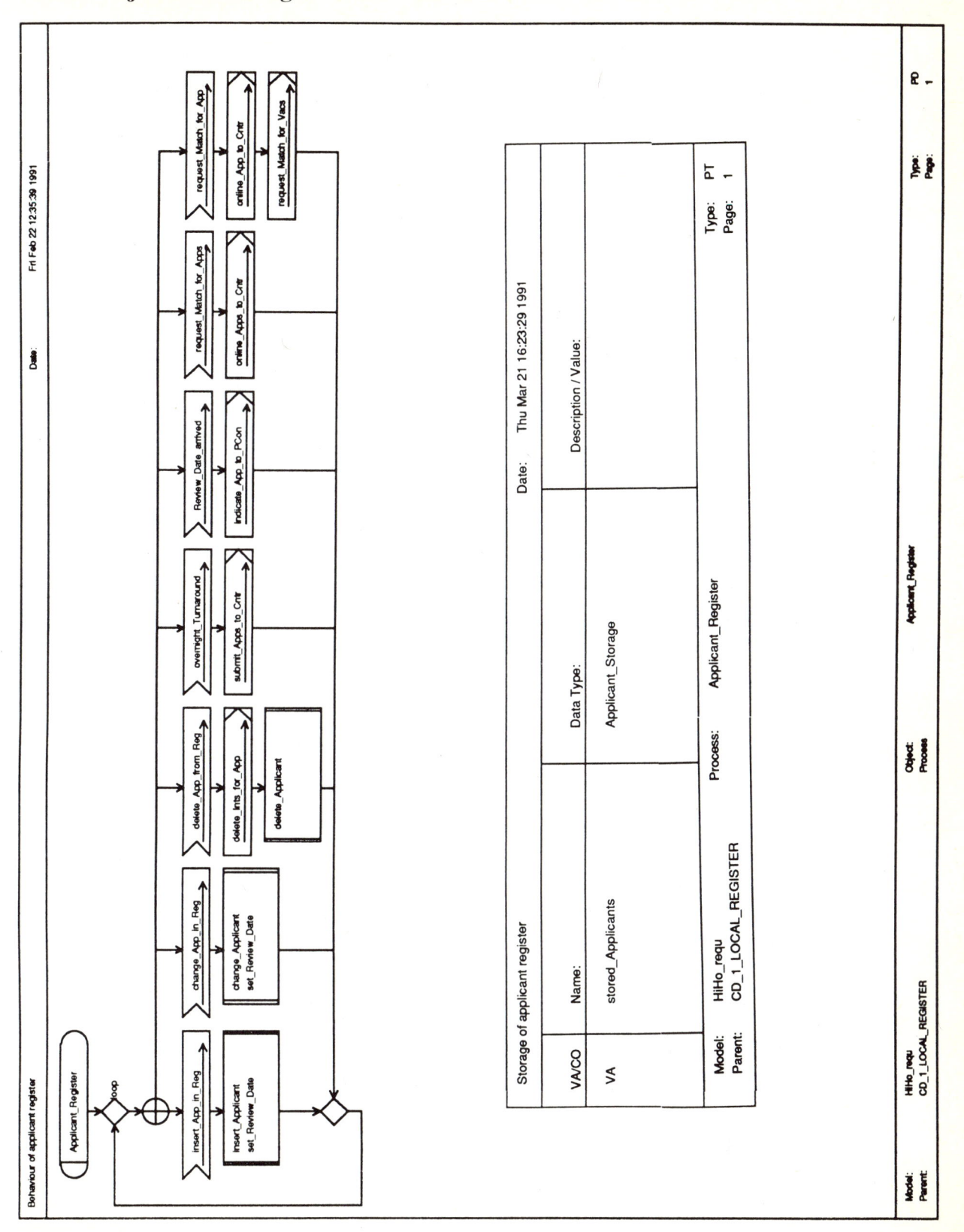

Behaviour of employer register

Date: Thu Mar 21 10:55:33 1991

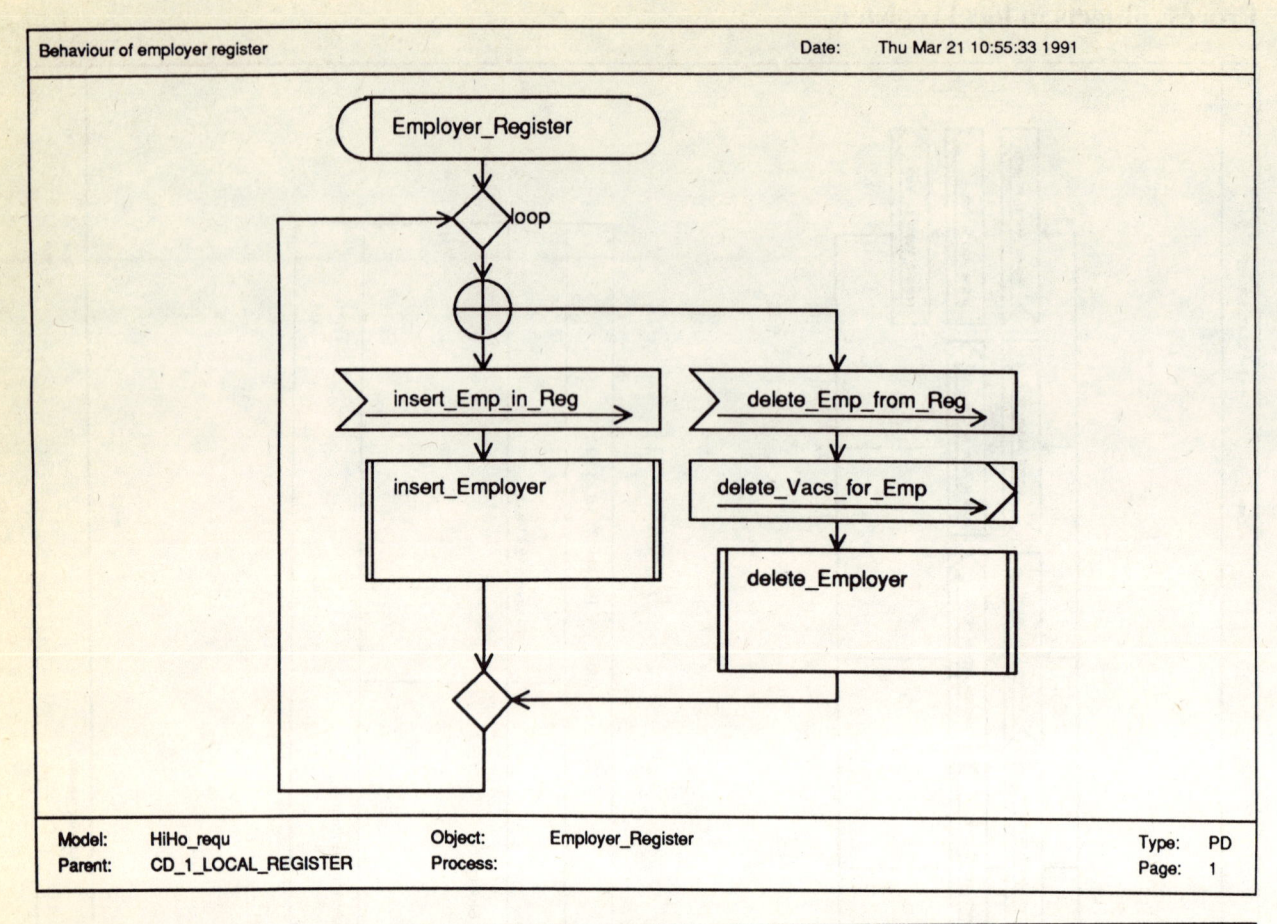

```
         Employer_Register
                │
                ▼
              ◇ loop
                │
                ⊕
         ┌──────┴──────┐
         ▼             ▼
   insert_Emp_in_Reg   delete_Emp_from_Reg
         │             │
         ▼             ▼
   insert_Employer     delete_Vacs_for_Emp
                       │
                       ▼
                       delete_Employer
         │             │
         └──────┬──────┘
                ▼
                ◇
```

Model:	HiHo_requ	Object:	Employer_Register	Type:	PD
Parent:	CD_1_LOCAL_REGISTER	Process:		Page:	1

Storage of employer register

Date: Thu Mar 21 16:22:39 1991

VA/CO	Name:	Data Type:	Description / Value:
VA	stored_Employers	Employer_Storage	

Model:	HiHo_requ	Process:	Employer_Register	Type:	PT
Parent:	CD_1_LOCAL_REGISTER			Page:	1

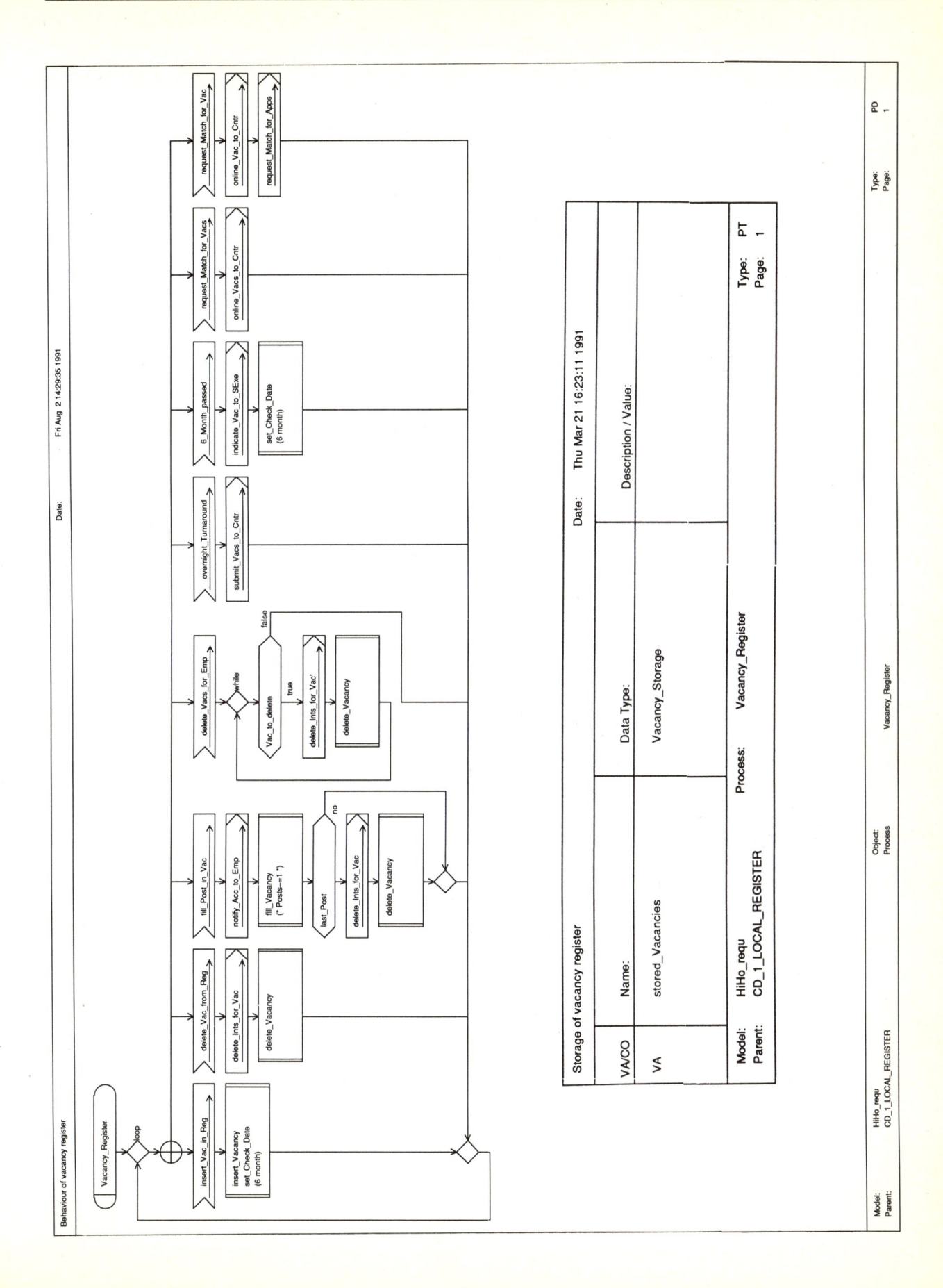

Behaviour

Behaviour of vacancy register

Date: Fri Aug 2 14:29:35 1991

- Vacancy_Register
- loop
- insert_Vac_in_Reg
- insert_Vacancy set_Check_Date (6 month)
- delete_Vac_from_Reg
- delete_Ints_for_Vac
- delete_Vacancy
- fill_Post_in_Vac
- notify_Acc_to_Emp
- fill_Vacancy (" Posts=1 ")
- last_Post — no
- delete_Ints_for_Vac
- delete_Vacancy
- delete_Vacs_for_Emp
- while Vac_to_delete — false / true
- delete_Ints_for_Vac'
- delete_Vacancy
- overnight_Turnaround
- submit_Vacs_to_Cntr
- 6_Month_passed
- indicate_Vac_to_SExe
- set_Check_Date (6 month)
- request_Match_for_Vacs
- online_Vacs_to_Cntr
- request_Match_for_Vac
- online_Vac_to_Cntr
- request_Match_for_Apps

Storage of vacancy register

Date: Thu Mar 21 16:23:11 1991

VA/CO	Name:	Data Type:	Description / Value:
VA	stored_Vacancies	Vacancy_Storage	

Model: HiHo_requ Process: Vacancy_Register Type: PT
Parent: CD_1_LOCAL_REGISTER Page: 1

Model: HiHo_requ Object: Process Type: PD
Parent: CD_1_LOCAL_REGISTER Vacancy_Register Page: 1

241

Annex 3 Hi-Ho GRAPES Development

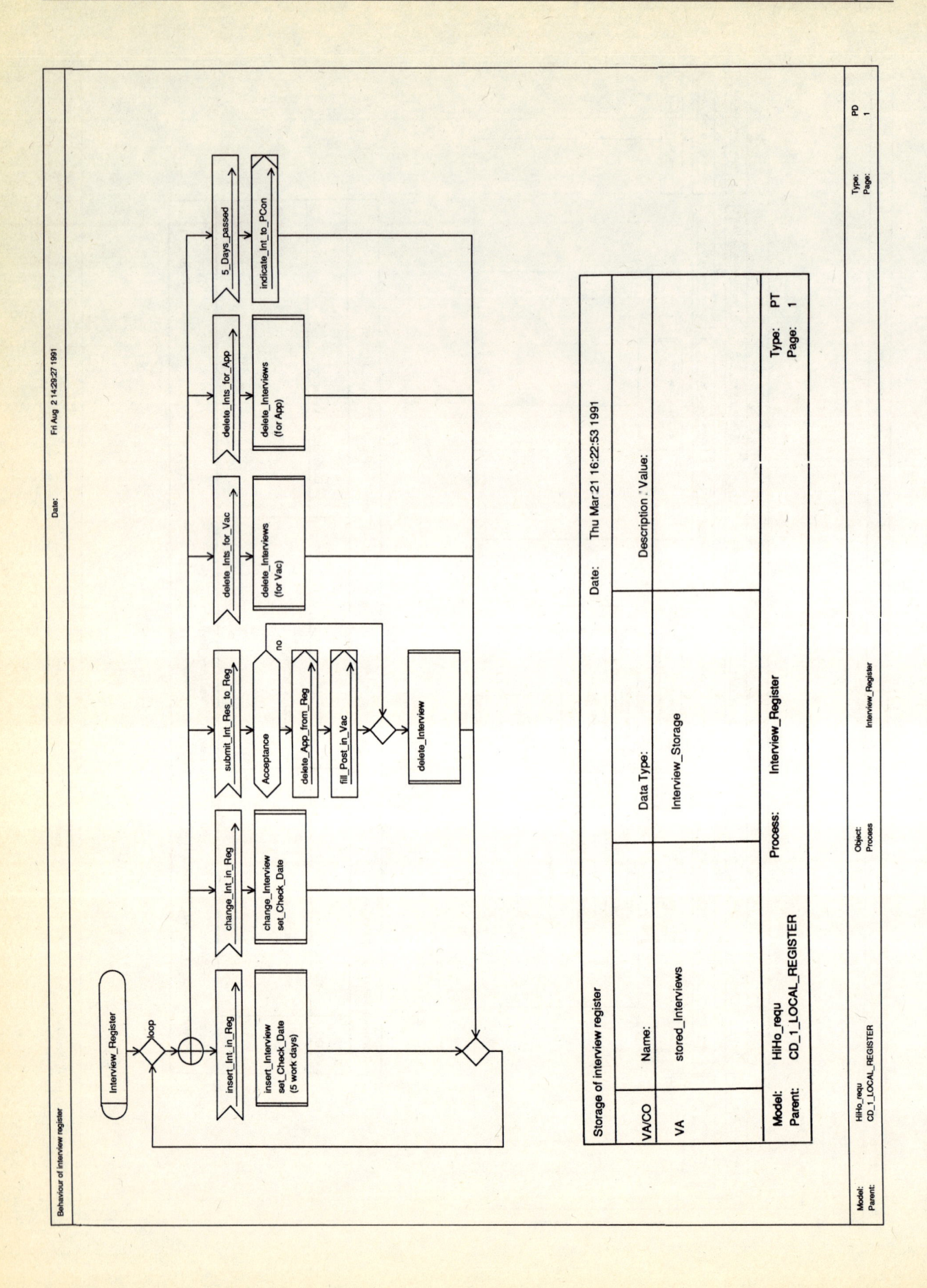

Process objects on Hi-Ho company level

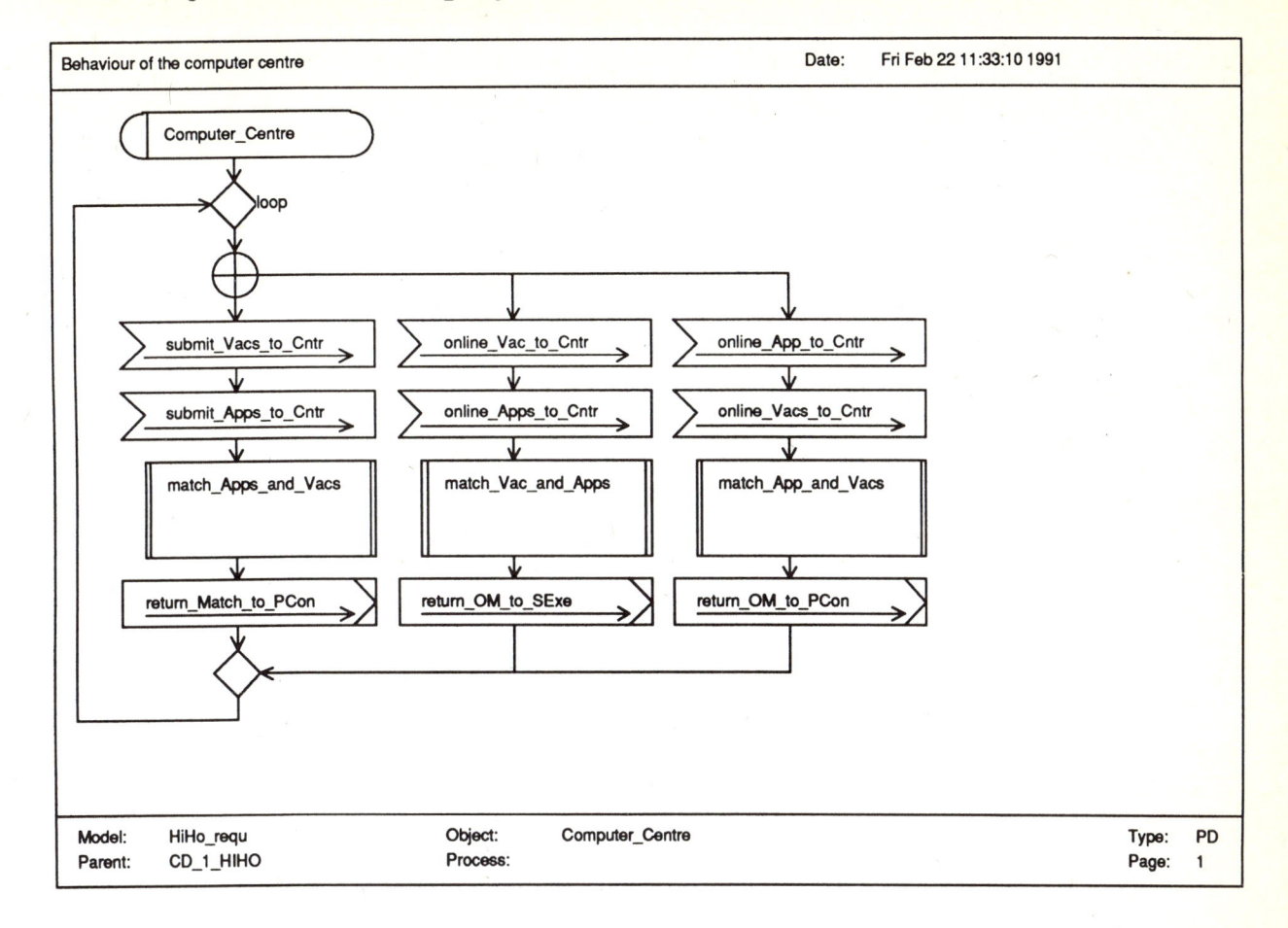

Behaviour of the computer centre Date: Fri Feb 22 11:33:10 1991

- Computer_Centre
  - loop
    - submit_Vacs_to_Cntr
    - submit_Apps_to_Cntr
    - match_Apps_and_Vacs
    - return_Match_to_PCon
  - online_Vac_to_Cntr
    - online_Apps_to_Cntr
    - match_Vac_and_Apps
    - return_OM_to_SExe
  - online_App_to_Cntr
    - online_Vacs_to_Cntr
    - match_App_and_Vacs
    - return_OM_to_PCon

Model: HiHo_requ
Parent: CD_1_HIHO
Object: Computer_Centre
Process:
Type: PD
Page: 1

Data structure of local register

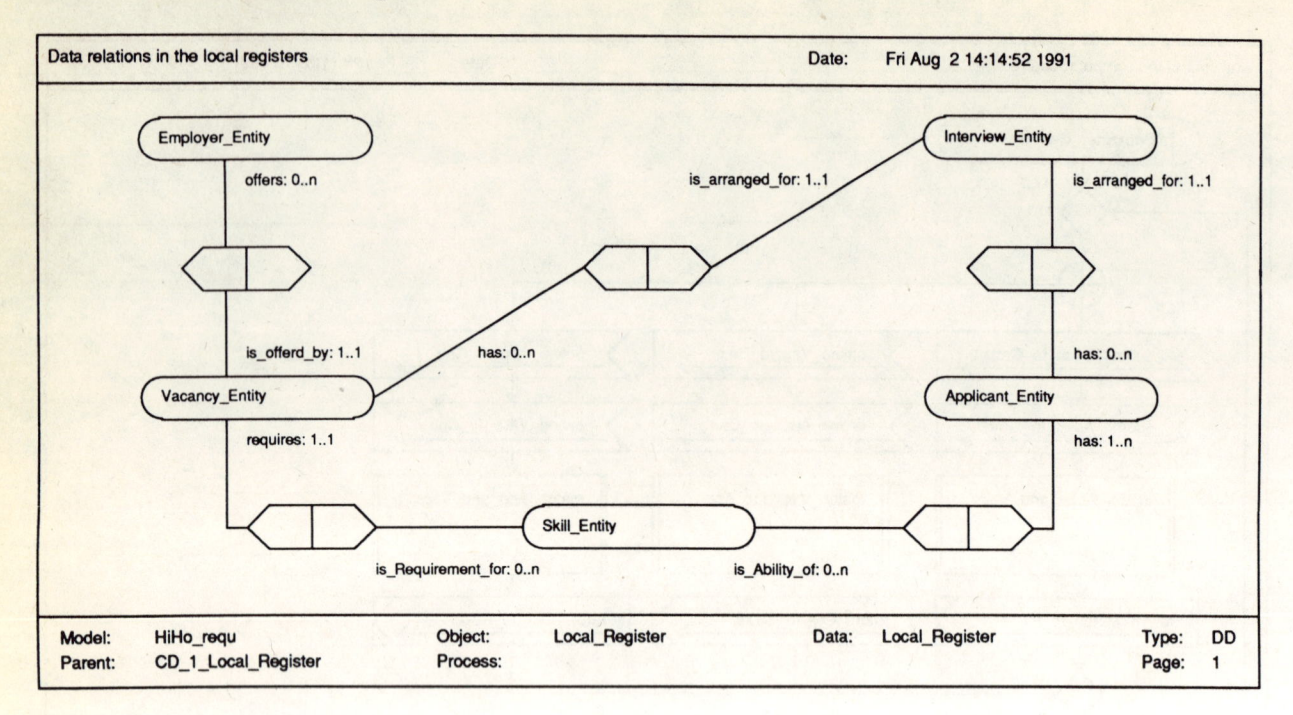

Data relations in the local registers			Date:	Fri Aug 2 14:14:52 1991	

- Employer_Entity — offers: 0..n
- Vacancy_Entity — is_offerd_by: 1..1
- Vacancy_Entity — has: 0..n — Interview_Entity (is_arranged_for: 1..1)
- Interview_Entity — is_arranged_for: 1..1 — Applicant_Entity
- Applicant_Entity — has: 0..n
- Vacancy_Entity — requires: 1..1 — Skill_Entity (is_Requirement_for: 0..n)
- Applicant_Entity — has: 1..n — Skill_Entity (is_Ability_of: 0..n)

Model:	HiHo_requ	Object:	Local_Register	Data:	Local_Register	Type:	DD
Parent:	CD_1_Local_Register	Process:				Page:	1

Data types for internal processing

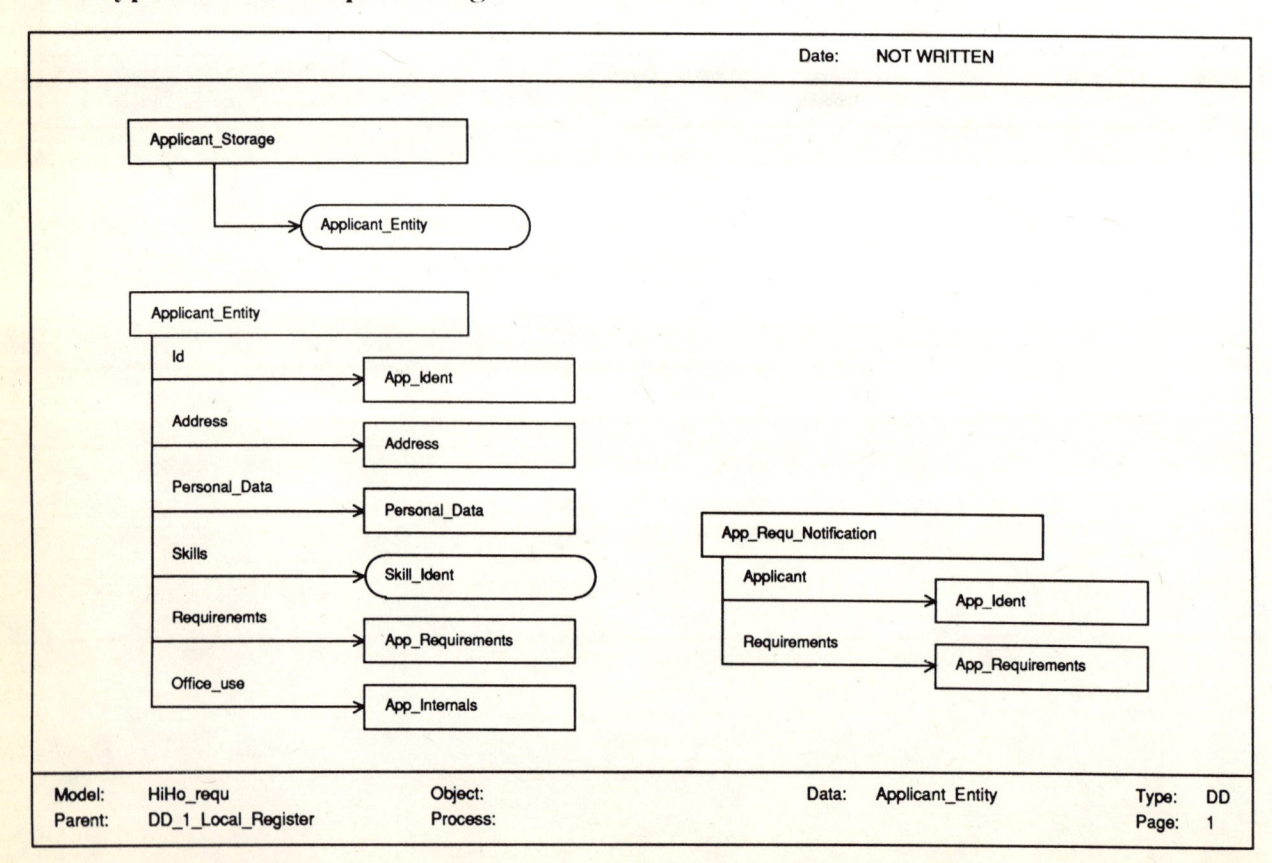

	Date:	NOT WRITTEN

Applicant_Storage → Applicant_Entity

Applicant_Entity
- Id → App_Ident
- Address → Address
- Personal_Data → Personal_Data
- Skills → Skill_Ident
- Requirenemts → App_Requirements
- Office_use → App_Internals

App_Requ_Notification
- Applicant → App_Ident
- Requirements → App_Requirements

Model:	HiHo_requ	Object:		Data:	Applicant_Entity	Type:	DD
Parent:	DD_1_Local_Register	Process:				Page:	1

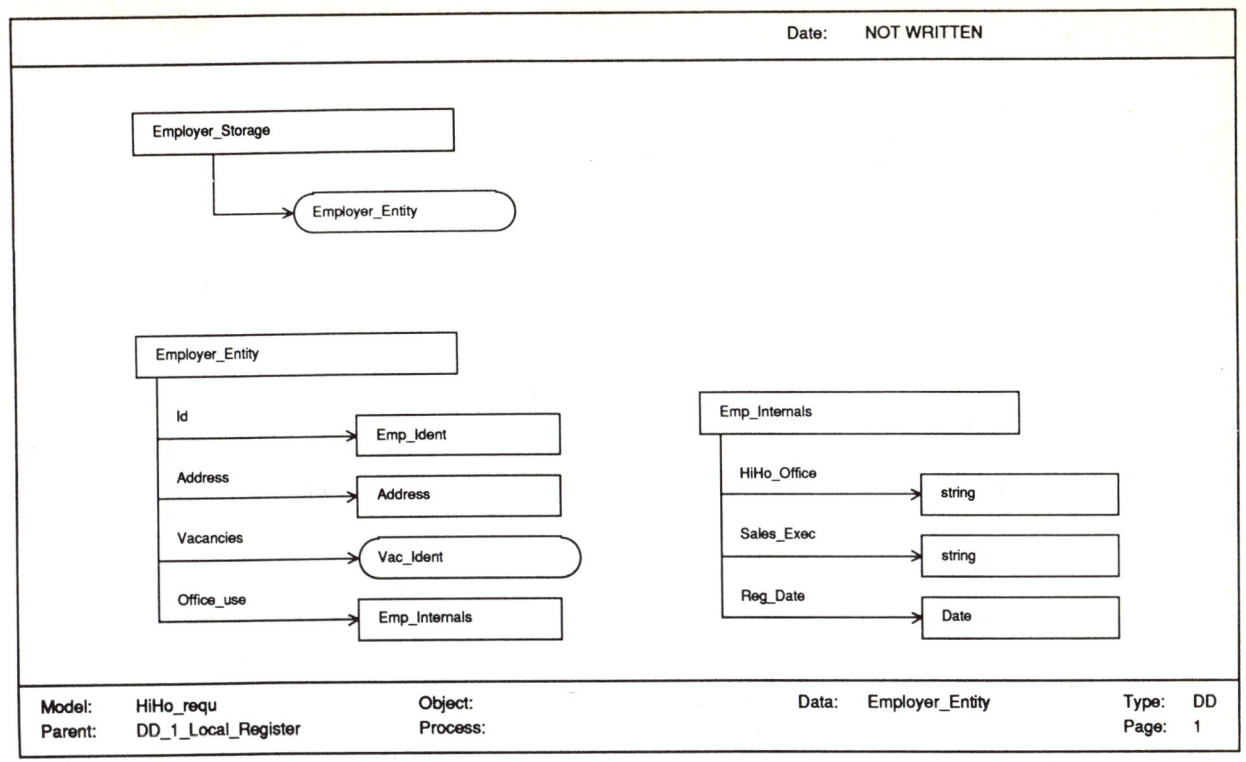

Behaviour

Date: NOT WRITTEN

Employer_Storage → Employer_Entity

Employer_Entity
- Id → Emp_Ident
- Address → Address
- Vacancies → Vac_Ident
- Office_use → Emp_Internals

Emp_Internals
- HiHo_Office → string
- Sales_Exec → string
- Reg_Date → Date

Model: HiHo_requ
Parent: DD_1_Local_Register
Object:
Process:
Data: Employer_Entity
Type: DD
Page: 1

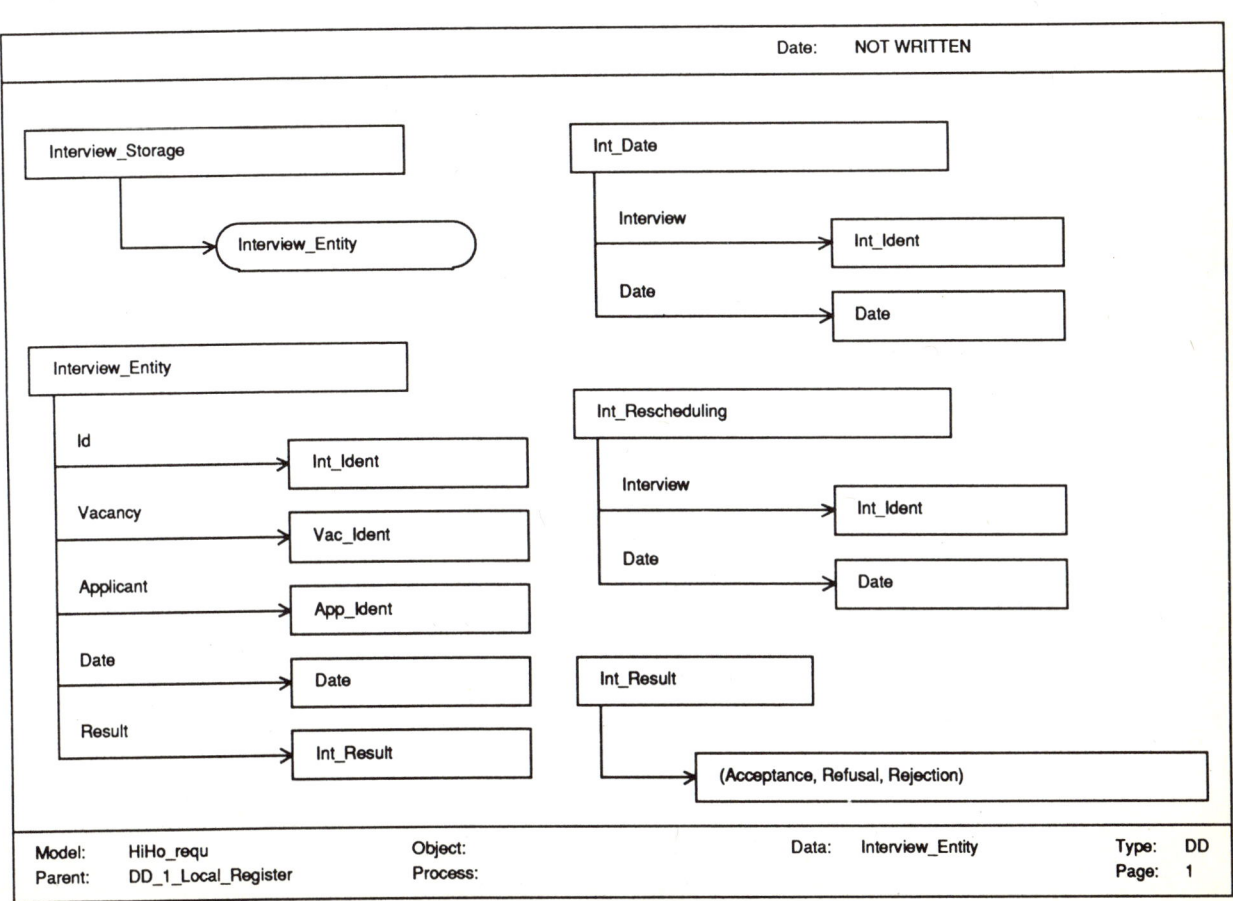

Date: NOT WRITTEN

Interview_Storage → Interview_Entity

Interview_Entity
- Id → Int_Ident
- Vacancy → Vac_Ident
- Applicant → App_Ident
- Date → Date
- Result → Int_Result

Int_Date
- Interview → Int_Ident
- Date → Date

Int_Rescheduling
- Interview → Int_Ident
- Date → Date

Int_Result
- → (Acceptance, Refusal, Rejection)

Model: HiHo_requ
Parent: DD_1_Local_Register
Object:
Process:
Data: Interview_Entity
Type: DD
Page: 1

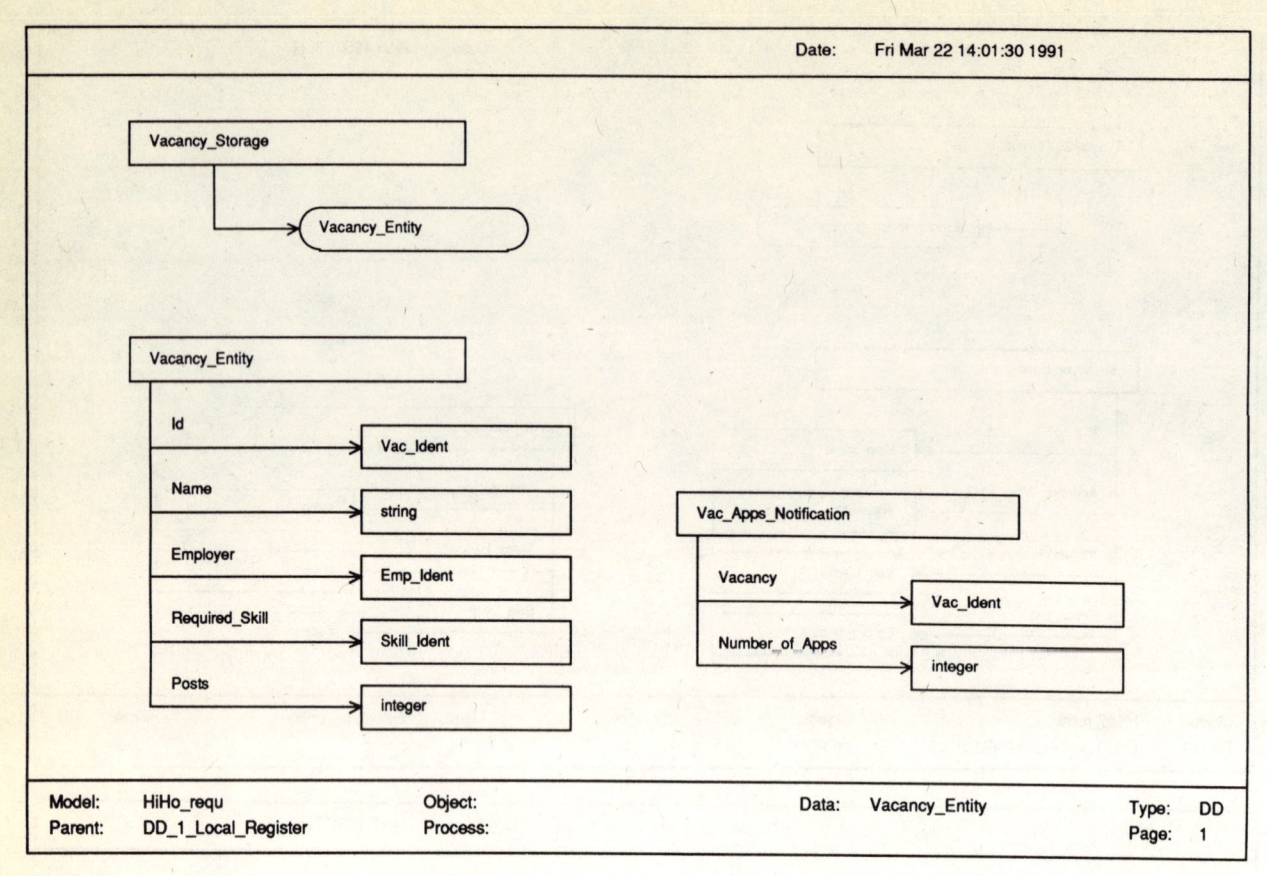

Diagram 1

Date: Fri Mar 22 14:01:30 1991

- Vacancy_Storage → Vacancy_Entity

- Vacancy_Entity
  - Id → Vac_Ident
  - Name → string
  - Employer → Emp_Ident
  - Required_Skill → Skill_Ident
  - Posts → integer

- Vac_Apps_Notification
  - Vacancy → Vac_Ident
  - Number_of_Apps → integer

Model:	HiHo_requ	Object:	Data: Vacancy_Entity	Type:	DD
Parent:	DD_1_Local_Register	Process:		Page:	1

Diagram 2

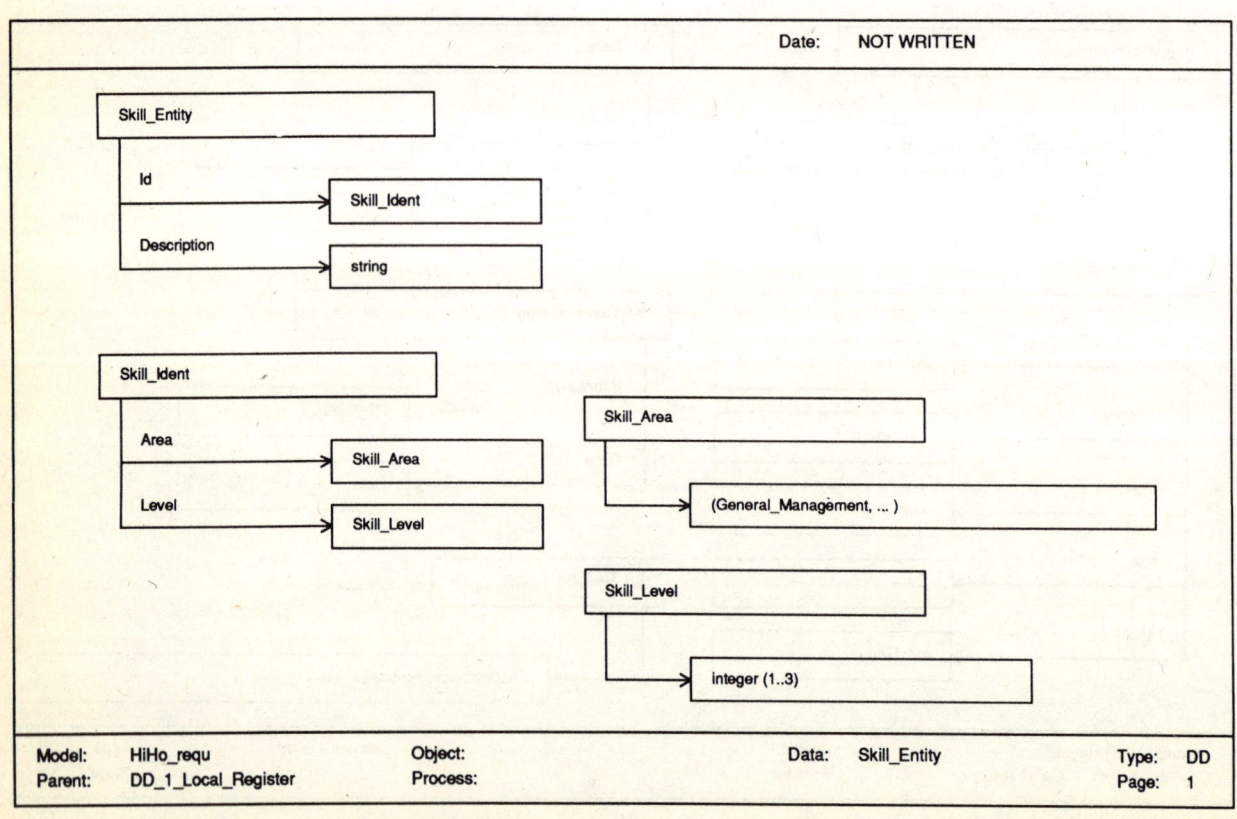

Date: NOT WRITTEN

- Skill_Entity
  - Id → Skill_Ident
  - Description → string

- Skill_Ident
  - Area → Skill_Area
  - Level → Skill_Level

- Skill_Area → (General_Management, ...)

- Skill_Level → integer (1..3)

Model:	HiHo_requ	Object:	Data: Skill_Entity	Type:	DD
Parent:	DD_1_Local_Register	Process:		Page:	1

Data types for external communication

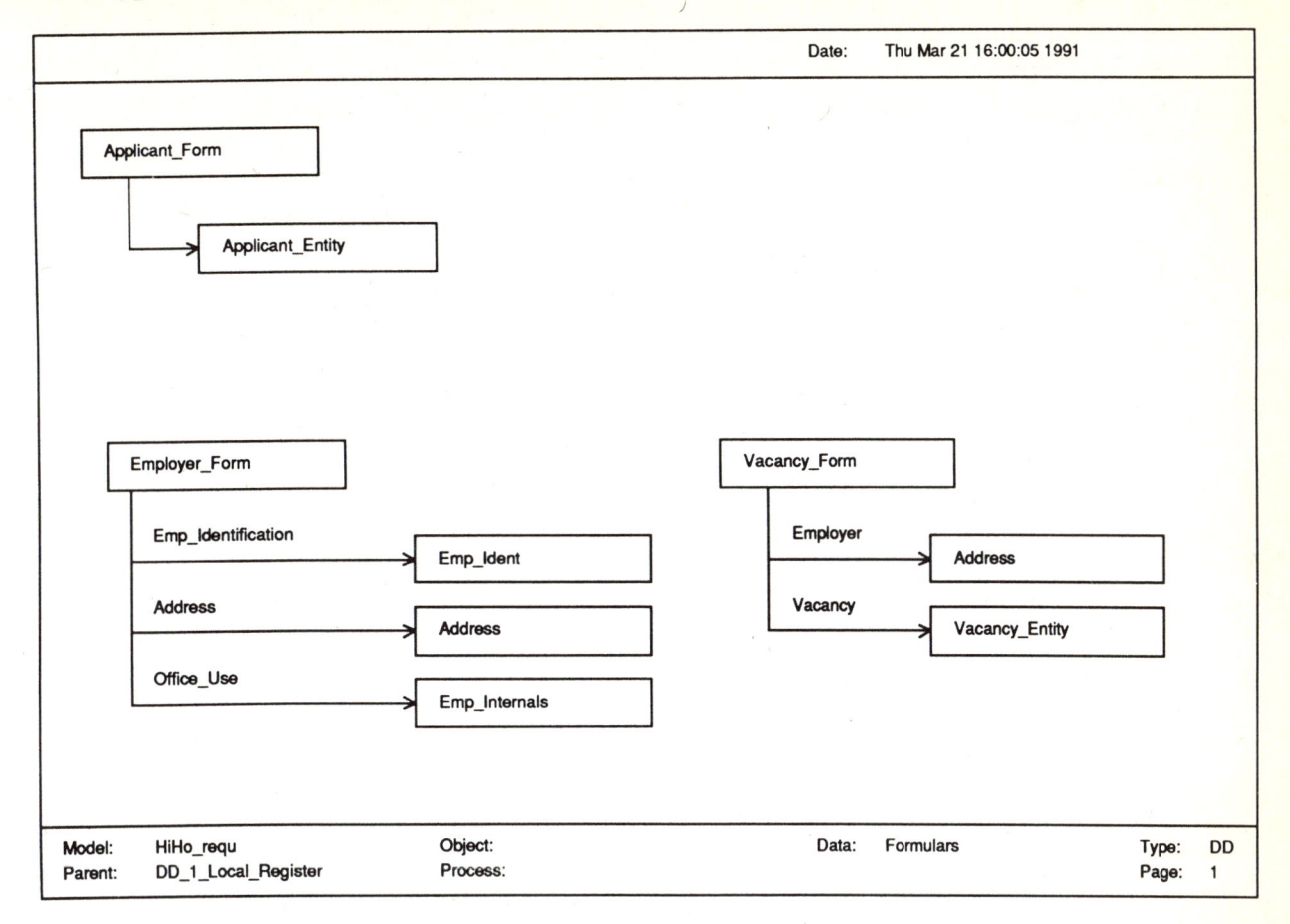

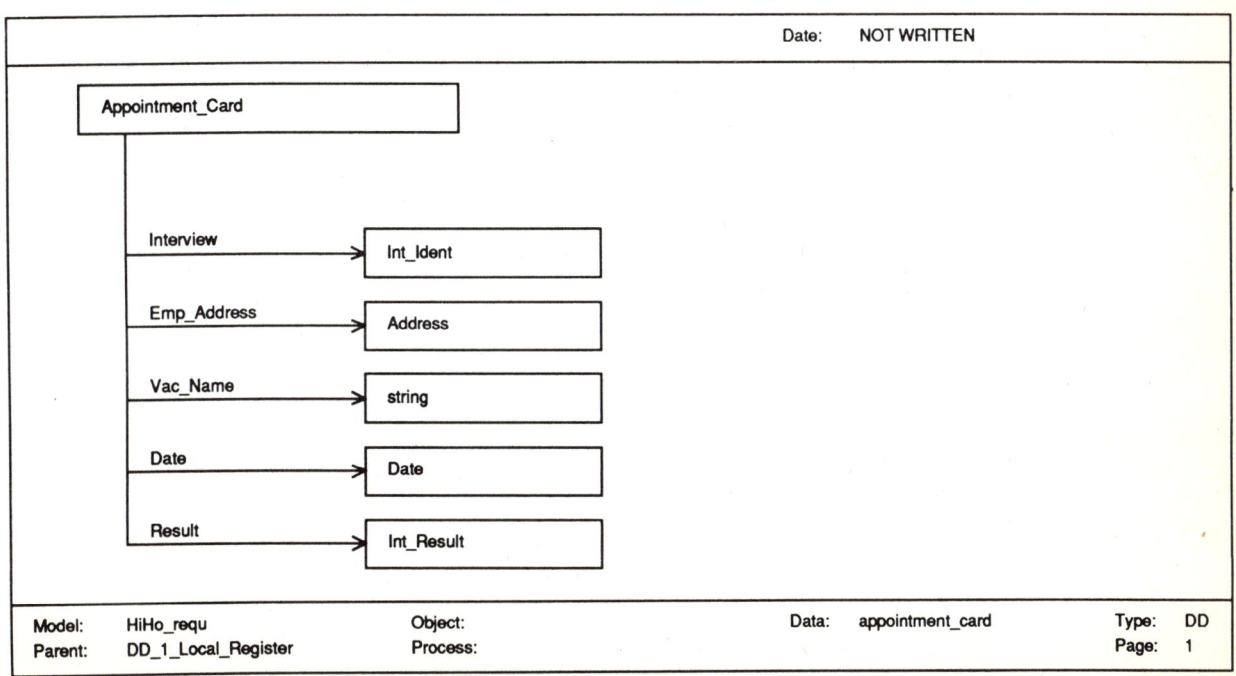

```
                                                    Date:    Thu Mar 21 16:00:05 1991

   ┌─────────────────┐
   │ Applicant_Form  │
   └────────┬────────┘
            └──▶ ┌─────────────────┐
                 │ Applicant_Entity│
                 └─────────────────┘


   ┌─────────────────┐                           ┌─────────────────┐
   │ Employer_Form   │                           │ Vacancy_Form    │
   └────────┬────────┘                           └────────┬────────┘
            │                                             │
      Emp_Identification                              Employer
            ├──▶ ┌─────────────┐                          ├──▶ ┌─────────────┐
            │    │  Emp_Ident  │                          │    │   Address   │
            │    └─────────────┘                          │    └─────────────┘
         Address                                       Vacancy
            ├──▶ ┌─────────────┐                          └──▶ ┌──────────────┐
            │    │   Address   │                               │Vacancy_Entity│
            │    └─────────────┘                               └──────────────┘
         Office_Use
            └──▶ ┌─────────────┐
                 │Emp_Internals│
                 └─────────────┘

Model:   HiHo_requ              Object:              Data:  Formulars     Type: DD
Parent:  DD_1_Local_Register    Process:                                   Page: 1
```

```
                                                    Date:    NOT WRITTEN

   ┌───────────────────┐
   │  Appointment_Card │
   └──────────┬────────┘
              │
         Interview
              ├──▶ ┌─────────────┐
              │    │   Int_Ident │
              │    └─────────────┘
         Emp_Address
              ├──▶ ┌─────────────┐
              │    │   Address   │
              │    └─────────────┘
         Vac_Name
              ├──▶ ┌─────────────┐
              │    │   string    │
              │    └─────────────┘
         Date
              ├──▶ ┌─────────────┐
              │    │    Date     │
              │    └─────────────┘
         Result
              └──▶ ┌─────────────┐
                   │  Int_Result │
                   └─────────────┘

Model:   HiHo_requ              Object:              Data:  appointment_card  Type: DD
Parent:  DD_1_Local_Register    Process:                                       Page: 1
```

```
┌─────────────────────────────────────────────────────────────────────────────┐
│                                          Date:    NOT WRITTEN               │
├─────────────────────────────────────────────────────────────────────────────┤
│                                                                             │
│   ┌───────────────────────────────┐      ┌───────────────────────────────┐  │
│   │ Personal_Data                 │      │ App_Internals                 │  │
│   └───────────────────────────────┘      └───────────────────────────────┘  │
│       Date_of_Birth                          HiHo_Office                    │
│                      ┌──────────────┐                      ┌──────────────┐ │
│                  ───▶│ Date         │                  ───▶│ string       │ │
│                      └──────────────┘                      └──────────────┘ │
│       Marital_Status                         Placement_Consultant           │
│                      ┌──────────────┐                      ┌──────────────┐ │
│                  ───▶│ Marital_Status│                 ───▶│ string       │ │
│                      └──────────────┘                      └──────────────┘ │
│       Advert_Ref                             Reg_Date                       │
│                      ┌──────────────┐                      ┌──────────────┐ │
│                  ───▶│ string       │                  ───▶│ Date         │ │
│                      └──────────────┘                      └──────────────┘ │
│                                                                             │
│   ┌───────────────────────────────┐                                         │
│   │ App_Requirements              │                                         │
│   └───────────────────────────────┘                                         │
│       Min_Salary                                                            │
│                      ┌──────────────┐                                       │
│                  ───▶│ integer      │                                       │
│                      └──────────────┘                                       │
│       willing_to_move                                                       │
│                      ┌──────────────┐                                       │
│                  ───▶│ boolean      │                                       │
│                      └──────────────┘                                       │
│                                                                             │
├─────────────────────────────────────────────────────────────────────────────┤
│  Model:   HiHo_requ         Object:            Data:  Applicant_Entity  Type: DD │
│  Parent:  DD_1_Local_Register  Process:                                 Page: 2  │
└─────────────────────────────────────────────────────────────────────────────┘
```

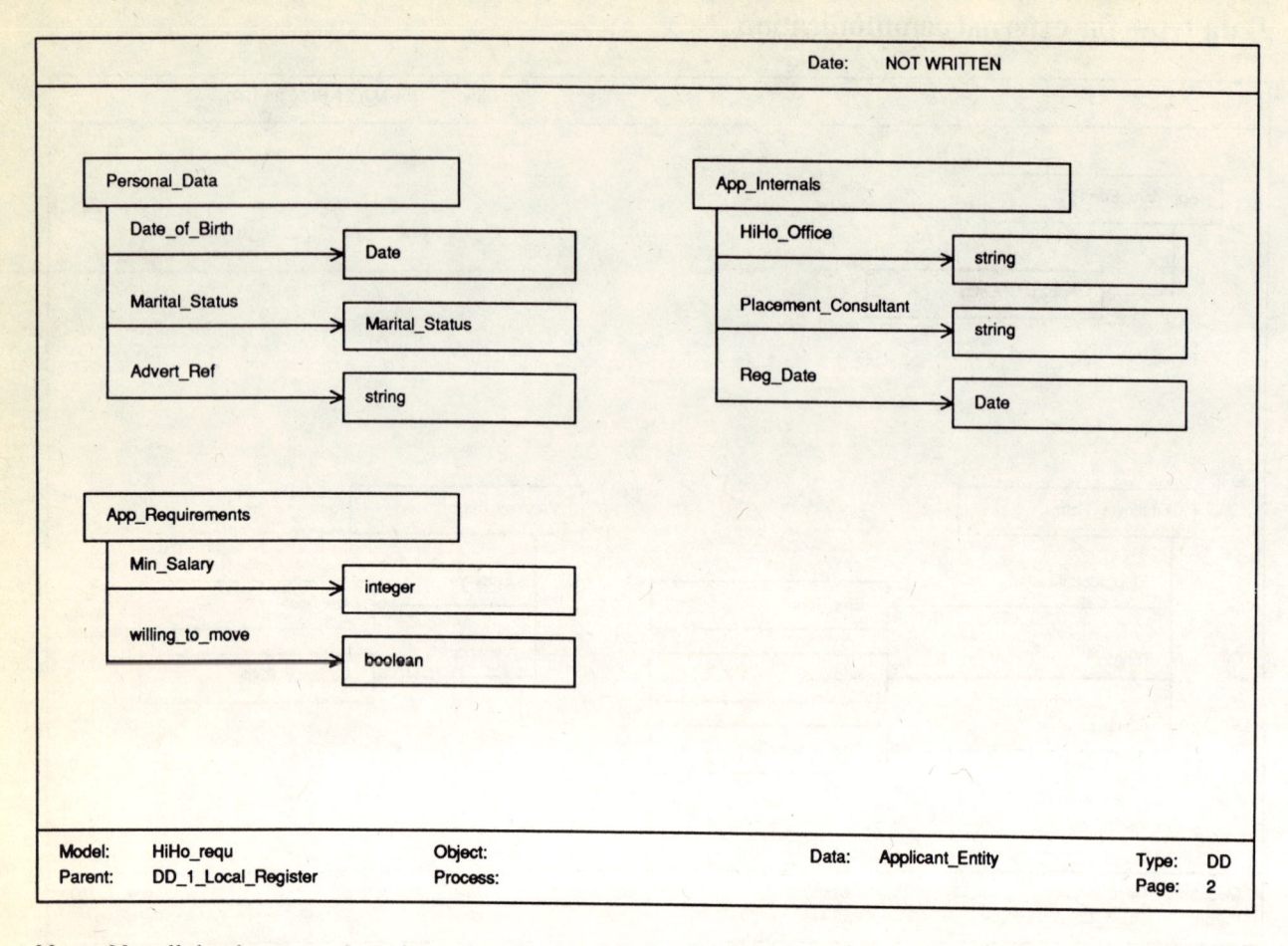

Note: Not all the data types have been described in full detail. The selected types were sufficient for the purposes of comparison.

Model hierarchy

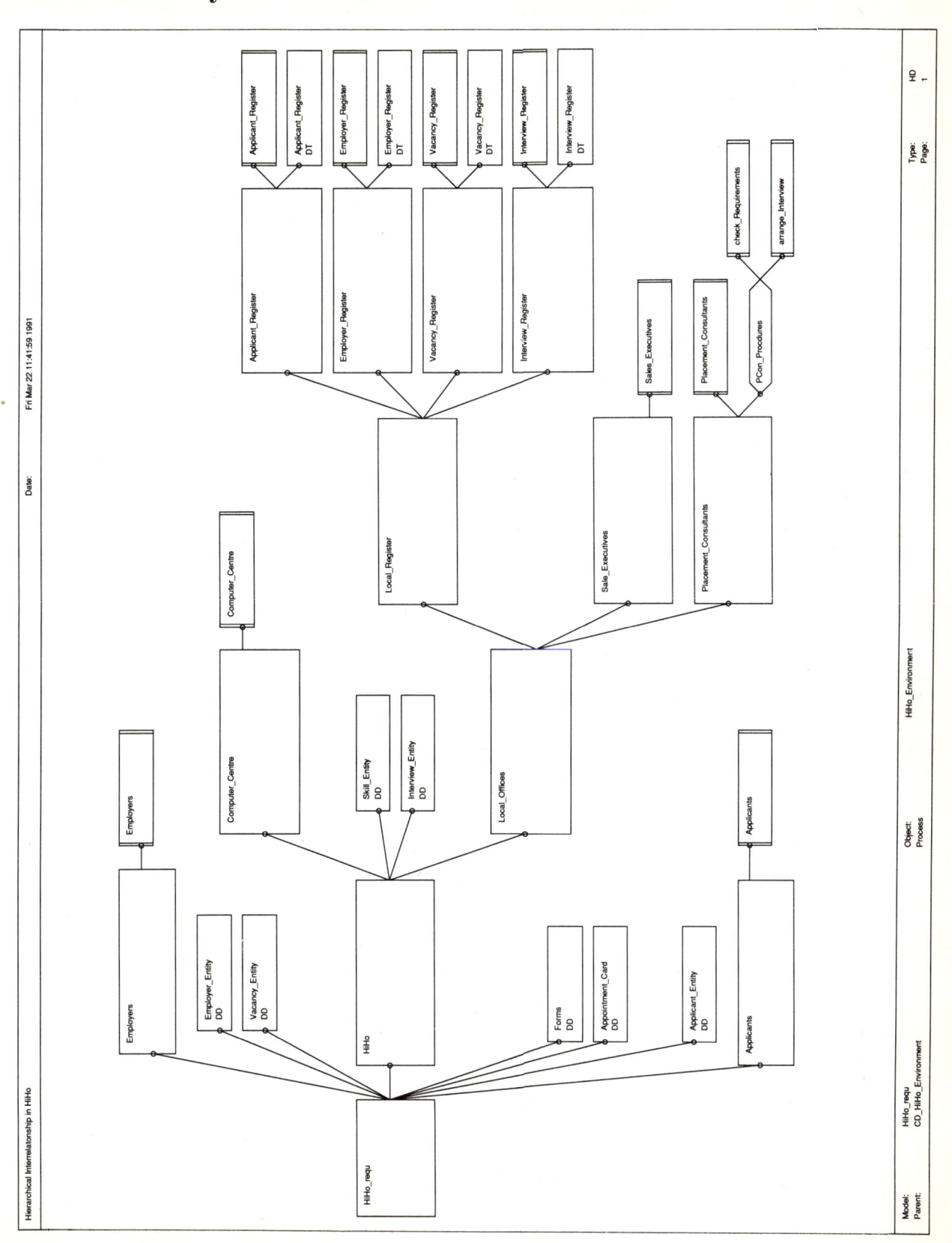

Reusable units

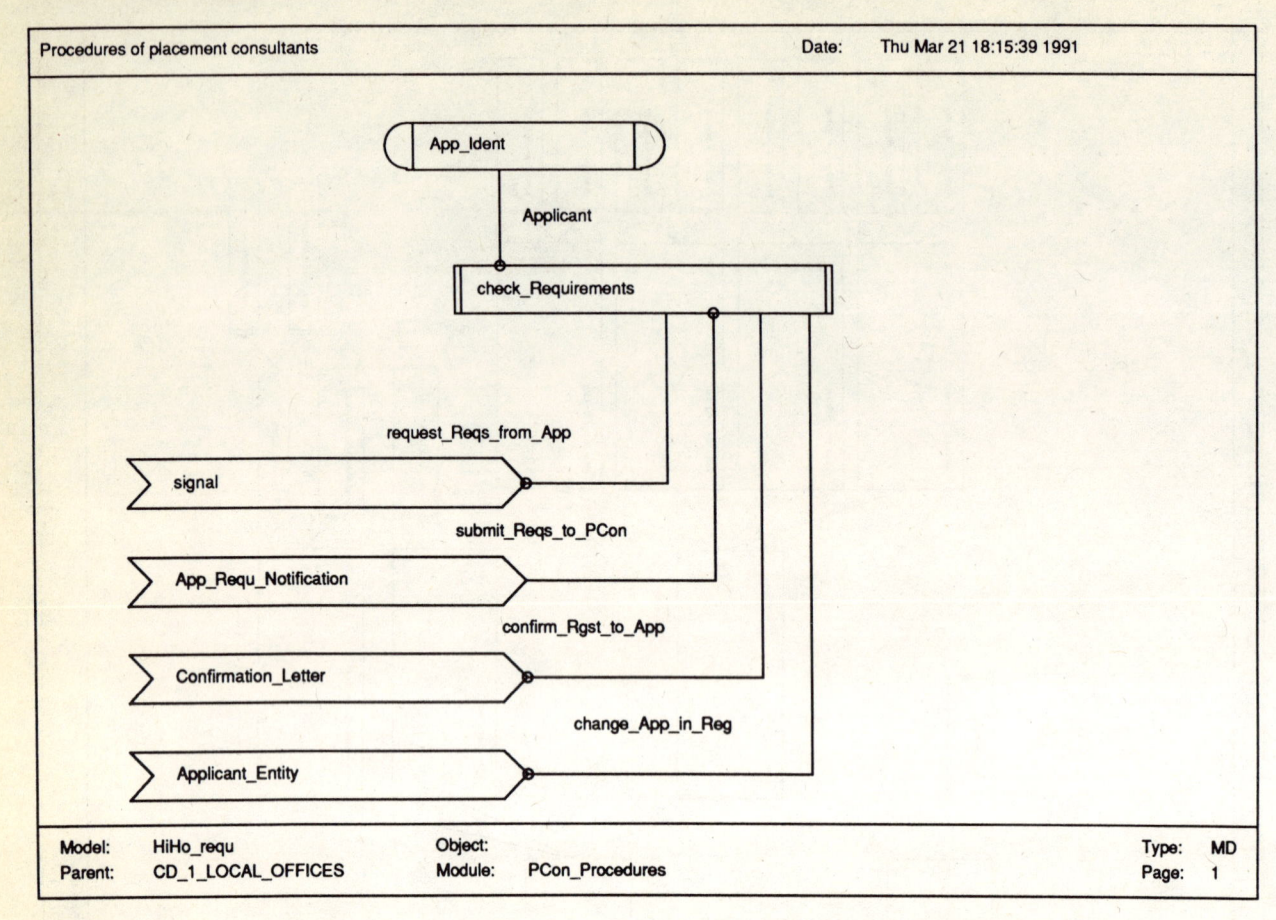

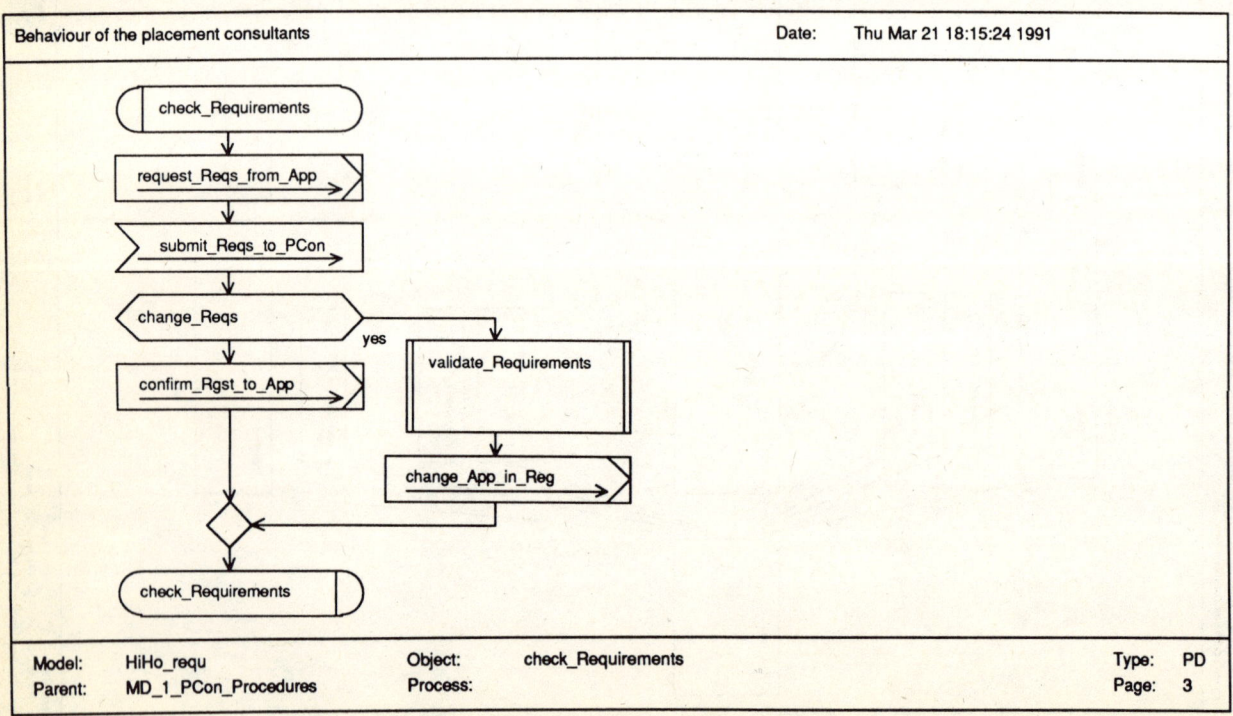

Procedures of placement consultants — Date: Thu Mar 21 18:15:39 1991

- App_Ident
- Applicant
- check_Requirements
- request_Reqs_from_App — signal
- submit_Reqs_to_PCon — App_Requ_Notification
- confirm_Rgst_to_App — Confirmation_Letter
- change_App_in_Reg — Applicant_Entity

Model:	HiHo_requ	Object:		Type:	MD
Parent:	CD_1_LOCAL_OFFICES	Module:	PCon_Procedures	Page:	1

Behaviour of the placement consultants — Date: Thu Mar 21 18:15:24 1991

- check_Requirements
- request_Reqs_from_App
- submit_Reqs_to_PCon
- change_Reqs — yes → validate_Requirements
- confirm_Rgst_to_App
- change_App_in_Reg
- check_Requirements

Model:	HiHo_requ	Object:	check_Requirements	Type:	PD
Parent:	MD_1_PCon_Procedures	Process:		Page:	3

Reusable units

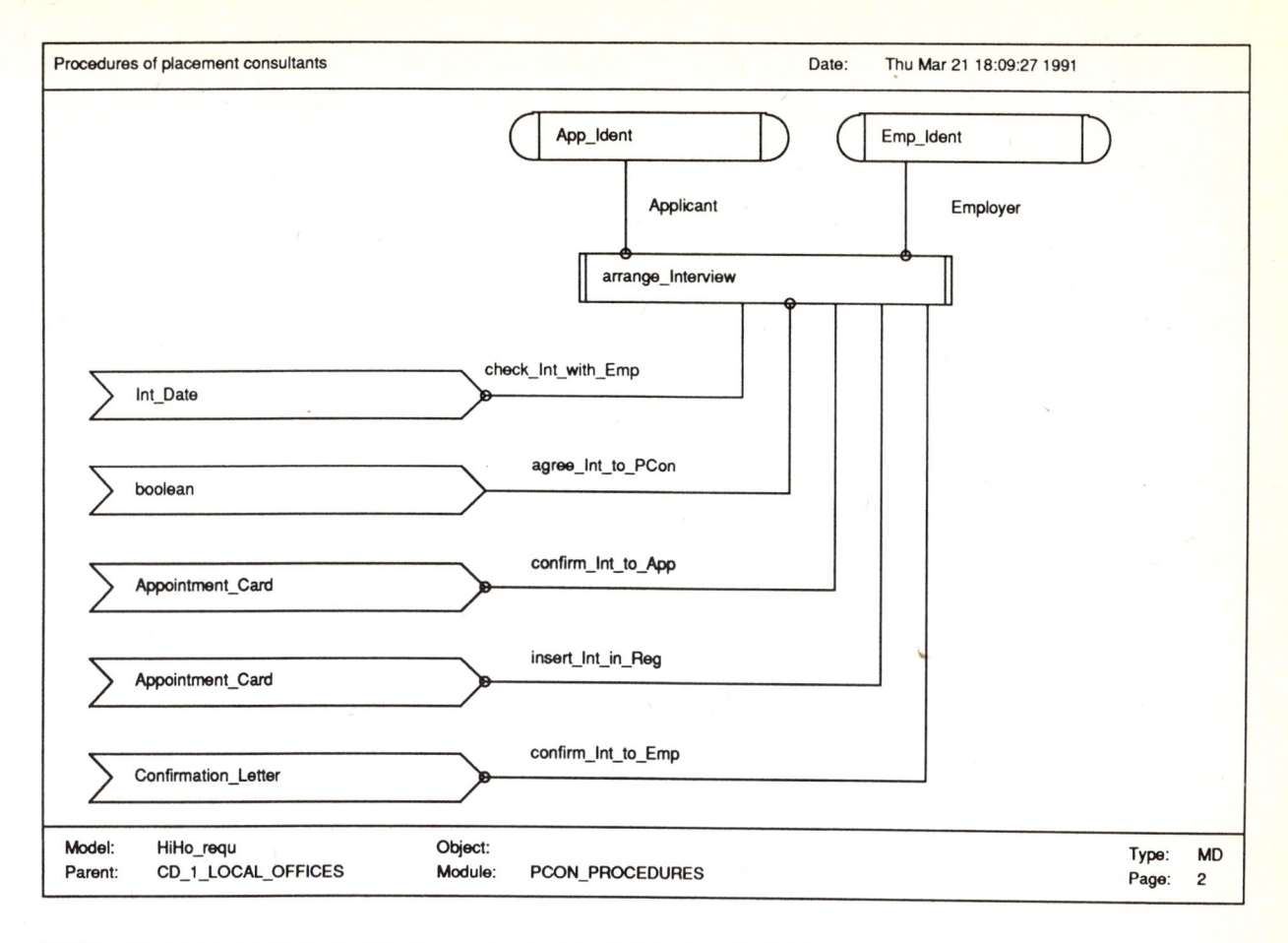

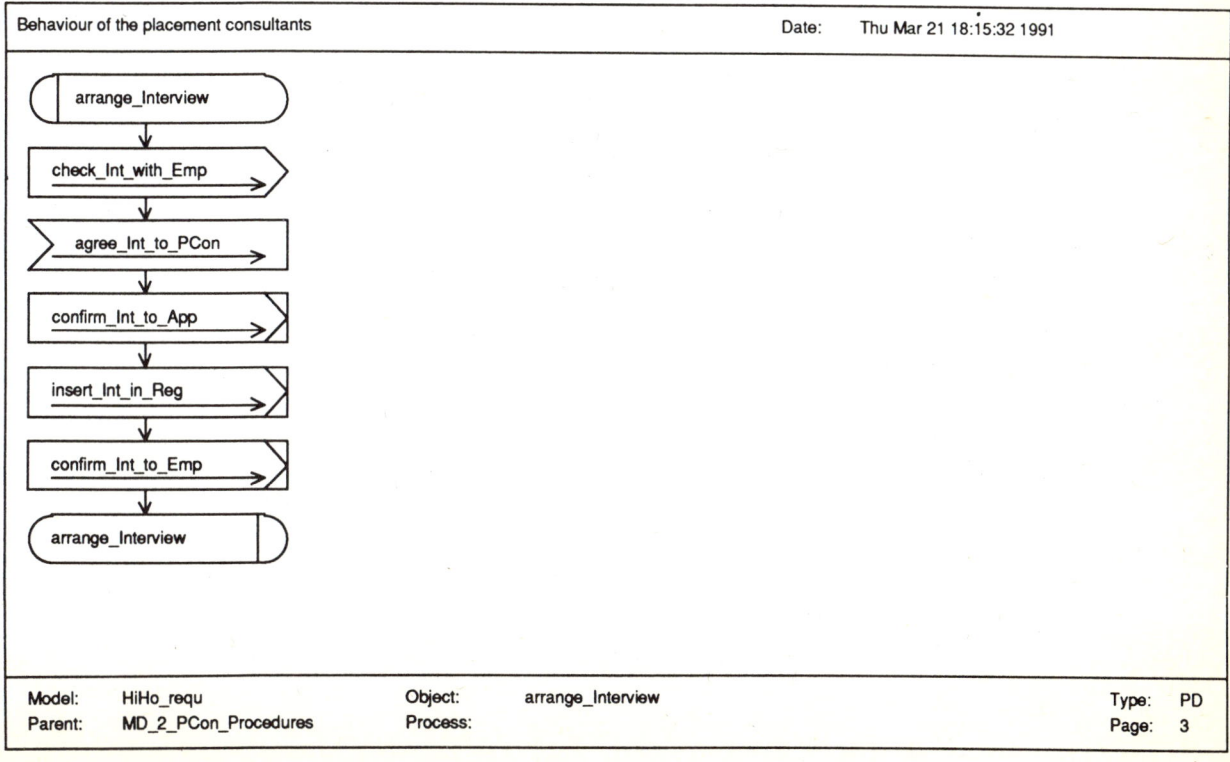

Procedures of placement consultants — Date: Thu Mar 21 18:09:27 1991

- App_Ident — Applicant
- Emp_Ident — Employer
- arrange_Interview
  - check_Int_with_Emp ← Int_Date
  - agree_Int_to_PCon ← boolean
  - confirm_Int_to_App ← Appointment_Card
  - insert_Int_in_Reg ← Appointment_Card
  - confirm_Int_to_Emp ← Confirmation_Letter

Model:	HiHo_requ	Object:	Type: MD
Parent:	CD_1_LOCAL_OFFICES	Module: PCON_PROCEDURES	Page: 2

Behaviour of the placement consultants — Date: Thu Mar 21 18:15:32 1991

arrange_Interview
→ check_Int_with_Emp
→ agree_Int_to_PCon
→ confirm_Int_to_App
→ insert_Int_in_Reg
→ confirm_Int_to_Emp
→ arrange_Interview

Model:	HiHo_requ	Object: arrange_Interview	Type: PD
Parent:	MD_2_PCon_Procedures	Process:	Page: 3

Model of
Current Situation

Static structure

External relations of the Hi-Ho company

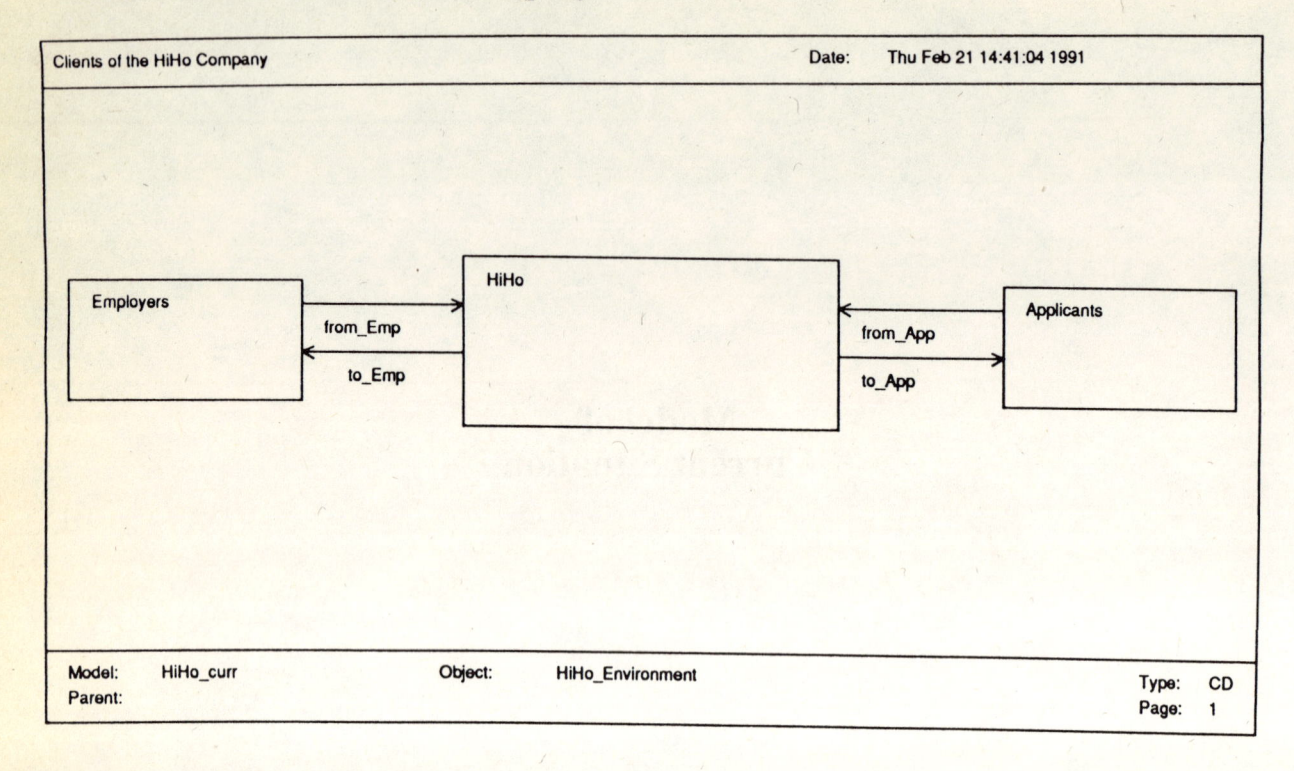

Clients of the HiHo Company		Date: Thu Feb 21 14:41:04 1991

```
                    from_Emp            from_App
  Employers  ─────────────────►  HiHo  ◄─────────────  Applicants
             ◄─────────────────        ─────────────►
                    to_Emp              to_App
```

Model: HiHo_curr   Object: HiHo_Environment   Type: CD
Parent: Page: 1

Structure of the Hi-Ho company

Structure of the HiHo Company		Date: Thu Feb 21 14:41:09 1991

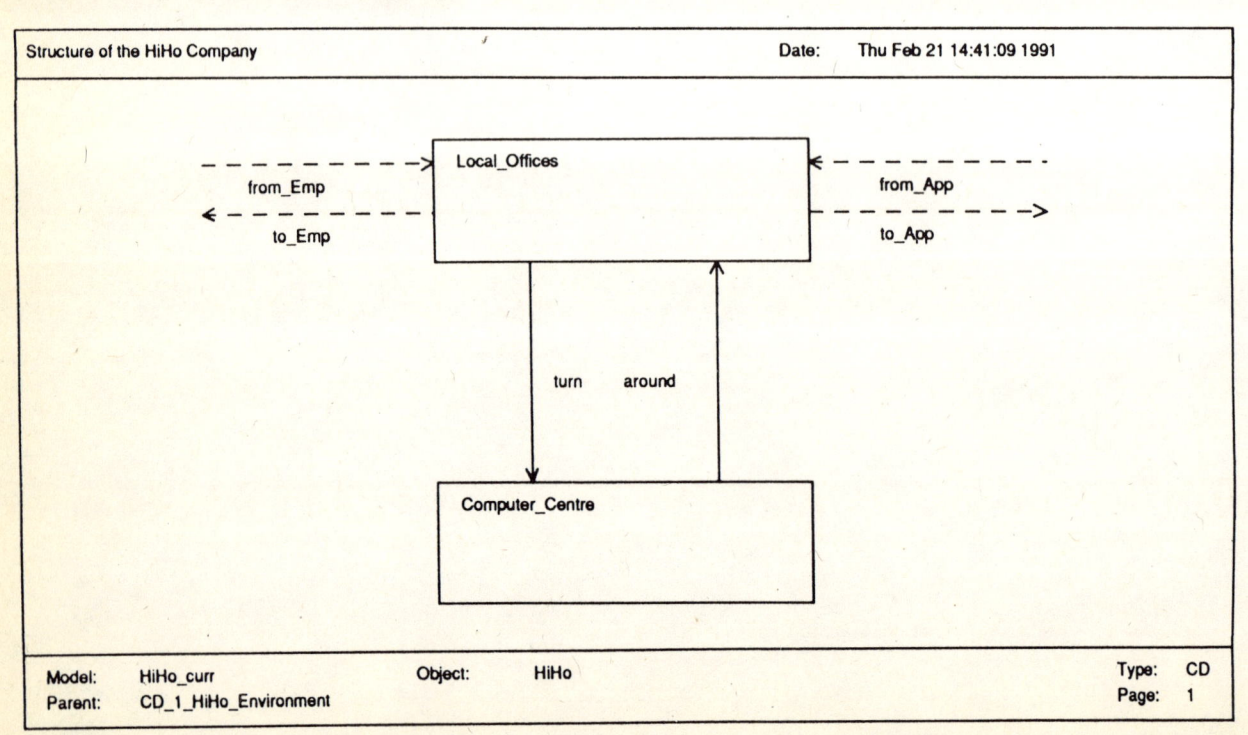

```
                    from_Emp                    from_App
       - - - - - - - - - - ►  Local_Offices  ◄- - - - - - - - - -
       ◄ - - - - - - - - - -                  - - - - - - - - - ►
                    to_Emp                      to_App
                                │   ▲
                                │   │
                              turn  around
                                ▼   │
                          Computer_Centre
```

Model: HiHo_curr   Object: HiHo   Type: CD
Parent: CD_1_HiHo_Environment   Page: 1

Static structure

Structure of local offices

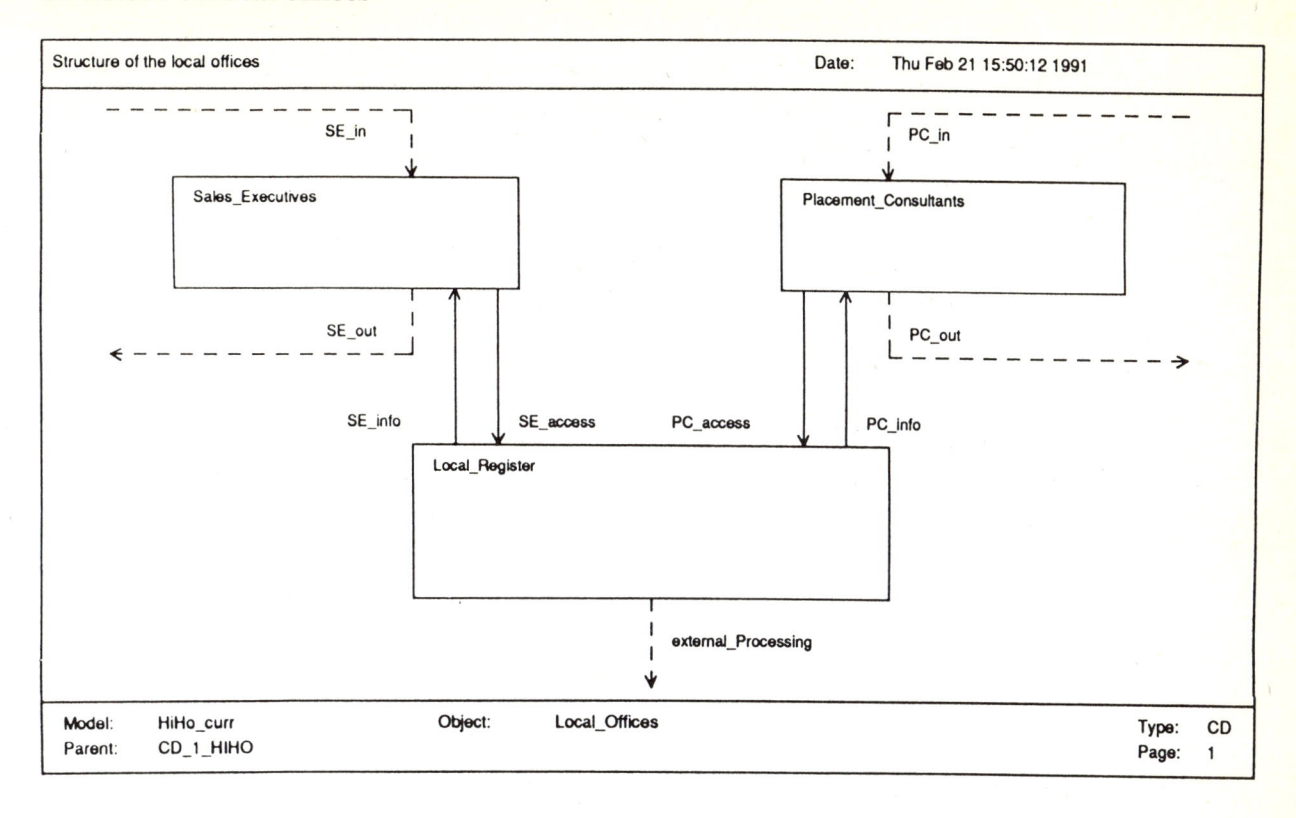

```
Structure of the local offices                    Date:   Thu Feb 21 15:50:12 1991

         SE_in                                        PC_in
      Sales_Executives                          Placement_Consultants
         SE_out                                        PC_out

      SE_info    SE_access        PC_access    PC_info
                        Local_Register
                              external_Processing

Model:  HiHo_curr        Object:  Local_Offices                Type:  CD
Parent: CD_1_HIHO                                              Page:  1
```

Structure of the local register

```
Structure of the local registers                  Date:   Thu Feb 21 14:08:15 1991

   Emp_Reg_In                                                  Int_Reg_In
      Employer_Register      Vac_Int_Cons      Interview_Register

                                               App_Int_Cons      Int_Reg_Out
   Vac_Reg_In    Emp_Vac_Cons                  Int_App_Cons      App_Reg_In
      Vacancy_Register                         Applicant_Register

   Vac_Reg_Out              Int_Vac_Cons                        App_Reg_Out

              There is no process-object
              for the skill-entity because
              there's no dynamic behaviour

Model:  HiHo_curr        Object:  Local_Register               Type:  CD
Parent: CD_1_LOCAL_OFFICES                                     Page:  1
```

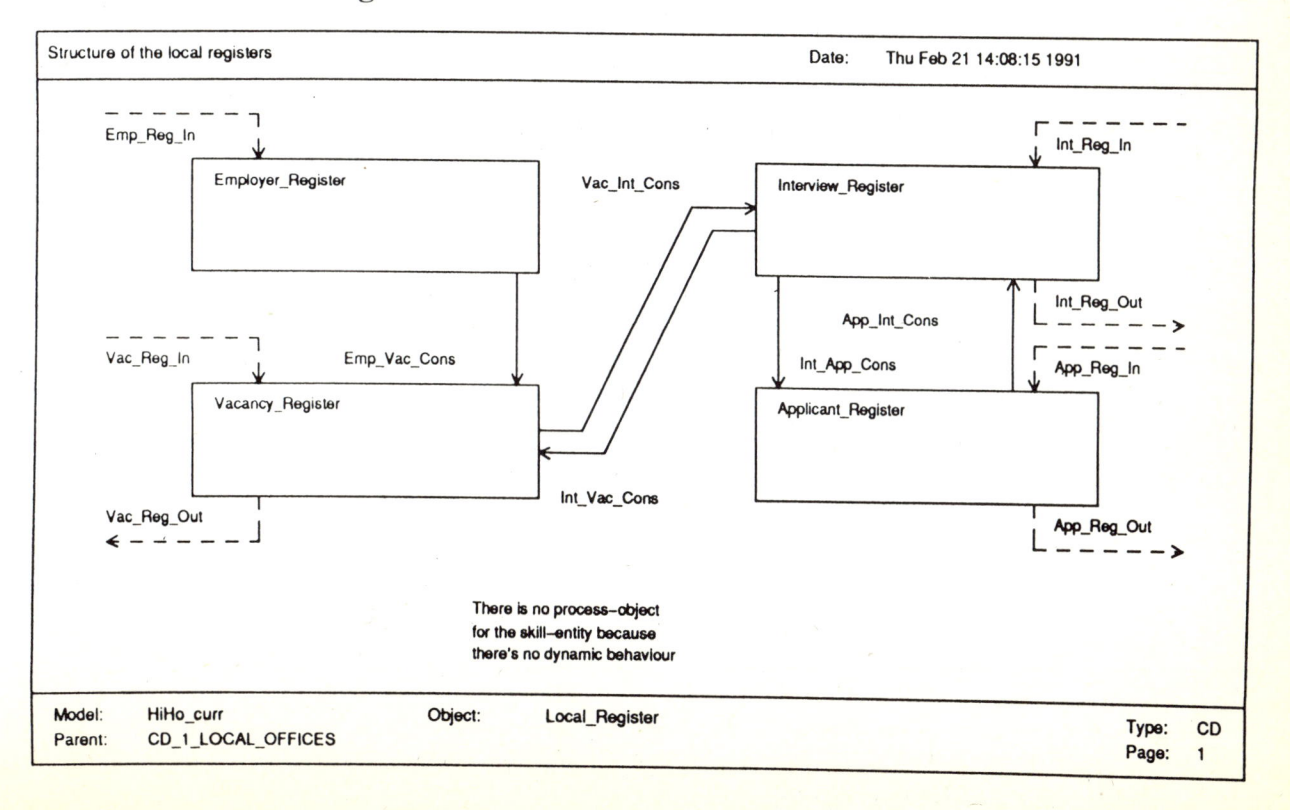

Communication

Input to local offices

Outside view

Name:	Data Type:	Description:
request_Conditions		
register_Employer		
withdraw_Employer		
notify_Vacancy		
withdraw_Vacancy		
reschedule_Interview		
return_Result		
agree_Int_to_PCon		
submit_Int_Res_to_PCon		
confirm_Vac_to_PCon		
confirm_Vac_to_SExe		

Date: Thu Feb 21 15:50:52 1991
Model: HiHo_curr Interface: from_emp Type: IT
Parent: CD_1_HIHO Page: 1

Name:	Data Type:	Description:
register_Applicant	Registration_Form	
withdraw_Applicant		
review_Requirements		
arrange_Interview		
reschedule_Interview		
submit_Reqs_to_PCon		

Date: Thu Feb 21 15:51:00 1991
Model: HiHo_curr Interface: from_App Type: IT
Parent: CD_1_HIHO Page: 1

Name:	Data Type:	Description:
return_Match_to_PCon	List_of_Matches	

Date: Thu Feb 21 15:51:17 1991
Model: HiHo_curr Interface: from_App Type: IT
Parent: CD_1_HIHO Page: 1

Inside view

Name:	Data Type:	Description:
request_Conditions		
register_Employer		
withdraw_Employer		
notify_Vacancy		
withdraw_Vacancy		
reschedule_Interview		
confirm_Vac_to_SExe		

Date: Thu Feb 21 15:51:21 1991

Model: HiHo_curr
Parent: CD_1_LOCAL_OFFICES
Interface: SE_in
Type: IT
Page: 1

Name:	Data Type:	Description:
register_Applicant		
review_Applicant		
withdraw_Applicant		
arrange_Interview		
reschedule_Interview		
return_Result		
submit_Reqs_to_PCon		
confirm_Vac_to_PCon		
agree_Int_to_PCon		
submit_Int_Res_to_PCon		
return_Match_to_PCon		

Date: Thu Feb 21 15:51:47 1991

Model: HiHo_curr
Parent: CD_1_LOCAL_OFFICES
Interface: PC_in
Type: IT
Page: 1

Output of local offices

Outside view

Name:	Data Type:	Description:
return_Cond_to_Emp	Condition_Letter	
check_Vac_with_Emp		
check_Int_with_Emp		SYN
confirm_Int_to_Emp		
request_Int_Res_from_Emp		
notify_Accout_to_Emp		

Date: Thu Feb 21 15:50:56 1991
Model: HiHo_curr Interface: to_Emp Type: IT
Parent: CD_1_HIHO Page: 1

Name:	Data Type:	Description:
confirm_Rgst_to_App		
request_Reqs_from_App		
invite_App_for_Rev		
invite_App_for_Argmt		
confirm_Int_to_App	Appointment_Card	

Date: Thu Feb 21 15:51:04 1991
Model: HiHo_curr Interface: to_App Type: IT
Parent: CD_1_HIHO Page: 1

Name:	Data Type:	Description:
submit_Vacs_to_Cntr		
submit_Apps_to_Cntr		

Date: Thu Feb 21 15:51:09 1991
Model: HiHo_curr Interface: turn Type: IT
Parent: CD_1_HIHO Page: 1

Communication 259

Inside view

		Date: Thu Feb 21 15:51:26 1991
Name:	**Data Type:**	**Description:**
return_Cond_to_Emp check_Vac_with_Emp		

Model: HiHo_curr   Interface: SE_out   Type: IT
Parent: CD_1_LOCAL_OFFICES   Page: 1

		Date: Thu Feb 21 15:51:42 1991
Name:	**Data Type:**	**Description:**
invite_App_for_Argmt invite_App_for_Rev confirm_Rgst_to_App confirm_Int_to_App request_Reqs_from_App check_Vac_with_Emp check_Int_with_Emp confirm_Int_to_Emp request_Int_Res_from_Emp	Appointment_Card	

Model: HiHo_curr   Interface: PC_out   Type: IT
Parent: CD_1_LOCAL_OFFICES   Page: 1

		Date: Thu Feb 21 15:50:25 1991
Name:	**Data Type:**	**Description:**
submit_Vacs_to_Cntr submit_Apps_to_Cntr notify_Account_to_Emp		

Model: HiHo_curr   Interface: external_Processing   Type: IT
Parent: CD_1_LOCAL_OFFICES   Page: 1

Input of local register

Outside view

	Date:	Thu Feb 21 15:50:21 1991
Name:	Data Type:	Description:
insert_Emp_in_Reg		
delete_Emp_from_Reg		
insert_Vac_in_Reg		
delete_Vac_from_Reg		
change_Int_in_Reg		

Model: HiHo_curr Interface: SE_access Type: IT
Parent: CD_1_LOCAL_OFFICES Page: 1

	Date:	Thu Feb 21 15:50:16 1991
Name:	Data Type:	Description:
insert_App_in_Reg		
change_App_in_Reg		
delete_App_from_Reg		
insert_Int_in_Reg		
change_Int_in_Reg		
submit_Int_Res_to_Reg		
delete_Vac_from_Reg		

Model: HiHo_curr Interface: PC_access Type: IT
Parent: CD_1_LOCAL_OFFICES Page: 1

Inside view

Date:	Thu Feb 21 15:51:53 1991

Name:	Data Type:	Description:
insert_Emp_in_Reg delete_Emp_from_Reg		

Model:	HiHo_curr	Interface:	Emp_Reg_In	Type:	IT
Parent:	CD_1_Local_Register			Page:	1

Date:	Wed Feb 20 11:36:33 1991

Name:	Data Type:	Description:
insert_App_in_Reg change_App_in_Reg delete_App_from_Reg		

Model:	HiHo_curr	Interface:	App_Reg_In	Type:	IT
Parent:	CD_1_Local_Register			Page:	1

Date:	Thu Feb 21 15:52:02 1991

Name:	Data Type:	Description:
insert_Int_in_Reg	Appointment_Card	
change_Int_in_Reg		
submit_Int_Res_to_Reg	Appointment_Card	

Model:	HiHo_curr	Interface:	Int_Reg_In	Type:	IT
Parent:	CD_1_LOCAL_REGISTER			Page:	1

Date:	Thu Feb 21 15:51:58 1991

Name:	Data Type:	Description:
insert_Vac_in_Reg delete_Vac_from_Reg		

Model:	HiHo_curr	Interface:	Vac_Reg_In	Type:	IT
Parent:	CD_1_Local_Register			Page:	1

Output of local register

Outside view

	Date: Thu Feb 21 15:51:33 1991

Name:	Data Type:	Description:
indicate_Vac_to_SExe		

Model: HiHo_curr Interface: SE_info Type: IT
Parent: CD_1_LOCAL_OFFICES Page: 1

	Date: Thu Feb 21 15:51:37 1991

Name:	Data Type:	Description:
indicate_App_to_PCon		
indicate_Int_to_PCon		

Model: HiHo_curr Interface: PC_info Type: IT
Parent: CD_1_LOCAL_OFFICES Page: 1

	Date: Thu Feb 21 15:50:25 1991

Name:	Data Type:	Description:
submit_Vacs_to_Cntr		
submit_Apps_to_Cntr		
notify_Account_to_Emp		

Model: HiHo_curr Interface: external_Processing Type: IT
Parent: CD_1_LOCAL_OFFICES Page: 1

Inside view

Name:	Data Type:	Description:
indicate_App_to_PCon		
submit_Apps_to_Cntr		

Date: Thu Feb 21 15:52:18 1991

Model: HiHo_curr Interface: App_Reg_Out Type: IT
Parent: CD_1_Local_Register Page: 1

Name:	Data Type:	Description:
indicate_Int_to_PCon		

Date: Thu Feb 21 15:52:12 1991

Model: HiHo_curr Interface: Int_Reg_Out Type: IT
Parent: CD_1_Local_Register Page: 1

Name:	Data Type:	Description:
submit_Vacs_to_Cntr		
notify_Account_to_Emp		
indicate_Vac_to_SExe		

Date: Thu Feb 21 15:52:07 1991

Model: HiHo_curr Interface: Vac_Reg_Out Type: IT
Parent: CD_1_Local_Register Page: 1

Internal communication of local register

		Date:	Wed Feb 20 11:36:28 1991
Name:	**Data Type:**	**Description:**	
delete_Ints_for_App			
Model: HiHo_curr **Parent:** CD_1_Local_Register	**Interface:** App_Int_Cons		**Type:** IT **Page:** 1

		Date:	Wed Feb 20 11:36:14 1991
Name:	**Data Type:**	**Description:**	
delete_Ints_for_Vac			
Model: HiHo_curr **Parent:** CD_1_Local_Register	**Interface:** Vac_Int_Cons		**Type:** IT **Page:** 1

		Date:	Wed Feb 20 11:36:24 1991
Name:	**Data Type:**	**Description:**	
delete_App_from_Reg			
Model: HiHo_curr **Parent:** CD_1_Local_Register	**Interface:** Int_App_Cons		**Type:** IT **Page:** 1

		Date:	Wed Feb 20 11:36:19 1991
Name:	**Data Type:**	**Description:**	
fill_Post_in_Vac			
Model: HiHo_curr **Parent:** CD_1_Local_Register	**Interface:** Int_Vac_Cons		**Type:** IT **Page:** 1

		Date:	Thu Feb 21 15:52:23 1991
Name:	**Data Type:**	**Description:**	
delete_Vacs_for_Emp			
Model: HiHo_curr **Parent:** CD_1_Local_Register	**Interface:** Emp_Vac_Cons		**Type:** IT **Page:** 1

Behaviour

Process objects of local offices

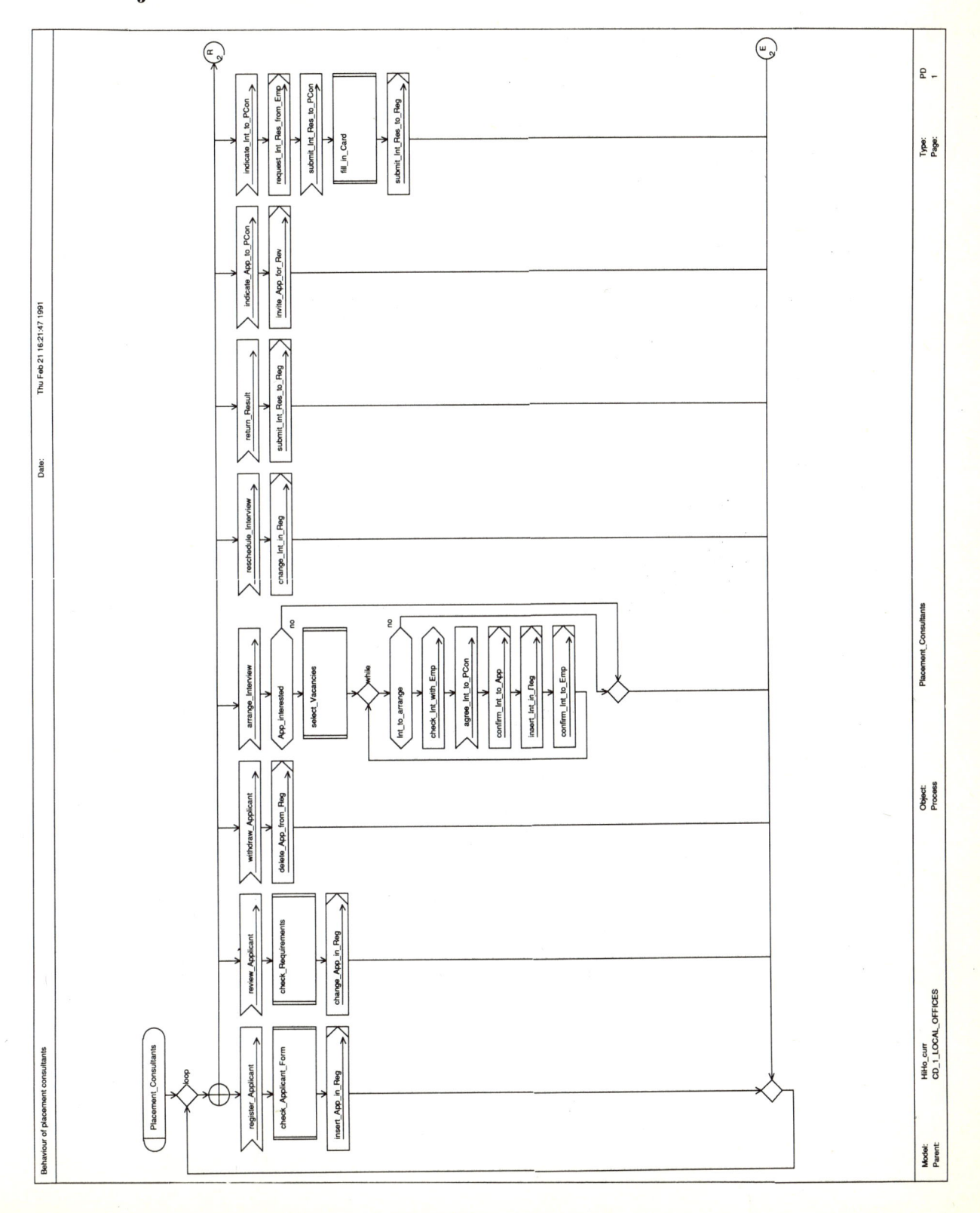

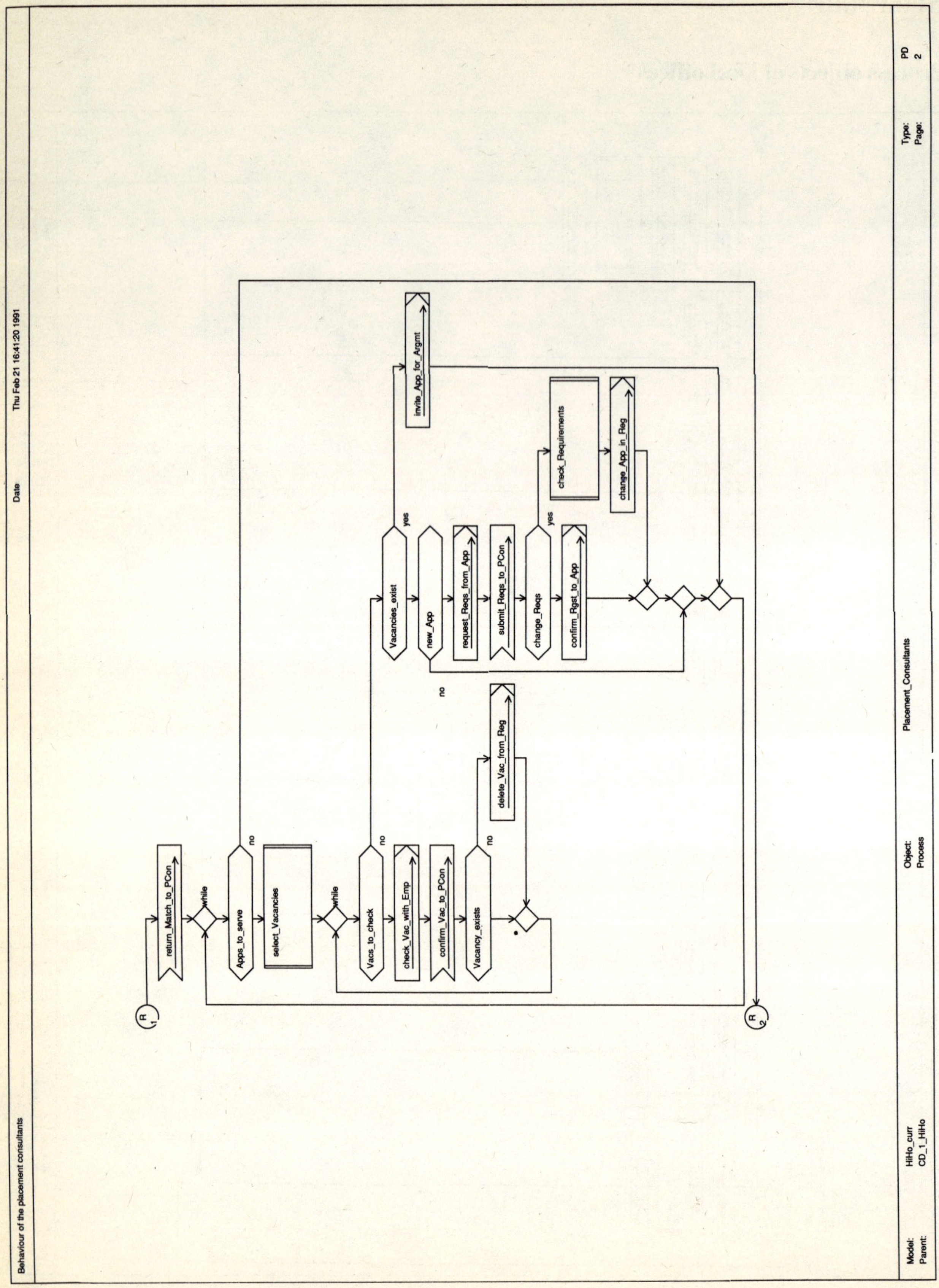
Behaviour of the placement consultants
Date
Thu Feb 21 16:41:20 1991
R 1
return_Match_to_PCon
while
Apps_to_serve
no
select_Vacancies
while
Vacs_to_check
no
check_Vac_with_Emp
confirm_Vac_to_PCon
Vacancy_exists
no
delete_Vac_from_Reg
Vacancies_exist
yes
invite_App_for_Argmt
new_App
no
request_Reqs_from_App
submit_Reqs_to_PCon
change_Reqs
yes
check_Requirements
change_App_in_Reg
confirm_Rgst_to_App
R 2
Model: HiHo_curr
Parent: CD_1_HiHo
Object: Process
Placement_Consultants
Type: PD
Page: 2

Behaviour

Behaviour of sales executives

Date: Thu Feb 21 16:22:10 1991

Model: HiHo_curr
Parent: CD_1_LOCAL_OFFICES
Object: Process
Sales_Executives
Type: PD
Page: 1

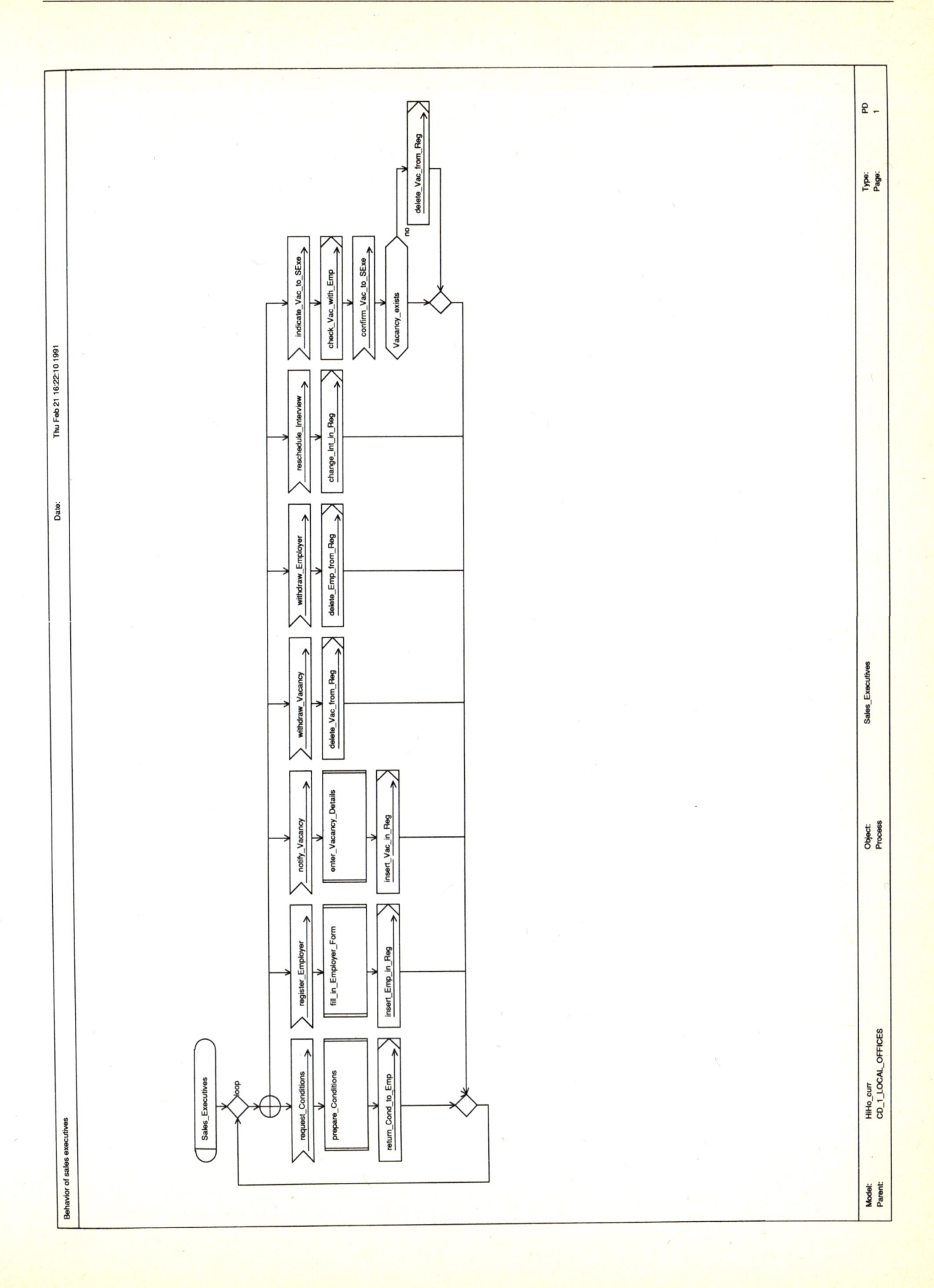

Process objects in local register

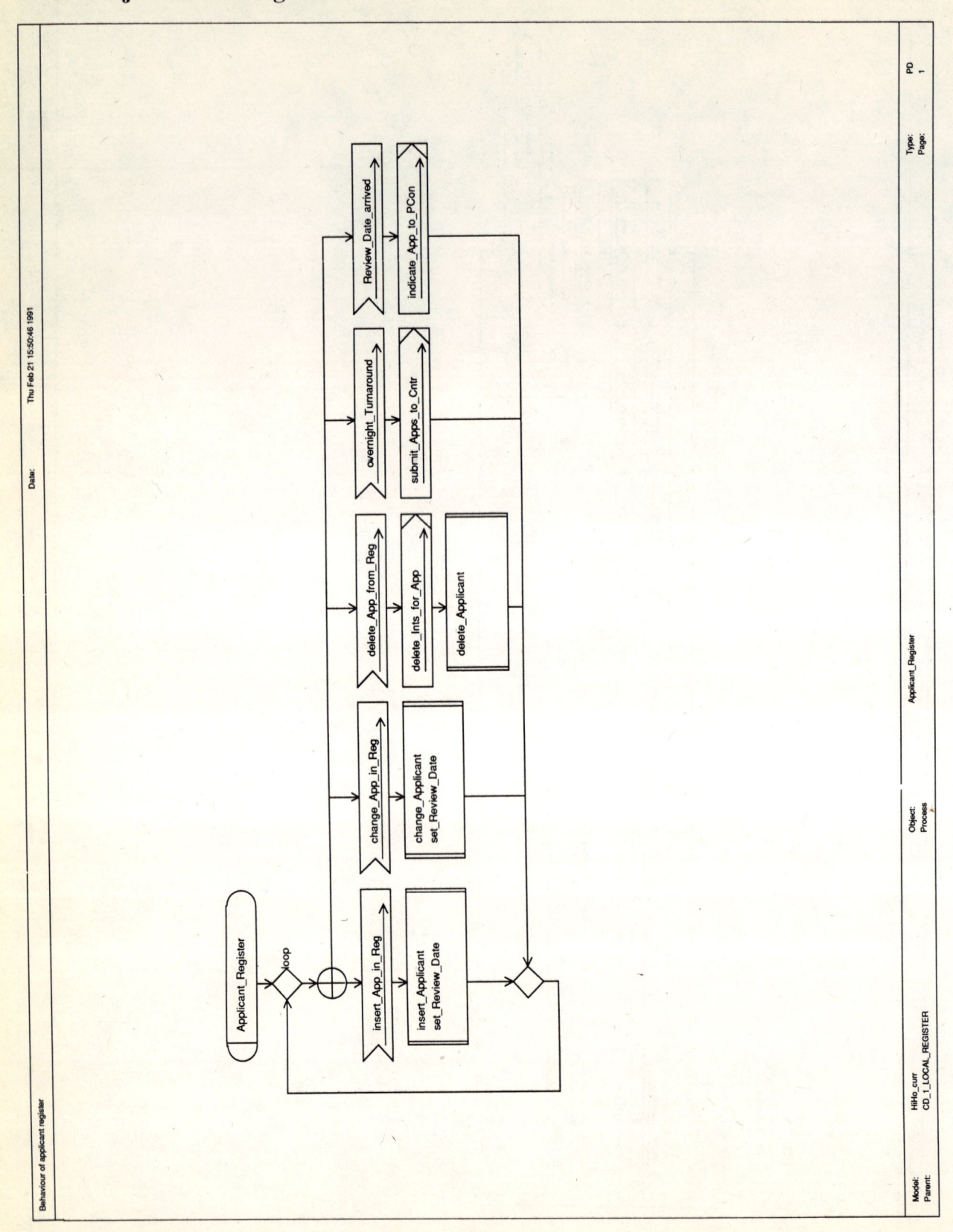

Behaviour of employer register

Date: Thu Feb 21 16:21:27 1991

```
                  ┌─────────────────┐
                  │ Employer_Register│
                  └────────┬────────┘
                           ◇ loop
          ┌────────────────┼────────────────┐
                           ⊕
                  ┌────────┴────────┬────────────────────┐
              insert_Emp_in_Reg          delete_Emp_from_Reg
              ┌─────────────────┐         
              │ insert_Employer │         delete_Vacs_for_Emp
              └────────┬────────┘         ┌─────────────────┐
                                          │ delete_Employer │
                                          └────────┬────────┘
                           ◇ ←──────────────────────┘
```

Model:	HiHo_curr	Object:	Employer_Register	Type: PD
Parent:	CD_1_LOCAL_REGISTER	Process:		Page: 1

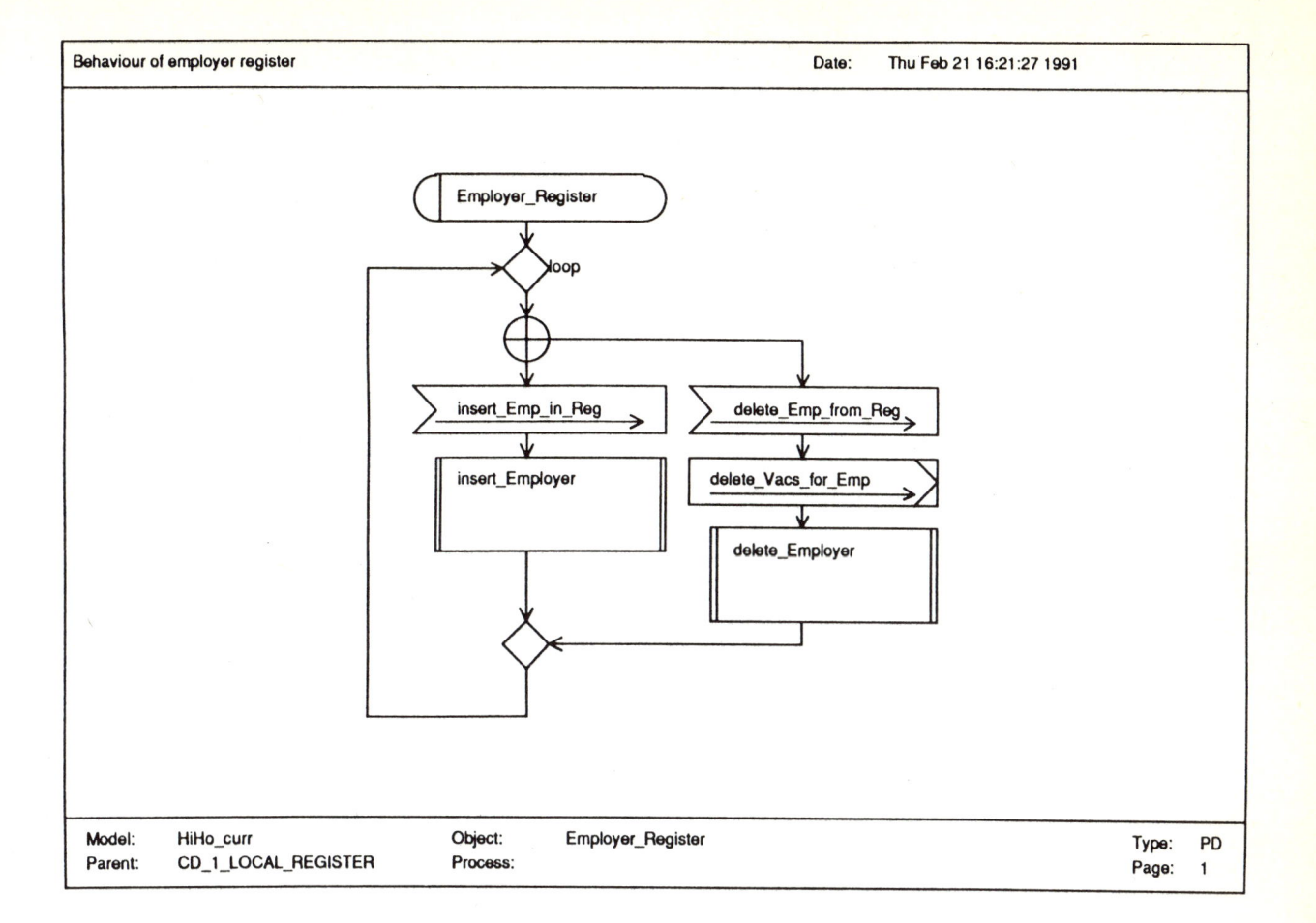

270 Annex 3 Hi-Ho GRAPES Development

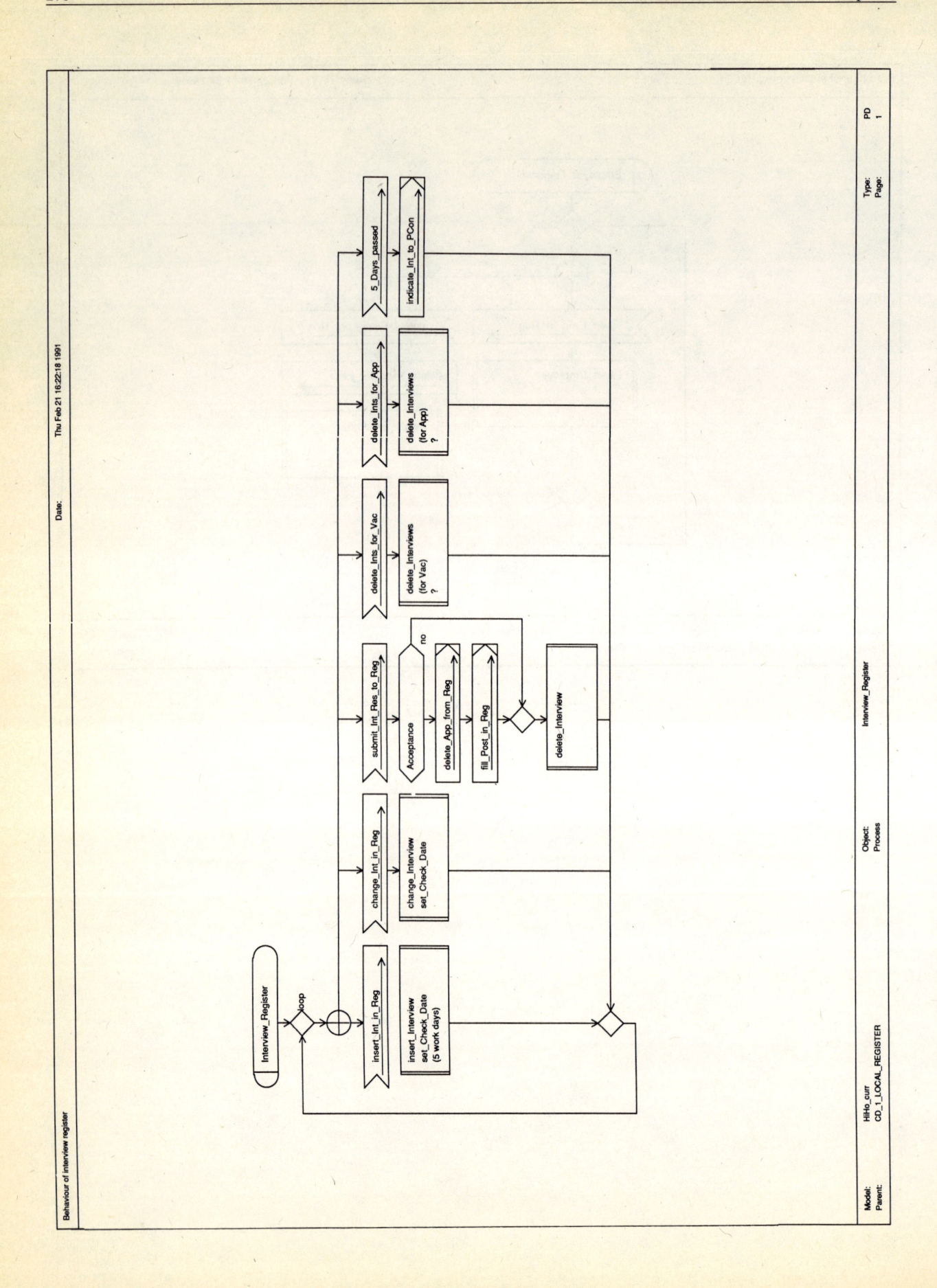

Behaviour

Behaviour of vacancy register

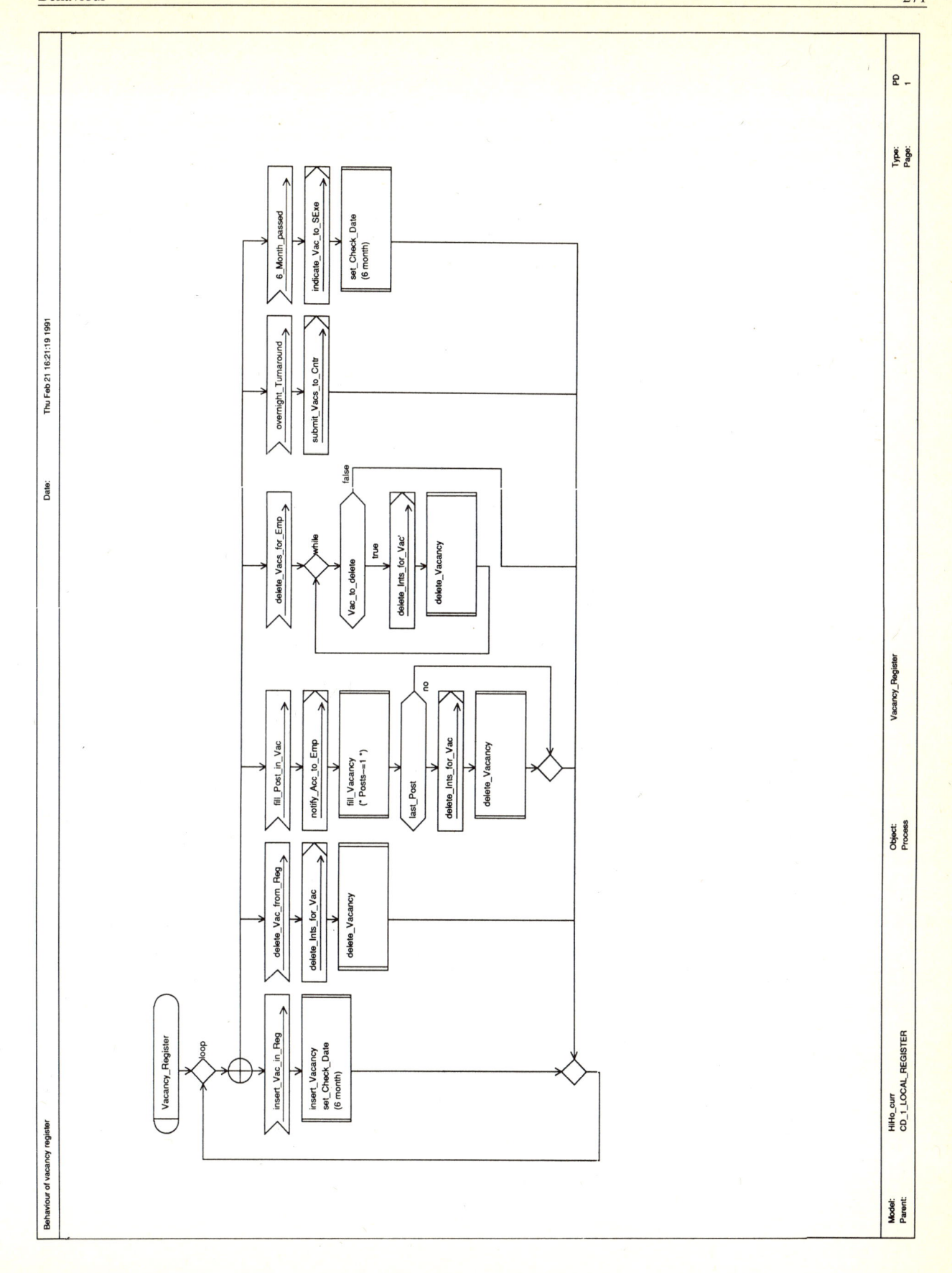

Process objects on Hi-Ho company level

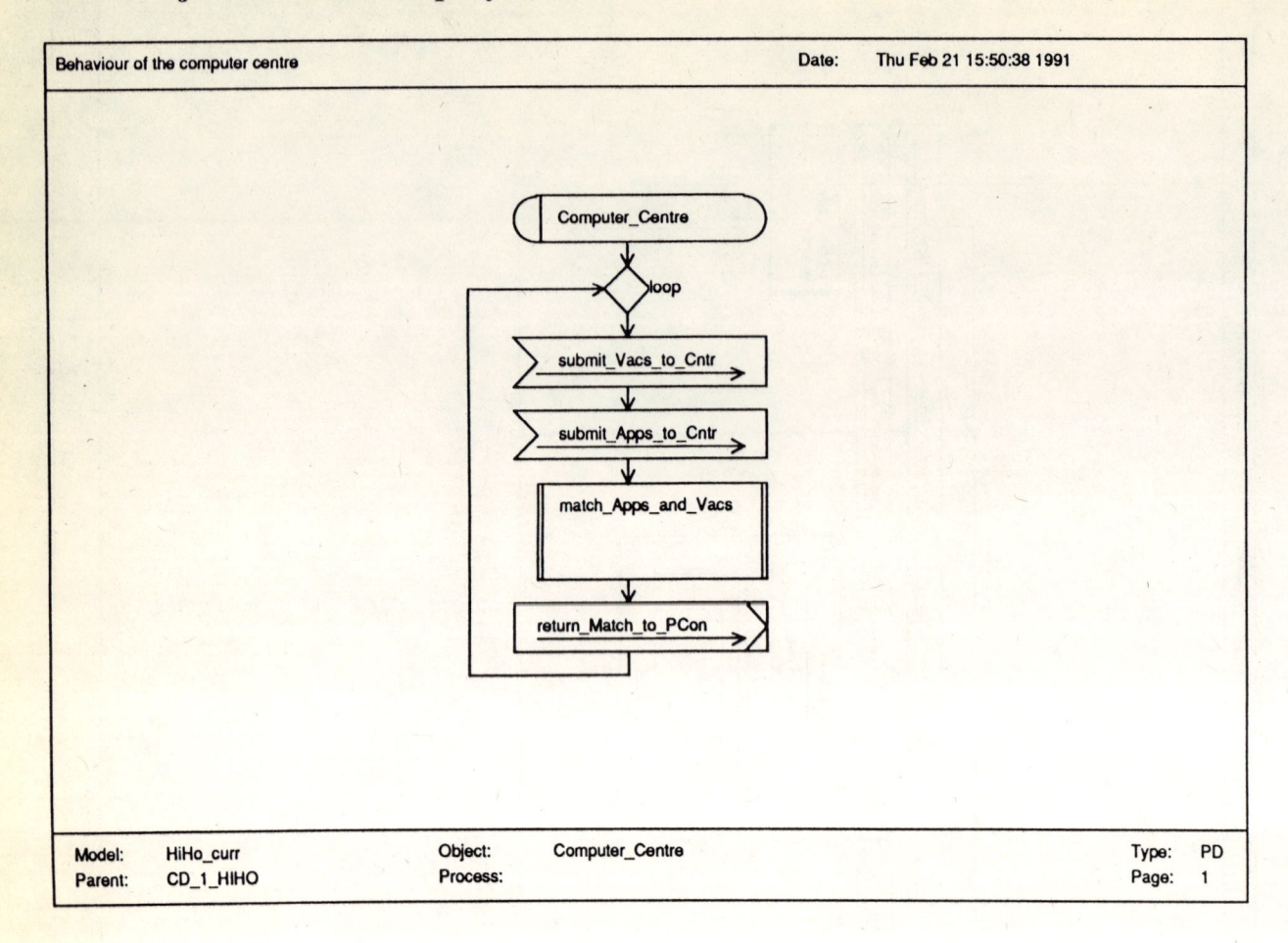

| Behaviour of the computer centre | | Date: | Thu Feb 21 15:50:38 1991 |

- Computer_Centre
  - loop
    - submit_Vacs_to_Cntr
    - submit_Apps_to_Cntr
    - match_Apps_and_Vacs
    - return_Match_to_PCon

| Model: | HiHo_curr | Object: | Computer_Centre | Type: | PD |
| Parent: | CD_1_HIHO | Process: | | Page: | 1 |

Data structure of local register

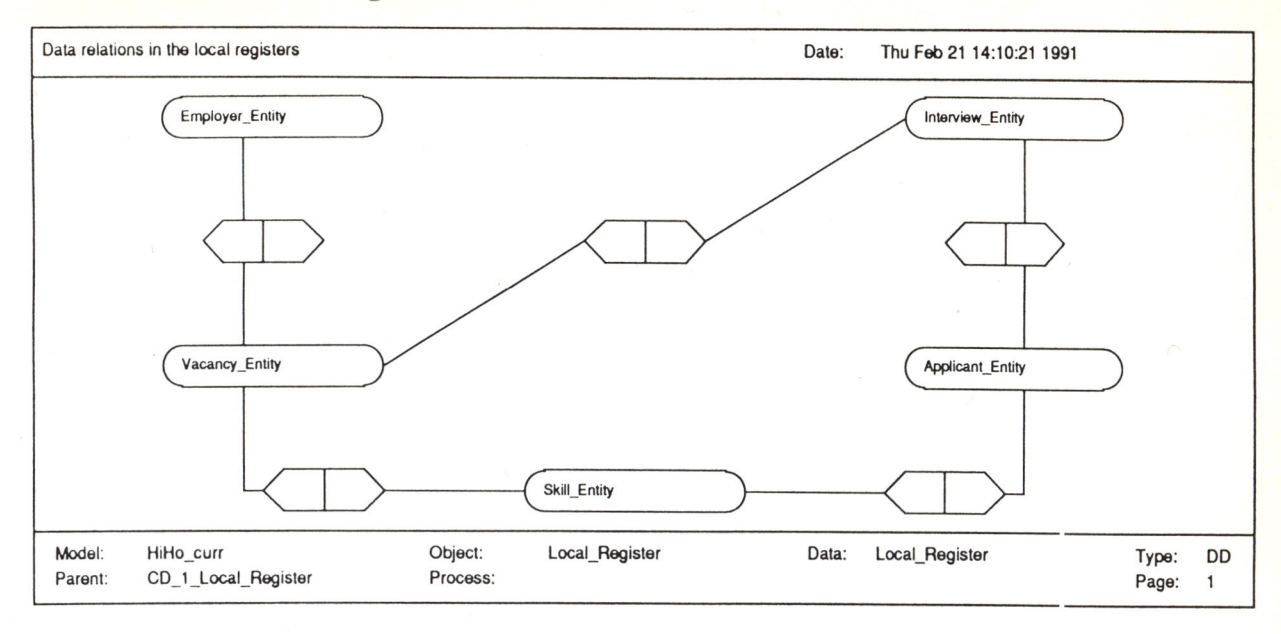

Data relations in the local registers Date: Thu Feb 21 14:10:21 1991

[Diagram showing entities: Employer_Entity, Interview_Entity, Vacancy_Entity, Applicant_Entity, Skill_Entity with relationship connectors between them]

Model:	HiHo_curr	Object:	Local_Register	Data:	Local_Register	Type:	DD
Parent:	CD_1_Local_Register	Process:				Page:	1

Data types for internal processing

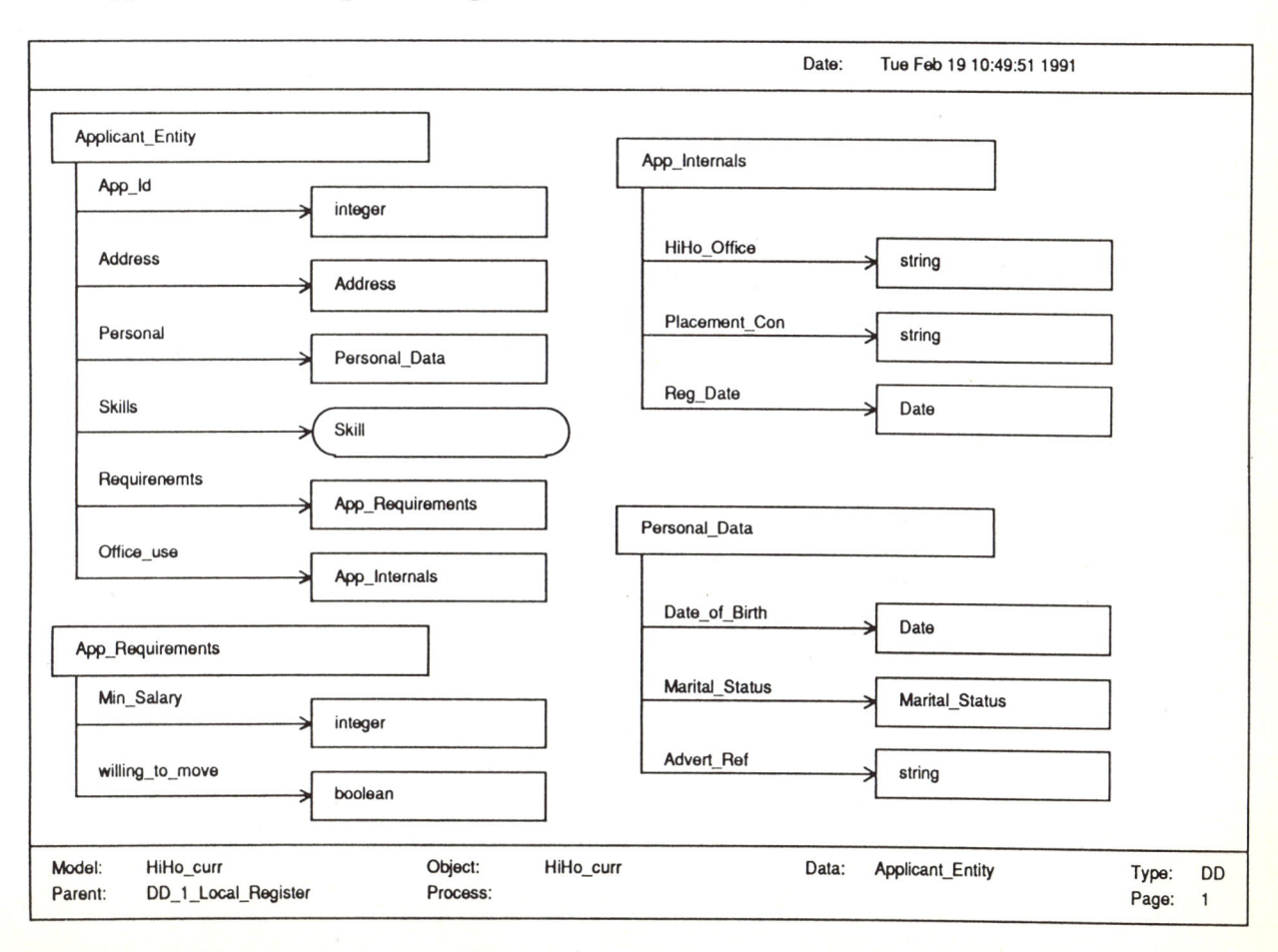

Date: Tue Feb 19 10:49:51 1991

Applicant_Entity
- App_Id → integer
- Address → Address
- Personal → Personal_Data
- Skills → Skill
- Requirenemts → App_Requirements
- Office_use → App_Internals

App_Requirements
- Min_Salary → integer
- willing_to_move → boolean

App_Internals
- HiHo_Office → string
- Placement_Con → string
- Reg_Date → Date

Personal_Data
- Date_of_Birth → Date
- Marital_Status → Marital_Status
- Advert_Ref → string

Model:	HiHo_curr	Object:	HiHo_curr	Data:	Applicant_Entity	Type:	DD
Parent:	DD_1_Local_Register	Process:				Page:	1

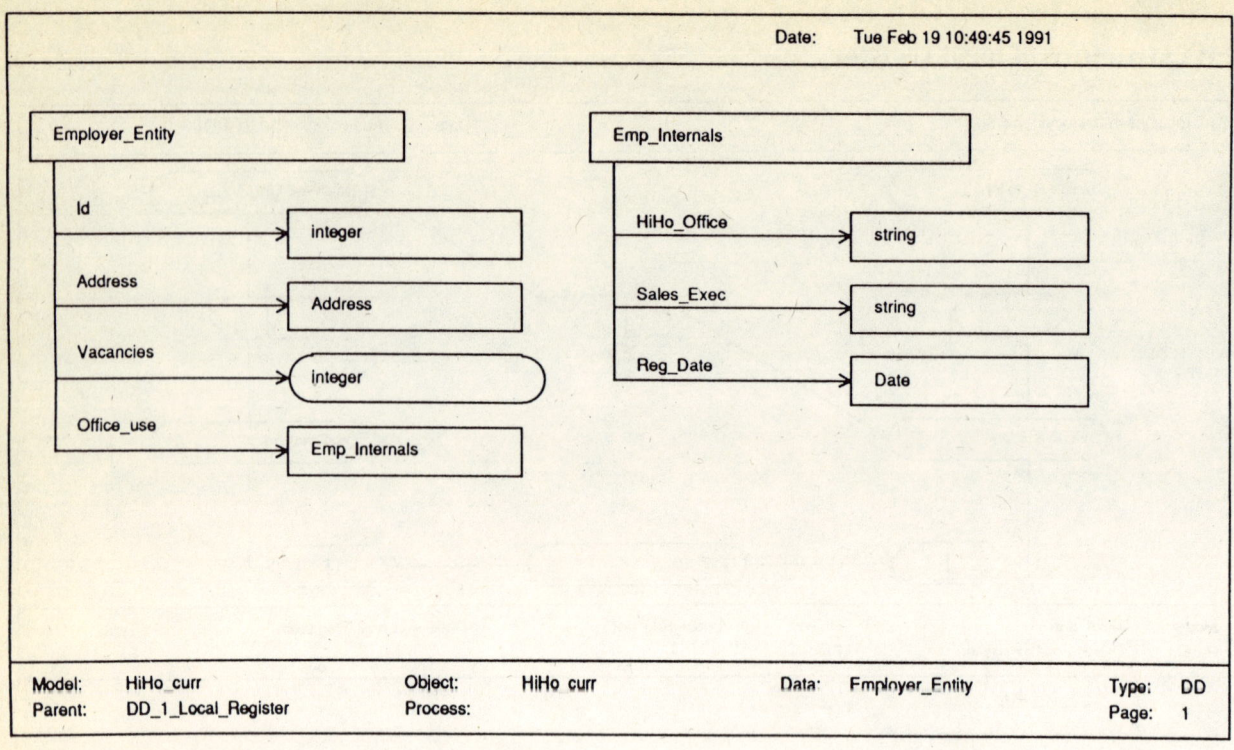

274　　　　　　　　　　　　　　　　　　　　　　　　　　Annex 3 Hi-Ho GRAPES Development

Date:　Tue Feb 19 10:49:45 1991

Employer_Entity
- Id → integer
- Address → Address
- Vacancies → integer
- Office_use → Emp_Internals

Emp_Internals
- HiHo_Office → string
- Sales_Exec → string
- Reg_Date → Date

Model:	HiHo_curr	Object:	HiHo_curr	Data:	Employer_Entity	Type: DD
Parent:	DD_1_Local_Register	Process:				Page: 1

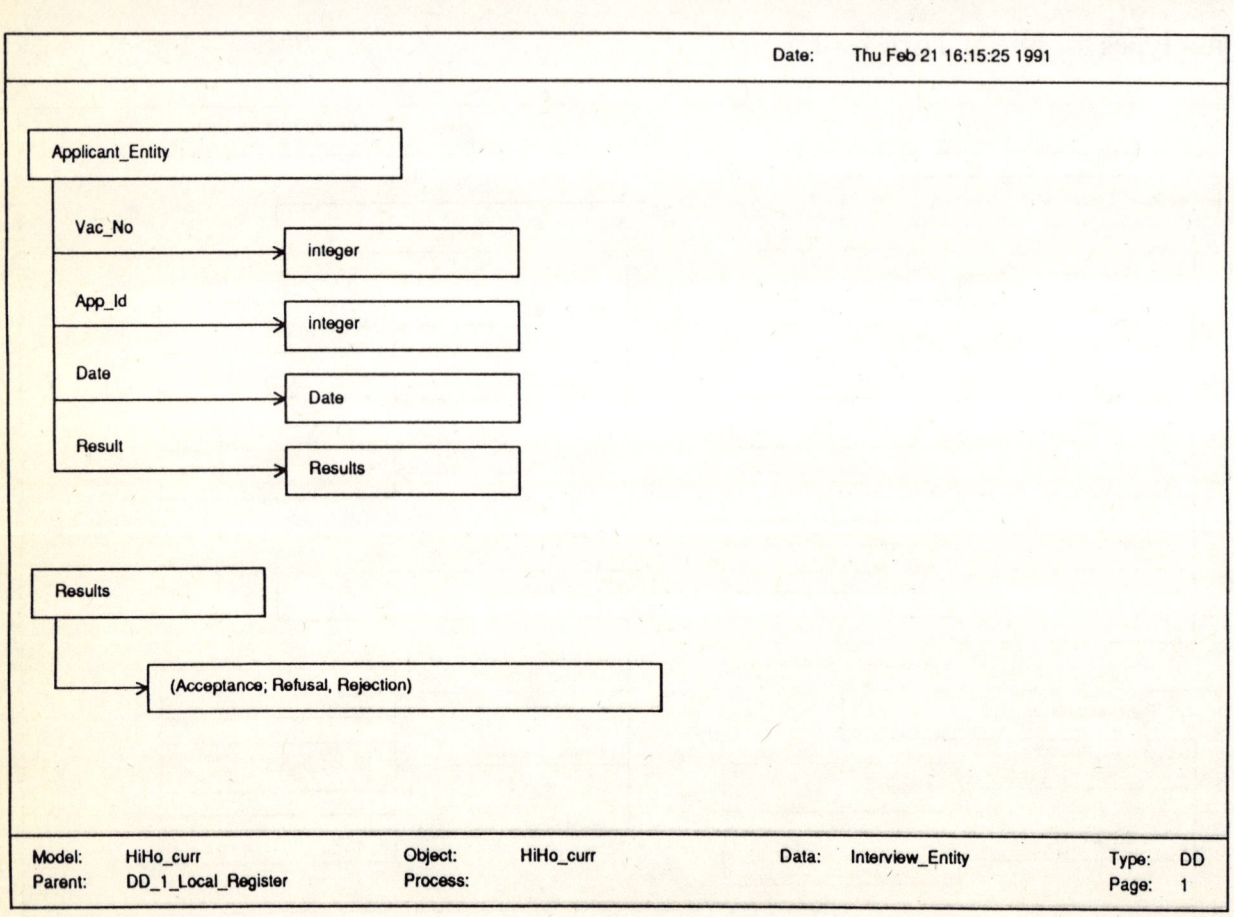

Date:　Thu Feb 21 16:15:25 1991

Applicant_Entity
- Vac_No → integer
- App_Id → integer
- Date → Date
- Result → Results

Results
- → (Acceptance; Refusal, Rejection)

Model:	HiHo_curr	Object:	HiHo_curr	Data:	Interview_Entity	Type: DD
Parent:	DD_1_Local_Register	Process:				Page: 1

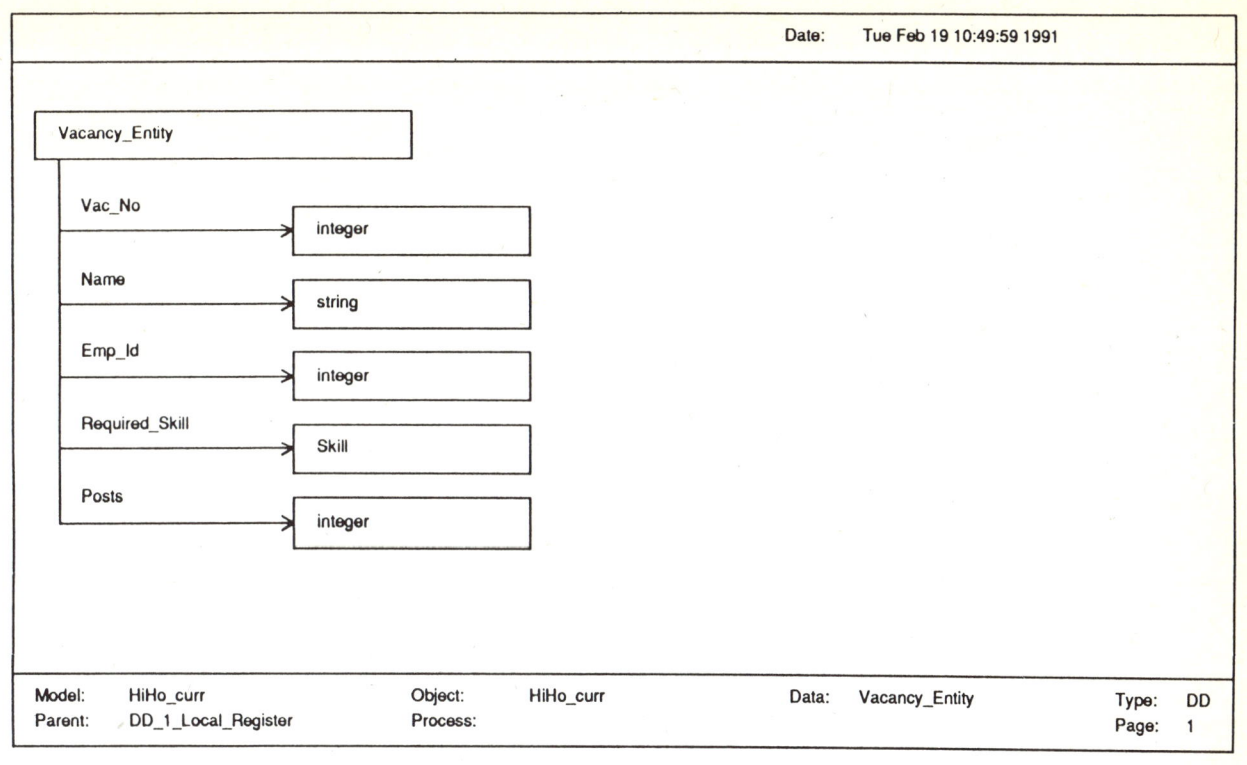

Behaviour 275

```
Vacancy_Entity

    Vac_No
       └──→ integer

    Name
       └──→ string

    Emp_Id
       └──→ integer

    Required_Skill
       └──→ Skill

    Posts
       └──→ integer
```

| Model: | HiHo_curr | Object: | HiHo_curr | Data: | Vacancy_Entity | Type: | DD |
| Parent: | DD_1_Local_Register | Process: | | | | Page: | 1 |

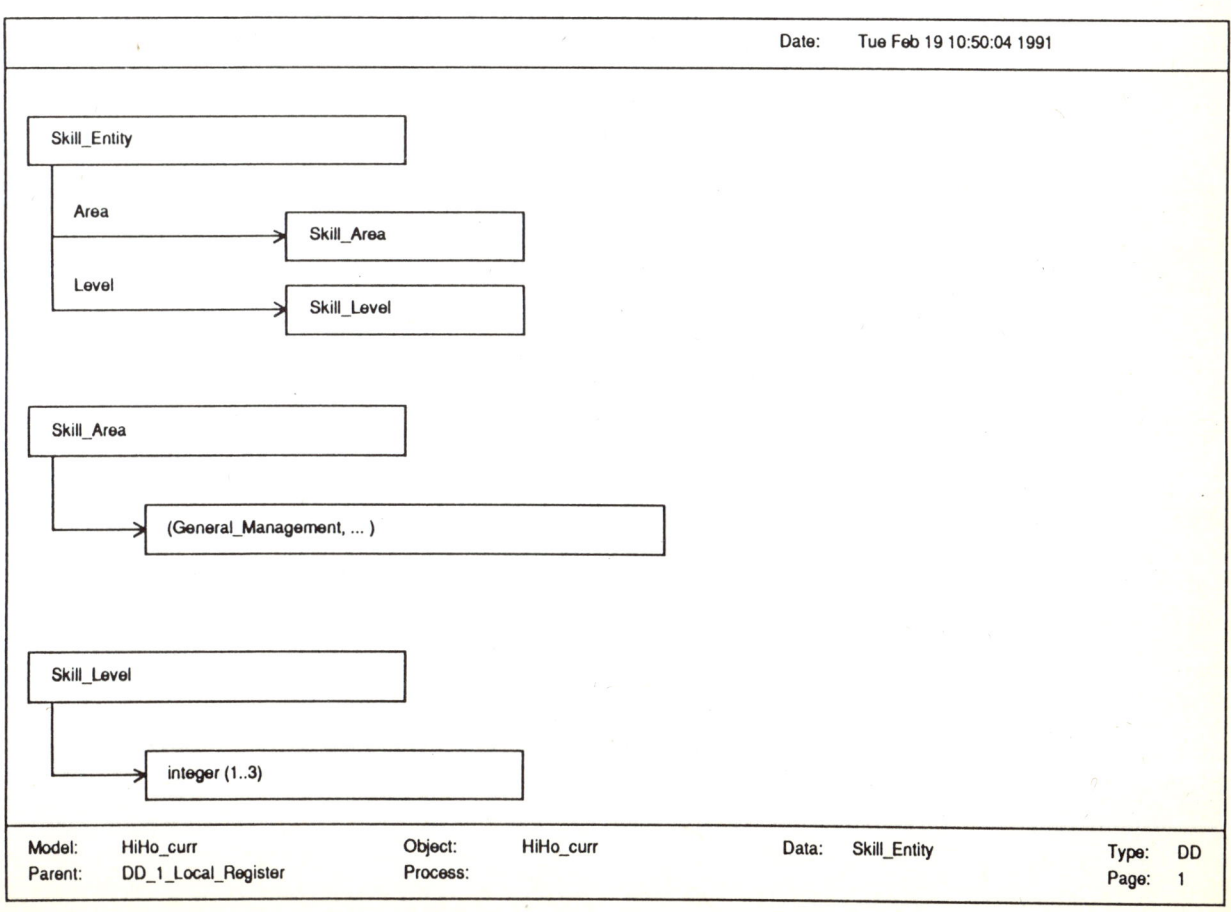

Date: Tue Feb 19 10:50:04 1991

```
Skill_Entity

    Area
       └──→ Skill_Area

    Level
       └──→ Skill_Level


Skill_Area
    └──→ (General_Management, ... )


Skill_Level
    └──→ integer (1..3)
```

| Model: | HiHo_curr | Object: | HiHo_curr | Data: | Skill_Entity | Type: | DD |
| Parent: | DD_1_Local_Register | Process: | | | | Page: | 1 |

Annex 4

Hi-Ho SSADM Development Using Super-Events

Entity life histories if super-events are used

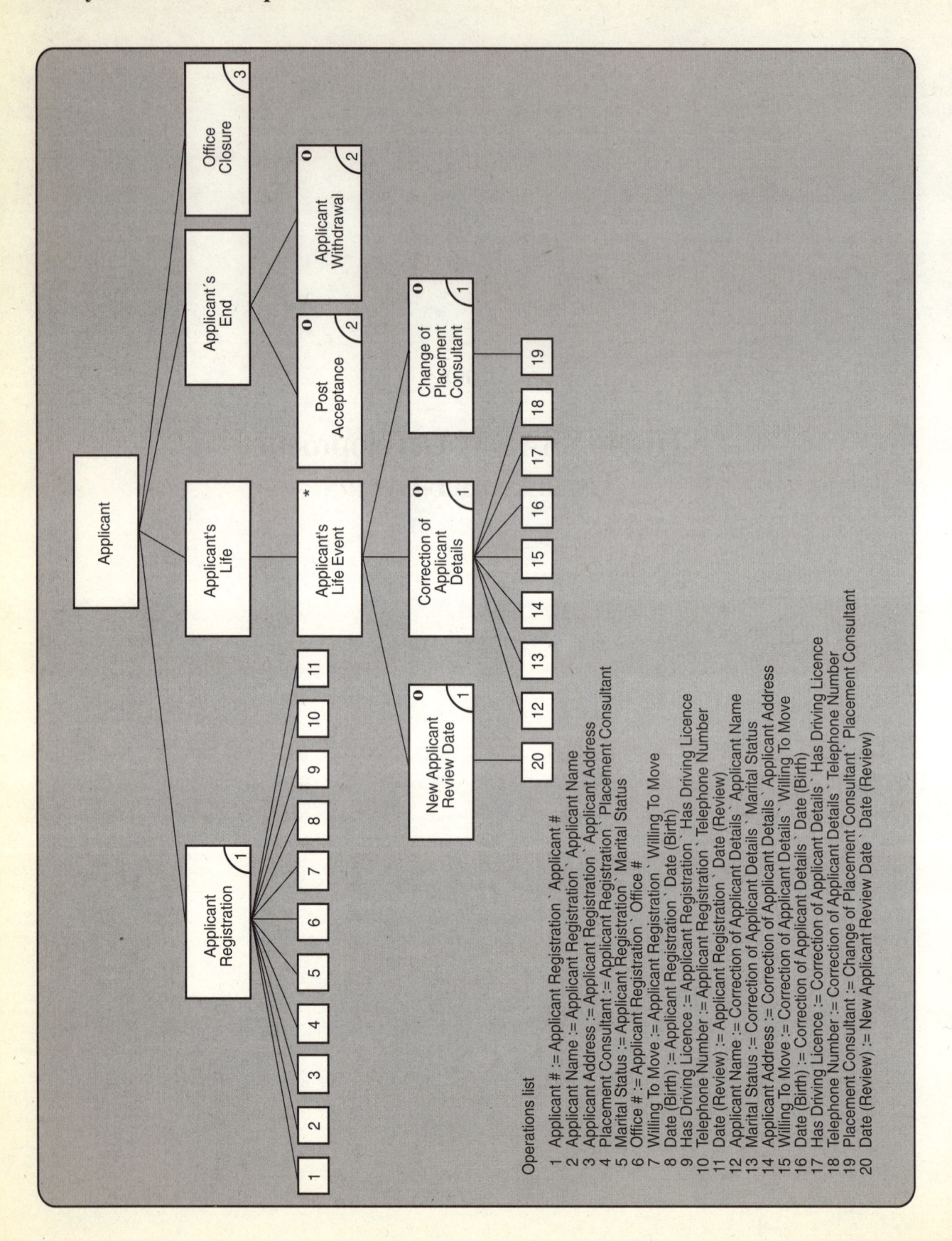

Operations list

1. Applicant # := Applicant Registration ` Applicant #
2. Applicant Name := Applicant Registration ` Applicant Name
3. Applicant Address := Applicant Registration ` Applicant Address
4. Placement Consultant := Applicant Registration ` Placement Consultant
5. Marital Status := Applicant Registration ` Marital Status
6. Office # := Applicant Registration ` Office #
7. Willing To Move := Applicant Registration ` Willing To Move
8. Date (Birth) := Applicant Registration ` Date (Birth)
9. Has Driving Licence := Applicant Registration ` Has Driving Licence
10. Telephone Number := Applicant Registration ` Telephone Number
11. Date (Review) := Applicant Registration ` Date (Review)
12. Applicant Name := Correction of Applicant Details ` Applicant Name
13. Marital Status := Correction of Applicant Details ` Marital Status
14. Applicant Address := Correction of Applicant Details ` Applicant Address
15. Willing To Move := Correction of Applicant Details ` Willing To Move
16. Date (Birth) := Correction of Applicant Details ` Date (Birth)
17. Has Driving Licence := Correction of Applicant Details ` Has Driving Licence
18. Telephone Number := Correction of Applicant Details ` Telephone Number
19. Placement Consultant := Change of Placement Consultant ` Placement Consultant
20. Date (Review) := New Applicant Review Date ` Date (Review)

Entity life histories if super-events are used

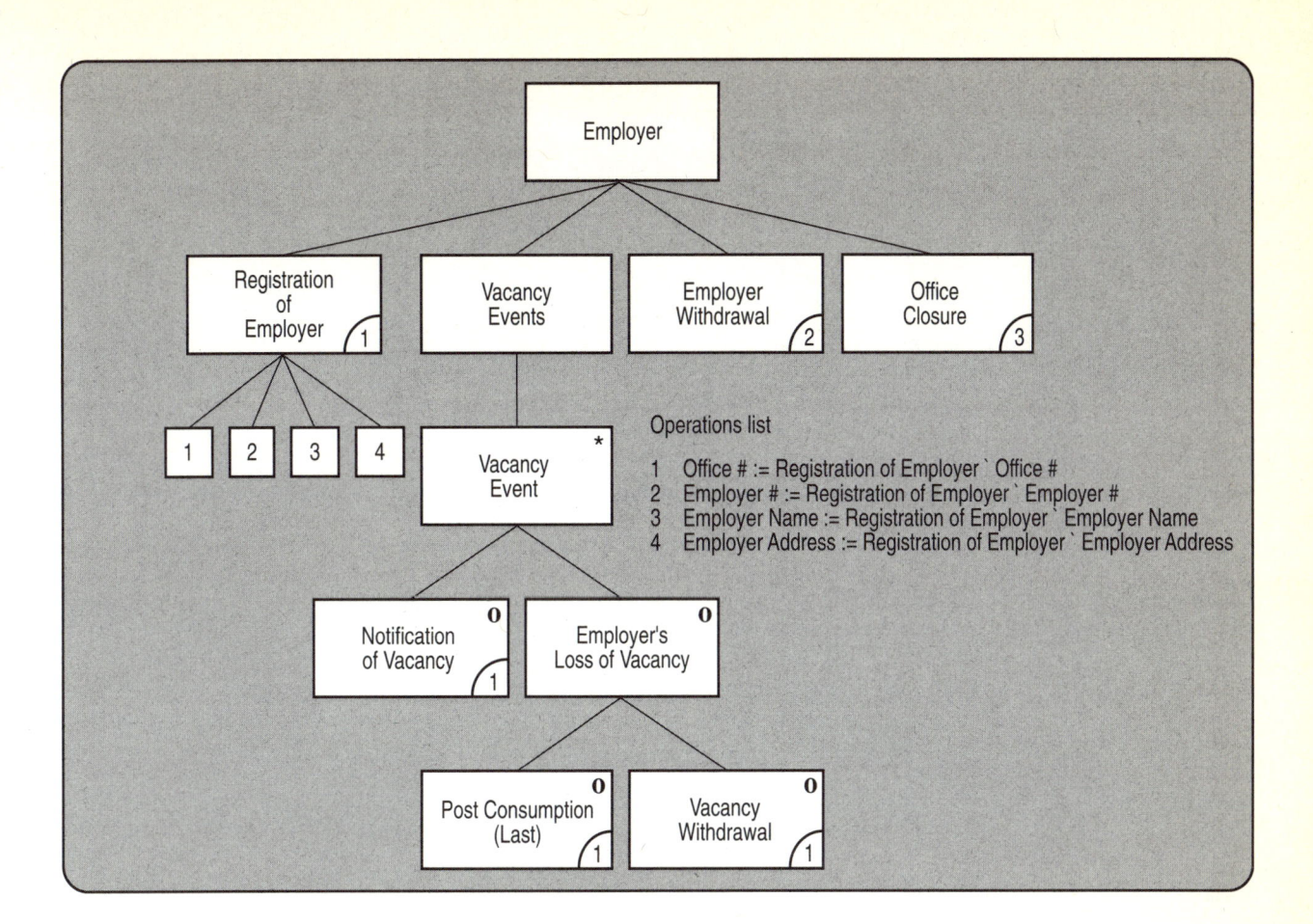

Employer

Structure:
- Employer
  - Registration of Employer (1)
    - 1, 2, 3, 4
  - Vacancy Events
    - Vacancy Event *
      - Notification of Vacancy (o) (1)
      - Employer's Loss of Vacancy (o)
        - Post Consumption (Last) (o) (1)
        - Vacancy Withdrawal (o) (1)
  - Employer Withdrawal (2)
  - Office Closure (3)

Operations list

1. Office # := Registration of Employer ` Office #
2. Employer # := Registration of Employer ` Employer #
3. Employer Name := Registration of Employer ` Employer Name
4. Employer Address := Registration of Employer ` Employer Address

280   Annex 4 Hi-Ho SSADM Development Using Super-Events

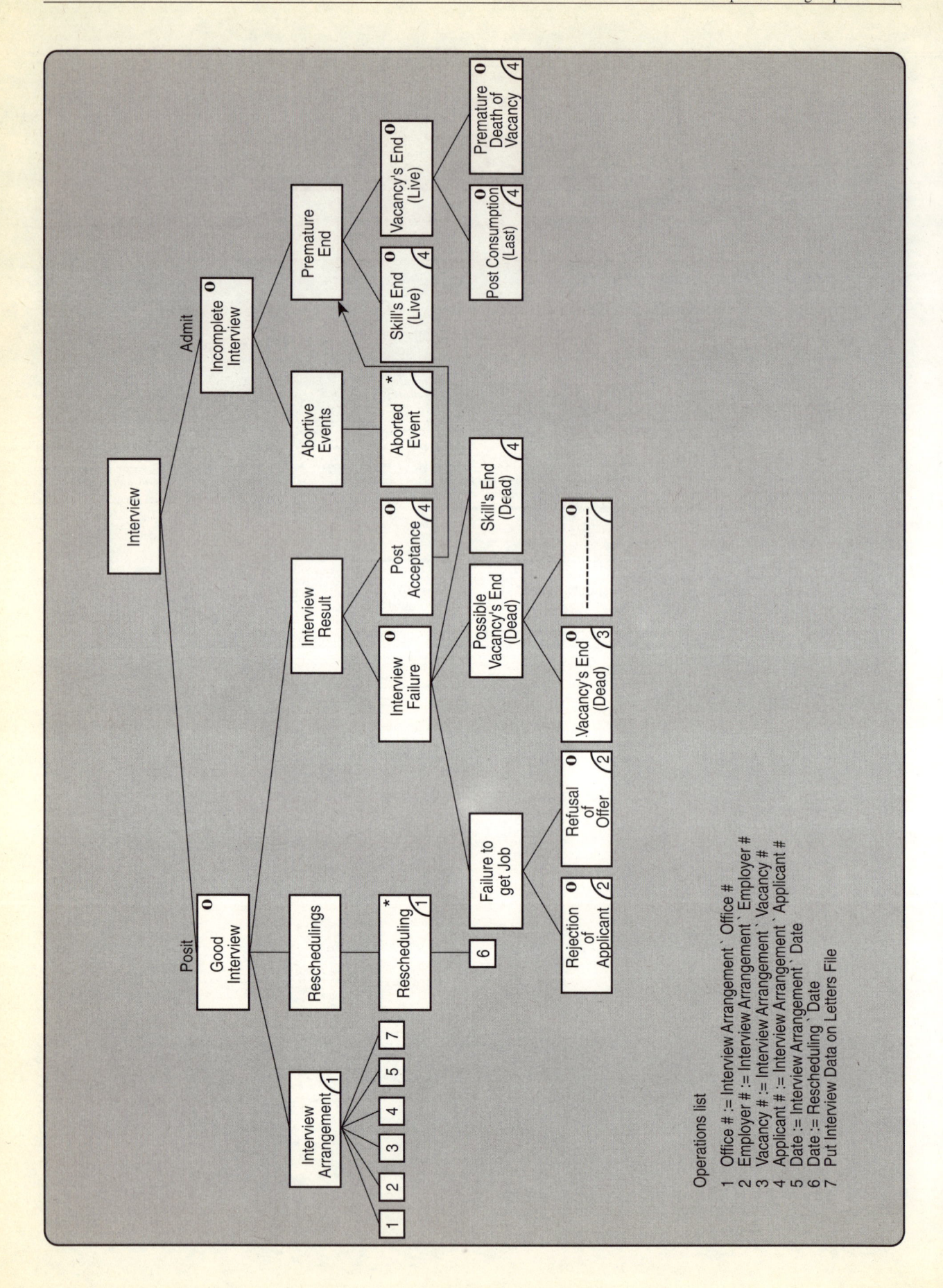

Interview

Posit: Good Interview
- Interview Arrangement (1)
  - 1, 2, 3, 4, 5, 7
- Reschedulings
  - Rescheduling *
    - 6
- Failure to get Job
  - Rejection of Applicant (2)
  - Refusal of Offer (2)

Interview Result
- Post Acceptance (4)
- Interview Failure
  - Skill's End (Dead) (4)
  - Possible Vacancy's End (Dead)
    - Vacancy's End (Dead) (3)
    - (─ ─ ─)

Admit: Incomplete Interview
- Abortive Events
  - Aborted Event *
- Premature End
  - Skill's End (Live) (4)
  - Vacancy's End (Live)
    - Post Consumption (Last) (4)
    - Premature Death of Vacancy (4)

Operations list

1 Office # := Interview Arrangement ` Office #
2 Employer # := Interview Arrangement ` Employer #
3 Vacancy # := Interview Arrangement ` Vacancy #
4 Applicant # := Interview Arrangement ` Applicant #
5 Date := Interview Arrangement ` Date
6 Date := Rescheduling ` Date
7 Put Interview Data on Letters File

Entity life histories if super-events are used   281

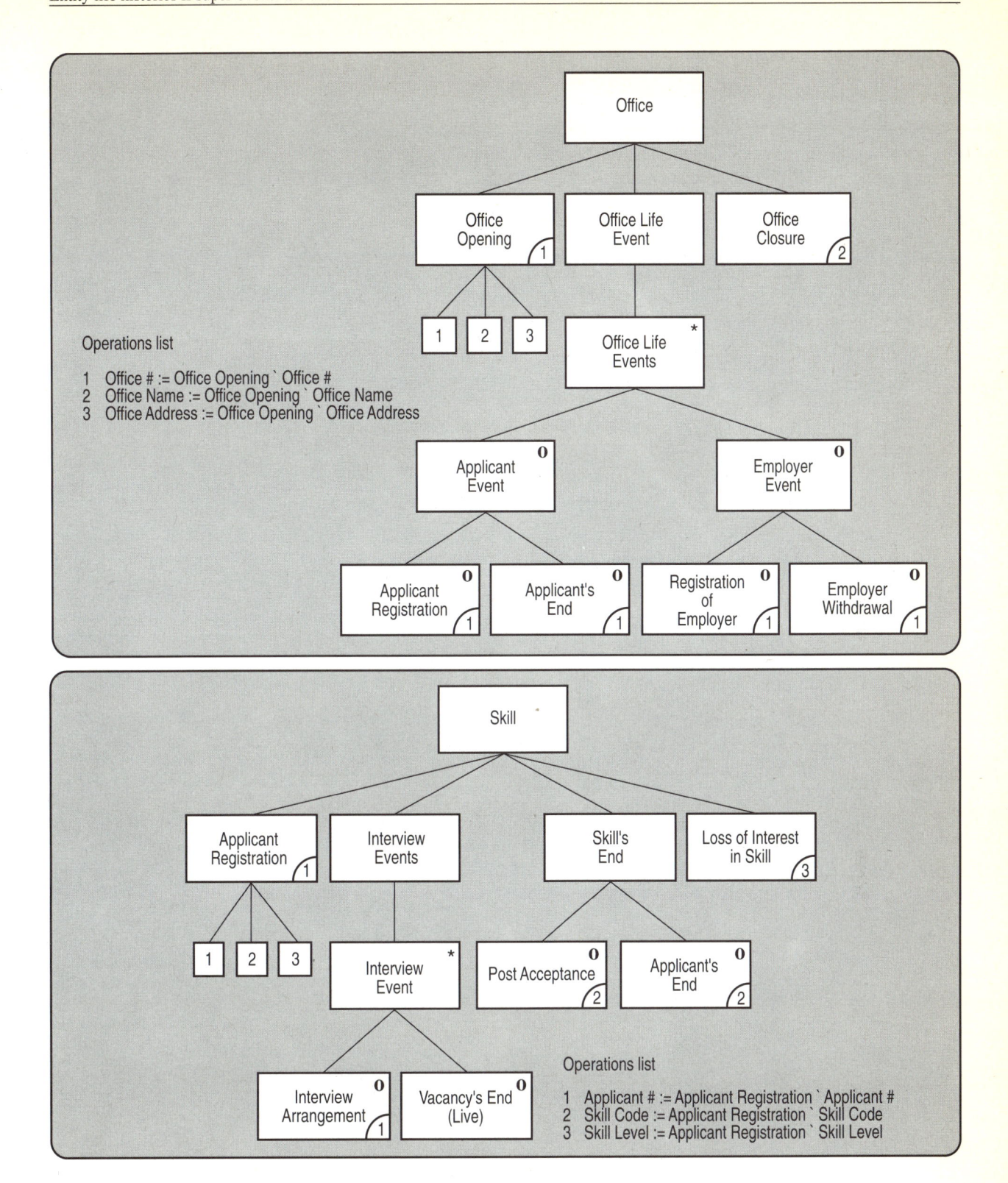

```
                            ┌─────────┐
                            │ Office  │
                            └────┬────┘
          ┌──────────────────────┼──────────────────────┐
    ┌───────────┐         ┌─────────────┐         ┌───────────┐
    │  Office   │         │ Office Life │         │  Office   │
    │  Opening 1│         │   Event     │         │ Closure  2│
    └─────┬─────┘         └──────┬──────┘         └───────────┘
       ┌──┼──┐                   │
      [1][2][3]          ┌───────────────┐
                         │  Office Life *│
                         │    Events     │
                         └───────┬───────┘
                    ┌────────────┴────────────┐
              ┌───────────┐ o           ┌───────────┐ o
              │ Applicant │             │ Employer  │
              │   Event   │             │   Event   │
              └─────┬─────┘             └─────┬─────┘
              ┌────┴────┐                ┌────┴────┐
      ┌───────────┐o ┌───────────┐o ┌───────────┐o ┌───────────┐o
      │ Applicant │  │Applicant's│  │Registration│ │ Employer  │
      │Registration│ │    End    │  │of Employer│  │Withdrawal │
      │          1│  │          1│  │          1│  │          1│
      └───────────┘  └───────────┘  └───────────┘  └───────────┘
```

Operations list

1 Office # := Office Opening ` Office #
2 Office Name := Office Opening ` Office Name
3 Office Address := Office Opening ` Office Address

```
                              ┌───────┐
                              │ Skill │
                              └───┬───┘
          ┌──────────────┬───────┴────────┬──────────────┐
    ┌───────────┐ ┌───────────┐    ┌───────────┐  ┌───────────┐
    │ Applicant │ │ Interview │    │  Skill's  │  │Loss of Interest│
    │Registration│ │  Events   │    │    End    │  │  in Skill │
    │          1│ └─────┬─────┘    └─────┬─────┘  │          3│
    └─────┬─────┘       │                │        └───────────┘
       ┌──┼──┐    ┌───────────┐          │
      [1][2][3]   │ Interview*│      ┌────┴────┐
                  │   Event   │  ┌───────────┐o ┌───────────┐o
                  └─────┬─────┘  │   Post    │  │Applicant's│
                 ┌──────┴──────┐ │ Acceptance│  │    End    │
          ┌───────────┐o ┌───────────┐o       │          2│  │          2│
          │ Interview │  │Vacancy's End│      └───────────┘  └───────────┘
          │Arrangement│  │   (Live)   │
          │          1│  └───────────┘
          └───────────┘
```

Operations list

1 Applicant # := Applicant Registration ` Applicant #
2 Skill Code := Applicant Registration ` Skill Code
3 Skill Level := Applicant Registration ` Skill Level

282 Annex 4 Hi-Ho SSADM Development Using Super-Events

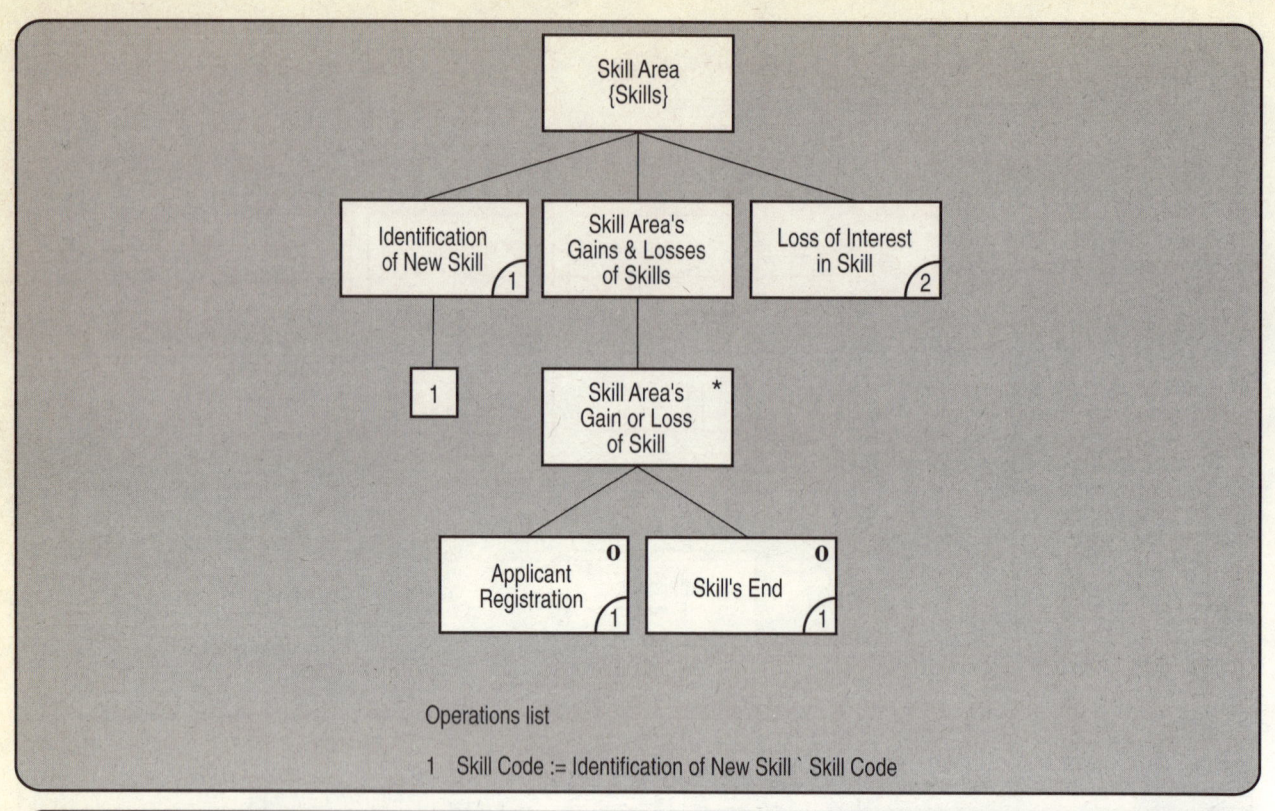

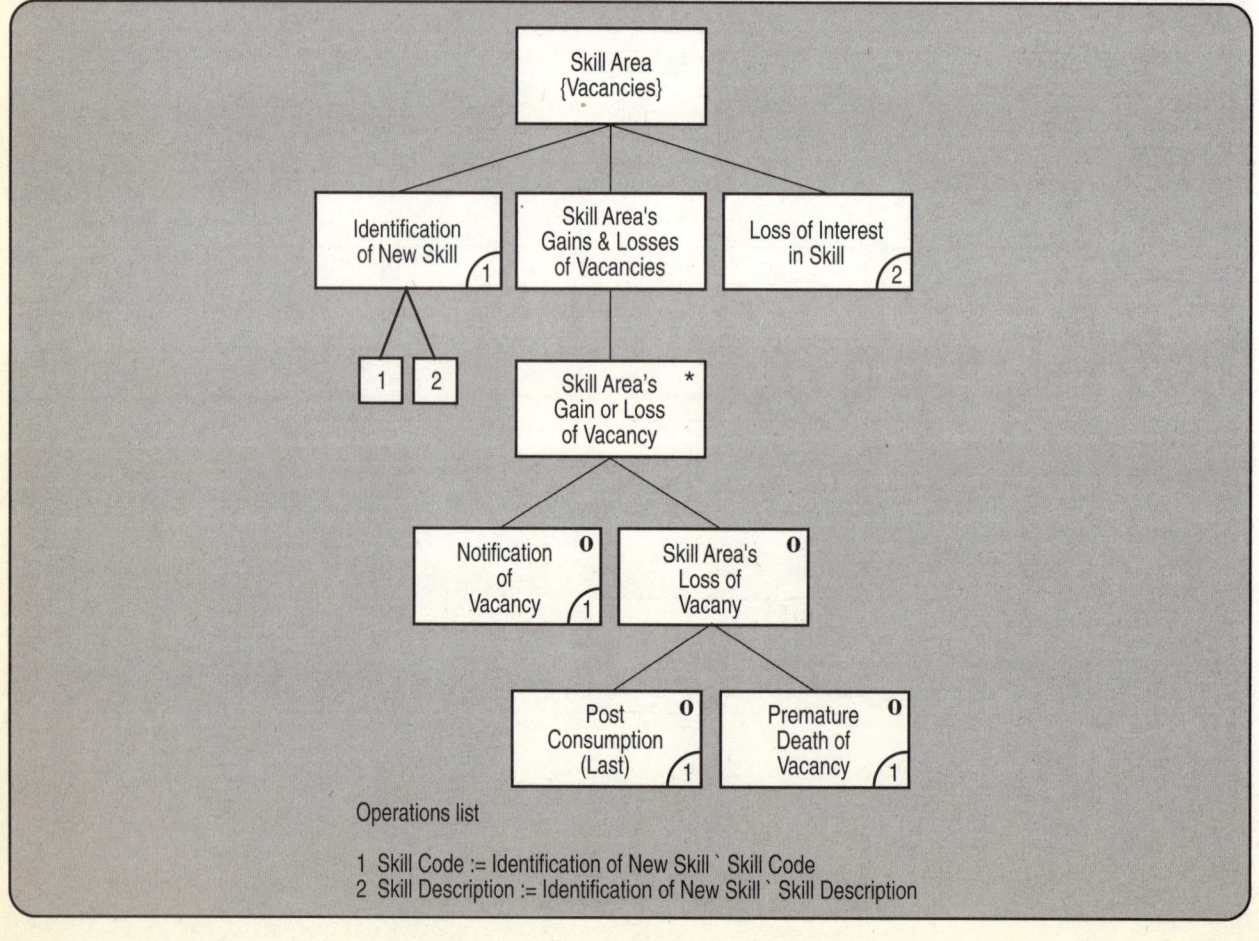

Diagram 1: Skill Area {Skills}

- Skill Area {Skills}
  - Identification of New Skill (1)
  - Skill Area's Gains & Losses of Skills
    - (1)
    - Skill Area's Gain or Loss of Skill *
      - Applicant Registration (o, 1)
      - Skill's End (o, 1)
  - Loss of Interest in Skill (2)

Operations list

1 Skill Code := Identification of New Skill ` Skill Code

Diagram 2: Skill Area {Vacancies}

- Skill Area {Vacancies}
  - Identification of New Skill (1)
  - Skill Area's Gains & Losses of Vacancies
    - (1) (2)
    - Skill Area's Gain or Loss of Vacancy *
      - Notification of Vacancy (o, 1)
      - Skill Area's Loss of Vacany (o)
        - Post Consumption (Last) (o, 1)
        - Premature Death of Vacancy (o, 1)
  - Loss of Interest in Skill (2)

Operations list

1 Skill Code := Identification of New Skill ` Skill Code
2 Skill Description := Identification of New Skill ` Skill Description

Entity life histories if super-events are used

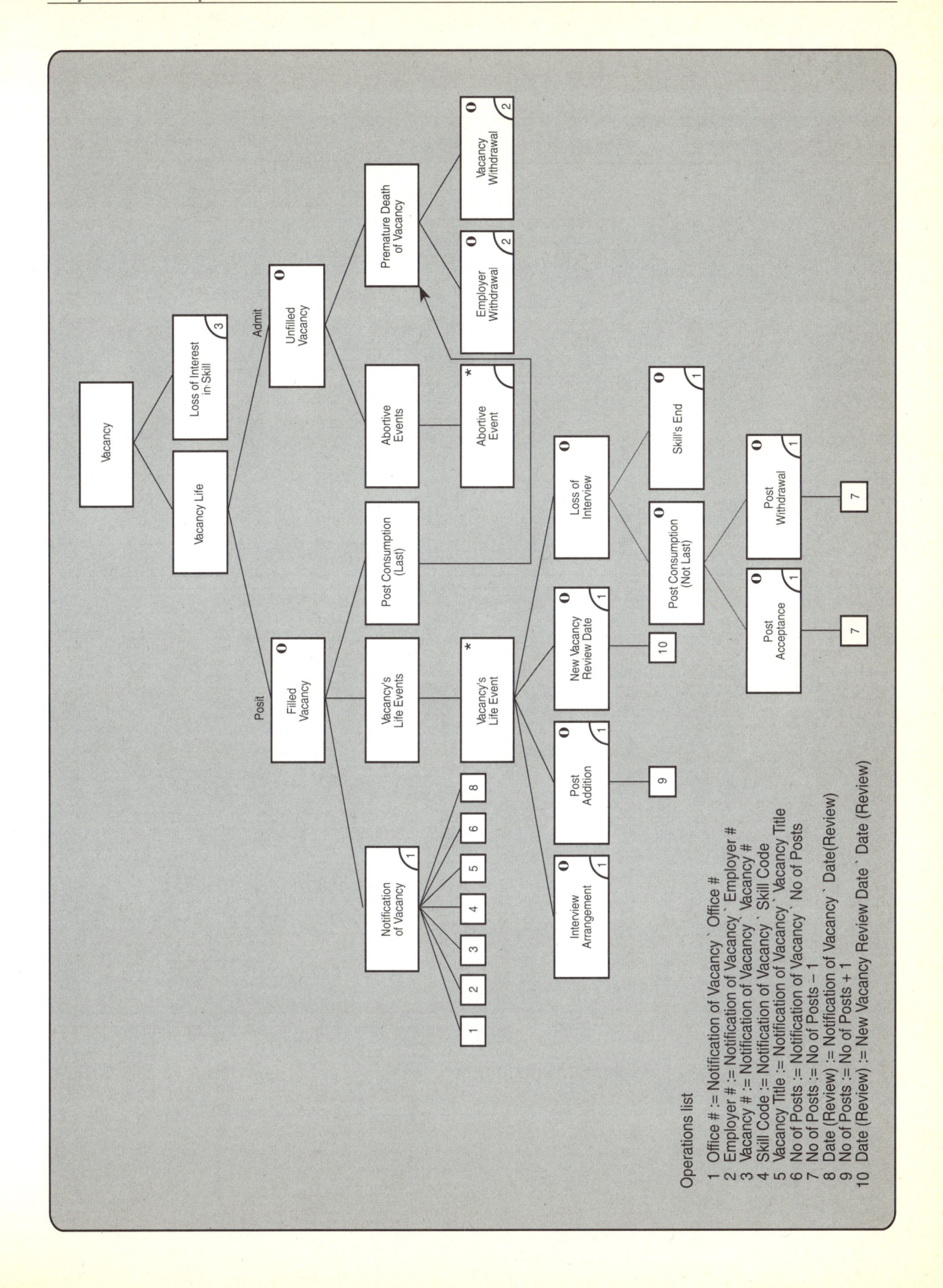

Operations list
1. Office # := Notification of Vacancy ` Office #
2. Employer # := Notification of Vacancy ` Employer #
3. Vacancy # := Notification of Vacancy ` Vacancy #
4. Skill Code := Notification of Vacancy ` Skill Code
5. Vacancy Title := Notification of Vacancy ` Vacancy Title
6. No of Posts := Notification of Vacancy ` No of Posts
7. No of Posts := No of Posts − 1
8. Date (Review) := Notification of Vacancy ` Date(Review)
9. No of Posts := No of Posts + 1
10. Date (Review) := New Vacancy Review Date ` Date (Review)

283

ECDs if super-events are used

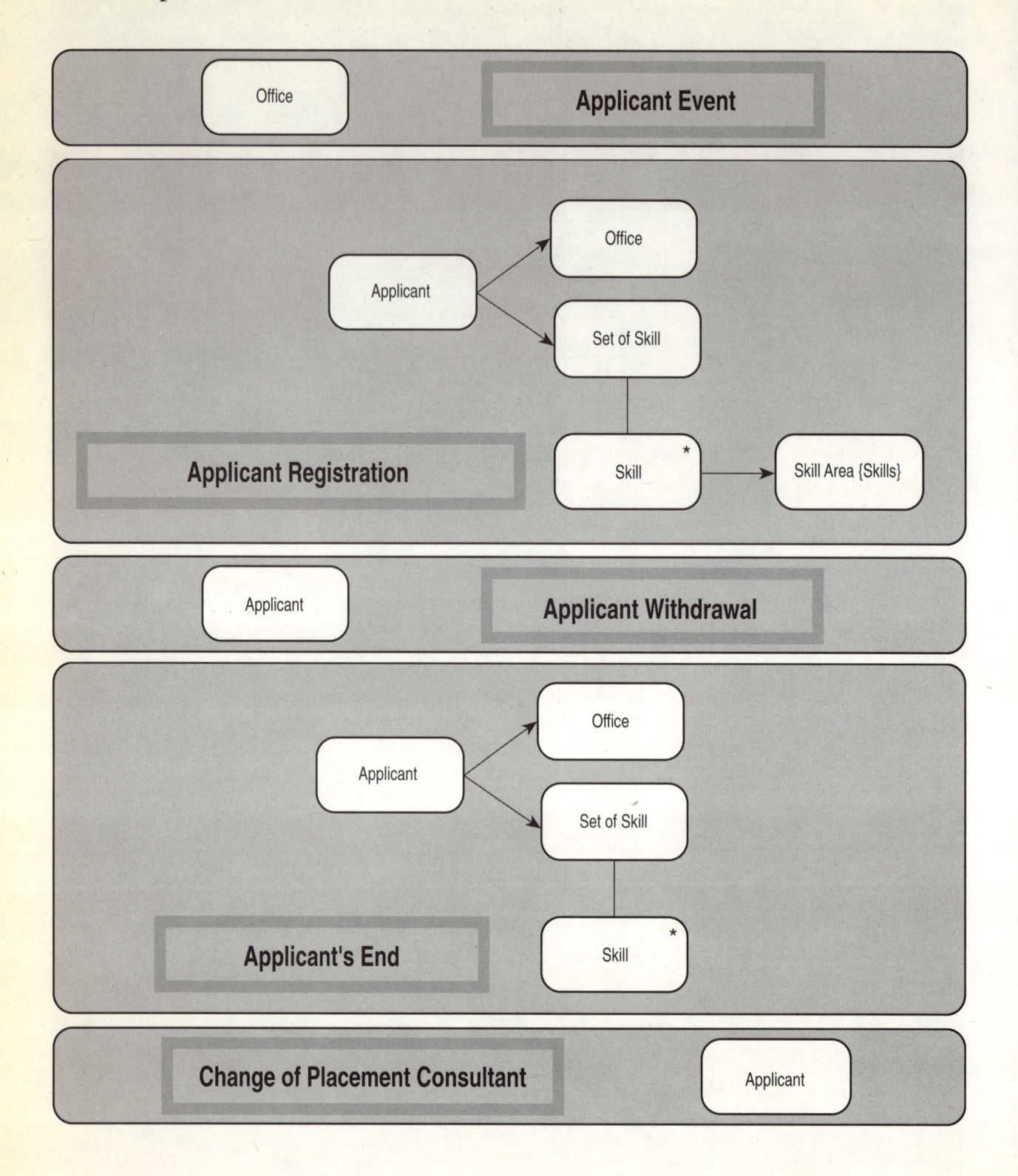

Applicant Event
- Office

Applicant Registration
- Applicant → Office
- Applicant → Set of Skill
- Set of Skill — Skill * → Skill Area {Skills}

Applicant Withdrawal
- Applicant

Applicant's End
- Applicant → Office
- Applicant → Set of Skill
- Set of Skill — Skill *

Change of Placement Consultant
- Applicant

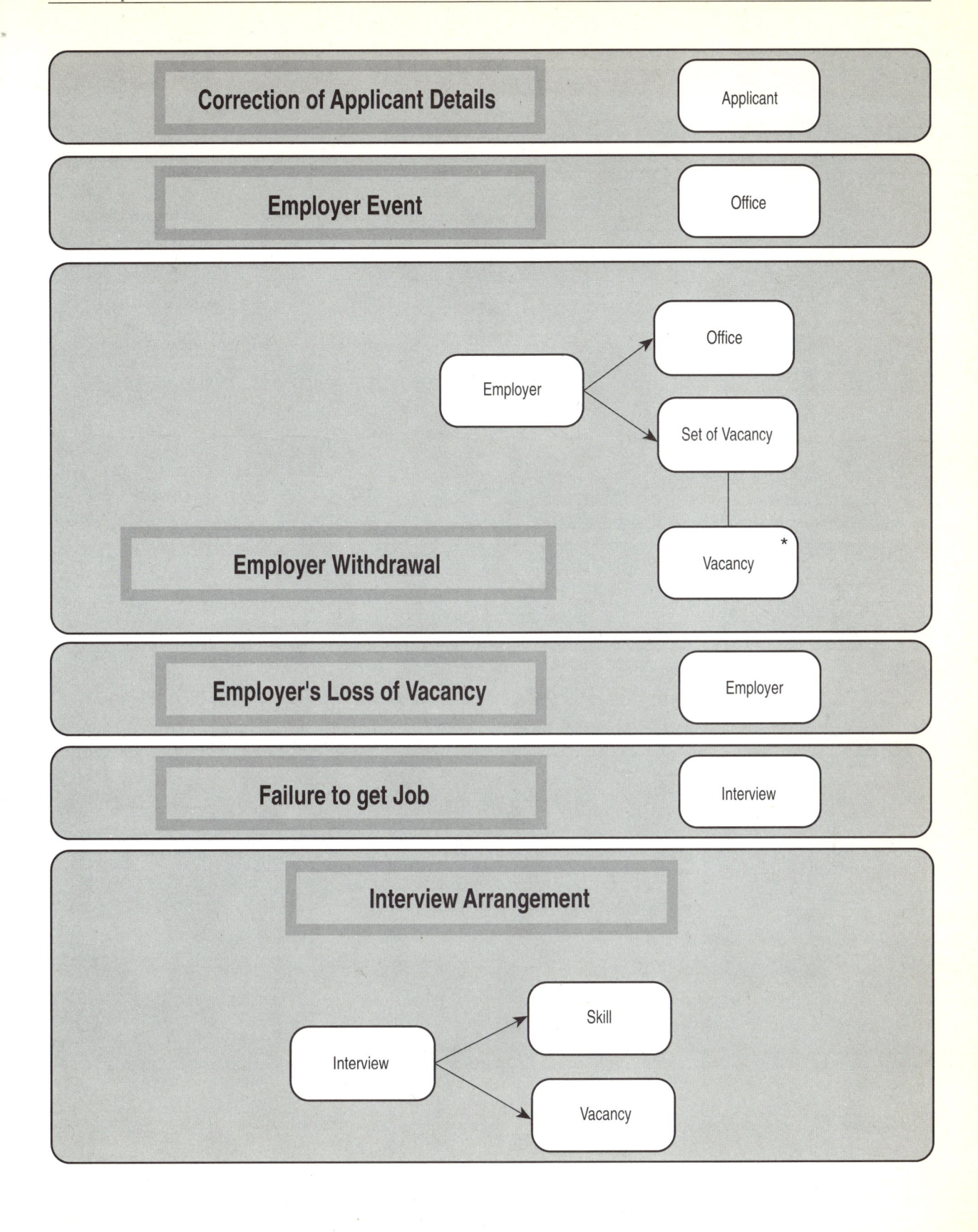

ECDs if super-events are used

Correction of Applicant Details

Applicant

Employer Event

Office

Employer Withdrawal

Employer → Office
Employer → Set of Vacancy
Set of Vacancy — Vacancy *

Employer's Loss of Vacancy

Employer

Failure to get Job

Interview

Interview Arrangement

Interview → Skill
Interview → Vacancy

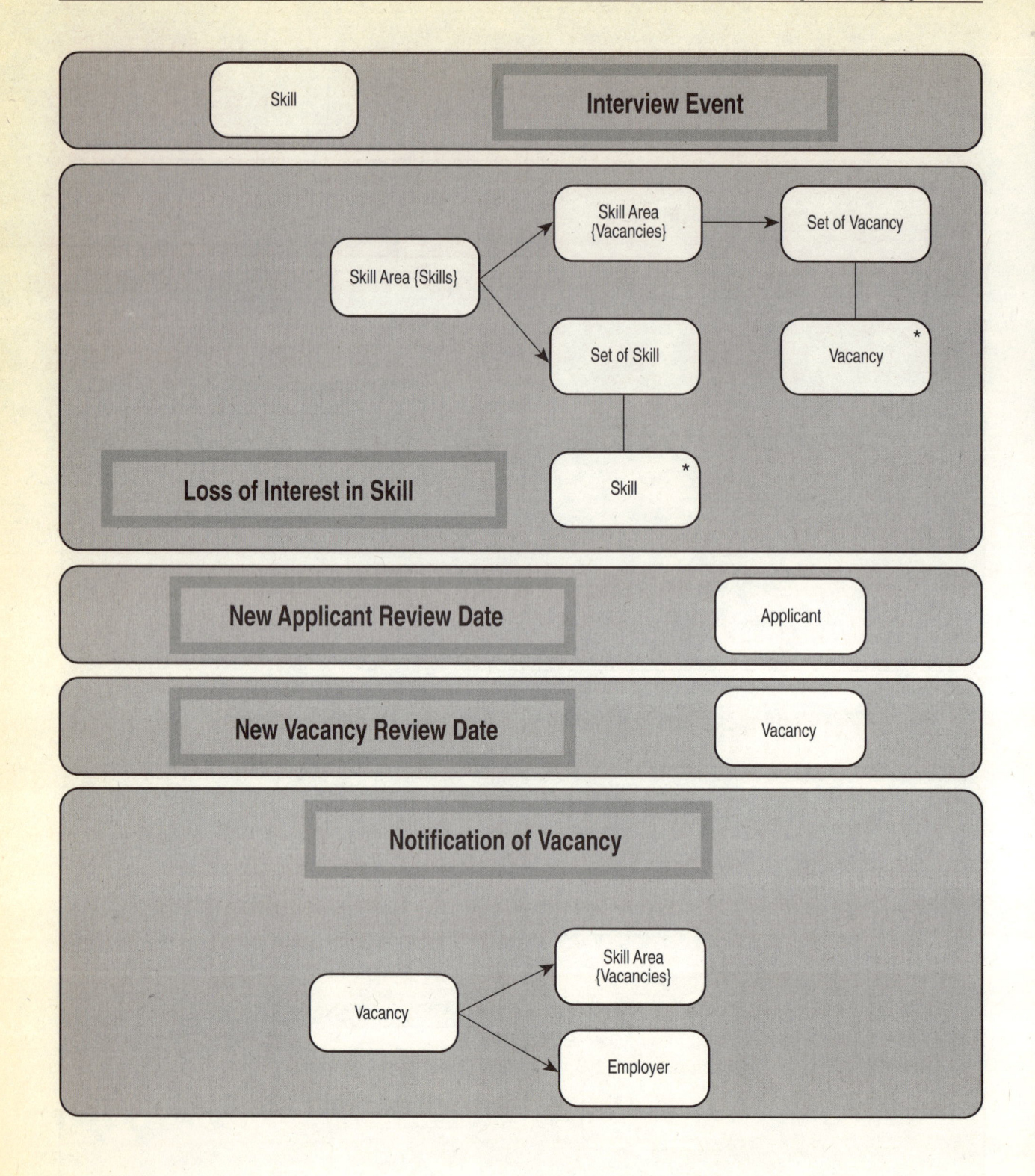

Interview Event

Skill

Loss of Interest in Skill

Skill Area {Skills} → Skill Area {Vacancies} → Set of Vacancy — Vacancy *

Skill Area {Skills} → Set of Skill → Skill *

New Applicant Review Date

Applicant

New Vacancy Review Date

Vacancy

Notification of Vacancy

Vacancy → Skill Area {Vacancies}

Vacancy → Employer

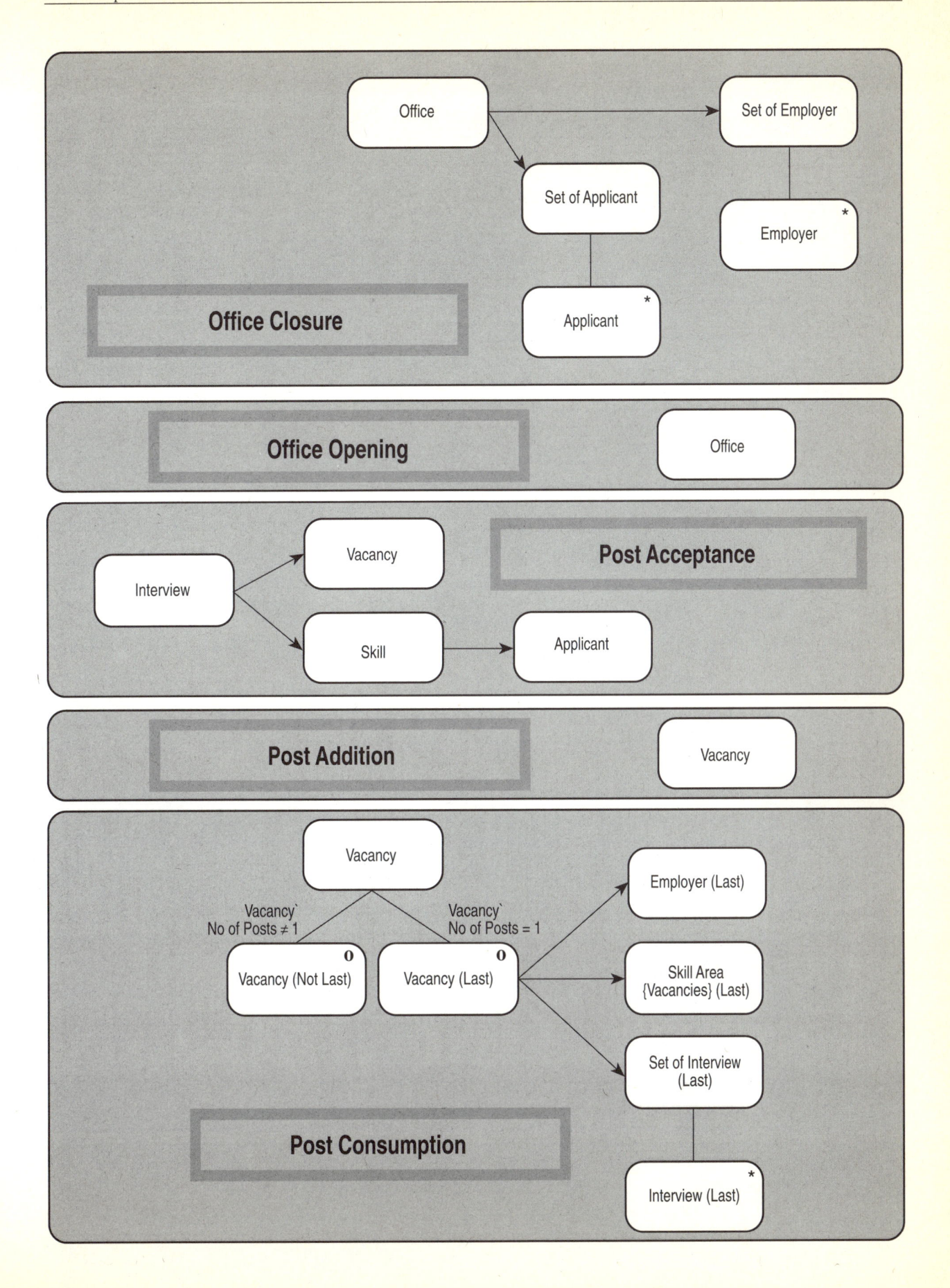

ECDs if super-events are used

Office Closure

- Office → Set of Employer
- Office → Set of Applicant
- Set of Employer → Employer *
- Set of Applicant → Applicant *

Office Opening

- Office

Post Acceptance

- Interview → Vacancy
- Interview → Skill → Applicant

Post Addition

- Vacancy

Post Consumption

- Vacancy
  - Vacancy` No of Posts ≠ 1 → Vacancy (Not Last) [o]
  - Vacancy` No of Posts = 1 → Vacancy (Last) [o]
- Vacancy (Last) → Employer (Last)
- Vacancy (Last) → Skill Area {Vacancies} (Last)
- Vacancy (Last) → Set of Interview (Last) → Interview (Last) *

287

Post Withdrawal

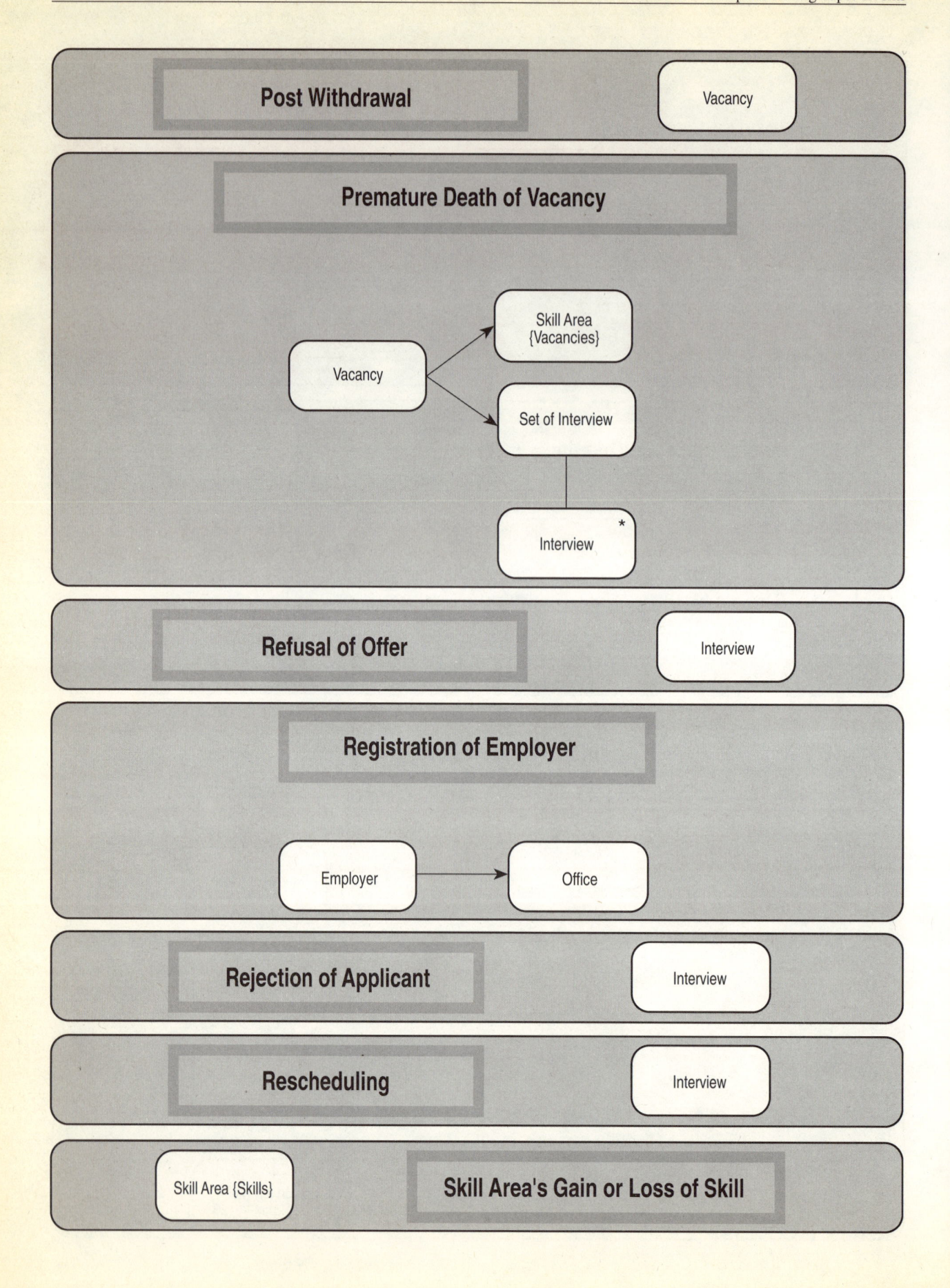

Vacancy

Premature Death of Vacancy

Vacancy → Skill Area {Vacancies}
Vacancy → Set of Interview — Interview *

Refusal of Offer

Interview

Registration of Employer

Employer → Office

Rejection of Applicant

Interview

Rescheduling

Interview

Skill Area's Gain or Loss of Skill

Skill Area {Skills}

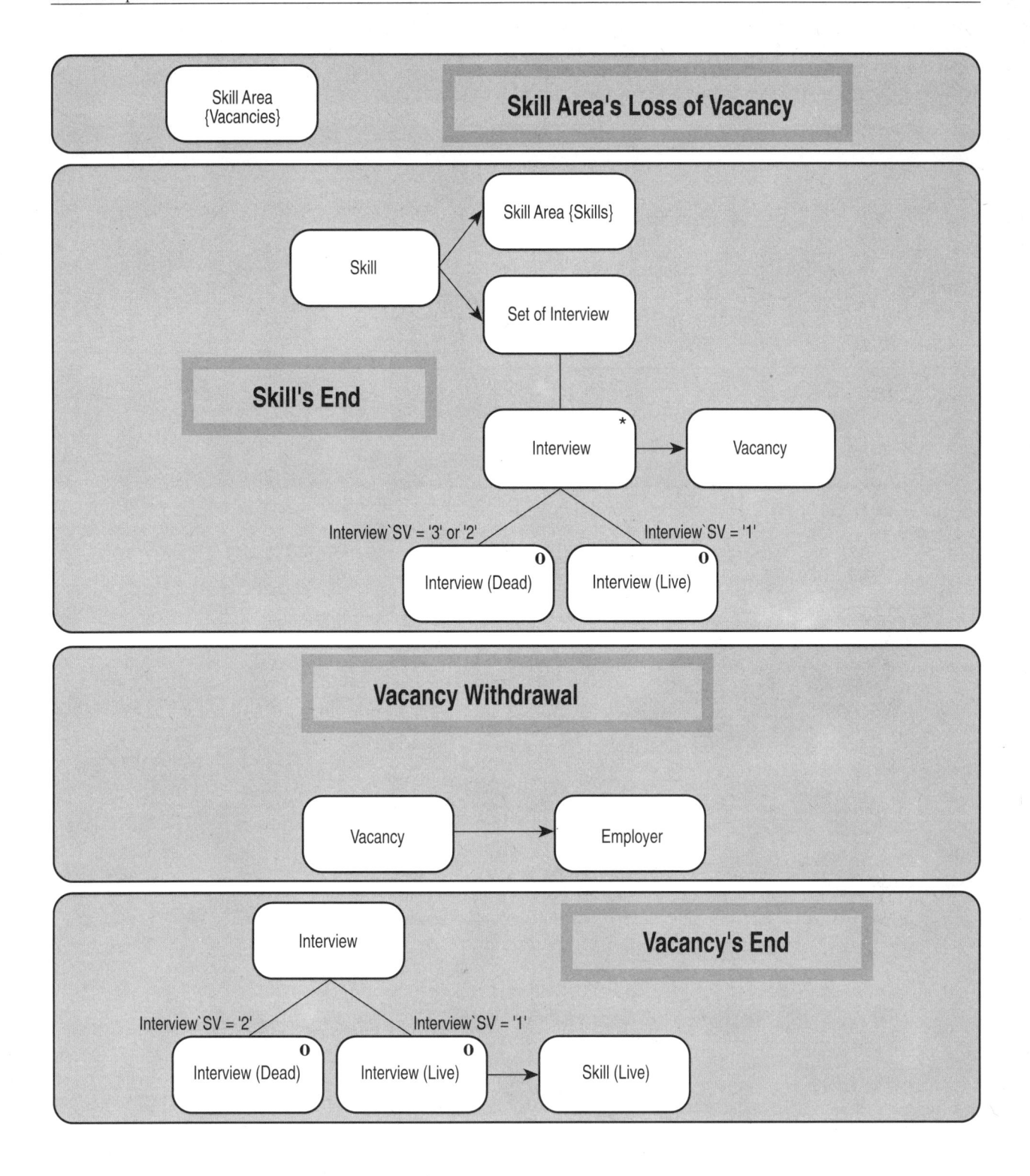

ECDs if super-events are used   289

Skill Area's Loss of Vacancy

Skill Area {Vacancies}

Skill's End

Skill → Skill Area {Skills}
Skill → Set of Interview
Set of Interview — Interview *→ Vacancy
Interview → Interview (Dead)   [Interview`SV = '3' or '2']   0
Interview → Interview (Live)   [Interview`SV = '1']   0

Vacancy Withdrawal

Vacancy → Employer

Vacancy's End

Interview → Interview (Dead)   [Interview`SV = '2']   0
Interview → Interview (Live)   [Interview`SV = '1']   0
Interview (Live) → Skill (Live)

Update process models if super-events are used

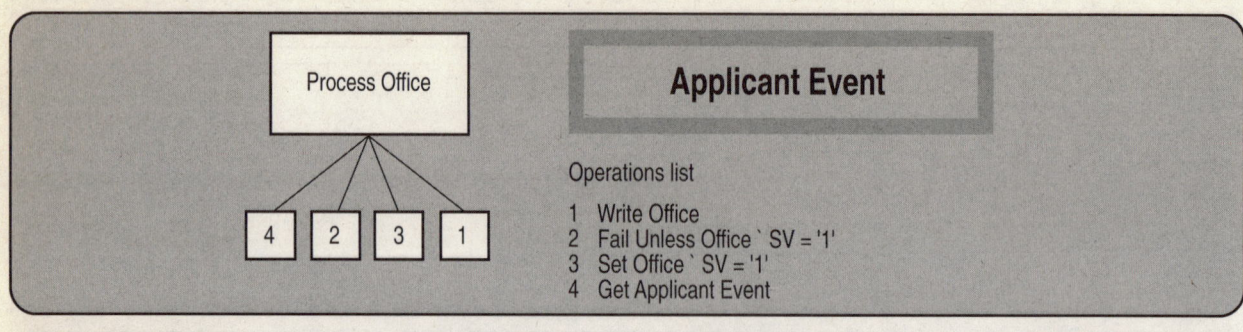

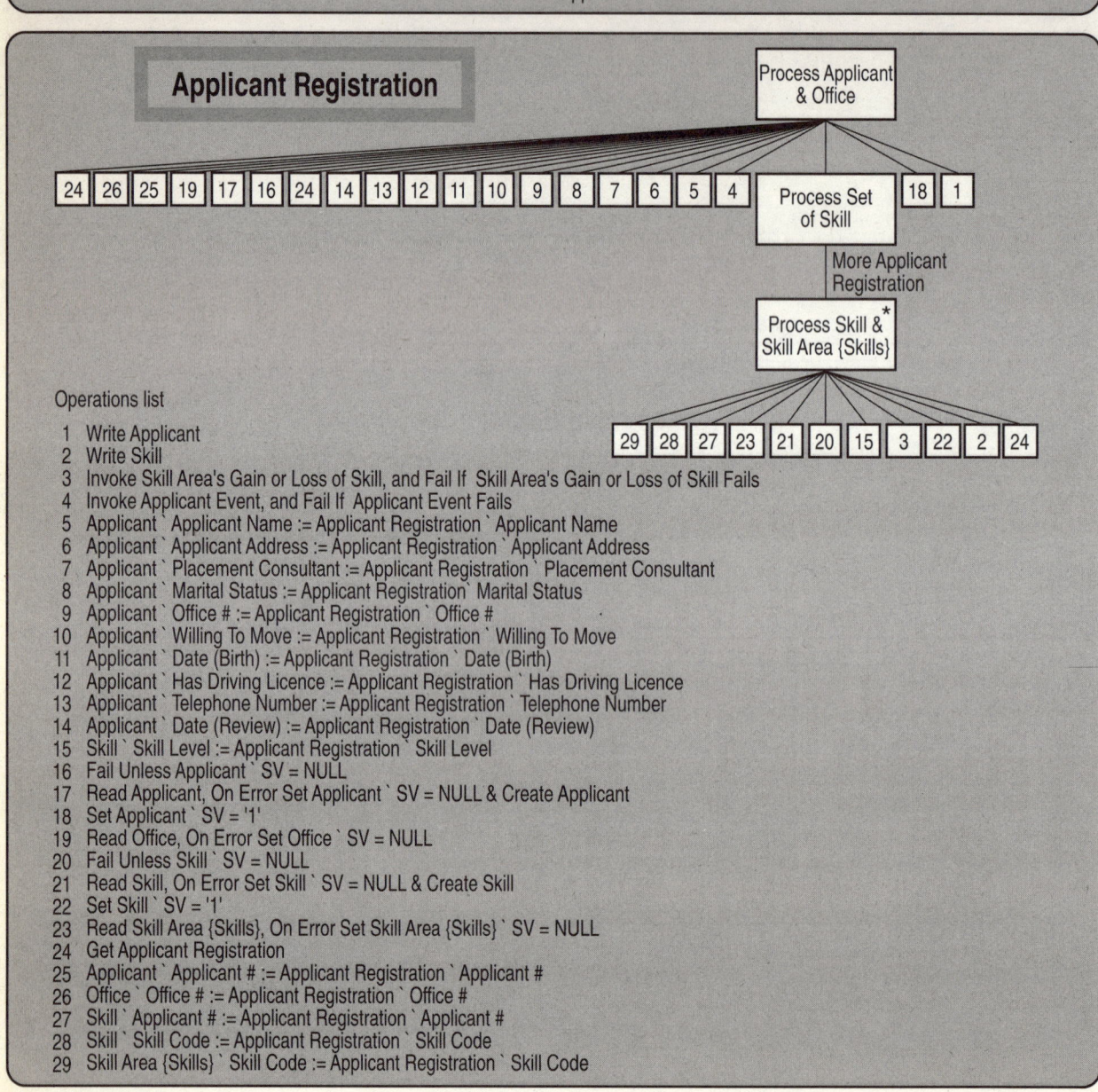

Process Office

```
        Process Office
       /    |    |   \
      4     2    3    1
```

Applicant Event

Operations list

1. Write Office
2. Fail Unless Office ` SV = '1'
3. Set Office ` SV = '1'
4. Get Applicant Event

Applicant Registration

```
                                          Process Applicant
                                              & Office
   ┌──┬──┬──┬──┬──┬──┬──┬──┬──┬──┬──┬──┬──┬──┤
   24 26 25 19 17 16 24 14 13 12 11 10 9 8 7 6 5 4    Process Set    18  1
                                                      of Skill
                                                          │
                                                     More Applicant
                                                      Registration
                                                          │
                                                    Process Skill &  *
                                                   Skill Area {Skills}
                                                          │
                                         ┌──┬──┬──┬──┬──┬──┬──┬──┬──┐
                                         29 28 27 23 21 20 15 3 22 2 24
```

Operations list

1. Write Applicant
2. Write Skill
3. Invoke Skill Area's Gain or Loss of Skill, and Fail If Skill Area's Gain or Loss of Skill Fails
4. Invoke Applicant Event, and Fail If Applicant Event Fails
5. Applicant ` Applicant Name := Applicant Registration ` Applicant Name
6. Applicant ` Applicant Address := Applicant Registration ` Applicant Address
7. Applicant ` Placement Consultant := Applicant Registration ` Placement Consultant
8. Applicant ` Marital Status := Applicant Registration ` Marital Status
9. Applicant ` Office # := Applicant Registration ` Office #
10. Applicant ` Willing To Move := Applicant Registration ` Willing To Move
11. Applicant ` Date (Birth) := Applicant Registration ` Date (Birth)
12. Applicant ` Has Driving Licence := Applicant Registration ` Has Driving Licence
13. Applicant ` Telephone Number := Applicant Registration ` Telephone Number
14. Applicant ` Date (Review) := Applicant Registration ` Date (Review)
15. Skill ` Skill Level := Applicant Registration ` Skill Level
16. Fail Unless Applicant ` SV = NULL
17. Read Applicant, On Error Set Applicant ` SV = NULL & Create Applicant
18. Set Applicant ` SV = '1'
19. Read Office, On Error Set Office ` SV = NULL
20. Fail Unless Skill ` SV = NULL
21. Read Skill, On Error Set Skill ` SV = NULL & Create Skill
22. Set Skill ` SV = '1'
23. Read Skill Area {Skills}, On Error Set Skill Area {Skills} ` SV = NULL
24. Get Applicant Registration
25. Applicant ` Applicant # := Applicant Registration ` Applicant #
26. Office ` Office # := Applicant Registration ` Office #
27. Skill ` Applicant # := Applicant Registration ` Applicant #
28. Skill ` Skill Code := Applicant Registration ` Skill Code
29. Skill Area {Skills} ` Skill Code := Applicant Registration ` Skill Code

Update process models if super-events are used
291

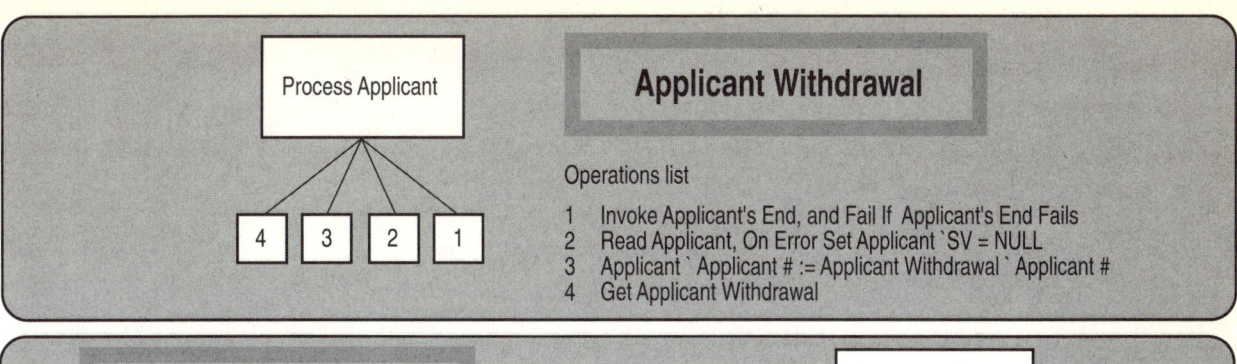

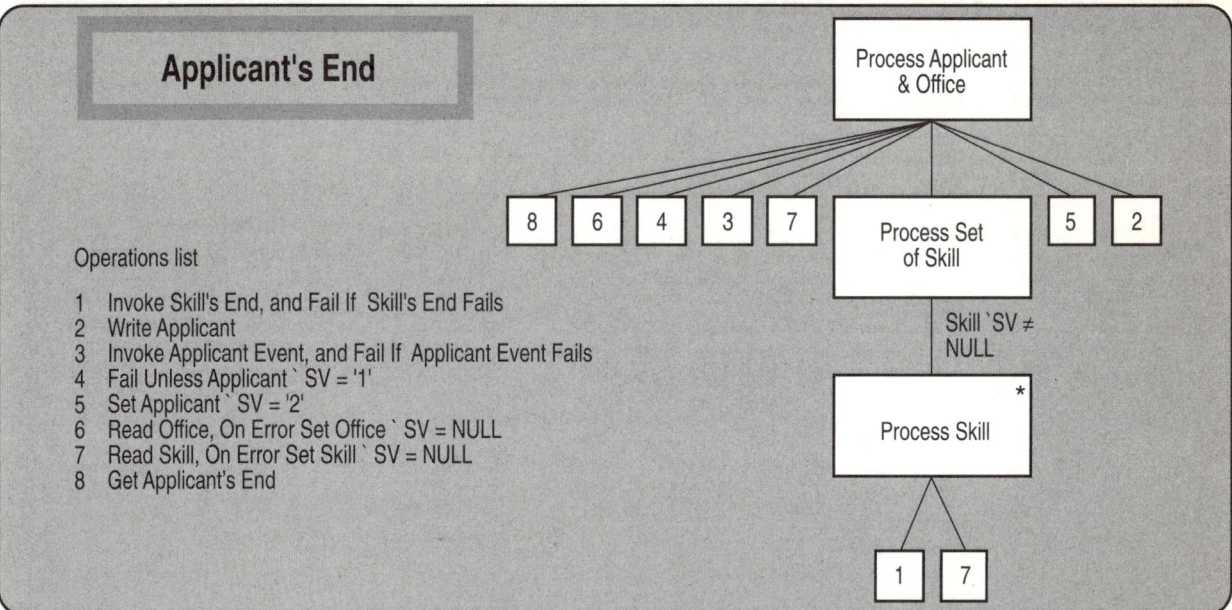

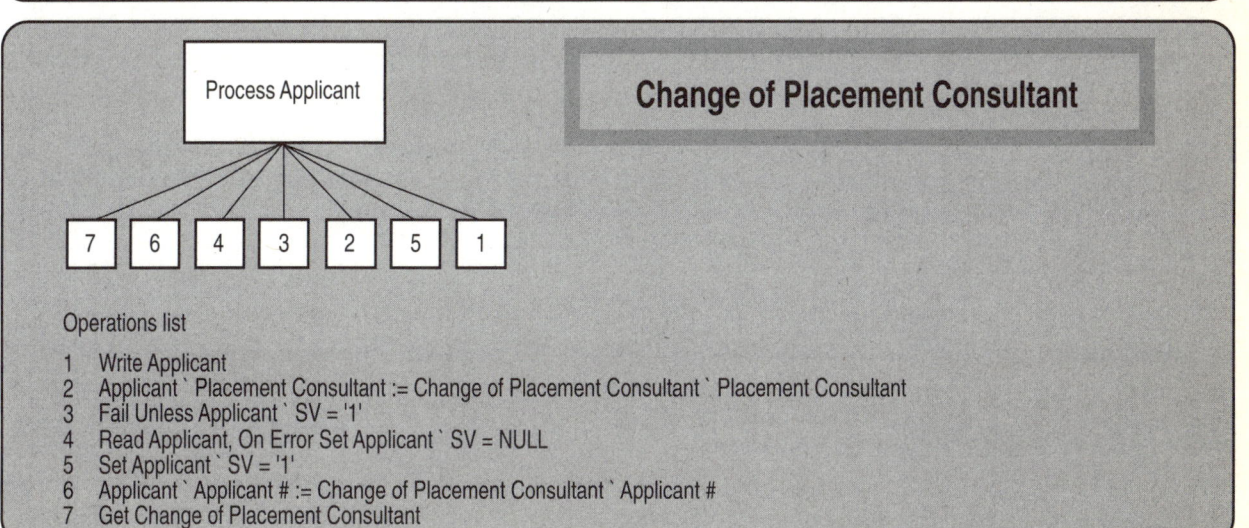

Applicant Withdrawal

Process Applicant
4 3 2 1

Operations list
1. Invoke Applicant's End, and Fail If Applicant's End Fails
2. Read Applicant, On Error Set Applicant ` SV = NULL
3. Applicant ` Applicant # := Applicant Withdrawal ` Applicant #
4. Get Applicant Withdrawal

Applicant's End

Process Applicant & Office
8 6 4 3 7   Process Set of Skill   5 2

Skill ` SV ≠ NULL

Process Skill *
1 7

Operations list
1. Invoke Skill's End, and Fail If Skill's End Fails
2. Write Applicant
3. Invoke Applicant Event, and Fail If Applicant Event Fails
4. Fail Unless Applicant ` SV = '1'
5. Set Applicant ` SV = '2'
6. Read Office, On Error Set Office ` SV = NULL
7. Read Skill, On Error Set Skill ` SV = NULL
8. Get Applicant's End

Change of Placement Consultant

Process Applicant
7 6 4 3 2 5 1

Operations list
1. Write Applicant
2. Applicant ` Placement Consultant := Change of Placement Consultant ` Placement Consultant
3. Fail Unless Applicant ` SV = '1'
4. Read Applicant, On Error Set Applicant ` SV = NULL
5. Set Applicant ` SV = '1'
6. Applicant ` Applicant # := Change of Placement Consultant ` Applicant #
7. Get Change of Placement Consultant

Correction of Applicant Details

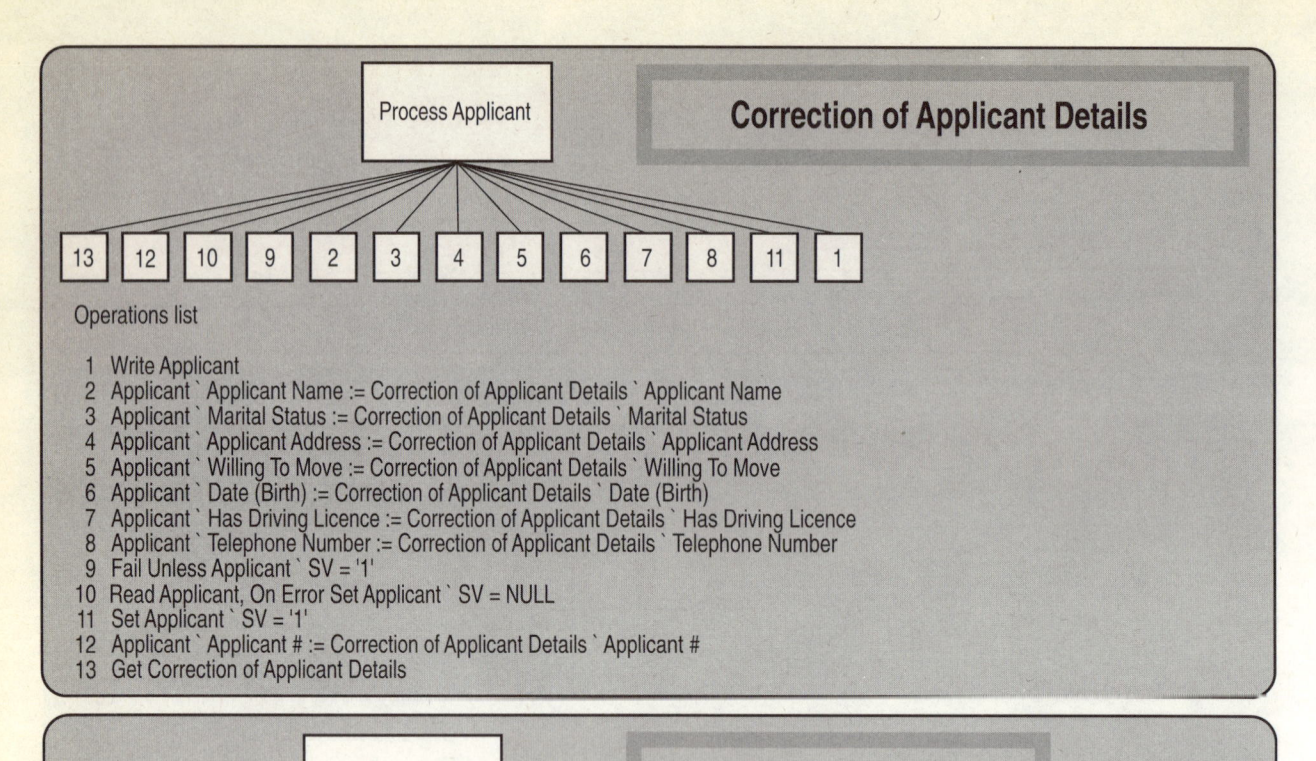

```
                 Process Applicant
                        |
    ┌────┬────┬───┬───┬─┴─┬───┬───┬───┬────┬───┐
   13   12   10   9   2   3   4   5   6   7   8   11   1
```

Operations list

1. Write Applicant
2. Applicant ` Applicant Name := Correction of Applicant Details ` Applicant Name
3. Applicant ` Marital Status := Correction of Applicant Details ` Marital Status
4. Applicant ` Applicant Address := Correction of Applicant Details ` Applicant Address
5. Applicant ` Willing To Move := Correction of Applicant Details ` Willing To Move
6. Applicant ` Date (Birth) := Correction of Applicant Details ` Date (Birth)
7. Applicant ` Has Driving Licence := Correction of Applicant Details ` Has Driving Licence
8. Applicant ` Telephone Number := Correction of Applicant Details ` Telephone Number
9. Fail Unless Applicant ` SV = '1'
10. Read Applicant, On Error Set Applicant ` SV = NULL
11. Set Applicant ` SV = '1'
12. Applicant ` Applicant # := Correction of Applicant Details ` Applicant #
13. Get Correction of Applicant Details

Employer Event

```
        Process Office
              |
       ┌───┬──┴┬───┐
       4   2   3   1
```

Operations list

1. Write Office
2. Fail Unless Office ` SV = '1'
3. Set Office ` SV = '1'
4. Get Employer Event

Employer Withdrawal

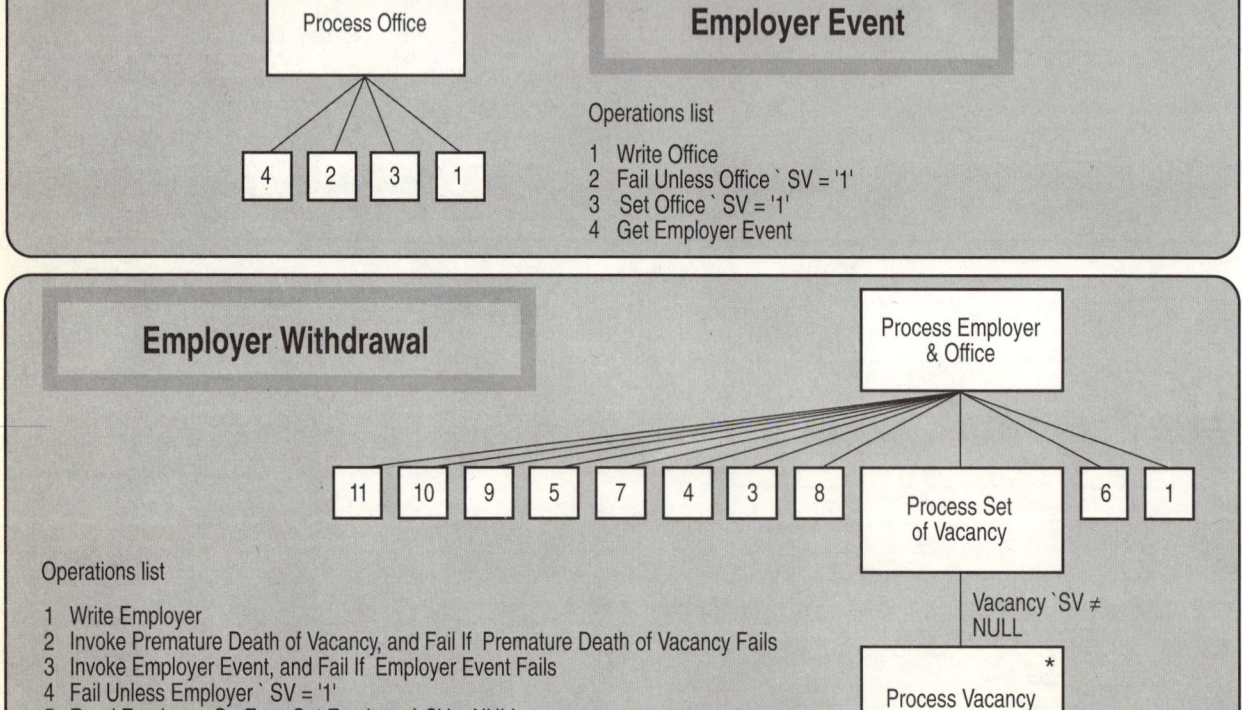

```
                         Process Employer
                             & Office
                                 |
      ┌────┬───┬───┬───┬───┬───┬─┴─────────────┬───┬───┐
     11   10   9   5   7   4   3   8    Process Set    6   1
                                         of Vacancy
                                              |
                                      Vacancy `SV ≠
                                         NULL
                                              |
                                        Process Vacancy *
                                              |
                                           ┌──┴──┐
                                           2     8
```

Operations list

1. Write Employer
2. Invoke Premature Death of Vacancy, and Fail If Premature Death of Vacancy Fails
3. Invoke Employer Event, and Fail If Employer Event Fails
4. Fail Unless Employer ` SV = '1'
5. Read Employer, On Error Set Employer ` SV = NULL
6. Set Employer ` SV = '2'
7. Read Office, On Error Set Office ` SV = NULL
8. Read Vacancy, On Error Set Vacancy ` SV = NULL
9. Employer ` Office # := Employer Withdrawal ` Office #
10. Employer ` Employer # := Employer Withdrawal ` Employer #
11. Get Employer Withdrawal

Update process models if super-events are used

Employer's Loss of Vacancy

[Process Employer with children: 4, 2, 3, 1]

Operations list
1. Write Employer
2. Fail Unless Employer ` SV = '1'
3. Set Employer ` SV = '1'
4. Get Employer's Loss of Vacancy

Failure to get Job

[Process Interview with children: 4, 2, 3, 1]

Operations list
1. Write Interview
2. Fail Unless Interview ` SV = '1'
3. Set Interview ` SV = '2'
4. Get Failure to get Job

Identification of New Skill

[Process Skill Area {Vacancies} & Skill Area {Skills} with children: 12, 11, 10, 8, 5, 7, 4, 3, 6, 9, 1, 2]

Operations list
1. Write Skill Area {Skills}
2. Write Skill Area {Vacancies}
3. Skill Area {Vacancies} ` Skill Description := Identitfication of New Skill ` Skill Description
4. Fail Unless Skill Area {Skills} ` SV = NULL
5. Read Skill Area {Skills}, On Error Set Skill Area {Skills} ` SV = NULL & Create Skill Area {Skills}
6. Set Skill Area {Skills} ` SV = '1'
7. Fail Unless Skill Area {Vacancies} ` SV = NULL
8. Read Skill Area {Vacancies}, On Error Set Skill Area {Vacancies} ` SV = NULL & Create Skill Area {Vacancies}
9. Set Skill are {Vacancies} ` SV = '1'
10. Skill Area {Skills} ` Skill code := Identification of New Skill ` Skill Code
11. Skill Area {Vacancies} ` Skill code := Identification of New Skill ` Skill Code
12. Get Identification of New Skill

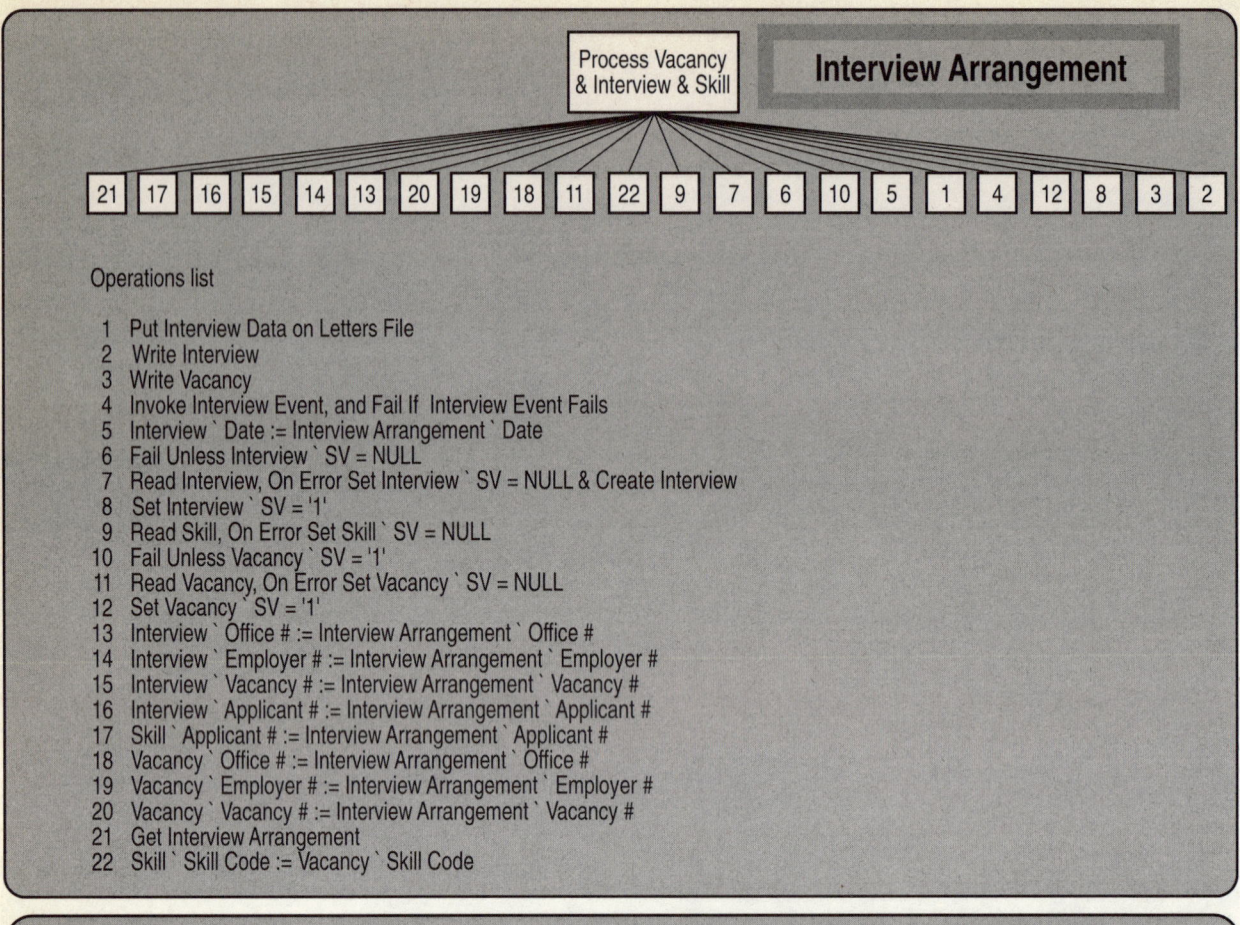

Interview Arrangement

Process Vacancy & Interview & Skill

21 17 16 15 14 13 20 19 18 11 22 9 7 6 10 5 1 4 12 8 3 2

Operations list

1. Put Interview Data on Letters File
2. Write Interview
3. Write Vacancy
4. Invoke Interview Event, and Fail If Interview Event Fails
5. Interview ` Date := Interview Arrangement ` Date
6. Fail Unless Interview ` SV = NULL
7. Read Interview, On Error Set Interview ` SV = NULL & Create Interview
8. Set Interview ` SV = '1'
9. Read Skill, On Error Set Skill ` SV = NULL
10. Fail Unless Vacancy ` SV = '1'
11. Read Vacancy, On Error Set Vacancy ` SV = NULL
12. Set Vacancy ` SV = '1'
13. Interview ` Office # := Interview Arrangement ` Office #
14. Interview ` Employer # := Interview Arrangement ` Employer #
15. Interview ` Vacancy # := Interview Arrangement ` Vacancy #
16. Interview ` Applicant # := Interview Arrangement ` Applicant #
17. Skill ` Applicant # := Interview Arrangement ` Applicant #
18. Vacancy ` Office # := Interview Arrangement ` Office #
19. Vacancy ` Employer # := Interview Arrangement ` Employer #
20. Vacancy ` Vacancy # := Interview Arrangement ` Vacancy #
21. Get Interview Arrangement
22. Skill ` Skill Code := Vacancy ` Skill Code

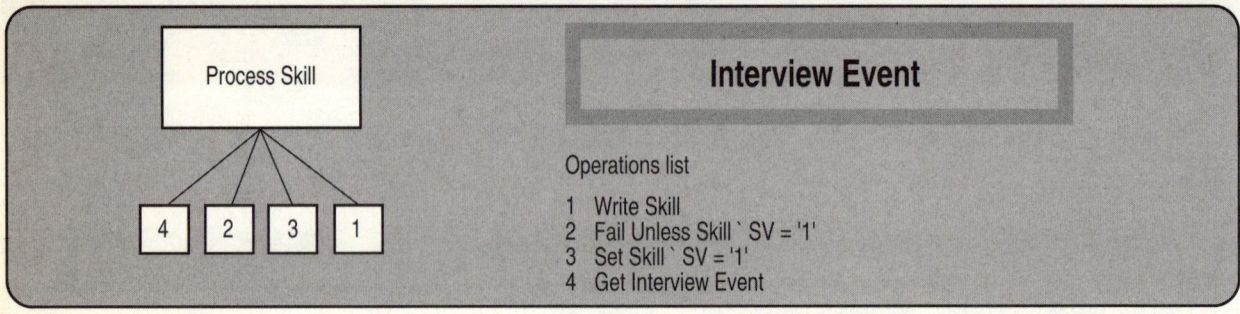

Interview Event

Process Skill

4 2 3 1

Operations list

1. Write Skill
2. Fail Unless Skill ` SV = '1'
3. Set Skill ` SV = '1'
4. Get Interview Event

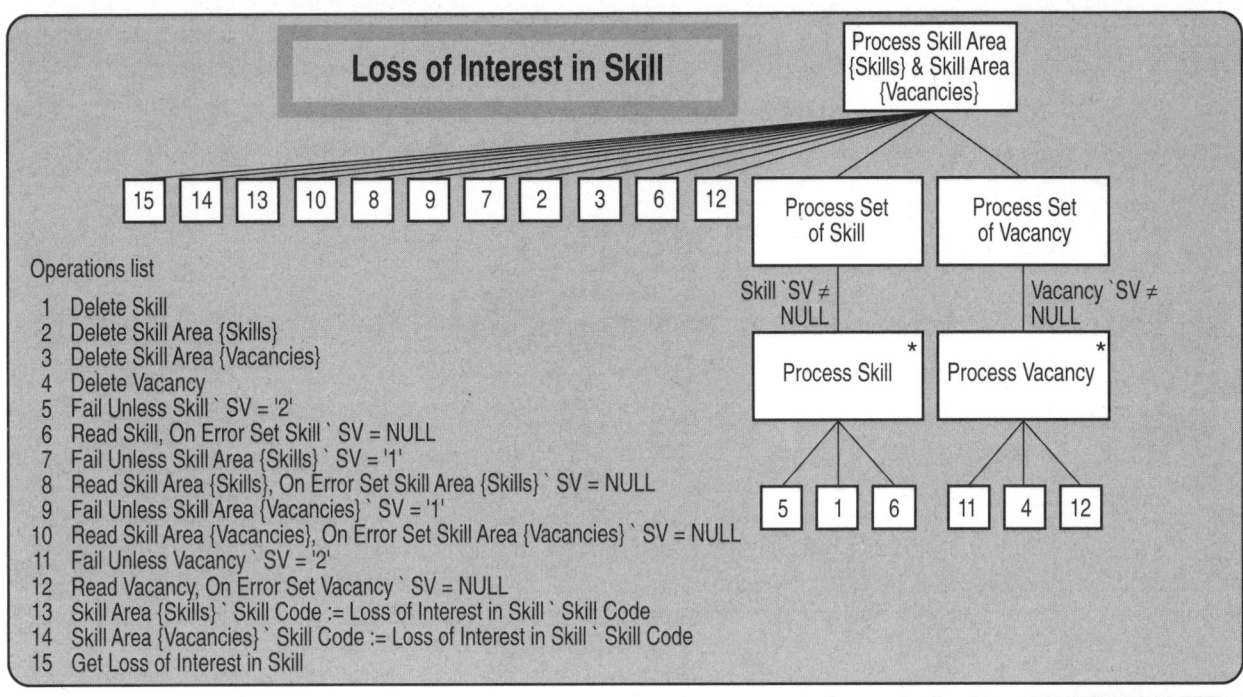

Loss of Interest in Skill

Operations list

1. Delete Skill
2. Delete Skill Area {Skills}
3. Delete Skill Area {Vacancies}
4. Delete Vacancy
5. Fail Unless Skill ` SV = '2'
6. Read Skill, On Error Set Skill ` SV = NULL
7. Fail Unless Skill Area {Skills} ` SV = '1'
8. Read Skill Area {Skills}, On Error Set Skill Area {Skills} ` SV = NULL
9. Fail Unless Skill Area {Vacancies} ` SV = '1'
10. Read Skill Area {Vacancies}, On Error Set Skill Area {Vacancies} ` SV = NULL
11. Fail Unless Vacancy ` SV = '2'
12. Read Vacancy, On Error Set Vacancy ` SV = NULL
13. Skill Area {Skills} ` Skill Code := Loss of Interest in Skill ` Skill Code
14. Skill Area {Vacancies} ` Skill Code := Loss of Interest in Skill ` Skill Code
15. Get Loss of Interest in Skill

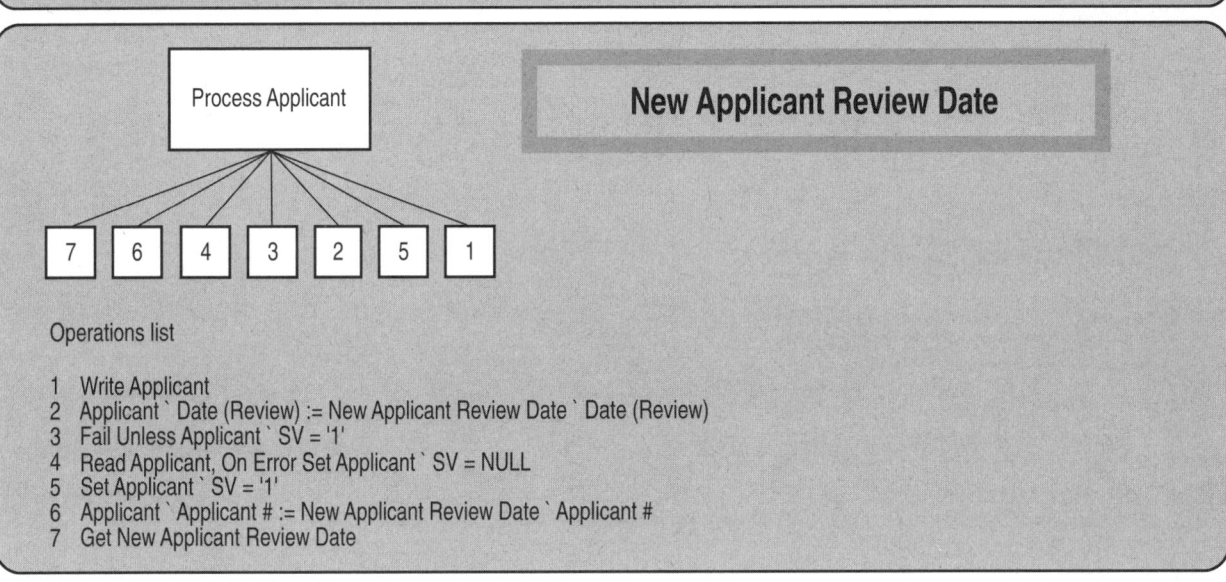

New Applicant Review Date

Operations list

1. Write Applicant
2. Applicant ` Date (Review) := New Applicant Review Date ` Date (Review)
3. Fail Unless Applicant ` SV = '1'
4. Read Applicant, On Error Set Applicant ` SV = NULL
5. Set Applicant ` SV = '1'
6. Applicant ` Applicant # := New Applicant Review Date ` Applicant #
7. Get New Applicant Review Date

New Vacancy Review Date

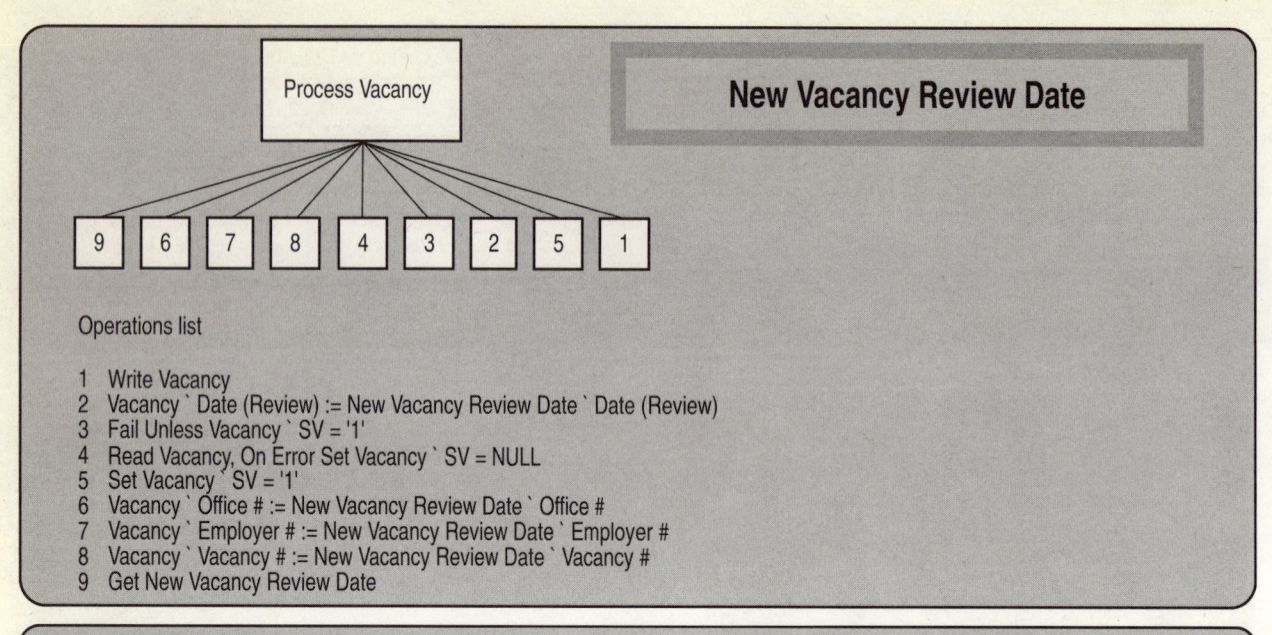

Process Vacancy
— 9, 6, 7, 8, 4, 3, 2, 5, 1

Operations list

1. Write Vacancy
2. Vacancy ` Date (Review) := New Vacancy Review Date ` Date (Review)
3. Fail Unless Vacancy ` SV = '1'
4. Read Vacancy, On Error Set Vacancy ` SV = NULL
5. Set Vacancy ` SV = '1'
6. Vacancy ` Office # := New Vacancy Review Date ` Office #
7. Vacancy ` Employer # := New Vacancy Review Date ` Employer #
8. Vacancy ` Vacancy # := New Vacancy Review Date ` Vacancy #
9. Get New Vacancy Review Date

Notification of Vacancy

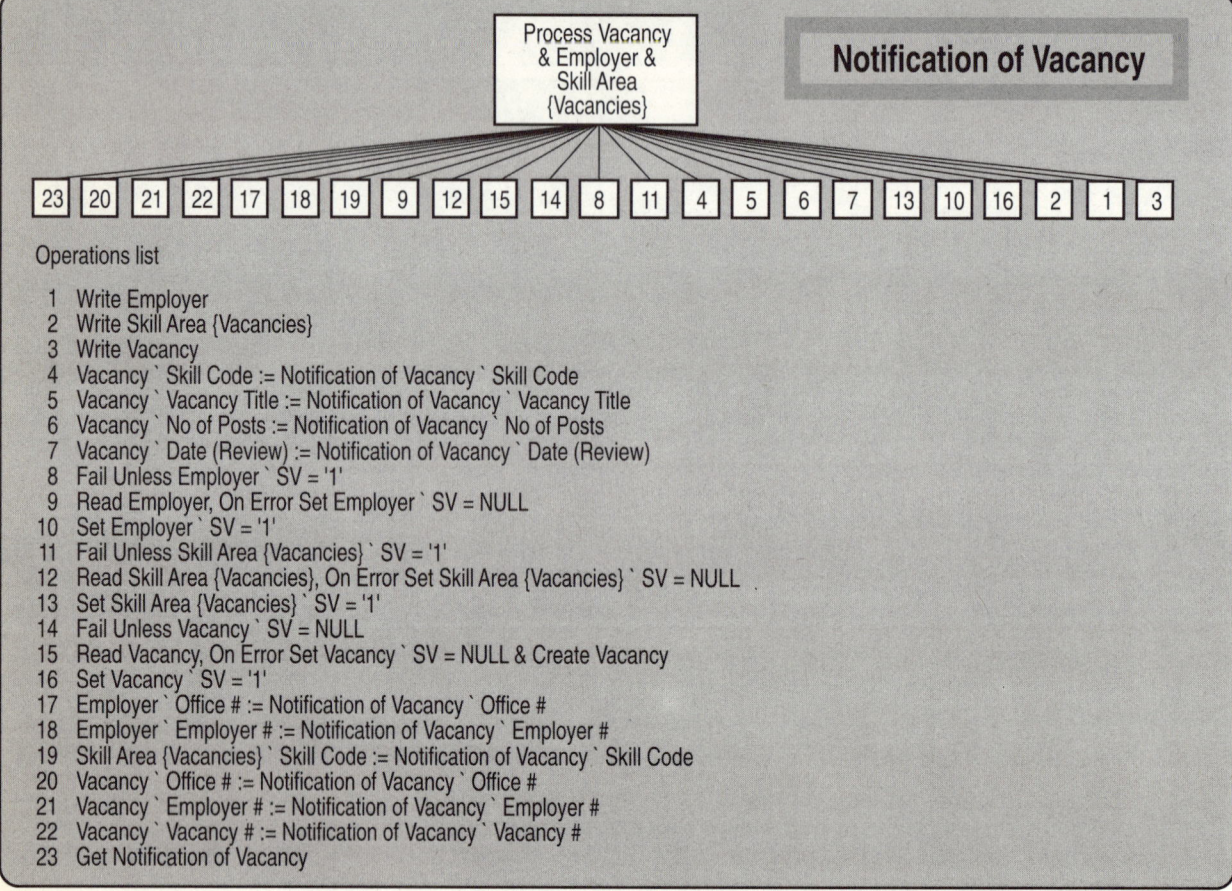

Process Vacancy & Employer & Skill Area {Vacancies}
— 23, 20, 21, 22, 17, 18, 19, 9, 12, 15, 14, 8, 11, 4, 5, 6, 7, 13, 10, 16, 2, 1, 3

Operations list

1. Write Employer
2. Write Skill Area {Vacancies}
3. Write Vacancy
4. Vacancy ` Skill Code := Notification of Vacancy ` Skill Code
5. Vacancy ` Vacancy Title := Notification of Vacancy ` Vacancy Title
6. Vacancy ` No of Posts := Notification of Vacancy ` No of Posts
7. Vacancy ` Date (Review) := Notification of Vacancy ` Date (Review)
8. Fail Unless Employer ` SV = '1'
9. Read Employer, On Error Set Employer ` SV = NULL
10. Set Employer ` SV = '1'
11. Fail Unless Skill Area {Vacancies} ` SV = '1'
12. Read Skill Area {Vacancies}, On Error Set Skill Area {Vacancies} ` SV = NULL
13. Set Skill Area {Vacancies} ` SV = '1'
14. Fail Unless Vacancy ` SV = NULL
15. Read Vacancy, On Error Set Vacancy ` SV = NULL & Create Vacancy
16. Set Vacancy ` SV = '1'
17. Employer ` Office # := Notification of Vacancy ` Office #
18. Employer ` Employer # := Notification of Vacancy ` Employer #
19. Skill Area {Vacancies} ` Skill Code := Notification of Vacancy ` Skill Code
20. Vacancy ` Office # := Notification of Vacancy ` Office #
21. Vacancy ` Employer # := Notification of Vacancy ` Employer #
22. Vacancy ` Vacancy # := Notification of Vacancy ` Vacancy #
23. Get Notification of Vacancy

Office Closure

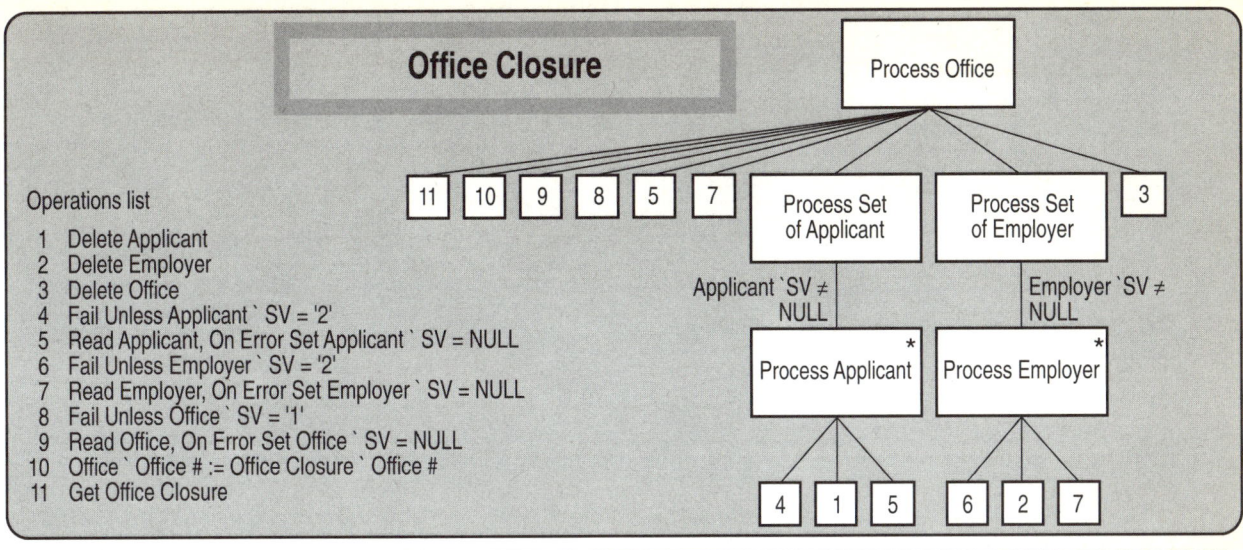

Operations list

1. Delete Applicant
2. Delete Employer
3. Delete Office
4. Fail Unless Applicant ` SV = '2'
5. Read Applicant, On Error Set Applicant ` SV = NULL
6. Fail Unless Employer ` SV = '2'
7. Read Employer, On Error Set Employer ` SV = NULL
8. Fail Unless Office ` SV = '1'
9. Read Office, On Error Set Office ` SV = NULL
10. Office ` Office # := Office Closure ` Office #
11. Get Office Closure

Office Opening

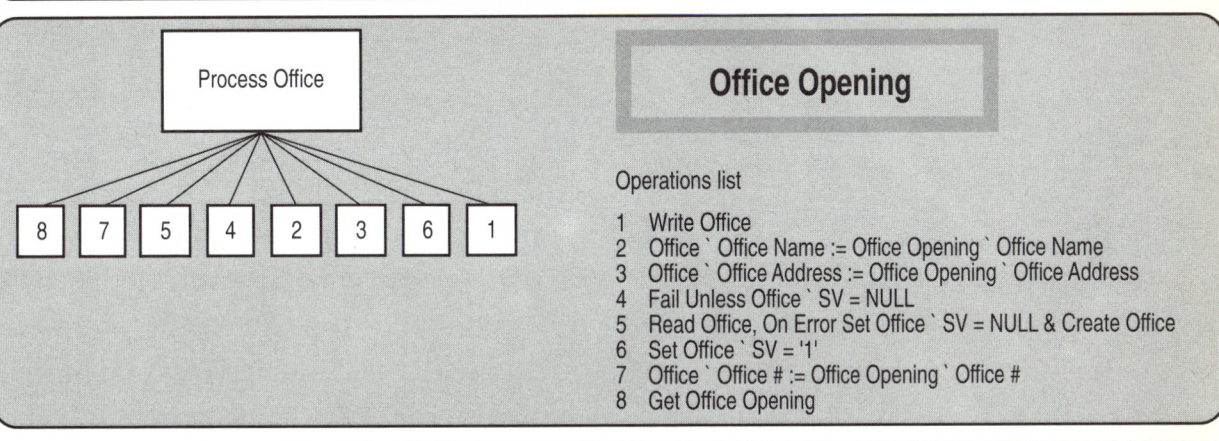

Operations list

1. Write Office
2. Office ` Office Name := Office Opening ` Office Name
3. Office ` Office Address := Office Opening ` Office Address
4. Fail Unless Office ` SV = NULL
5. Read Office, On Error Set Office ` SV = NULL & Create Office
6. Set Office ` SV = '1'
7. Office ` Office # := Office Opening ` Office #
8. Get Office Opening

Post Acceptance

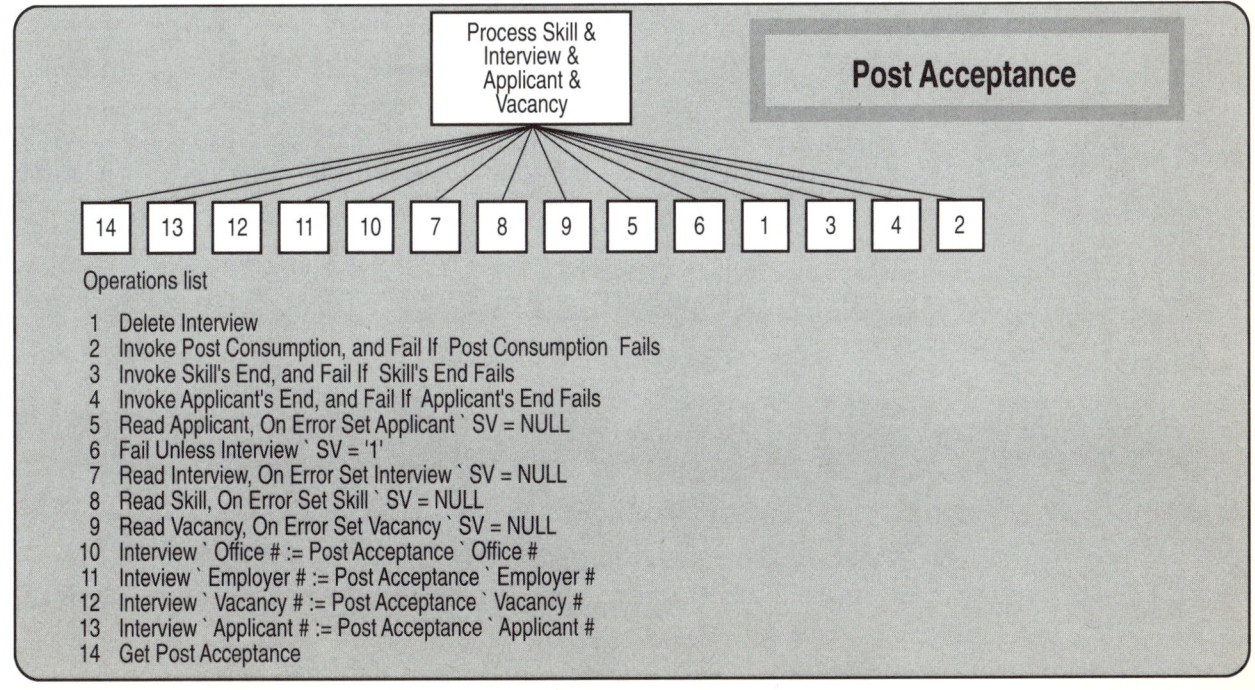

Operations list

1. Delete Interview
2. Invoke Post Consumption, and Fail If Post Consumption Fails
3. Invoke Skill's End, and Fail If Skill's End Fails
4. Invoke Applicant's End, and Fail If Applicant's End Fails
5. Read Applicant, On Error Set Applicant ` SV = NULL
6. Fail Unless Interview ` SV = '1'
7. Read Interview, On Error Set Interview ` SV = NULL
8. Read Skill, On Error Set Skill ` SV = NULL
9. Read Vacancy, On Error Set Vacancy ` SV = NULL
10. Interview ` Office # := Post Acceptance ` Office #
11. Inteview ` Employer # := Post Acceptance ` Employer #
12. Interview ` Vacancy # := Post Acceptance ` Vacancy #
13. Interview ` Applicant # := Post Acceptance ` Applicant #
14. Get Post Acceptance

Post Addition

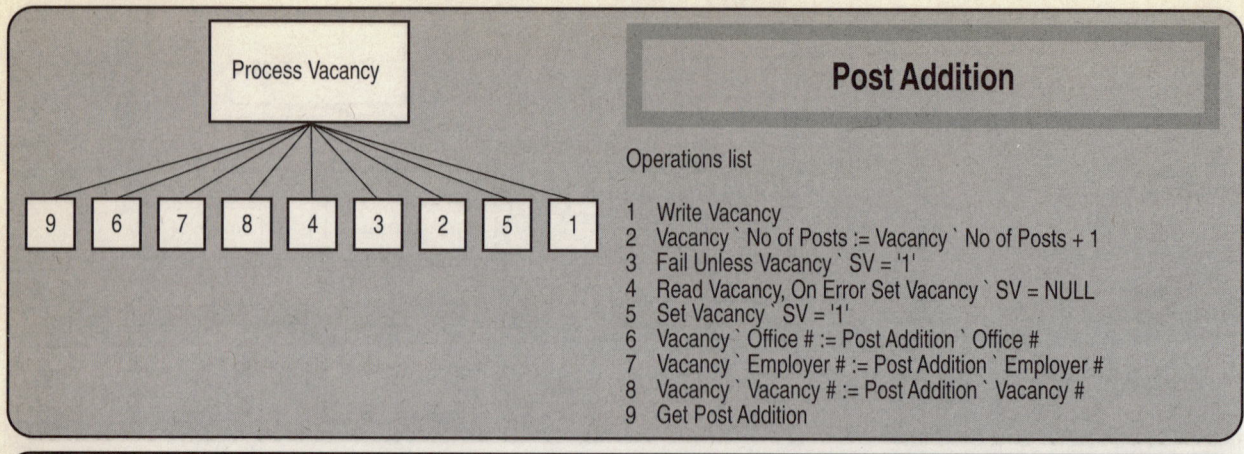

Process Vacancy
├── 9
├── 6
├── 7
├── 8
├── 4
├── 3
├── 2
├── 5
└── 1

Operations list

1. Write Vacancy
2. Vacancy ` No of Posts := Vacancy ` No of Posts + 1
3. Fail Unless Vacancy ` SV = '1'
4. Read Vacancy, On Error Set Vacancy ` SV = NULL
5. Set Vacancy ` SV = '1'
6. Vacancy ` Office # := Post Addition ` Office #
7. Vacancy ` Employer # := Post Addition ` Employer #
8. Vacancy ` Vacancy # := Post Addition ` Vacancy #
9. Get Post Addition

Post Consumption

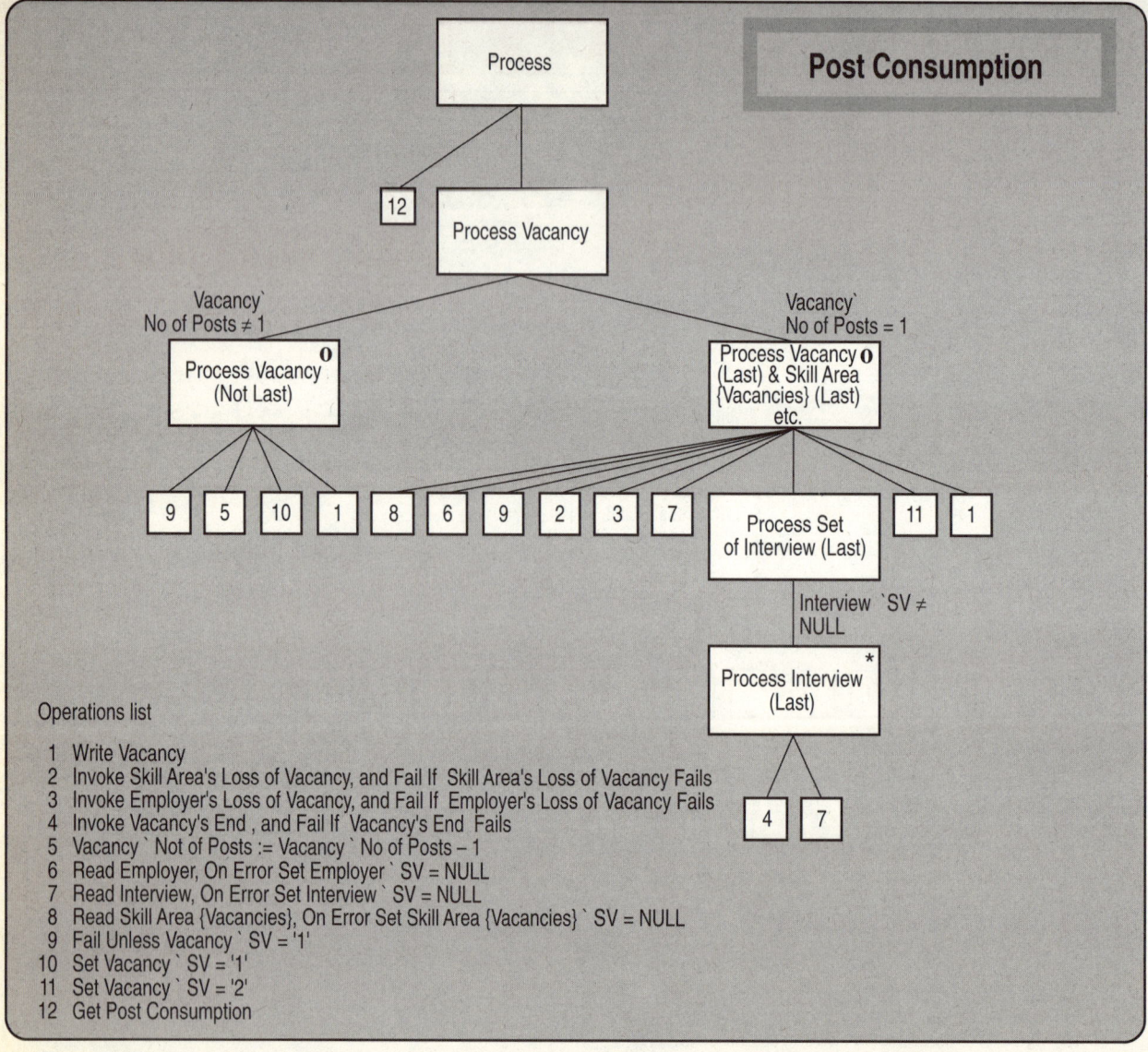

Process
└── [12] Process Vacancy
    ├── Vacancy ` No of Posts ≠ 1: Process Vacancy (Not Last) ○
    │   └── 9, 5, 10, 1, 8, 6, 9, 2, 3, 7
    └── Vacancy ` No of Posts = 1: Process Vacancy (Last) & Skill Area {Vacancies} (Last) etc. ○
        ├── 11, 1
        └── Process Set of Interview (Last)
            └── Interview ` SV ≠ NULL: Process Interview (Last) *
                └── 4, 7

Operations list

1. Write Vacancy
2. Invoke Skill Area's Loss of Vacancy, and Fail If Skill Area's Loss of Vacancy Fails
3. Invoke Employer's Loss of Vacancy, and Fail If Employer's Loss of Vacancy Fails
4. Invoke Vacancy's End, and Fail If Vacancy's End Fails
5. Vacancy ` Not of Posts := Vacancy ` No of Posts − 1
6. Read Employer, On Error Set Employer ` SV = NULL
7. Read Interview, On Error Set Interview ` SV = NULL
8. Read Skill Area {Vacancies}, On Error Set Skill Area {Vacancies} ` SV = NULL
9. Fail Unless Vacancy ` SV = '1'
10. Set Vacancy ` SV = '1'
11. Set Vacancy ` SV = '2'
12. Get Post Consumption

Update process models if super-events are used 299

Post Withdrawal

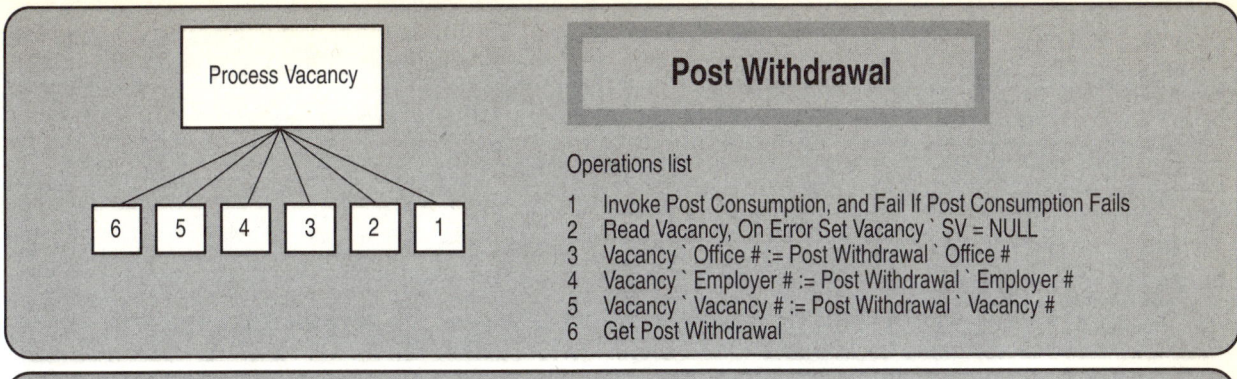

Process Vacancy
6 5 4 3 2 1

Operations list

1. Invoke Post Consumption, and Fail If Post Consumption Fails
2. Read Vacancy, On Error Set Vacancy ` SV = NULL
3. Vacancy ` Office # := Post Withdrawal ` Office #
4. Vacancy ` Employer # := Post Withdrawal ` Employer #
5. Vacancy ` Vacancy # := Post Withdrawal ` Vacancy #
6. Get Post Withdrawal

Premature Death of Vacancy

Process Vacancy & Skill Area {Vacancies}
8 5 6 2 4 7 1

Process Set of Interview

Interview ` SV ≠ NULL

Process Interview *
3 4

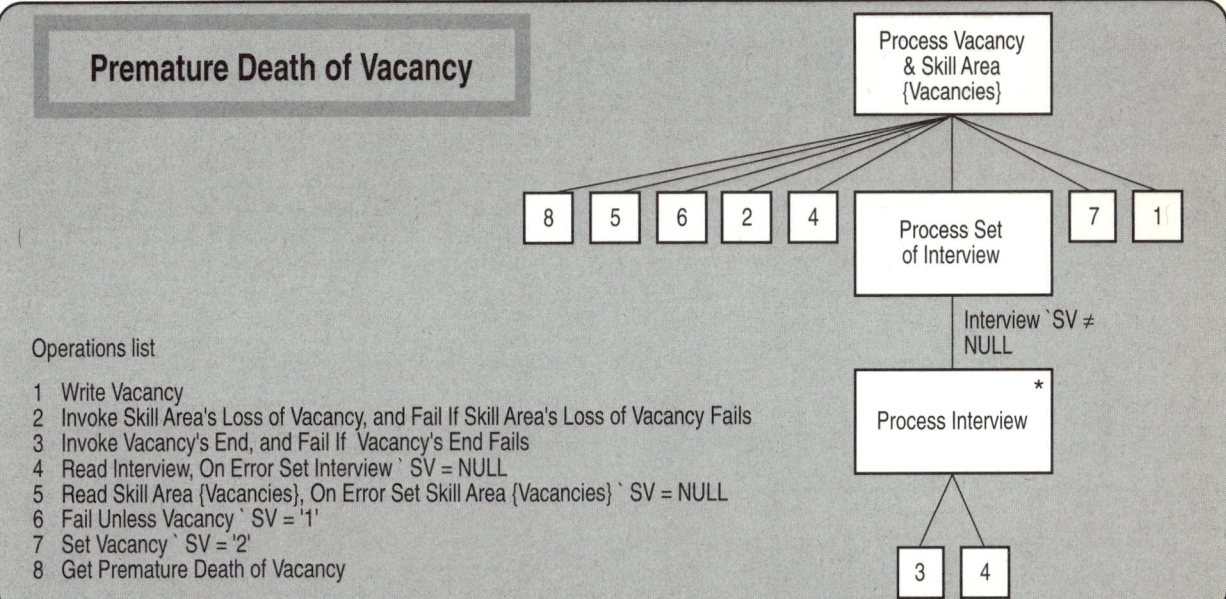

Operations list

1. Write Vacancy
2. Invoke Skill Area's Loss of Vacancy, and Fail If Skill Area's Loss of Vacancy Fails
3. Invoke Vacancy's End, and Fail If Vacancy's End Fails
4. Read Interview, On Error Set Interview ` SV = NULL
5. Read Skill Area {Vacancies}, On Error Set Skill Area {Vacancies} ` SV = NULL
6. Fail Unless Vacancy ` SV = '1'
7. Set Vacancy ` SV = '2'
8. Get Premature Death of Vacancy

Refusal of Offer

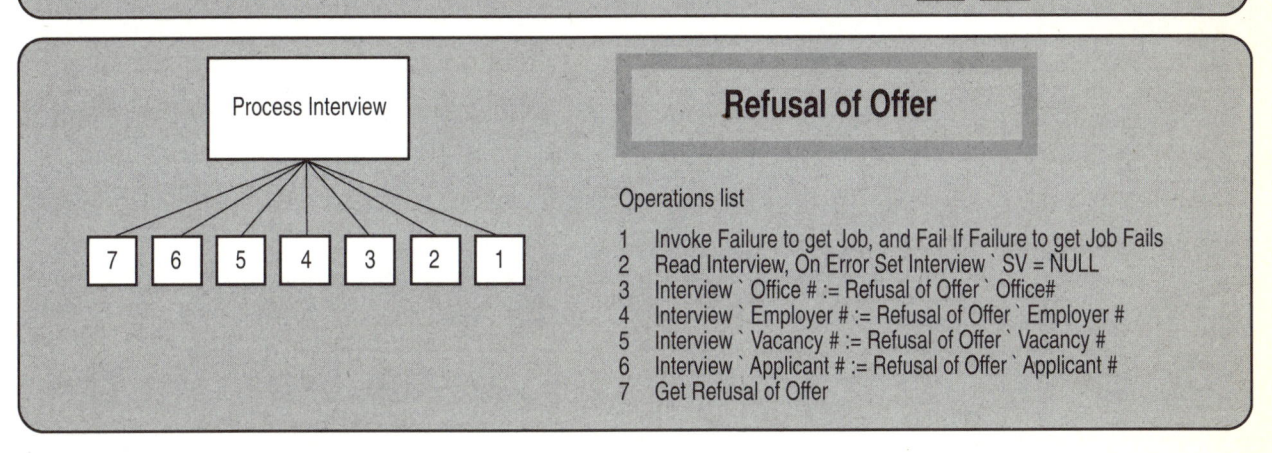

Process Interview
7 6 5 4 3 2 1

Operations list

1. Invoke Failure to get Job, and Fail If Failure to get Job Fails
2. Read Interview, On Error Set Interview ` SV = NULL
3. Interview ` Office # := Refusal of Offer ` Office#
4. Interview ` Employer # := Refusal of Offer ` Employer #
5. Interview ` Vacancy # := Refusal of Offer ` Vacancy #
6. Interview ` Applicant # := Refusal of Offer ` Applicant #
7. Get Refusal of Offer

Registration of Employer

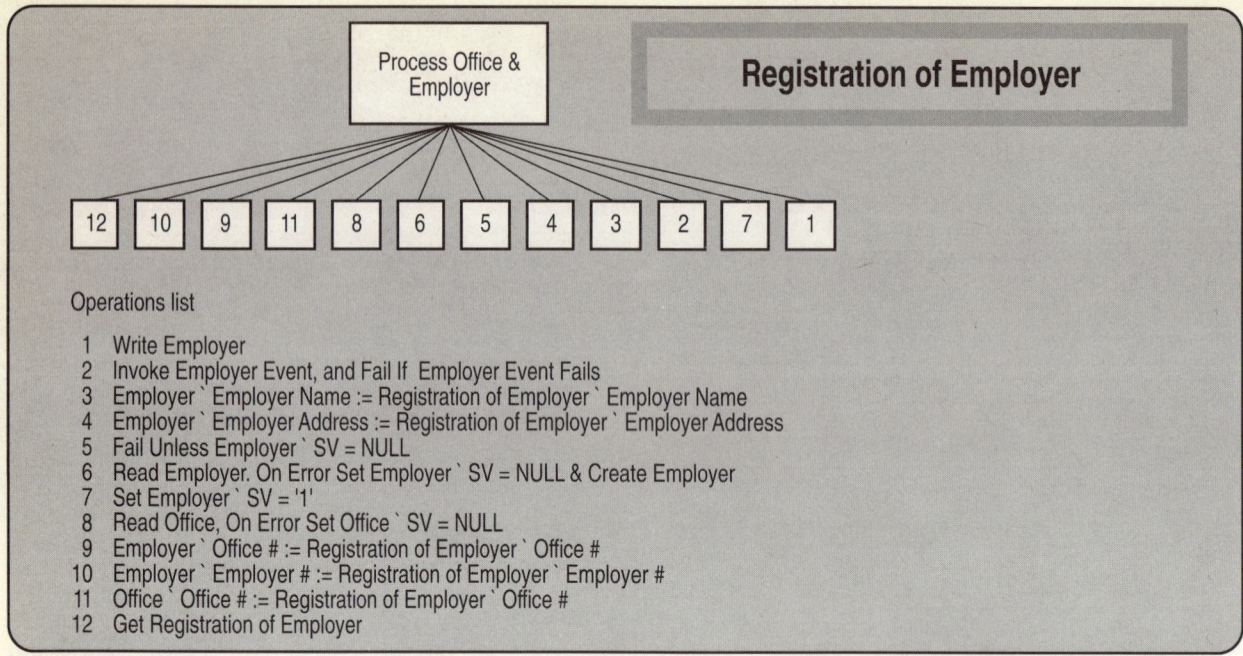

[Diagram: Process Office & Employer → 12, 10, 9, 11, 8, 6, 5, 4, 3, 2, 7, 1]

Operations list

1. Write Employer
2. Invoke Employer Event, and Fail If Employer Event Fails
3. Employer ` Employer Name := Registration of Employer ` Employer Name
4. Employer ` Employer Address := Registration of Employer ` Employer Address
5. Fail Unless Employer ` SV = NULL
6. Read Employer. On Error Set Employer ` SV = NULL & Create Employer
7. Set Employer ` SV = '1'
8. Read Office, On Error Set Office ` SV = NULL
9. Employer ` Office # := Registration of Employer ` Office #
10. Employer ` Employer # := Registration of Employer ` Employer #
11. Office ` Office # := Registration of Employer ` Office #
12. Get Registration of Employer

Rejection of Applicant

[Diagram: Process Interview → 7, 6, 5, 4, 3, 2, 1]

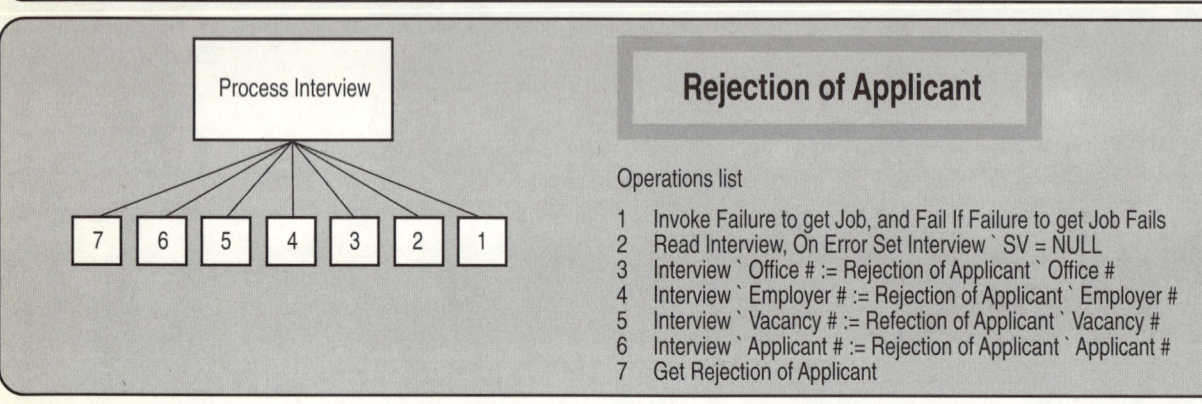

Operations list

1. Invoke Failure to get Job, and Fail If Failure to get Job Fails
2. Read Interview, On Error Set Interview ` SV = NULL
3. Interview ` Office # := Rejection of Applicant ` Office #
4. Interview ` Employer # := Rejection of Applicant ` Employer #
5. Interview ` Vacancy # := Refection of Applicant ` Vacancy #
6. Interview ` Applicant # := Rejection of Applicant ` Applicant #
7. Get Rejection of Applicant

Rescheduling

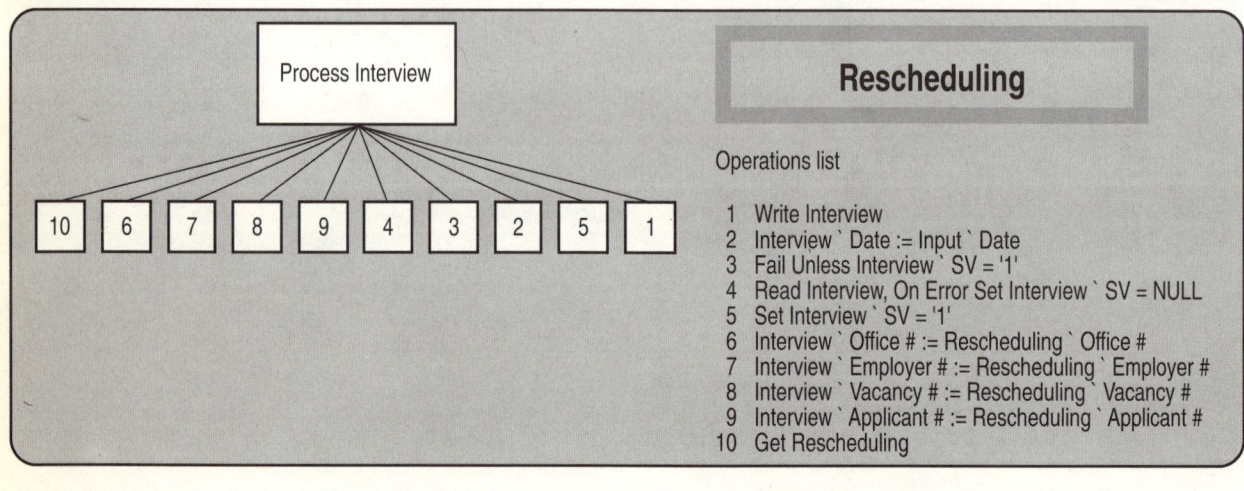

[Diagram: Process Interview → 10, 6, 7, 8, 9, 4, 3, 2, 5, 1]

Operations list

1. Write Interview
2. Interview ` Date := Input ` Date
3. Fail Unless Interview ` SV = '1'
4. Read Interview, On Error Set Interview ` SV = NULL
5. Set Interview ` SV = '1'
6. Interview ` Office # := Rescheduling ` Office #
7. Interview ` Employer # := Rescheduling ` Employer #
8. Interview ` Vacancy # := Rescheduling ` Vacancy #
9. Interview ` Applicant # := Rescheduling ` Applicant #
10. Get Rescheduling

Update process models if super-events are used

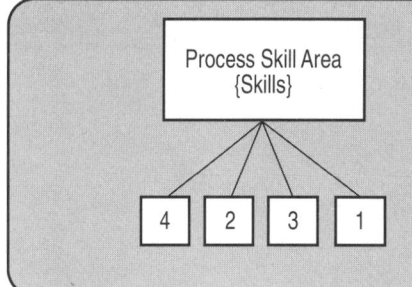

Skill Area's Gain or Loss of Skill

Operations list

1. Write Skill Area {Skills}
2. Fail Unless Skill Area {Skills} ` SV = '1'
3. Set Skill Area {Skills} ` SV = '1'
4. Get Skill Area's Gain or Loss of Skill

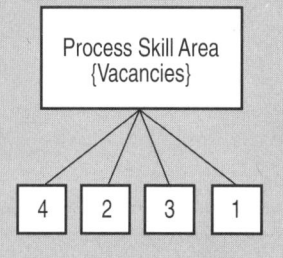

Skill Area's Loss of Vacancy

Operations list

1. Write Skill Area {Vacancies}
2. Fail Unless Skill Area {Vacancies} ` SV = '1'
3. Set Skill Area {Vacancies} ` SV = '1'
4. Get Skill Area's Loss of Vacancy

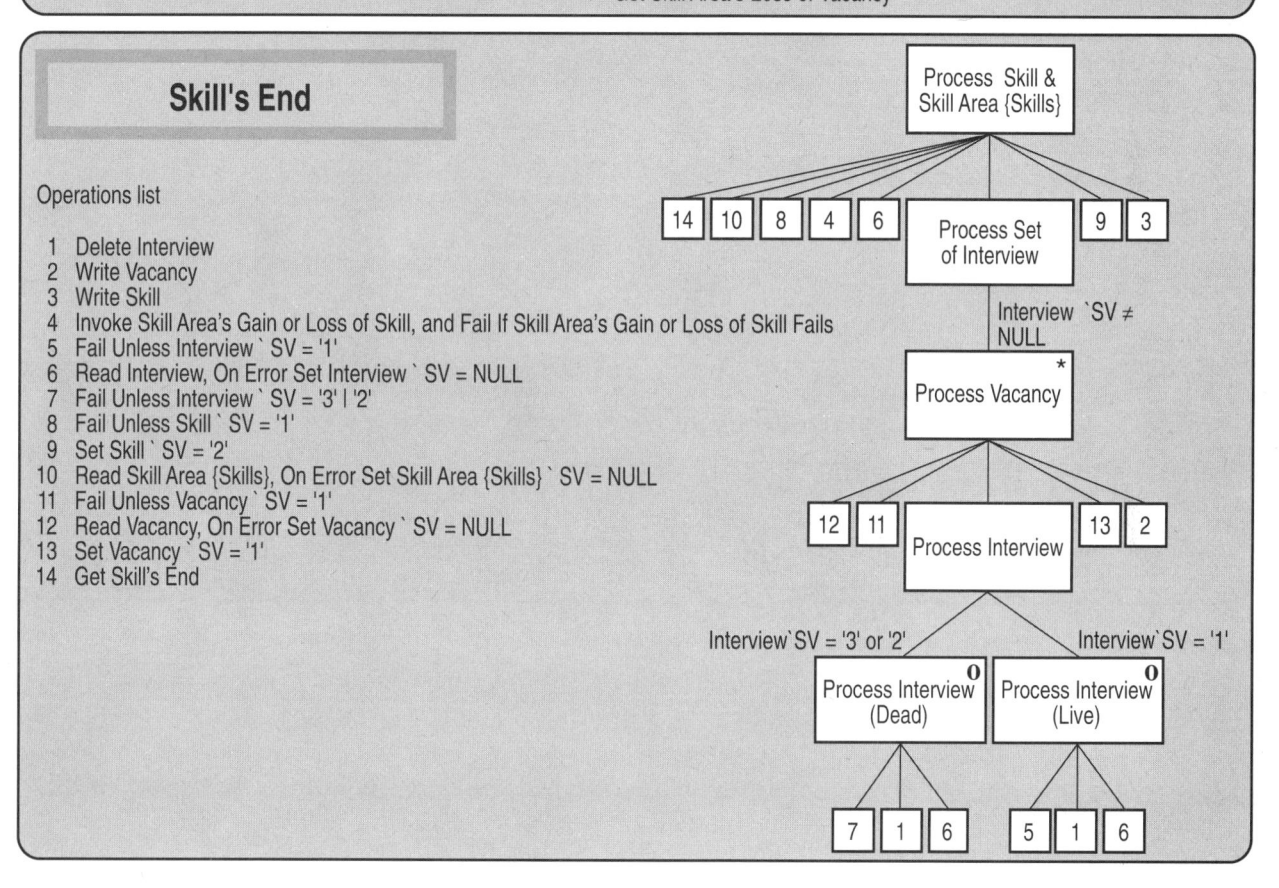

Skill's End

Operations list

1. Delete Interview
2. Write Vacancy
3. Write Skill
4. Invoke Skill Area's Gain or Loss of Skill, and Fail If Skill Area's Gain or Loss of Skill Fails
5. Fail Unless Interview ` SV = '1'
6. Read Interview, On Error Set Interview ` SV = NULL
7. Fail Unless Interview ` SV = '3' | '2'
8. Fail Unless Skill ` SV = '1'
9. Set Skill ` SV = '2'
10. Read Skill Area {Skills}, On Error Set Skill Area {Skills} ` SV = NULL
11. Fail Unless Vacancy ` SV = '1'
12. Read Vacancy, On Error Set Vacancy ` SV = NULL
13. Set Vacancy ` SV = '1'
14. Get Skill's End

Vacancy Withdrawal

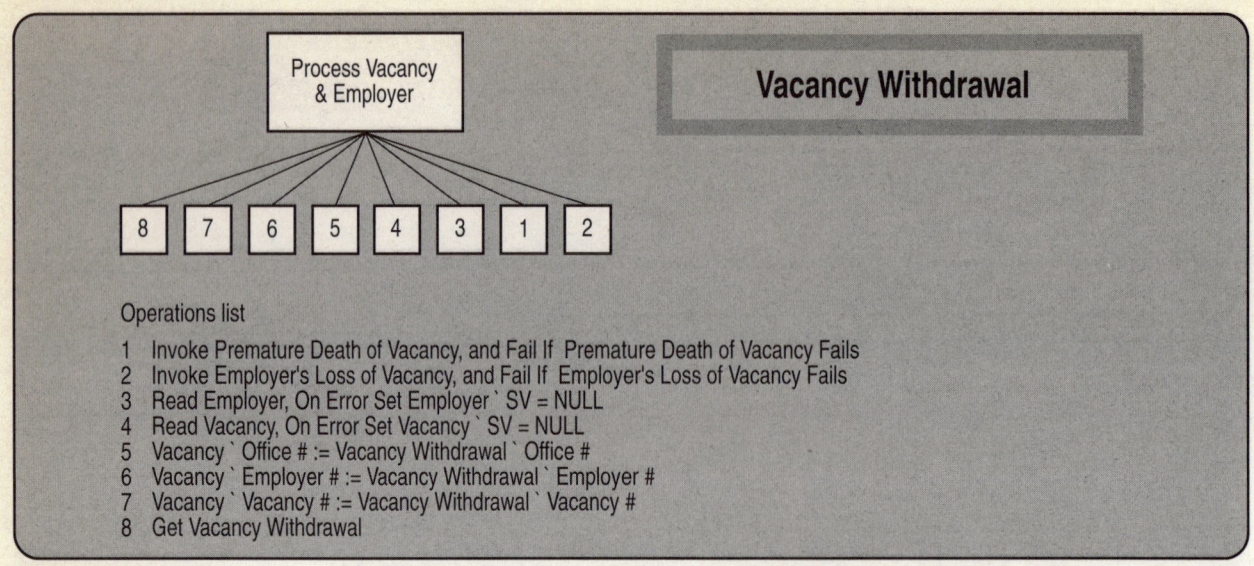

Structure: Process Vacancy & Employer → 8, 7, 6, 5, 4, 3, 1, 2

Operations list

1. Invoke Premature Death of Vacancy, and Fail If Premature Death of Vacancy Fails
2. Invoke Employer's Loss of Vacancy, and Fail If Employer's Loss of Vacancy Fails
3. Read Employer, On Error Set Employer ` SV = NULL
4. Read Vacancy, On Error Set Vacancy ` SV = NULL
5. Vacancy ` Office # := Vacancy Withdrawal ` Office #
6. Vacancy ` Employer # := Vacancy Withdrawal ` Employer #
7. Vacancy ` Vacancy # := Vacancy Withdrawal ` Vacancy #
8. Get Vacancy Withdrawal

Vacancy's End

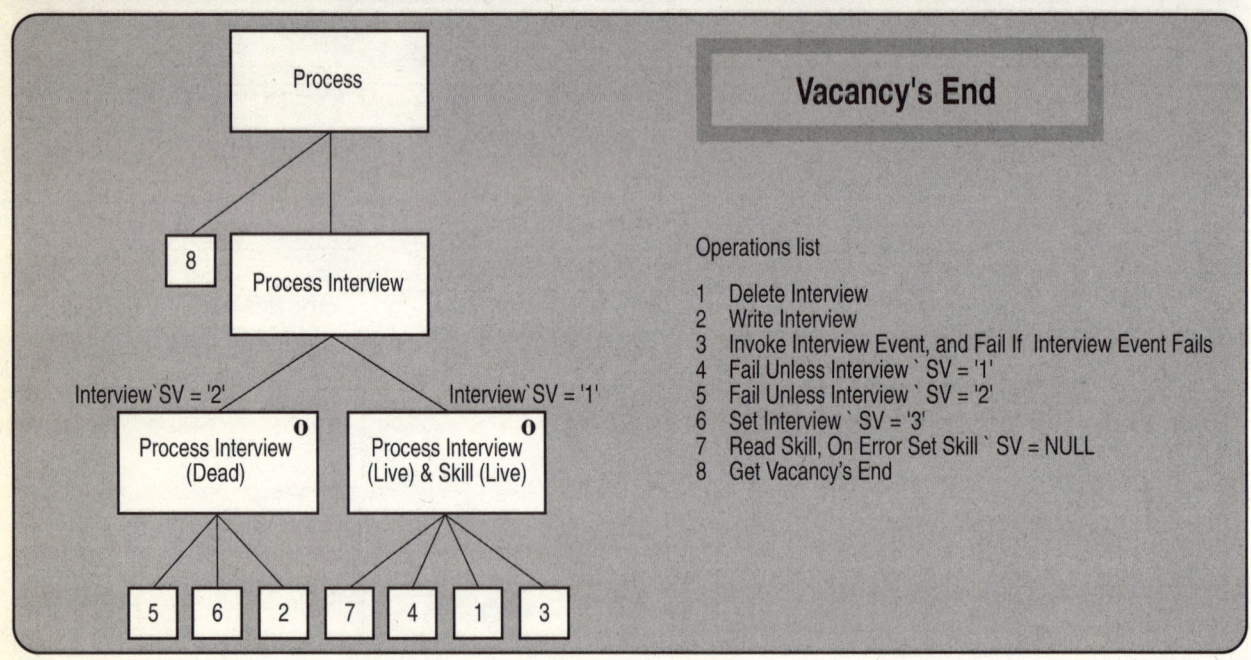

Structure:
- Process
  - 8
  - Process Interview
    - Process Interview (Dead) [o] — Interview`SV = '2' — 5, 6, 2
    - Process Interview (Live) & Skill (Live) [o] — Interview`SV = '1' — 7, 4, 1, 3

Operations list

1. Delete Interview
2. Write Interview
3. Invoke Interview Event, and Fail If Interview Event Fails
4. Fail Unless Interview ` SV = '1'
5. Fail Unless Interview ` SV = '2'
6. Set Interview ` SV = '3'
7. Read Skill, On Error Set Skill ` SV = NULL
8. Get Vacancy's End

Super-event invocation structure

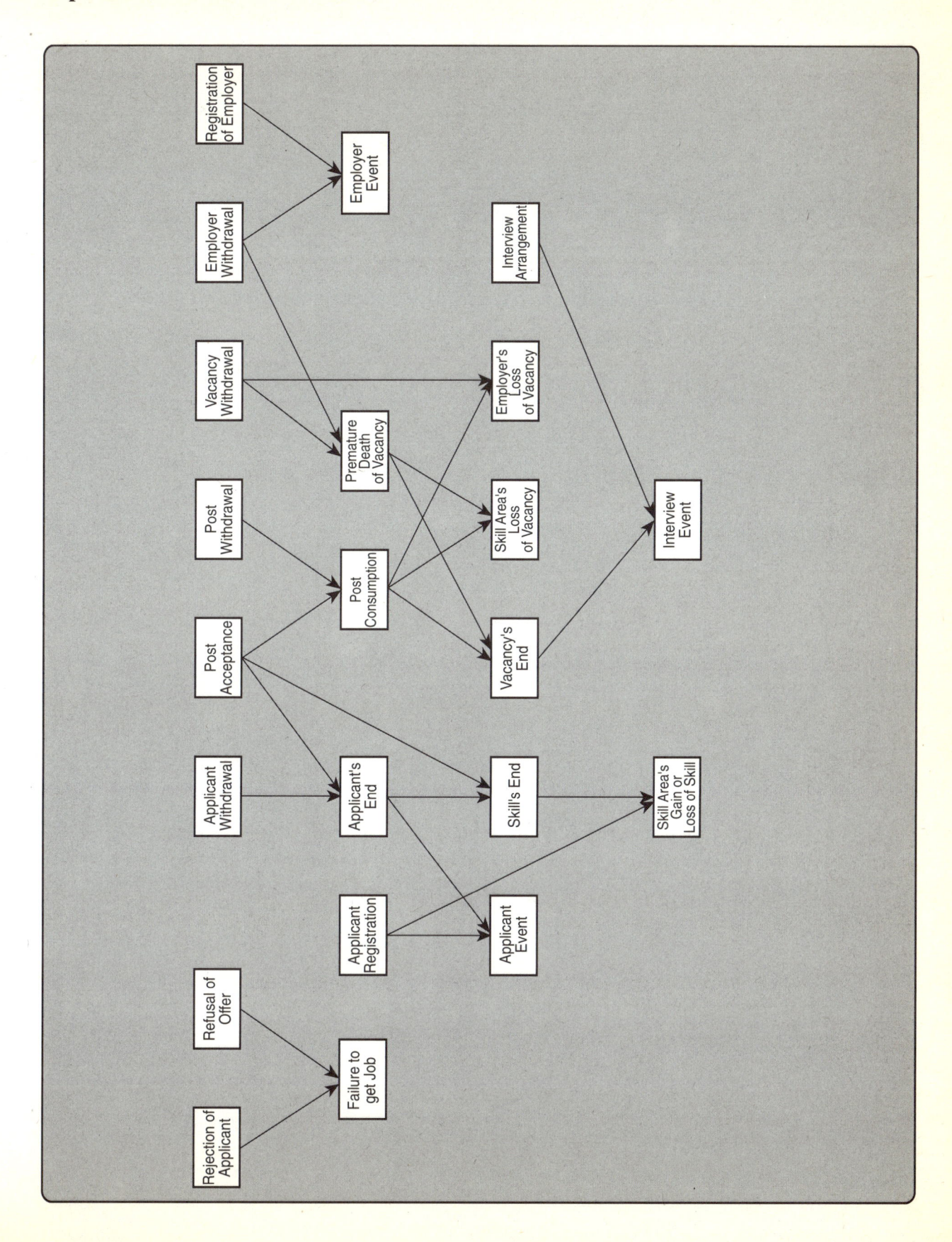

Annex 5

**Hi-Ho GRAPES
Intermediate Diagrams
(Sequence Charts)**

Model of
Required System

Annex 5 Hi-Ho GRAPES Intermediate Diagrams (Sequence Charts)

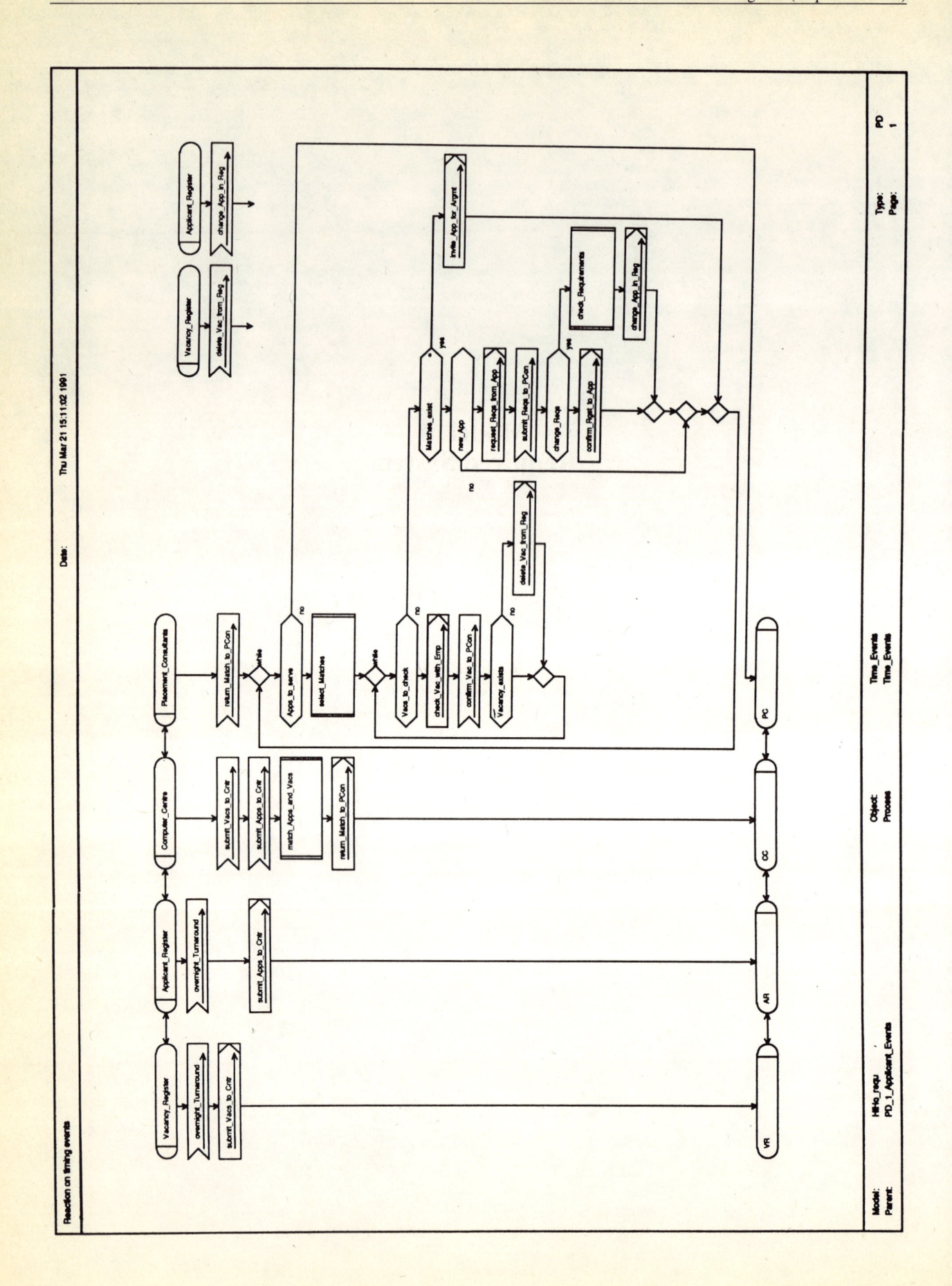

Annex 5 Hi-Ho GRAPES Intermediate Diagrams (Sequence Charts) 309

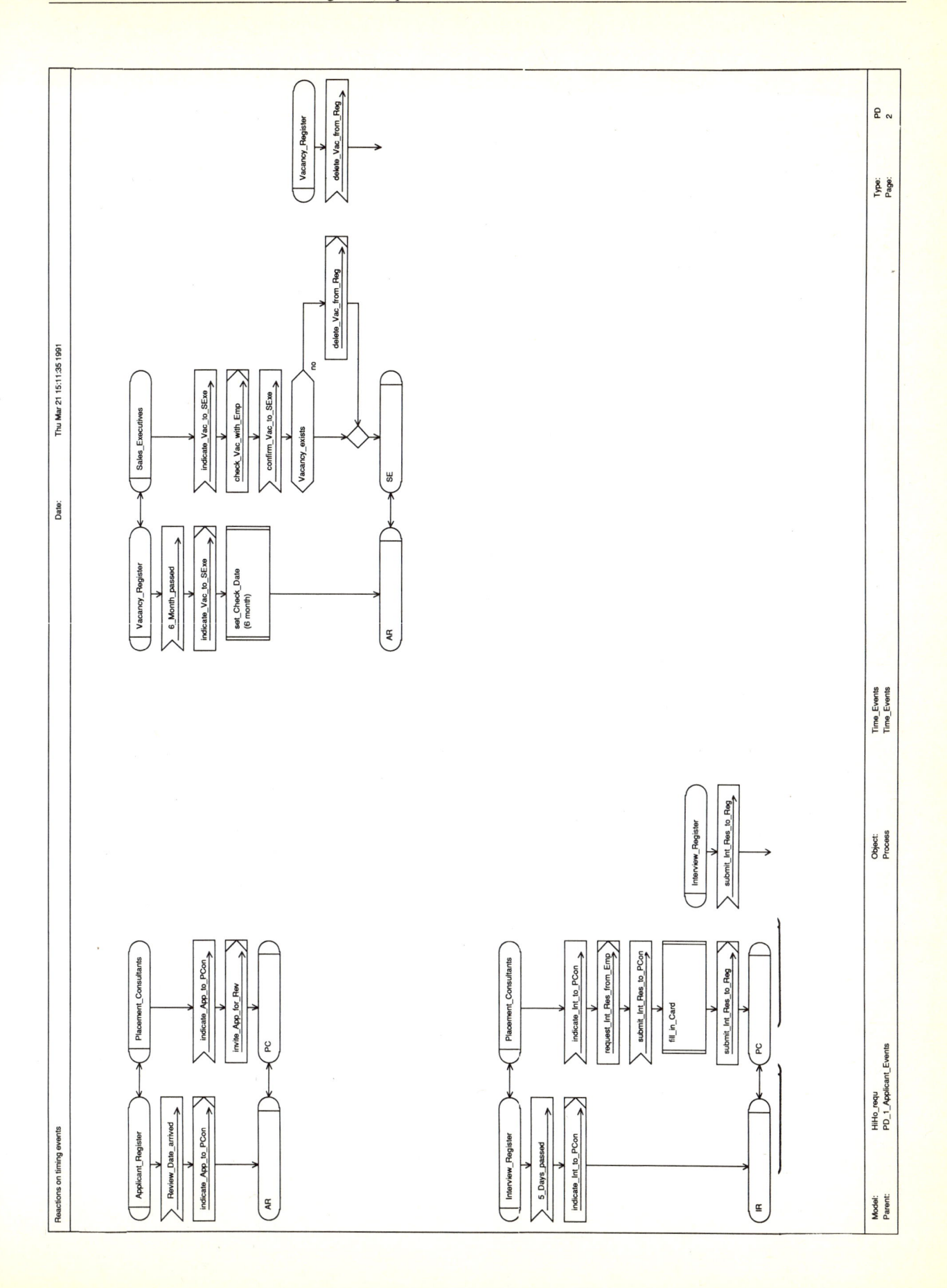

310   Annex 5 Hi-Ho GRAPES Intermediate Diagrams (Sequence Charts)

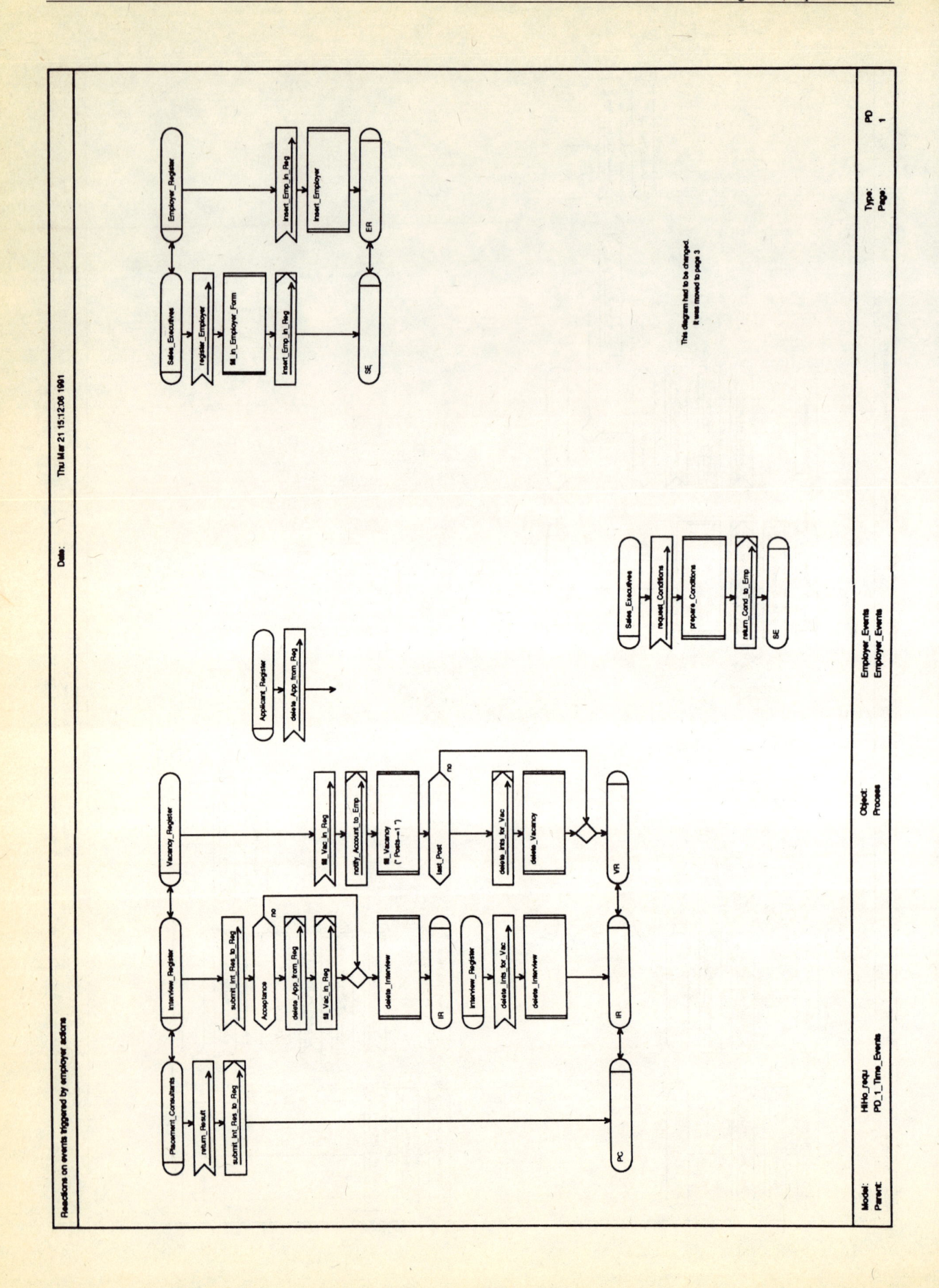

Annex 5 Hi-Ho GRAPES Intermediate Diagrams (Sequence Charts) 311

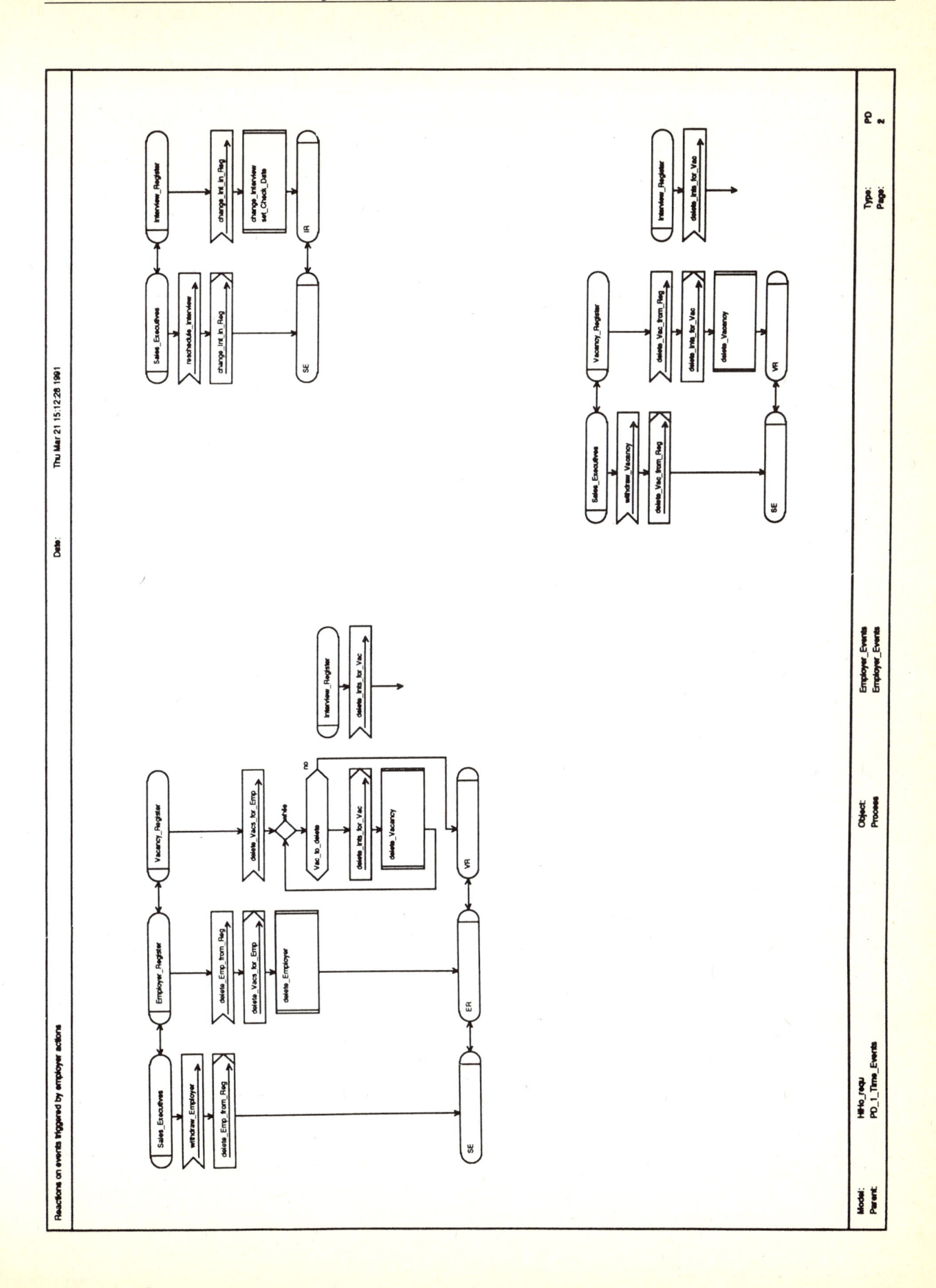

Annex 5 Hi-Ho GRAPES Intermediate Diagrams (Sequence Charts)

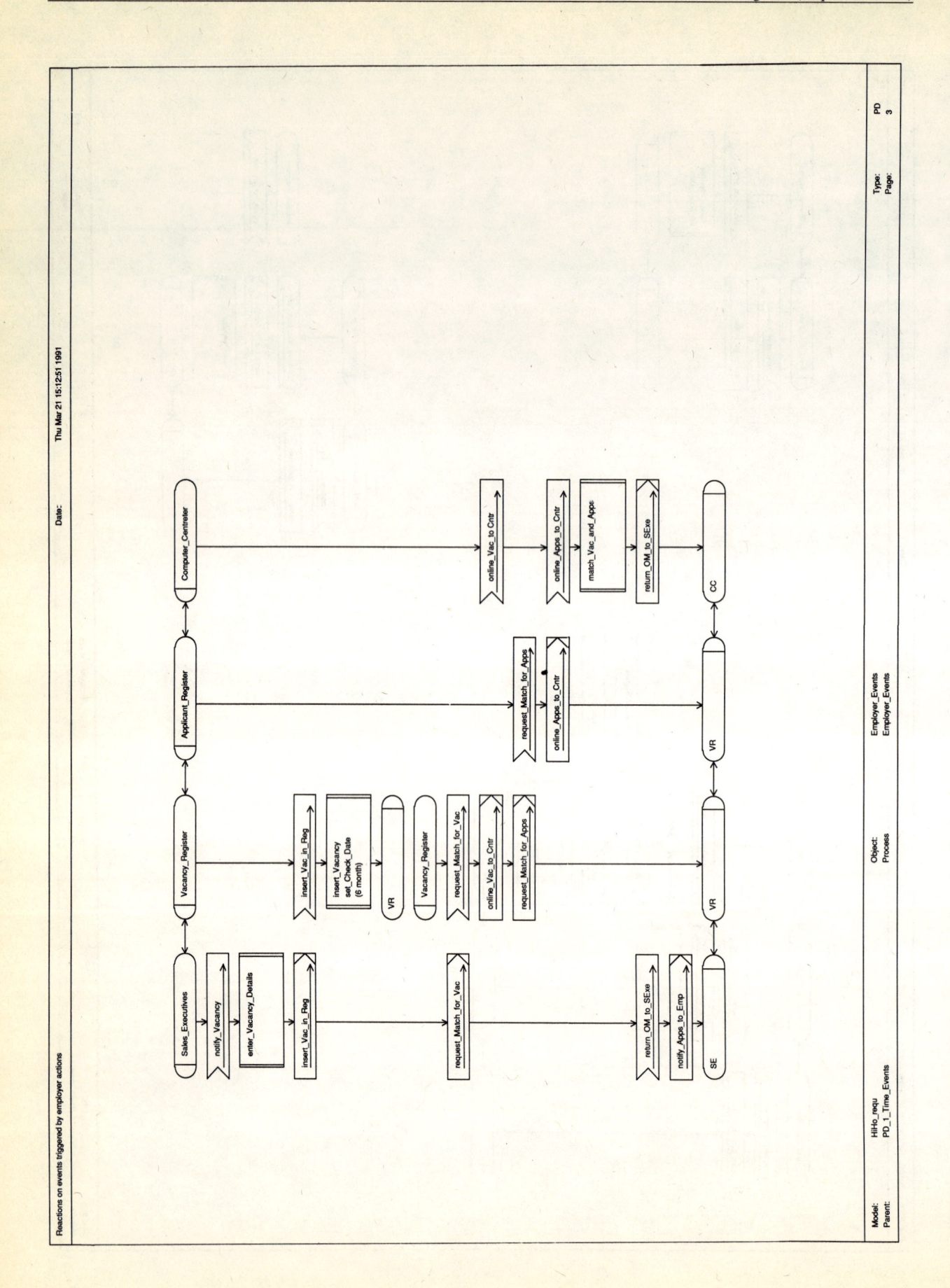

Annex 5 Hi-Ho GRAPES Intermediate Diagrams (Sequence Charts)

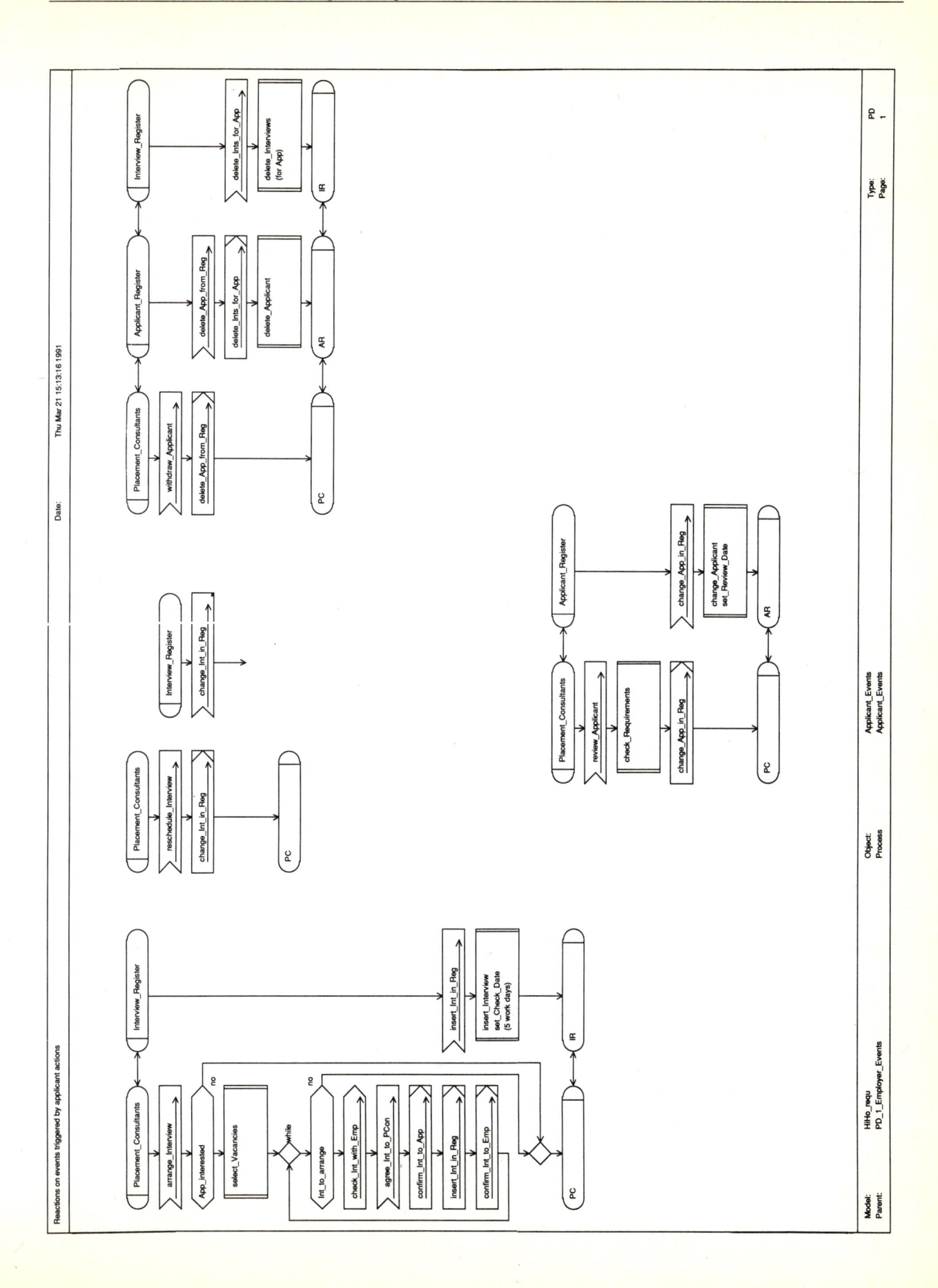

Annex 5 Hi-Ho GRAPES Intermediate Diagrams (Sequence Charts)

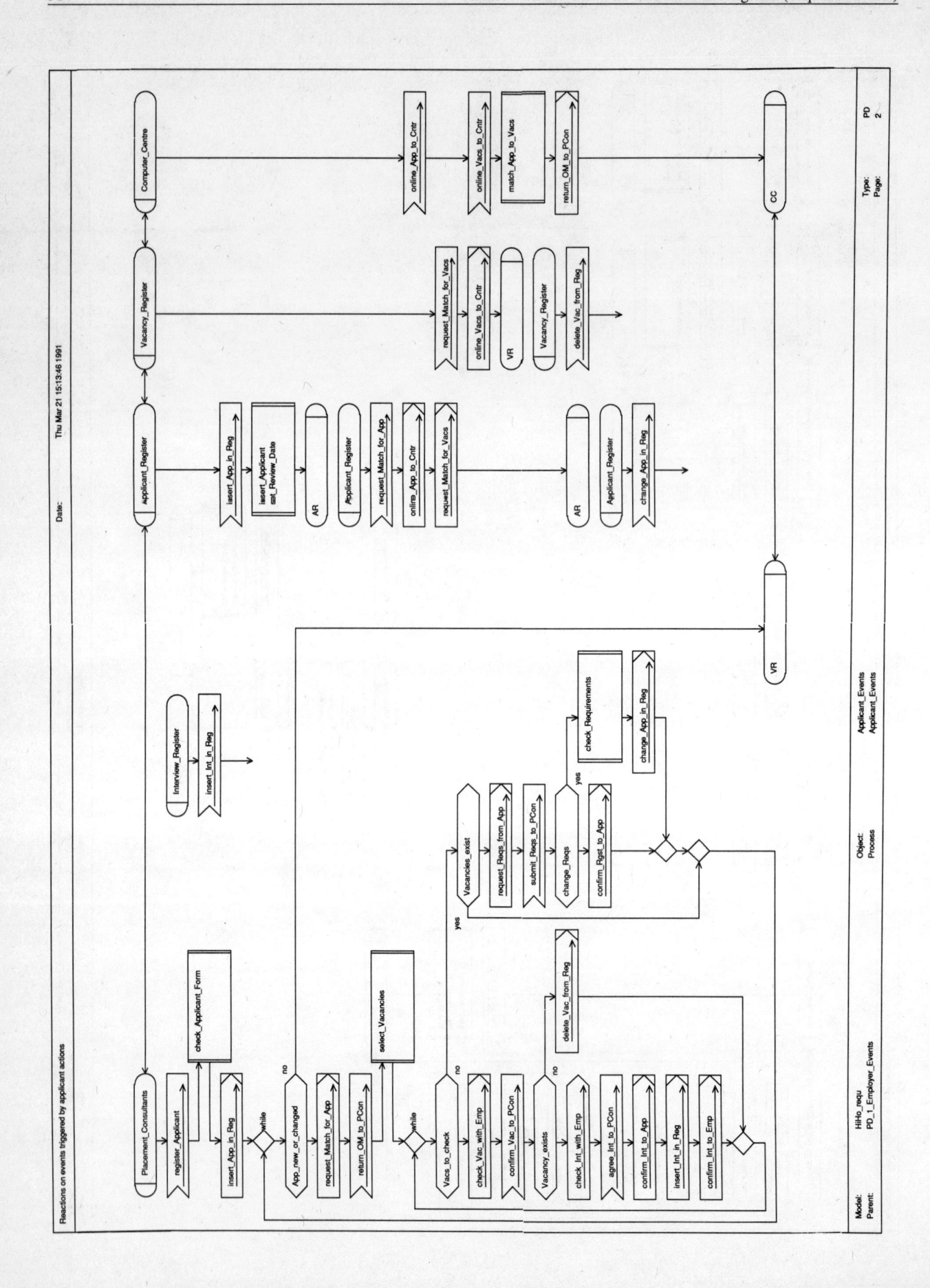

Model of
Current Situation

Annex 5 Hi-Ho GRAPES Intermediate Diagrams (Sequence Charts)

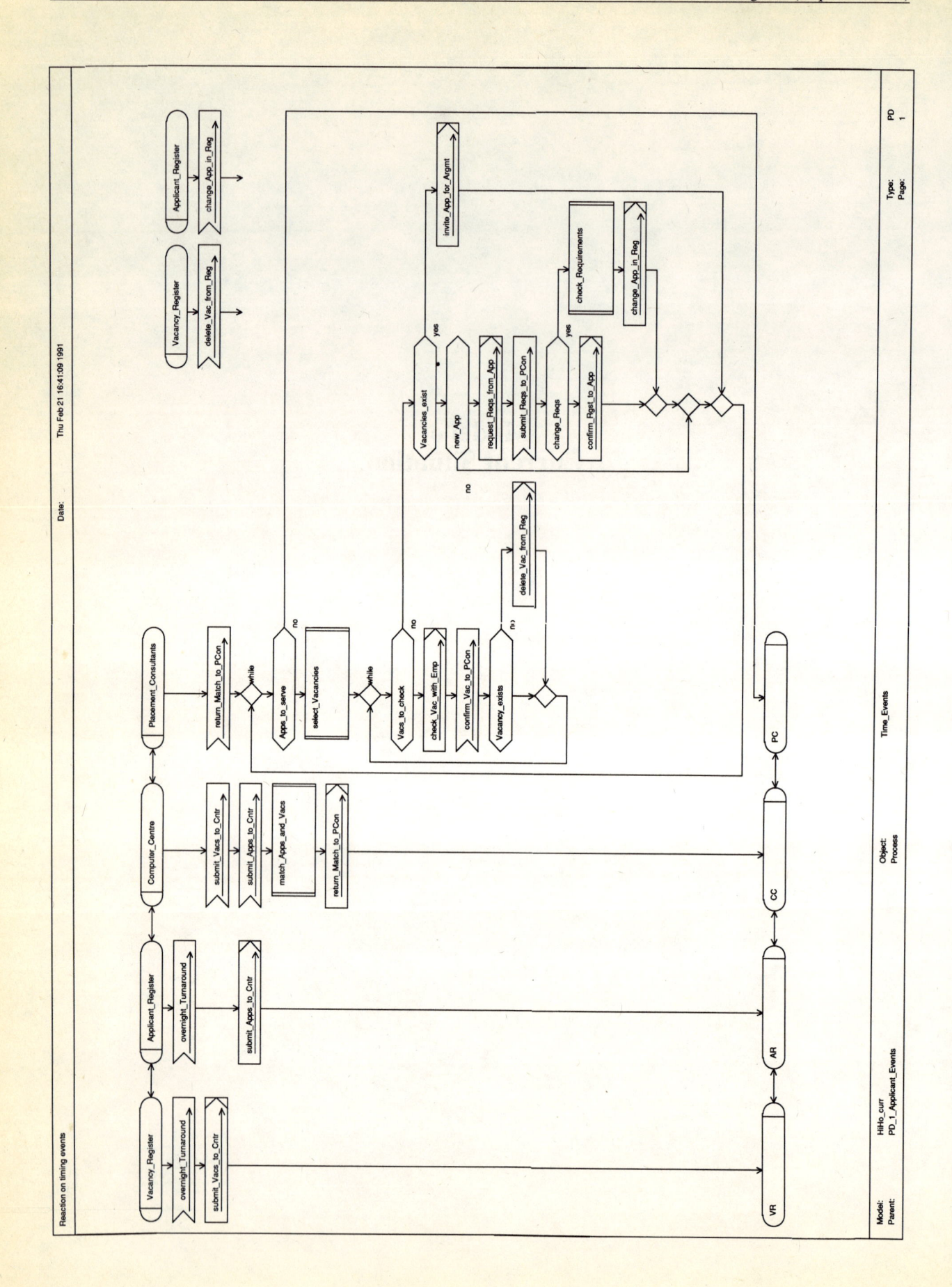

Annex 5 Hi-Ho GRAPES Intermediate Diagrams (Sequence Charts)

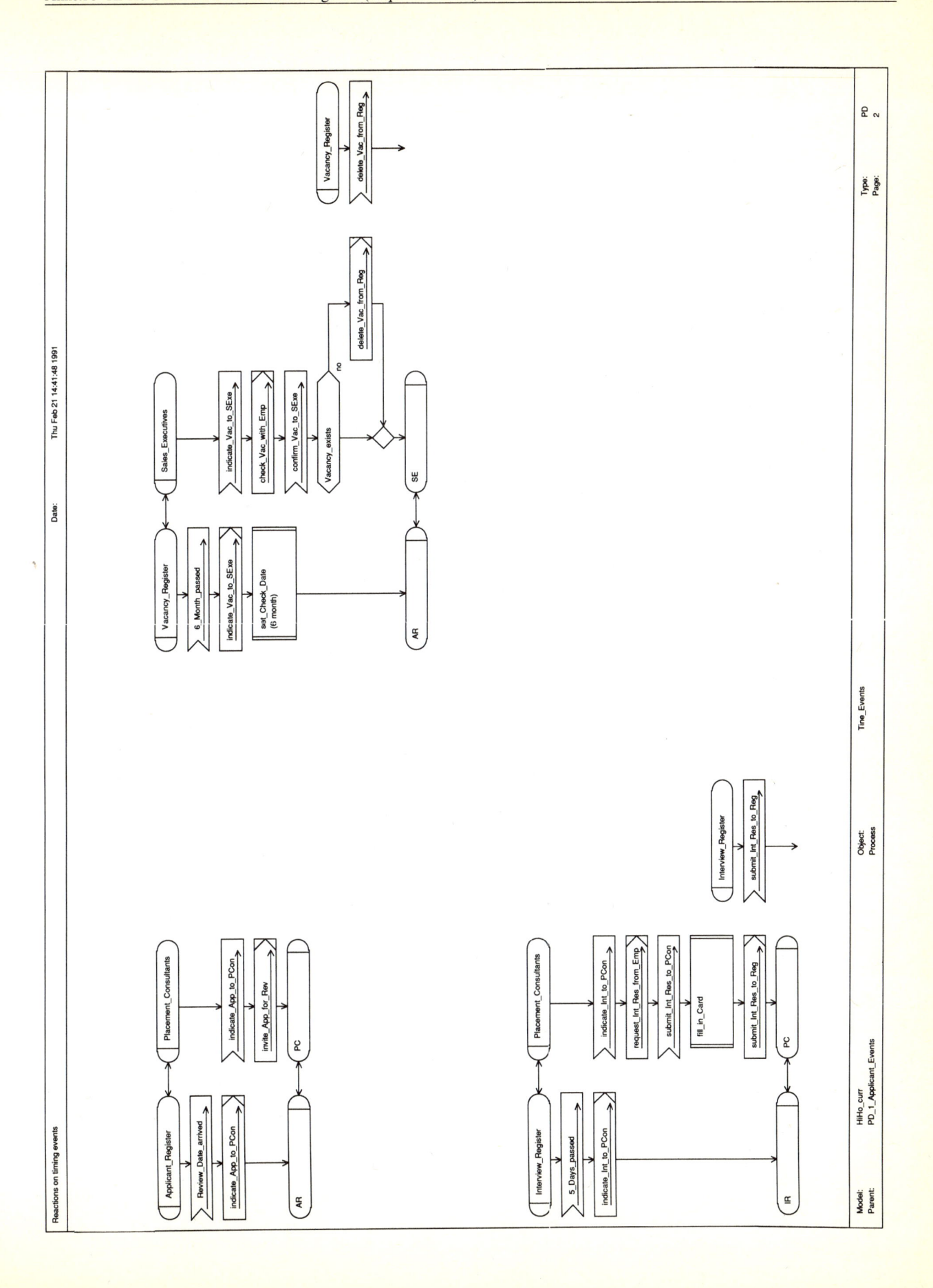

318　　　　　　　　　　　　　　　　　Annex 5 Hi-Ho GRAPES Intermediate Diagrams (Sequence Charts)

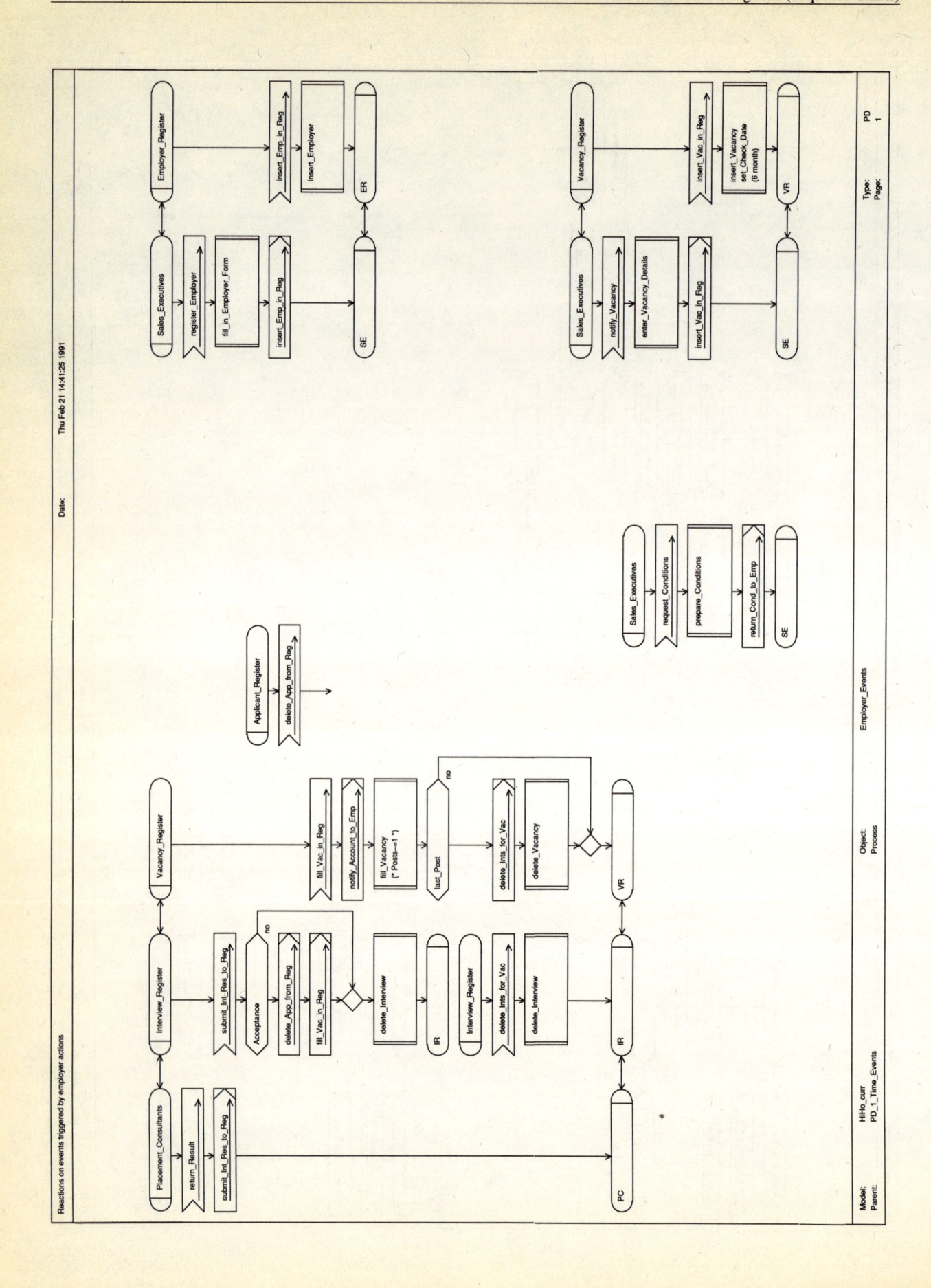

Annex 5 Hi-Ho GRAPES Intermediate Diagrams (Sequence Charts)

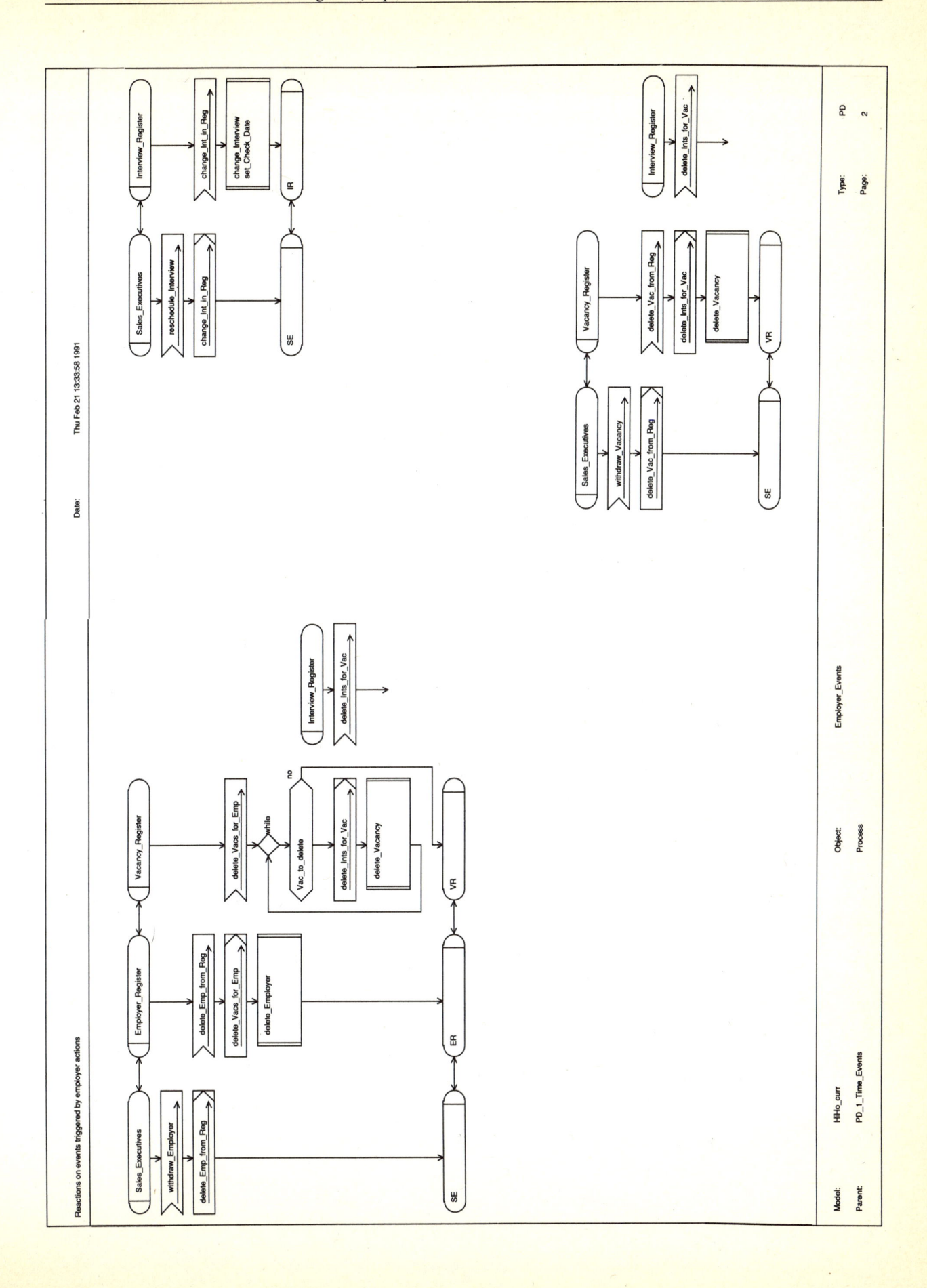

320 Annex 5 Hi-Ho GRAPES Intermediate Diagrams (Sequence Charts)

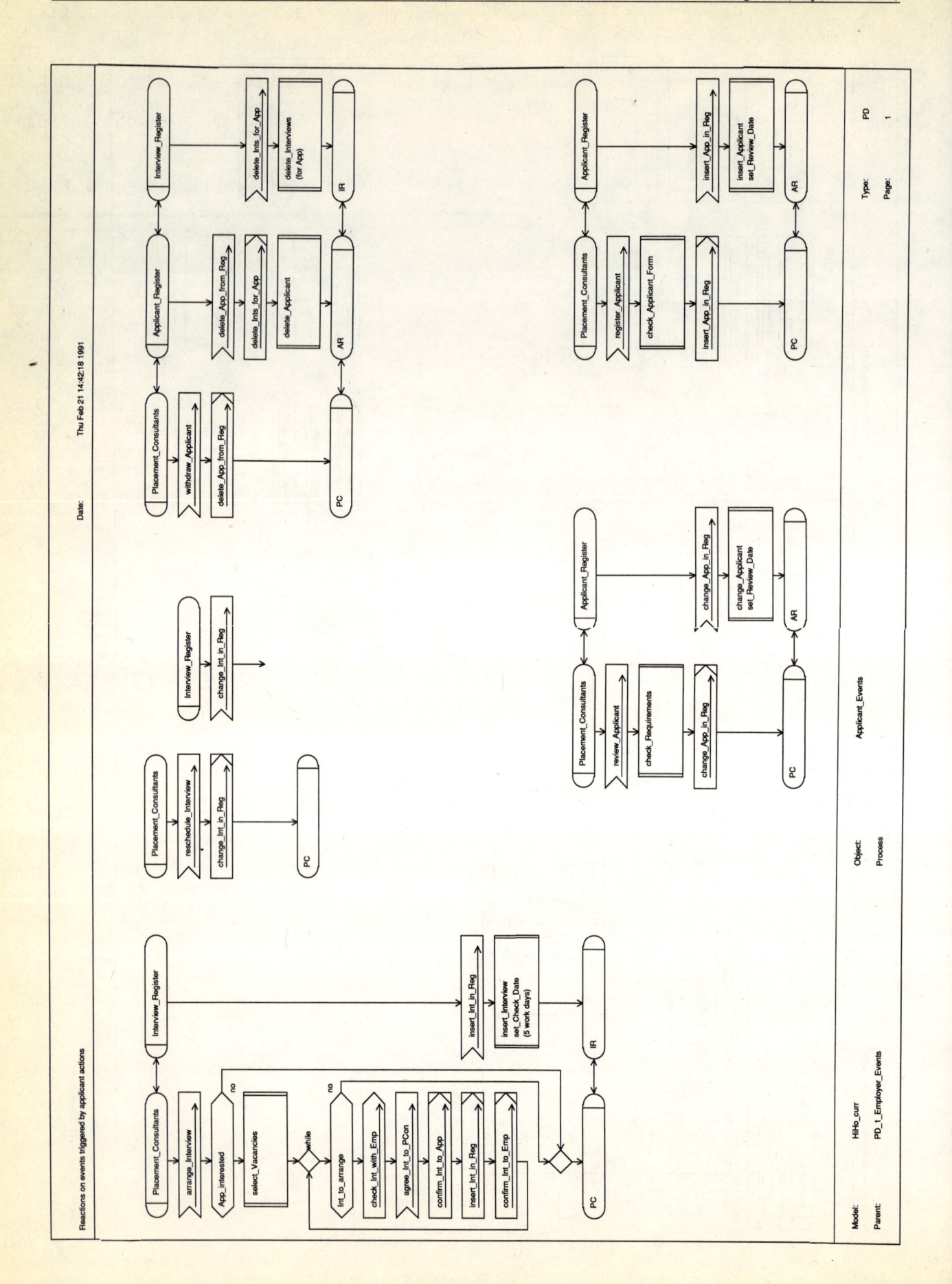

References

Bachman C. W., Data Structure Diagrams, Data Base, The Quarterly Newsletter of the Special Interest Group on Business Data Processing of the ACM, vol. 1, no. 2, Summer 1969, pp. 4-10

CCITT, Specification and Description Language SDL, Recommendation Z. 100, 1988

Commision of the European Communities, Euromethod Phase 3a: Information Pack, Brussels, 1991

CCTA, SSADM Version 4 Reference Manual, Oxford, 1990

CCTA, SSADM 3 GL Interface Guide, to be published

CCTA, SSADM Guide on Distributed Systems, to be published

CCTA, SSADM Soft Systems Guide, to be published

Chen P., The Entity-Relationship-Model, ACM Transactions on Database Systems, vol. 1, no. 1, 1976

DeMarco T., Structured Analysis and Systems Specifications, Englewood Cliffs., 1978

Gane C. and Sarson T., Structured Systems Analysis and Design, New York, 1977

Held G. (ed.), GRAPES Language Description, Munich, 1991

Held G. (ed), Informations- und Funktionsmodellierung mit GRAPES, 1991

Martin J., Information Engineering (3 Volumes) Englewood Cliffs., 1989 – 1990

Jackson M. A., Systems Development, Englewood Cliffs., 1983

Siemens AG, DOMINO Comprehensive Process Technology, Introduction, Munich, 1988

SNI AG, Mosaik: Brief Description, to be published

SNI AG, Process Engineering Handbook for Application Software Manufacture and Project Implementation, Munich, 1991

Wegner P., Conceptual Evolution of object-oriented Programming, Technical Report nr. CS-89-48, Brown University, 1989

Wirth N., Strukturiertes Programmieren, Stuttgart, 1983

Zachman J., A framework for Information Systems Architecture, IBM Systems Journal, vol. 26, no. 3, 1987

Index

3-schema architecture 31, 35, 36

A10 milestone 51, 53, 56
A20 milestone 51, 53, 56
A30 milestone 53, 56
Access path 57, 77
ANSI-SPARC 35, 36
Areas of application 117

Business systems options 49

CADOS 30
Communication diagram (CD) 24, 59, 60, 77, 79, 102, 109, 110, 119, 120
Corporate data modelling 81

Data diagram (DD) 25
Data flow diagram (DFD) 19, 31, 32, 37, 57, 58, 77, 79, 97, 120
Data flow model (DFM) 16, 119
Data structure diagram 59, 60, 78
Data table (DT) 24, 59, 78
Data volumes 19
Definition of requirements 49
DOMINO 20
DOMINO life cycle 21, 51

Effect correspondence diagram (ECD) 19, 58, 77
Entity-event analysis 37
Entity life history (ELH) 16, 19, 31, 58, 77, 79, 82, 96, 120
Entity relationship diagram 59, 60, 78, 79, 102, 110, 119, 120
Entity relationship modelling (ERM) 21

Functions 19, 57, 77

Hierarchy diagram (HD) 25, 60, 79, 102, 110, 112

I/O structure 57, 58, 77
Information hiding 59
Input/Output requirements language 21, 30
Interface table (IT) 24, 59, 78, 102, 109, 112
Intermediate diagram 96, 112
Investigation of current environment 49
IORL 21, 30

Life cycle coverage 49
Life Cycles 49
Logical data model (LDM) 16, 18, 31, 37, 57, 58, 77, 79, 98, 119, 120
Logical data modelling 81
Logical design 17, 49

Maintenance 51
Message passing 32
Milestones 51, 53, 56

Operation 51

P20 milestone 53
P30 milestone 53
Phases 51
Physical design 16, 17, 49
PRINCE 28
Problem analysis 51
Process diagram (PD) 24, 59, 60, 73, 78, 79, 102, 112, 120
Process models 19
Process object 33
Prototyping 37

Quasi-conceptual scenario 45, 47
Quasi-external scenario 45

Relational data analysis (RDA) 19, 37, 82
Relational data analysis models 19
Requirements definition 51
Resource / Service flow 81, 82
Reusability 33

SDM 28
Specification and description language (SDL) 21
Specification diagram (SD) 24, 59, 78
SPU 20
SSADM life cycle 49
SSADM life history 73
Stages 49
Structure object 33
Structured analysis (SA) 21
Structured analysis and design technique (SADT) 21
Super-events 80

T20 milestone 53
Technical implementation 51
Technical systems options 49
Third normal form (TNF) 18, 32, 84, 98
Tool support 21, 117

Update process models 19

Weaknesses 58, 60

Zachman framework 50, 53